Ecological Studies, Vol. 186

Analysis and Synthesis

Ecological Studies

Volumes published since 2001 are listed at the end of this book.

R.B. Allen W.G. Lee (Eds.)

Biological Invasions in New Zealand

With 38 Figures and 28 Tables

Springer

Dr. Robert B. Allen
Landcare Research
PO Box 69
Lincoln 8152
New Zealand

Dr. William G. Lee
Landcare Research
Private Bag 1930
Dunedin
New Zealand

Cover illustration: A diverse range of species have invaded New Zealand, largely from the Northern Hemisphere, and now there is an increasing number of New Zealand species that are invasive in other parts of the world. Cover design by Ruth Guthrie and Jenny Hurst.

ISSN 0070-8356
ISBN-10 3-540-30022-8 Springer Berlin Heidelberg New York
ISBN-13 978-3-540-30022-9 Springer Berlin Heidelberg New York

Springer is a part of Springer Science+Business Media
springer.com

Editor: Dr. Dieter Czeschlik, Heidelberg, Germany
Desk editor: Dr. Andrea Schlitzberger, Heidelberg, Germany
Cover design: *design & production* GmbH, Heidelberg, Germany
Typesetting and production: Friedmut Kröner, Heidelberg, Germany

31/3152 YK – 5 4 3 2 1 0 – Printed on acid free paper

Foreword

Rob Allen and William Lee, along with a comprehensive cast of experts, have produced a book on the "Biological Invasions in New Zealand" that will be welcomed by the growing community of scholars around the world who are studying the spread and impacts of alien invasive species on the land and in the waters of the world.

In Charles Darwin's visit to New Zealand in 1835, during the voyage of the Beagle, he noted after passing through miles of "uninhabited useless country", the joy of coming to an English farmhouse, with their large gardens offering "every fruit and vegetable that England produces" and gorse for fences.

Fast forward a little over a hundred years later to Charles Elton's seminal book on invasive species, "The Ecology of Invasions by Animals and Plants". In this book, Elton too calls attention to New Zealand but with a very different message than Darwin. Elton notes that "No place in the world has received for such a long time such a steady stream of aggressive invaders ..." In Elton's book is found a quaint figure, produced by hand, depicting the sources of the many animal invaders to New Zealand. It is this drawing, and of course the masterful accompanying material, that first drew my attention to the burning issue of invasive species. Here was a place, New Zealand, with no native mammals except bats, but which by Elton's time had already acquired some 29 different "problem" mammals originating from other lands, to say nothing of a host of other groups of animals and plants. Clearly, a major biological revolution was taking place in New Zealand that has become an important test bed for what is happening also in the rest of the world, although perhaps to a lesser extent.

In his book, Elton came to a number of conclusions about the nature of biological rearrangements that are exploding across the face of the earth. His main thesis was that the complexity of ecosystems affords resistance to invaders. In support of this, he points to an overwhelming abundance of invaders in certain island ecosystems that he highlights, and which are often species poor. Further, he observes that it is disturbed communities, again species poor, which are most receptive to introductions. These arguments have subsequently been the subject of considerable debate and clarification as we have accumulated more data, and in particular, as experimental approaches have been applied to the issue.

This volume revisits these earlier, and many contemporary views pertaining to the success of invaders. It is now clear that some of Elton's arguments have not stood the test of extensive new findings. The chapters in this book probe the fates and impacts of diverse groups of organisms that have entered into New Zealand from foreign lands. The overriding importance of the human effort to introduce new biota to the islands, rather than characteristics of the islands per se promoting invasion success, is clearly documented. As for the contention that natural systems are inherently resistant to invasions, the editors of this volume note that "even small patches of relatively unmodified native forest may now contain six alien mammal herbivores, five alien mammal predators, two alien fish, numerous alien plants, and an unknown number of alien invertebrate, fungi and bacteria species"!

The political response to the flood of invasive species into New Zealand has been one of the most aggressive among the nations of the world. The Hazardous Substance and New Organisms Act controls the importation of alien species not presently in the country, and similarly there is a national Act directed toward eradication and management of unwanted alien organisms. New Zealand has successfully extirpated damaging alien species from small islands. Yet, new invertebrates, and no doubt also microbial organisms, continue to enter into the New Zealand biota. Those organisms already resident will continue their dynamic interaction with the native biota, in time transforming this fascinating landscape into a new biotic world.

This book will prove to be an invaluable record of the biological state of New Zealand in the early 21st Century. Importantly, it provides invasion biologists throughout the world a test of current theories on those factors leading to success of invaders, and a solid basis for new approaches for understanding the drivers, nature, consequences, and management of biological invasions everywhere.

Stanford University *Harold Mooney*

Preface

Biological invasions are an important global change issue and are currently the focus for considerable ecological research and conservation management. Determining the level and rate of alien invasions, and understanding what controls the success of invaders and their consequences for ecosystems are research topics central to the effective management of biological invasions. Although increasing knowledge on the significance of biological invasions should help mitigate their consequences, it will plainly be countered, to some degree, by the entrenched patterns of increasing travel and trade between countries, and human pressures on natural resources.

Colonization of New Zealand has dramatically altered the resident biota and resulted in the introduction of numerous alien organisms. In our view, an in-depth review of biological invasions in New Zealand was long overdue, coupled with the international contribution such a synthesis could make. Historically, New Zealand was considered an exemplar of island vulnerability to biological invasions reflecting its insularity, temperate maritime climate, and lack of significant native herbivores and predators. However, increasingly questions are being raised about the level of New Zealand's historical biotic isolation, and also whether its current high levels of invasion are a consequence of introduction effort and the expanding areas of human-transformed habitat, rather than of geography. Clearly, a synthesis of biological invasions in New Zealand would contribute to the ongoing debate about the vulnerability of island ecosystems.

This book was designed to present a wide range of New Zealand studies and experiences on biological invasions: for example, chapters were selected to represent marine, freshwater and terrestrial ecosystems. As such, we consider the book will be of interest to a broad group of researchers, managers and students concerned about biological invasions. In producing the concept for such a book, we were initially encouraged by Lars Walker and Richard Mack. Colleagues at Lincoln and Dunedin assisted in numerous ways. Although chapter contributions, and their review, were funded by various organizations, a major input to the preparation of this book was funded by the New Zealand Foundation for Research, Science and Technology. We appreciate the contribution made by all the individuals and organizations supporting the production of this book, and express our special thanks to Melissa Brignall-Theyer and Christine Bezar for their assistance in preparing this volume.

Lincoln, New Zealand *Rob Allen*
Dunedin, New Zealand *William Lee*

January 2006

Contents

Contributors

ALLEN, ROBERT B.
Landcare Research, P.O. Box 69, Lincoln 8152, New Zealand,
e-mail: allenr@landcareresearch.co.nz

ANDERSON, SANDRA H.
School of Geography and Environmental Sciences, University of Auckland,
Private Bag 92019, Auckland, New Zealand

ARMSTRONG, TRISTAN
Ecological Genetics Laboratory, Landcare Research, Private Bag 92170,
Auckland, New Zealand

ATKINSON, IAN A.E.
Ecological Research Associates of New Zealand Inc., P.O. Box 48-147, Silverstream, Upper Hutt 6430, New Zealand, e-mail: i.atkinson@paradise.net.nz

BEE, JENNIE
Department of Plant Sciences, University of Cambridge, Downing Street,
Cambridge CB2 3EA, UK, and Landcare Research, P.O. Box 69, Lincoln 8152,
New Zealand

BEGGS, JACQUELINE R.
School of Biological Sciences, University of Auckland, Private Bag 92019,
Auckland, New Zealand, e-mail: j.beggs@auckland.ac.nz

BELLINGHAM, PETER J.
Landcare Research, P.O. Box 69, Lincoln 8152, New Zealand,
e-mail: bellinghamp@landcareresearch.co.nz

Blackburn, Tim M.
School of Biosciences, University of Birmingham, Edgbaston, Birmingham B15 2TT, UK

Byrom, Andrea E.
Landcare Research, P.O. Box 69, Lincoln 8152, New Zealand

Cameron, Ewen K.
Auckland Museum, Private Bag 92018, Auckland, New Zealand

Cassey, Phillip
Ecole Normale Supérieure, Laboratoire d'Ecologie, 46 Rue d'Ulm, 75230 Paris cedex 05, France, present address: School of Biosciences, University of Birmingham, Edgbaston, Birmingham BIS 2TT, UK

Choquenot, David
Landcare Research, Private Bag 92170, Auckland, New Zealand

Clarkson, Bruce D.
Centre for Biodiversity and Ecology Research, Department of Biological Sciences, The University of Waikato, Private Bag 3105, Hamilton, New Zealand, e-mail: b.clarkson@waikato.ac.nz

Clout, Mick N.
Centre for Biodiversity and Biosecurity, School of Biological Sciences, University of Auckland, Private Bag 92019, Auckland, New Zealand, e-mail: m.clout@auckland.ac.nz

Coomes, David A.
Department of Plant Sciences, University of Cambridge, Downing Street, Cambridge CB2 3EA, UK, and Landcare Research, P.O. Box 69, Lincoln 8152, New Zealand, e-mail: dac18@cam.ac.uk

Craine, Joseph M.
Landcare Research, Private Bag 1930, Dunedin, New Zealand
e-mail: crain010@umn.edu

Duckworth, Janine A.
Landcare Research, P.O. Box 69, Lincoln 8152, New Zealand, e-mail: duckworthj@landcareresearch.co.nz

DUNCAN, RICHARD P.

Bio-Protection and Ecology Division, P.O. Box 84, Lincoln University, Canterbury, New Zealand, and Landcare Research, P.O. Box 69, Lincoln 8152, New Zealand, e-mail: DuncanR@Lincoln.ac.nz, DuncanR@landcareresearch.com

FISHER, PENNY

Landcare Research, P.O. Box 69, Lincoln 8152, New Zealand

FORREST, BARRIE M.

Cawthron Institute, Private Bag 2, Nelson, New Zealand, e-mail: barrie.forrest@cawthron.org.nz

FORSYTH, DAVID M.

Landcare Research, P.O. Box 69, Lincoln 8152, New Zealand, present address: Arthur Rylah Institute for Environmental Research, P.O. Box 137, 123 Brown Street, Heidelberg, Victoria 3084, Australia, e-mail: dave.forsyth@dse.vic.gov.au

FOWLER, SIMON V.

Landcare Research, P.O. Box 69, Lincoln 8152, New Zealand, e-mail: fowlers@landcareresearch.co.nz

GLEESON, DIANNE

Ecological Genetics Laboratory, Landcare Research, Private Bag 92170, Auckland, New Zealand, e-mail: gleesond@landcareresearch.co.nz,

HARMAN, HELEN M.

Ecological Genetics Laboratory, Landcare Research, Private Bag 92170, Auckland, New Zealand

HAYDEN, BARBARA J.

National Centre for Aquatic Biodiversity and Biosecurity, NIWA, P.O. Box 8602, Christchurch, New Zealand

HORN, CHRYS

Landcare Research, P.O. Box 69, Lincoln 8152, New Zealand

INGLIS, GRAEME J.

National Centre for Aquatic Biodiversity and Biosecurity, NIWA, P.O. Box 8602, Christchurch, New Zealand, e-mail: g.inglis@niwa.co.nz

Kelly, Dave

School of Biological Sciences, University of Canterbury, Private Bag 4800, Christchurch, New Zealand, e-mail: dave.kelly@canterbury.ac.nz

Ladley, Jenny J.

School of Biological Sciences, University of Canterbury, Private Bag 4800, Christchurch, New Zealand

Lee, William G.

Landcare Research, Private Bag 1930, Dunedin, New Zealand, e-mail: leew@landcareresearch.co.nz

Mark, Alan F.

Botany Department, University of Otago, P.O. Box 56, Dunedin, New Zealand

McGlone, Matt S.

Landcare Research, P.O. Box 69, Lincoln 8152, New Zealand, e-mail: mcglonem@landcareresearch.co.nz

McKenzie, Robert J.

School of Biological Sciences, University of Canterbury, Private Bag 4800, Christchurch, New Zealand

McLennan, John A.

Landcare Research, 33 Simla Avenue, Havelock North, New Zealand, e-mail: mclennanj@landcareresearch.co.nz

McQueen, Joanna C.

Centre for Biodiversity and Ecology Research, Department of Biological Sciences, The University of Waikato, Private Bag 3105, Hamilton, New Zealand, e-mail: Joanna.McQueen@scionresearch.com

Nelson, Wendy A.

National Centre for Aquatic Biodiversity and Biosecurity, NIWA, P.O. Box 14901, Wellington, New Zealand

Norbury, Grant

Landcare Research, P.O. Box 282, Alexandra, New Zealand

NORRIS, REBECCA
School of Biological Sciences, University of Bristol, Woodland Road,
Bristol BS8 1UG, UK

PARKES, JOHN P.
Landcare Research, P.O. Box 69, Lincoln 8152, New Zealand,
e-mail: parkesj@landcareresearch.co.nz

POULIN, ROBERT
Department of Zoology, University of Otago, P.O. Box 56, Dunedin,
New Zealand

ROBERTSON, ALASTAIR W.
Ecology, Institute of Natural Resources, Massey University,
Private Bag 11222, Palmerston North, New Zealand

RUSCOE, WENDY A.
Landcare Research, P.O. Box 69, Lincoln 8152, New Zealand,
e-mail: ruscoew@landcareresearch.co.nz

SIMON, KEVIN S.
Department of Biology, James Madison University, Harrisonburg, VA 22807,
USA

SINNER, JIM
Ecologic Foundation, P.O. Box 756, Nelson, New Zealand

TAYLOR, MICHAEL D.
Cawthron Institute, Private Bag 2, Nelson, New Zealand

TOMPKINS, DANIEL M.
Landcare Research, Private Bag 1930, Dunedin, New Zealand,
e-mail: tompkinsd@landcareresearch.co.nz

TOWNSEND, COLIN R.
Department of Zoology, University of Otago, P.O. Box 56, Dunedin,
New Zealand, e-mail: colin.townsend@stonebow.otago.ac.nz

Tozer, Wade C.

Centre for Biodiversity and Ecology Research, Department of Biological Sciences, The University of Waikato, Private Bag 3105, Hamilton, New Zealand

Walker, Susan

Landcare Research, Private Bag 1930, Dunedin, New Zealand

Ward, Darren

Landcare Research, Private Bag 92170, Auckland, New Zealand

Wardle, David A.

Landcare Research, P.O. Box 69, Lincoln 8152, New Zealand, and Department of Forest Vegetation Ecology, Swedish University of Agricultural Sciences, 90183 Umeå, Sweden,
e-mail: wardled@landcareresearch.co.nz, david.wardle@svek.slu.se

Williams, Peter A.

Landcare Research, Private Bag 6, Nelson, New Zealand,
e-mail: williamsp@landcareresearch.co.nz

Wiser, Susan K.

Landcare Research, P.O. Box 69, Lincoln 8152, New Zealand,
e-mail: wisers@landcareresearch.co.nz

Withers, Toni M.

Forest Biosecurity and Protection, Ensis – a joint venture between SCION (the Forest Research Institue New Zealand Ltd) and CSIRO, 49 Sala Street, Private Bag 3020, Rotorua, New Zealand

Yeates, Gregor W.

Landcare Research, Private Bag 11052, Palmerston North, New Zealand,
e-mail: yeatesg@landcareresearch.co.nz

1 Paradise Lost – the Last Major Colonization

W.G. Lee, R.B. Allen and D.M. Tompkins

1.1 Introduction

Non-native or alien species have invaded New Zealand in a manner that is rarely equaled elsewhere in the world. For over a century, and long before biological invasions surfaced on the global research agenda in the late 20th century, alien plants and animals have been the focus of often intense public, legislative, management, and research activity and debate. In the early 20th century, New Zealand provided one of the earliest and most comprehensive national accounts of alien invasions (Thomson 1922), and New Zealand examples of invasion impacts and management commonly feature in the international literature on alien biota. To this day, New Zealand ecologists in general maintain a distinction between native and alien species, which is frequently viewed with bemusement by those from other countries where such categorizations, until relatively recently, have seldom been considered in research (Bannister 1994).

This heightened awareness of alien species reflects both the timing of European settlement, and the stark contrast between the alien and New Zealand resident flora and fauna. When Europeans began to settle in New Zealand in the early 19th century, they found an unfamiliar biota unsuitable as quality pasture for stock, edible fruits, or trees for short-rotation plantation forestry. The aspiration to recreate the resources and recreational opportunities of their largely European homelands saw the introduction of thousands of alien plant and animal species (McDowall 1994). The incidental arrival of plants and animals was also enormous, particularly as the colony looked to provide resources for a growing population and sought to develop primary industries for foreign trade (Guthrie-Smith 1921). In 1870, a few decades after European settlement, the explorer Julius von Haast could claim that he had over 700 non-native plant species to establish (McDowall 1994). A passion for the introduction of alien plants and animals continued until the late-20th century. Introductions were initially from Eurasia and North America. However,

Ecological Studies, Vol. 186
R.B. Allen and W.G. Lee (Eds.)
Biological Invasions in New Zealand

with increasing trade and transportation links to other countries, the need for commercial crops, pasture and forestry species, and an insatiable desire for novel garden plants, biota were sought from across the globe.

Facilitated by a temperate maritime climate, the lack of significant native herbivores and predators, and expanding areas of human-transformed habitat, an increasing proportion of alien species naturalized. Naturalized alien seed plant species now outnumber natives (Wilton and Breitwiesser 2000), and amongst vertebrates several trophic levels (predators, herbivores) have been dominated by introduced species (King 1990) for nearly two centuries. New Zealand's native biodiversity is renowned for high levels of endemism, ancient lineages, local radiations, and notable taxonomic and functional absences (Diamond 1990). Hence, the alien biota often represents novel functional groups, and species biogeographical and evolutionary histories vastly different from those experienced by the indigenous biota.

Research on alien invasive species in New Zealand was initially driven by concern over economic threats to agro-ecosystems arising from the early spread of pasture weeds. However, the early and rapid decline of many native terrestrial birds created apprehension about the impacts of carnivorous mammals in forest ecosystems (Worthy and Holdaway 2002), and presaged legislation to declare some alien species as noxious pests, to be controlled and eliminated where possible. Although agribusiness and forestry issues continue to drive much of our biosecurity legislation that is focused on the detection and elimination of new and unwanted alien species at the border, environmental issues (particularly the protection of biodiversity) are becoming increasingly important.

In this chapter, we address some of the major themes and debates that have developed through research in New Zealand on alien species, which provide a context for the book. We then introduce the structure and content of the book. The long focus on invasive species in New Zealand provides unique insights that may not be apparent in places that have recently begun to investigate invasions, or where alien species are not long established.

1.2 Themes and Debates on Alien Species

Research on biological invasions in New Zealand has undergone major shifts in approach and emphasis that have been driven by increased ecological understanding, changes in the operational management requirements of agencies controlling alien species, expansion of the diversity of alien organisms being detected, and the need to quantify impacts for risk assessment. Here, we outline some of the major issues, and how they have affected the development of research on biological invasions.

1.2.1 Human Dimension of Alien Invasions

Alien species introductions are obviously dependent on human-assisted transportation to New Zealand, but the important role of humans in the naturalization, establishment and spread of species is only beginning to be understood. Historically in New Zealand, nearly all introduced organisms are selected for surviving outdoors, and generally arrive in sufficient quantity for commercial use. Changing agribusiness requirements and fashions strongly influence alien species introductions (Mack 2001). However, for alien plants, the number of naturalized species is positively related to regional human population densities (Allan 1937). Proximity to human settlement also accounts for the level of alien plant species in native forest fragments (Sullivan et al. 2005), and the pattern and rate of spread of many vertebrate animals are largely a function of the intensity of human activity (Fraser et al. 2000). Habitat modification, coupled with the direct and indirect effects of human disturbance are also important factors maintaining numerical dominance of invasive alien species. What is becoming clear is that the invasion process involves multiple phases (importation, establishment, naturalization, spread, etc.), and that each stage is governed by a different set of key processes. Predicting invasion success requires integration of research on these different phases, and incorporation of a range of key processes.

1.2.2 Ecological Superiority of Alien Species

Since Darwin's (1845) assertion that the New Zealand vegetation was defenseless against the new and competitively superior alien flora from the Northern Hemisphere, assessment of the susceptibility of the native biota to introduced alien species has been controversial, especially for plants where impacts were less obvious and slower to unfold. Cockayne (1910) and Thomson (1922) challenged the supposed inferiority of the native flora, concluding that alien plant species only prevailed in natural ecosystems where indigenous plant species were weakened by human activity. They considered that local evolutionary processes had produced a native flora strongly adapted to New Zealand conditions and resistant to invasion by Northern Hemisphere plants. However, history and ecological research are showing that a verdict about the outcome of ecological interactions between native and alien plant species is context-dependent, and not necessarily favorable to native species. Although many communities currently comprise mixtures of native and exotic plant species, there is no evidence of stable states, and compositional and structural shifts continue in relation to management and environmental changes (Wilson 1989). For example, in the semi-arid regions of the South Island (which has one of the longest histories in New Zealand of human disturbance and weed invasion), native and exotic species continue to change within communities, and are poorly correlated with environmental variables (Wilson et al. 1989). Most natural habitats in New Zealand have at least one clear example of a successful invasive alien species, suggesting there are enough safe sites or windows of opportunity, created by natural processes, to allow invasion.

The initial assessment regarding the resistance of New Zealand vegetation and communities to invasion now appears increasingly premature, for three reasons.

- First, early authors greatly underestimated the number, range and intensity of plant introductions, not knowing that in little more than 100 years over 20,000 alien plant species would be imported to New Zealand, and over 2,100 would become naturalized (Wilton and Breitwiesser 2000).
- Second, they failed to consider the invasion process over ecologically realistic time frames. In European landscapes, alien woody species show an average time lag of 147 years between introduction and the initiation of invasion (Kowarik 1995). The onset of alien woody species invasion may well be more rapid in New Zealand than in Europe, due to the absence of original predators, less intensive forms of land use, and milder climatic conditions. However, natural spread for long-lived, slow-growing, woody species is usually incremental, and is further impeded by the availability of safe sites or regeneration opportunities. In forests, for example, entry points for introduced plant species are usually gaps created by disturbance or tree senescence that occur at century-level frequencies (Stewart et al. 1991). Hence, the invasion potential of most species will not be recognized for many decades, and a residence time of at least a century will be required to adequately assess the ability of an alien woody species to spread into natural communities.
- Third, assumptions about the relative resistance of native vegetation focused on widespread forest communities, which frequently have an inherent resistance due to low irradiance levels and slow canopy turnover times. Such a prognosis is less convincing when we consider non-forested terrestrial habitats and aquatic systems that clearly demonstrate the potential for alien plants to invade and modify ecosystems, often displacing indigenous plant species (Scott et al. 2001; Closs et al. 2004).

The obvious impact of alien animal species is in sharp contrast to that observed for plants. While plant introductions (in terms of number of species) have been far more numerous, animal introductions have more frequently represented new functional groups, notably for introduced mammals (King 1990). The opportunities available to introduced mammals, particularly through predation of native fauna with few co-evolved defenses, have allowed a large and rapid impact by the comparatively few species introduced. In general, the large effect of alien animals on native New Zealand plants and animals is not in question. Rather, the key debates involving alien animal species have centered on whether they are the only cause of native species decline (and the relative culpability of different alien species), or whether other factors (e.g., habitat loss, climate change) are also involved (Worthy and Holdaway 2002; McDowall 2003). Ecological systems are frequently complex, and causal factors driving historical versus recent trends may be different. Thus, there are numerous factors potentially involved in native species' decline, providing some justification for diverse perspectives on which particular alien animal, if any, should be targeted for control.

1.2.3 Invasion Impacts Are Complex

The impact measurement for invasive weeds or herbivorous mammals in agricultural systems is relatively simple: loss of pasture production. Both the impact criteria and the consequences of alien species activity are often not clear in natural systems dominated by native species. Here, research has struggled because both the goals and the interactions are more complex.

Some alien species in New Zealand have clearly extended their niche breadth compared with that in their native range, and we have been slow to detect this expansion. For example, the perspective of the Australian brushtail possum (*Trichsurus vulpecula*) as primarily herbivorous in New Zealand forests was challenged by video footage of the species eating native bird eggs and pre-fledged young. For several decades, possum research focused on vegetation impacts associated with canopy dieback, ignorant of the more immediate threat on native birds (Brown et al. 1996).

The application of ecosystem-level studies has further expanded our perspective on invasive species impacts in New Zealand. For example, traditionally, mammalian herbivore effects were assessed primarily in relation to vegetation composition and regeneration. However, in the last decade, research on multiple-trophic-level interactions and above- and below-ground processes has extended our understanding and demonstrated significant, if somewhat idiosyncratic, influences of deer herbivory on the composition and diversity of soil organisms through the nonrandom selection of vascular plant traits, which in turn influences decomposer processes (Wardle et al. 2001). The long-term significance of deer herbivory on these ecosystem features has yet to be determined, but this is a compelling indication of how alien species may have subtle impacts far beyond those generally measured. Similar multi-trophic and ecosystem-level impacts of alien plant species are also emerging (Standish et al. 2001, 2004; Standish 2004). Deliberate or accidental introductions of cryptic alien organisms causing diseases are also being implicated in multiple trophic changes (Beggs 2001; Norbury et al. 2002), although for some groups such as introduced plant viruses the consequences of transfer to native species are largely unknown (Davis and Guy 2001).

Studies at the community level, rather than at the scale of the individual species, are also increasing our understanding of how synergistic interactions among alien species can impact on native biodiversity. The term invasional meltdown, coined by Simberloff and von Holle (1999), describes the potential for such positive interactions among alien species, consolidating their abundance, and thereby enhancing their impacts on native biota. The clearest examples in New Zealand are predator–prey systems in forests and induced grassland ecosystems, where mustelids are maintained and enhanced by the presence of their primary prey (introduced rodent and rabbit populations, respectively; Norbury 2001). Similarly, grazing by cattle and sheep frequently maintains the dominance of alien sward grasses that in turn sustain high rabbit numbers in many areas. In many lowland ecosystems a suite of alien species, including plants, herbivores, competitors, and predators, interact positively to have increasingly negative effects on threatened native species (Brockie 1992).

Alien invasive species with both serious agribusiness and conservation impacts have invariably been priorities for research. The outstanding recent example is the Australian brushtail possum, which is a known major vector of livestock disease, and a threat to some native forest plant and animal species. The possum is universally ranked as a major pest, occurring in ca. 98 % of the country, and possum research in New Zealand is funded to a level of NZ$ 24 million per annum, which is an order of magnitude above the level of expenditure on any other single invasive alien species (Montague 2000).

1.2.4 Management Objectives

Management of alien animals in New Zealand traditionally focused on mitigating impacts of a single species (Parkes and Murphy 2003). Hence, for decades, ecologists have investigated single alien species (often in isolation) when attempting to establish the cause of selected native species' decline, and determine threshold pest densities that would allow target species' persistence (e.g., Basse et al. 1999; Innes et al. 1999). However, it has become increasingly apparent in recent years that interactions among alien species, combined with considerable ecological redundancy in multi-pest systems, frequently ensures that there are very few circumstances where individual alien species singly depress native species. This has driven a wider, more community-scale perspective on alien species management (Ramsey and Veltman 2005). Hence, in tandem with this development, the scope of management for biodiversity protection is broadening beyond the traditional targeted control of a single large predator/herbivore to the control of pest communities, creating new challenges for control techniques.

Eradication of alien species was for many decades a conceptually attractive but practically unattainable goal, even on offshore islands. The difficulties were both technical and policy-related. However, after early success with several mammal eradication programs, particularly on small offshore islands (Climo 1973; Taylor and Thomas 1989), perceptions have changed markedly, and nowadays larger islands and smaller mammals are being considered for total alien mammal extirpation. Successfully used on offshore islands, total alien mammal removal is now being attempted on the mainland using predator-proof fences and ongoing predator control (Saunders and Norton 2001). These approaches, using zero-pest levels, will undoubtedly reveal the level of alien species control required for the restoration of many biodiversity components.

For natural communities, weeds were largely ignored until the 1970s when the spread and impacts of several alien plants became too obvious to disregard (Kelly and Skipworth 1984). In recent years, the whole process of alien plant invasion, from importation across the border to spread in the wild, has been considered and research initiated to identify the major driving factors at each stage. This has led to more research on the process of naturalization and the role of human activities (Sullivan et al. 2005). A similar perspective has been adopted for many cryptic alien species and largely concealed habitats (e.g., marine), where elimination at or near the point of entry in New Zealand is the only realistic approach to control (Hewitt et al. 2004). Research priorities have therefore shifted from studying large-scale spread and impacts to early detection and surveillance issues. Furthermore, the geographic center of attention has been transferred from natural and rural areas to urban and peri-urban environments. Local extermination of alien plant species outside of the agro-ecosystems has been slower in developing, reflecting the perceived level of ecological impact, and the transient role of alien plant species in many natural systems.

In a broad sense, management objectives of regulatory and land management agencies reflect national government legislation and policy. These have undergone significant changes with the development of a more coherent suite of biosecurity laws aimed at protecting agri-industry and the natural environment, as well as providing a clear allocation of responsibilities among agencies (Storey and Clayton 2002). The Hazardous Substance and New Organisms Act (1996) restricts the importation of alien species currently not in New Zealand, unless approved by the Environmental Risk Management Authority. Because of the costs involved, very few new plant and animal species have been considered, which has markedly reduced the rate of increase in the national pool of potentially invasive alien species. The Biosecurity Act (1993) aims to eradicate and effectively manage unwanted alien organisms, by border surveillance and control, early eradication of founding populations of new pests, and the development of national and regional pest management strategies by government agencies. These laws strongly influence government-funded research on alien invasive species.

1.3 Outline of the Book

It is now over 50 years since a scientific book was published on alien invasive species in New Zealand (Clark 1949), and an update encompassing modern research is long overdue. This book covers a range of biological invasions occurring in New Zealand's natural or extensively managed ecosystems, which are usually dominated by native species. We have selected topics to encompass the diversity of studies that have focused on alien species invasions. Most have received considerable attention in New Zealand, and as such have contributed significantly to our knowledge of biological invasions. We have attempted to include familiar invaders (e.g., plants and herbivores) as well as lesser known examples (e.g., invertebrates and algae). The book is divided into four sections, which examine the magnitude of alien invasions (Sect. A), the controls on invasion success (Sect. B), the consequences of invasions (Sect. C), and finally, the management of biological invasions (Sect. D).

1.3.1 Magnitude of Alien Invasions

The scale and magnitude of some biological invasions are detailed in Chapters 2–5. Chapter 2 provides a biogeographic and evolutionary perspective on the influence of prehuman migration of plants and animals in the formation of the New Zealand biota. Subsequent chapters describe the introductions brought about by Polynesian migration and European colonization (Chaps. 3–5). Polynesian migrants introduced a relatively small suite of organisms to New Zealand, and many of the plant species were only suitable to more northern climates. However, debate continues about the timing of Polynesian settlement and consequences of the animal introductions, particularly the impacts of the kiore (*Rattus exulans*). In contrast, Europeans introduced a plethora of plant and animal species into an anthropogenically disturbed landscape over a relatively short time scale (ca. 150 years). The basis for alien plant introductions is outlined in Chapter 3. The arrival of domestic and feral mammals initiated a new type of disturbance for components of native ecosystems (Chap. 4), and mammals as predators and grazers have long been the major invasion research focus in New Zealand. Introductions of cryptic alien species, especially those that cause diseases, have only recently emerged as important components of invasion research. In Chapter 5, the changes in host–parasite relationships that frequently result from the independent introduction of novel parasites, new hosts, or exotic vectors are described, and their actual and potential impact on native biodiversity explored. In this book and elsewhere, New Zealand's insularity and unique evolutionary history are often cited as the primary reasons for the successful establishment of non-native plants and animals. However, Chapter 6 demonstrates that ecological susceptibility may not be an intrinsic property of all our biota, and provides case studies of selected native species that have been exported to other regions where they have become major pests.

1.3.2 Controls on Invasion Success

Chapters 7–13 consider factors that contribute to a successful invasion, including introduction effort, attributes of invaders and community invasibility, based on analysis within the New Zealand context. Chapter 7 explores the evidence for the importance of genetic factors as determinants of invasion success, particularly the genetic diversity and evolutionary potential of founder populations. Marine invasions have only recently been researched in New Zealand, and Chapter 8 compares the relative vulnerability of island and continental marine biotas. Chapter 9 examines invasibility in avian communities, using extensive acclimatization society records of bird introductions throughout the 19th and 20th centuries to evaluate invasion success in relation to the timing and nature of introduction effort, as well as interspecific competition. Attempts at biological control have a long history in New Zealand, and the factors that influence the establishment and extinction of small populations of mainly alien plant and insect species are discussed in Chapter 10. Predicting invasive species and vulnerable plant communities is a central issue in invasion research worldwide, and Chapter 11 suggests that an understanding of the evolutionary history and the extent and duration of habitats may assist in determining where alien plants will ultimately succeed. Introduced large mammalian herbivores in New Zealand occupy a predator-free environment, and Chapter 12 examines what it is that controls population dynamics in the absence of top-down controls. Chapter 13 uses time-series data to examine factors controlling the invasibility of forest understories by alien herbaceous plant species, including variation in canopy disturbance and community structure along environmental gradients.

1.3.3 Consequences of Alien Invasions

Ecosystem consequences of non-native species invasions form the basis of Chapters 14–21, and these represent overviews of specific ecosystems or examinations of the impacts of single species or guilds. For over a century, introduced brown trout have been liberated widely in freshwater ecosystems in New Zealand, and currently sustain a world-renowned recreational salmonid fishery. Chapter 14 examines the impact of brown trout on native fishes and invertebrates, and ecosystem processes. The next four chapters (Chaps. 15–18) explore what is known about competition between native and invading animal species for high-quality food resources produced by plants.

Chapter 15 considers the ability of introduced birds and insects to pollinate flowers and disperse seeds of native plant species, relative to the level of services currently provided by native animal species.

Alien species introductions have occurred at both the predator and herbivore trophic levels, and in many ecosystems exhibit dynamic populations that fluctuate seasonally and inter-annually depending on variations in plant growth and/or mast seeding of dominant plants species. Chapter 16 outlines the linkages between resources, herbivore populations, and predators in forest and dryland systems managed to maintain native biodiversity values. Because of their widespread impacts, some non-native invasive species have achieved keystone status in natural ecosystems. Evidence for impacts on native vegetation and birds, and as a vector for serious animal diseases, is reviewed in Chapter 17 for the Australian brushtail possum. The functional role of native keystone species may also be significantly affected by invasive alien species, and Chapter 18 outlines the multi-trophic impacts of non-native invertebrates (wasps) capturing most of the carbon- and energy-rich honeydew produced by native scale insects on *Nothofagus* trees in warm-temperate forests.

Nitrogen-fixing shrubs are amongst the most successful invasive plant species in New Zealand, and Chapter 19 compares their traits and successional roles with those of the native nitrogen-fixing shrubs on potentially forested sites. In natural forests, introduced mammalian herbivores exhibit nonrandom feeding preferences that constrain the spectrum of plant traits that remain. The possible effects of shifts in litter quality induced by non-native browsers on below-ground decomposers are discussed in Chapter 20. Mechanistic understanding of the ecological consequences of introduced mammalian herbivores is often confounded by natural disturbance regimes, the presence of multiple herbivores, and natural successional trends. Chapter 21 discusses several notable impacts long attributed to non-native herbivores, and the challenge of distinguishing natural processes from invasive species impacts.

1.3.4 Management of Biological Invasions

New Zealand's experience of attempting to eradicate and control invasive introduced species, as well as preventing the entry of further potentially invasive species, is well known, and forms the theme for the six end chapters (Chaps. 22–27). Ecosystem recovery following the removal or reduction of introduced herbivorous mammal populations was widely expected, but Chapter 22 examines the extent to which this has occurred in forest and grassland vegetation, and discusses why ecosystem restoration goals are often unrealistic ecologically. Deliberate alien introductions are frequently undertaken as part of biological control programs, most often to suppress non-native weed and insect pests in agricultural and natural systems, and Chapter 23 evaluates the environmental outcomes and risks associated with this approach. Protection of our declining native avifauna continues to drive developments in predator management in New Zealand, and Chapter 24 describes the different approaches to predator control, from total extermination to manipulating functional responses to reduce predation impacts, to maintain and restore bird populations. The control of introduced species nowadays largely depends on standardized assessments of the invasive potential and impact of the species, and the relative vulnerability of different resident communities or environments. In this light, Chapter 25 outlines an invasive species risk-management system for marine ecosystems. Chapter 26 examines whether commercial exploitation of introduced feral animals in New Zealand provides an effective method for reducing their densities, and therefore their impacts on native biota. Molecular biology offers new technologies for the control of pests and weeds, and Chapter 27 considers the technical, operational, and societal challenges in developing and utilizing this approach.

The final chapter (Chap. 28) concludes the book with an evaluation and synthesis of the New Zealand contribution to understanding biological invasions now and potentially in the future.

1.4 Conclusion

The ecology, impacts and control of invasive non-native species has been a dominant research focus for over a century in New Zealand. This has coincided with the mass arrival of introduced plant and animal species, and we have attempted to provide an overview of aspects of that research. We have particularly focused on situations where researchers have sought to develop predictions about the characteristics of invasive species, vulnerable ecosystems, and general impacts. Biological invasions will continue to drive much of the ecological research in New Zealand as we attempt to reduce the importation of new invasive species, and assess and manage those already present. As contributions to this book indicate, successful control of alien invasive species in New Zealand will continue to depend on the development of robust ecological theory, improved understanding of natural communities and how they function, effective management approaches, and a strong societal commitment to the issue.

Acknowledgements Comments from Kelvin Lloyd, Jake Overton, Joe Craine, Peter Bellingham, and David Wardle improved the chapter, and Christine Bezar assisted with editing. Preparation of this text was financially supported by the New Zealand Foundation for Research Science and Technology.

References

Allan HH (1937) The origin and distribution of the naturalized plants of New Zealand. Proc Linn Soc Lond 150th session (1):25–46

Bannister PB (1994) Upon this blasted heath. Guest Editorial. NZ J Ecol 18:83–86

Basse B, McLennan JA, Wake GC (1999) Analysis of the impact of stoats, *Mustela erminea*, on northern brown kiwi, *Apteryx mantelli*, in New Zealand. Wildl Res 26:227–237

Beggs J (2001) The ecological consequences of social wasps (*Vespula* spp.) invading an ecosystem that has an abundant carbohydrate resource. Biol Conserv 99:17–28

Brockie B (1992) A living New Zealand forest. Bateman, Auckland, New Zealand

Brown KP, Moller H, Innes J (1996) Sign left by brushtail possums after feeding on bird eggs and chicks. NZ J Ecol 20:277–284

Clark AH (1949) The invasion of New Zealand by people, plants and animals: the South Island. Rutgers Univ Press, Piscataway, NJ

Climo FM (1973) The systematics, biology and zoogeography of the landsnail fauna of Great Island, Three Kings group, New Zealand. J R Soc NZ 3:565–628

Closs G, Dean T, Champion P, Hofstra D (2004) Aquatic invaders and pest species in lakes. In: Harding J, Mosley P, Pearson C, Sorrell B (eds) Freshwaters of New Zealand, chap 27. Caxton Press, Christchurch, New Zealand

Cockayne L (1910) New Zealand plants and their story. Government Printer, Wellington, New Zealand

Darwin C (1845) Journal of researches into the natural history and geology of the countries visited during the voyage of H.M.S. "Beagle" round the world: under the command of Capt. Fitz Roy. Murray, London

Davis LT, Guy PL (2001) Introduced plant viruses and the invasion of a native grass flora. Biol Invasions 3:89–95

Diamond JM (1990) New Zealand as an archipelago: an international perspective. In: Towns DR, Daugherty CH, Atkinson IAE (eds) Ecological restoration of New Zealand islands. Conservation Sciences Publications no 2. Department of Conservation, Wellington, New Zealand, pp 3–8

Fraser KW, Cone JM, Whitford EJ (2000) A revision of the established ranges and new populations of 11 introduced ungulate species in New Zealand. J R Soc NZ 30:419–437

Guthrie-Smith H (1921) Tutira: the story of a New Zealand sheep station. Blackwood, London

Hewitt CL, Willing J, Bauckham A, Cassidy AM, Cox CMS, Jones L, Wotton DM (2004) New Zealand marine biosecurity: delivering outcomes in a fluid environment. NZ J Mar Freshwater Res 38:429–438

Innes J, Hay R, Flux I, Bradfield P, Speed H, Jansen P (1999) Successful recovery of North island kokako *Callaeas cinerea wilsoni* populations, by adaptive management. Biol Conserv 87:201–214

Kelly D, Skipworth JP (1984) *Tradescantia fluminensis* in a Manawatu (New Zealand) forest. II. Management by herbicides. NZ J Bot 22:399–402

King CM (ed) (1990) The handbook of New Zealand mammals. Oxford Univ Press in association with the Mammal Society, New Zealand branch

Kowarik I (1995) Time lags in biological invasions with regard to the success and failure of alien species. In: Py_ek P, Prach K, Rejmánek M, Wade M (eds) Plant invasions: general aspects and special problems. SBP Academic Publishing, Amsterdam, pp 15–38

Mack RN (2001) Motivations and consequences of the human dispersal of plants. In: McNeely JA (ed) The great reshuffling: human dimensions in invasive alien species. IUCN, Gland, Switzerland and Cambridge, UK, pp 23–24

McDowall RM (1994) Gamekeepers for the nation: the story of New Zealand's acclimatisation societies, 1861–1990. Canterbury Univ Press, Christchurch, New Zealand
McDowall RM (2003) Impacts of introduced salmonids on native Galaxiids in New Zealand upland streams: a new look at an old problem. Trans Am Fish Soc 132:229–238
Montague TL (ed) (2000) The brushtail possum: biology, impact and management of an introduced marsupial. Manaaki Whenua Press, Lincoln, New Zealand
Norbury G (2001) Conserving dryland lizards by reducing predator-mediated apparent competition and direct competition with introduced rabbits. J Appl Ecol 38:1350–1361
Norbury G, Heyward R, Parkes J (2002) Short-term ecological effects of rabbit haemorrhagic disease in the short-tussock grasslands of the South Island, New Zealand. Wildl Res 29:599–604
Parkes J, Murphy E (2003) Management of introduced mammals in New Zealand. NZ J Zool 30:335–359
Ramsey D, Veltman C (2005) Predicting the effects of perturbations on ecological communities: what can qualitative models offer? J Anim Ecol 74:905–916
Saunders A, Norton DA (2001) Ecological restoration at mainland islands in New Zealand. Biol Conserv 99:109–119
Scott NA, Saggar S, McIntosh PD (2001) Biogeochemical impact of *Hieracium* invasion in New Zealand's grazed tussock grasslands: sustainability implications. Ecol Appl 11:1311–1322
Simberloff D, von Holle B (1999) Positive interactions of nonindigenous species: invasional meltdown? Biol Invasions 1:21–32
Standish RJ (2004) Impact of an invasive clonal herb on epigaeic invertebrates in forest remnants in New Zealand. Biol Conserv 116:49–58
Standish RJ, Robertson AW, Williams PA (2001) The impact of the invasive weed *Tradescantia fluminensis* on native forest regeneration. J Appl Ecol 38:1253–1263
Standish RJ, Williams PA, Robertson AW, Scott NA, Hedderley DI (2004) Invasion by a perennial herb increases decomposition rate and alters nutrient availability in warm temperate lowland forest remnants. Biol Invasions 6:71–82
Stewart GH, Rose AB, Veblen TT (1991) Forest development in canopy gaps in old-growth beech (*Nothofagus*) forests, New Zealand. J Veg Sci 2:679–690
Storey T, Clayton B (2002) Biosecurity law in New Zealand. NZ J For 47(2):28–30
Sullivan JJ, Timmins SM, Williams PA (2005) Movement of exotic plants into coastal native forests from gardens in northern New Zealand. NZ J Ecol 29:1–10
Taylor RH, Thomas BW (1989) Eradication of Norway rats (*Rattus norvegicus*) from Hawea Island, Fiordland, using brodifacoum. NZ J Ecol 12:23–32
Thomson GM (1922) The naturalisation of animals and plants in New Zealand. Cambridge Univ Press, Cambridge, UK
Wardle DA, Barker GM, Yeates GW, Bonner KI, Ghani A (2001) Introduced browsing mammals in New Zealand natural forests: aboveground and belowground consequences. Ecol Monogr 71:587–614
Wilson JB (1989) Relations between native and exotic plant guilds in the Upper Clutha, New Zealand. J Ecol 77:223–235
Wilson JB, Williams PA, Lee WG (1989) Vegetation composition and segregation in relation to the environment at low altitudes in the Upper Clutha Basin, New Zealand. NZ J Ecol 12:103–116
Wilton AD, Breitwiesser I (2000) Composition of the New Zealand seed plant flora. NZ J Bot 38:537–549
Worthy TH, Holdaway RN (2002) The lost world of the moa: prehistoric life of New Zealand. Canterbury Univ Press, Christchurch, New Zealand

A Magnitude of Alien Invasion

2 Becoming New Zealanders: Immigration and the Formation of the Biota

M.S. McGlone

2.1 Introduction

Plants and animals have been flying, drifting, swimming and floating to the New Zealand archipelago over the millions of years in which it has been in existence. The continuing interaction of invader and resident has had a powerful influence on the formation of the biota. The current fauna and flora therefore represent the cumulative outcome of many cycles of invasion, speciation and extinction extending back 80 Ma to Gondwana.

Invasion is not simply about dispersal and establishment. Many factors come into play – the biotic legacy from the remote Gondwanic past, the constantly altering geology and climate of the archipelago, and the makeup of the flora and fauna of New Zealand's closest neighbours – all have had a strong influence on the outcome. Successful invaders ultimately become an integral part of the biota, changed by their environment and, in turn, changing it.

The focus in this chapter is the history and consequences of these prehuman biotic invasions. I first survey the post-Gondwanic history of New Zealand, and review the evidence for ancient or dispersal origins of the biota. I then show how the makeup of the biota has been strongly influenced by the ever-changing climate and physical geography of the archipelago, as well as the long-term consequences of the absence of functional groups common elsewhere in the world, but rarely found on oceanic islands.

2.2 Dispersal and Vicariance: Late Cretaceous to the Pleistocene

There has been intense controversy over the extent to which the current New Zealand biota derives from ancestors that have been continuously in the

Ecological Studies, Vol. 186
R.B. Allen and W.G. Lee (Eds.)
Biological Invasions in New Zealand

Table 2.1 Geoclimatic history (after Fleming 1979; McGlone et al. 1996; Lee et al. 2001)

Geological epoch	Time range (Ma)	Location	Latitude	Geology and topography	Climate
Early Cretaceous	100–85	Joined to West Gondwana, but shallow seas forming	>80°	Mountainous	Cool and equable
Late Cretaceous	85–65	First oceanic seafloor formed; seaways of a few hundred kilometres	80°	Peneplanation begins	Cool and equable
Cretaceous–late Palaeocene	65–55	Rapid movement to current 2,000-km separation; New Caledonia, Lord Howe rise and Norfolk ridge submerged	70–60°	Reduction in land area; low to moderate relief; peat swamps and coastal plains. Global bolide impact affects vegetation and causes extinctions	Cool-temperate
Early Eocene–mid-Oligocene	55–30	Continued northward movement; volcanic island chain to north; re-emergence of land in New Caledonia region ca. 40 Ma ago	60–50°	Further subsidence and flooding by shallow seas; leached soils	Maximum warmth (subtropical) in the Eocene; early Oligocene cool-temperate
Mid-Oligocene–early Miocene	30–20	Continued northward movement	50–45°	Minimum land area, split into several islands, possibly as little as 10% of present	Warm-temperate
Early–mid Miocene	20–10	Island chain linkage formed to New Caledonia region	50–45°	Uplift, increased land size; volcanism in north	Maximum warmth (subtropical)
Late Miocene–Plio-Pleistocene	10–2	Present location; Lord Howe and Norfolk islands erupted 7 and 3 Ma ago	45–35°	Alpine fault active; mountain uplift; central North Island volcanism	Cooling climates in the Pliocene; onset of Plio-Pleistocene ice ages ca. 3.0 Ma ago

region since it was part of Gondwana. At one extreme, some have argued that nearly all extant biota have been continuously present since before separation (Craw et al. 1999; but see Cox 1998). Others have suggested the opposite view that all or nearly all of the biota is post-separation long-distance dispersal (Pole 1994). Although most would follow Fleming (1979) in accepting that both dispersal and vicariance (splitting of taxon ranges by geomorphic change) have had an influence, global evidence for frequent trans-oceanic dispersal is now increasingly strong (de Queiroz 2005).

The history of post-Gondwanic New Zealand has been summarised from a biogeographic perspective several times in recent years, most comprehensively by Lee et al. (2001). A synopsis of present understanding is presented in Table 2.1.

2.2.1 Gondwanic Pre-Separation Biota

Mid-Cretaceous environments (120–80 Ma) were radically different to today's. The region ancestral to the present New Zealand landmass experienced 3 months of complete darkness in winter, and near-continuous sunlight during the brief summer because of its high latitude (ca. 70°S). However, because warm water penetrated to high latitudes, climates were cool, but not cold (mean annual temperatures 10–15 °C), and humid (Quilty 1994; Parrish et al. 1998; Kennedy et al. 2002). The mid-Cretaceous Gondwanan vegetation was dominated by forests of tall trees of diverse podocarps, araucarians, *Ginkgo* and *Bennettites* over understories of ferns, horsetails, lycopods and bryophytes (Dettman 1989; Parrish et al. 1998; Falcon-Lang and Cantrill 2002). By 100 Ma, shrubby chloranthaceous angiosperms had appeared, and rapid spread and differentiation of angiosperms occurred by the latest Cretaceous. A striking feature of the mid to late Cretaceous flora is the predominance of deciduous trees, including many with leaves strongly reminiscent of those in current temperate Northern Hemisphere forests (see illustrations in Pole 1992 and Parrish et al. 1998). The Cretaceous fossil record includes dinosaurs, but *Sphenodon*, fish, geckos and freshwater crododilians were almost certainly present because of their occurrence in early Miocene deposits (Worthy et al. 2002). Monotremes (early mammals) were likely present, as they were found in Australia by that time (Archer et al. 1994).

Plant macrofossil evidence from Gondwana indicates the presence of Atherospermataceae, Cunoniaceae, Winteraceae, Lauraceae, Nothofagaceae, Proteaceae, Podocarpaceae and Auracariaceae (Poole and Gottwald 2001; Vajda et al. 2001), and even some extant genera such as *Agathis* (Stockler et al. 2002). However, this evidence is rarely conclusive, as extinction of one lineage in a family or genus and reinvasion by another is always possible. For instance, although Gondwanic pollen fossils of the Nothofagaceae indicate the presence of the family, they belong to groups not extant in the modern New Zealand flora (McGlone et al. 1996). Molecular clock investigations do not, in general, support the concept of the New Zealand flora being ancient. One of the oldest pollen types in New Zealand is the early Cretaceous *Clavatipollenites* (near identical to *Ascarina lucida*), and this pollen type has been continuously present ever since (Mildenhall 1980). However, extant *Ascarina lucida* in New Zealand can be no more than 18–9 Ma old, and is in all likelihood a relatively recent arrival from the subtropics (Zhang and Renner 2003). Ancient genera such as *Laurelia* (Atherospermataceae) appear to have dispersed from southern South America to New Zealand ca. 38 Ma ago (Renner et al. 2000; Wardle et al. 2001), in this case confirmed by the pollen record (Mildenhall 1980). The extant *Lophozonia* subgenus of *Nothofagus* (*N. menziesii* ancestor) and subgenus *Fuscospora* (*N. fusca* group) dispersed from Australia to New Zealand, perhaps as recently as 5–10 Ma ago (Swenson et al. 2001; Knapp et al. 2005), despite the presence of their pollen fossils earlier in the record. Extinction and reinvasion of the same taxon thus seems probable.

It is possible that animal groups have a greater Gondwanan component. The invertebrate fossil record is very poor, but some ancient elements are present such as the gradungulid spiders, the Dipluran *Heterojapyx* (little differentiated from the 300-Ma-old fossil *Testajapyx*), and the Jacobsoniid beetle (*Saphophagus minutus*: Coleopteran, Dermestoidea) regarded as coeval with tuatara (Daugherty et al. 1993). The high levels of endemism at family and generic level in invertebrates is indirect evidence for ancient and probably vicariant origins for many. For instance, in the beetles (Coleoptera) there are several endemic subfamilies and tribes and many endemic genera (Watt 1975; Klimaszewski 1997; Larochelle and Larivière 2001). There is uncertainty, however, over some other putatively ancient groups, such as the Onychophorans (peripatus) where molecular clock studies suggest recent migration for at least some species whereas phylogenetic and biogeographic studies tend to support Gondwanic vicariance (Gleeson et al. 1998). For vertebrate groups, recent molecular work suggests that the moa and acanthisittid wrens (sister taxon to all other extant passerines) and geckos were part of the original Gondwanic fauna (Cooper et al. 2001; Haddrath and Baker 2001; Ericson et al. 2002), and the phylogenetic isolation of the endemic families Aptornithidae (extinct adzebill *Aptornis*) and Callaeatidae (wattlebirds) suggests the same (Worthy and Holdaway 2002).

2.2.2 Late Cretaceous to Pliocene: Post-Separation Change

As the New Zealand continental fragment moved further from the Antarctic–Australian landmass, it lost heat and subsided, becoming an oceanic archipelago of low-lying islands making up perhaps ~20 % of the present landmass or less in the mid-Oligocene (Cooper and Cooper 1995). Warm-temperate, moist climates tended to prevail throughout the Tertiary, although there were intervals of somewhat cooler, drier climates in the early Oligocene and the middle to late Miocene (Mildenhall 1980; Pole and Douglas 1998). Steady northward movement into lower-latitude, warmer oceans helped counteract global cooling in the middle to late Miocene (Nelson and Cooke 2001).

For a prolonged period after separation, possibly until the end of the Eocene, New Zealand retained land connections north to the New Caldedonian region (Herzer et al. 1997; Herzer 1998). From then until the mid-Oligocene, New Zealand, although surrounded by ocean, faced an immense, near-continuous Gondwanan coastline consisting of South America, Antarctica and Australia. Until the late Miocene, island chains to the north and successive dispersal along ocean ridges may have provided a pathway for taxa incapable of long dispersal leaps (Herzer et al. 1997). Biotic interchange of all but the most immobile groups appears to have continued unabated until the Pliocene.

Generic and family-level endemism is low among plants (Godley 1975; Fenner et al. 1997), which suggests the flora as a whole is of relatively recent origin. Macrofossil and Tertiary pollen evidence supports this view, as these data demonstrate a constant influx of new taxa from the mid-Cretaceous to the present (Mildenhall 1980; Pole 1993). Australia and New Zealand had closely similar floras that underwent near-synchronous change throughout the Cenozoic, possible only if there was repeated invasion of New Zealand by taxa from Australia, or of both regions from another source (Pole 1994; Macphail 1997). Taxa belonging to families evolving after the end of the Cretaceous have always been regarded as having dispersed to New Zealand, and recent molecular work strongly supports this assumption (e.g. *Hebe*; Wagstaff et al. 2002). Of the 22 related vascular plant species pairs between southern South America and New Zealand examined by Wardle et al. (2001), only one (*Lepidothamnus laxifolius–L. fonckii*) indisputably originated before separation.

Terrestrial animals have a poor fossil record, and it is therefore difficult to demonstrate post-separation dispersal, but molecular phylogenies and clocks indicate many distributions that must have been achieved in this way (Arensburger et al. 2004). For instance, a number of vertebrates made transoceanic dispersal leaps, including skinks (Hickson et al. 2000), kiwi and parakeets (Chambers et al. 2001; Haddrath and Baker 2001). The five modern cicada genera in New Zealand probably arrived after the Oligocene period (and possibly as late as 12 Ma ago) in two waves of colonization, one from Australia and the other from New Caledonia (Chambers et al. 2001; Arensburger et al. 2004).

Until the early Miocene, New Zealand and southern Australia lay at equivalent latitudes (50–40°S) and shared similar moist, temperate to subtropical climates and low-lying landscapes. Only when the climate of New Zealand, Australia and Antarctica started diverging significantly in the late Miocene and Pliocene did the biotas begin to differentiate. The first Antarctic ice sheet formed 34 Ma ago in the early Oligocene and, at around 24 Ma, Antarctic coastal *Nothofagus* forest was replaced by tundra (Raine and Askin 2001). A permanent ice sheet formed 15 Ma ago in the middle Miocene (Barrett 2003). Antarctica was therefore lost as a source of immigrants and as a dispersal pathway from South America to Australasia. In the middle Miocene, Australia, as a result of steady northward movement, became increasingly warmer, more arid and monsoonal. New Zealand remained to the south, falling increasingly under the influence of the developing circum-Antarctic current system and the intensifying westerly wind belts, accompanied by a cooler and moister climate. At the same time, the New Zealand archipelago began to change radically as the Alpine Fault became active. This resulted in a main landmass that was more mountainous and compact, island chains to the north being submerged, and the total land area much reduced (McGlone et al. 2001).

2.2.3 Pleistocene

Pronounced global cooling and the onset of glacial–interglacial cycles (with periodicities ranging between 40,000 and 100,000 years) at about 3 Ma markedly increased climatic variability in the late Pliocene and Pleistocene (Head and Nelson 1994; Carter and Naish 1998). Near-complete forest cover during brief, warm, moist interglacials alternated with cool, windy, glacial grassland and shrub landscapes (McGlone et al. 1993). Permanent ice fields, glaciers and alpine habitats appeared, which spread massively during glacials; steep mountains, consisting for the most part of weak sandstones and schists, developed extensive unstable scree and boulderfield habitats; in addition, dry, warm-summer/cool-winter climate regimes established in the lee of the rising mountains.

Most of the speciose radiations that characterize the New Zealand biota involve these recently formed alpine and subalpine or rainshadow environments, in particular alpine to subalpine herb field, grassland and shrublands. Although some post-Pliocene taxa evolved from long-established groups, introduction by long-distance transport of groups (many ultimately from Northern Hemisphere sources) pre-adapted to cool climates and steep terrain provided most novelties (Raven 1973; Wardle 1978; Mildenhall 1980; Winkworth et al. 2002). In the case of the speciose plant radiations, nearly all appear to involve relatively recent immigrants (e.g. *Epilobium*, *Hebe*, *Aciphylla*, *Ranunculus* and *Carmichaelia*). This does not seem to apply to the fauna, in which speciose adaptive radiations appear to have occurred in groups long established in the country, such as snails, weta, stoneflies, skinks, cicadas, wrens and moa.

2.3 Dispersal and Formation of the Modern Biota

2.3.1 Independent Dispersal During the Human Era

Recent independent trans-Tasman dispersal has been observed in a number of groups (see Close et al. 1978 for a discussion), mediated largely by the persistent westerly winds that transport low-level air from Australia to New Zealand in 3–5 days (Sturman et al. 1997). A range of plants have arrived unaided (McGlone et al. 2001), most notably dust-seeded orchids from Australia. Flighted animals are, as might be expected, relatively common immigrants. A substantial number of non-breeding vagrant birds occur from time to time, and some 16 species (e.g. silvereye *Zosterops lateralis*) have established breeding populations unaided after human settlement (Holdaway et al. 2001). Compared with this, some 37 human-introduced bird species have established (see Chap. 9). A number of recent arrivals of butterflies (e.g. the blue moon, *Bolina nerina*, from Australia) have been observed (some 36 in the eight seasons between 1968 and 1976; Fox 1978), and some have become established, most notably the monarch butterfly. Several dragonflies have colonized from the southwest Pacific in recent years: *Hemianax papuensis* (ca. 1918 A.D.), *Ischnura aurora* (ca. 1926 A.D.) and *Hemicordulia australiae* (ca. 1900 A.D.; Rowe 1987).

Changing environments with clearance of forest by humans appear to have rendered the landscape more invasible for some open-habitat species such as the welcome swallow (*Hirundo tahitica*) and the spur-winged plover (*Vanellus miles*). Extinction of endemic competitors by introduced mammals or humans may be responsible for the success of some immigrants, e.g. the Australasian harrier (*Circus approximans*) after the elimination of Eyle's harrier (*Circus eylesi*), and the pukeko (*Porphyrio melanotus*) with the functional extinction of the North Island and South Island takahe (*P. mantelli* and *P. hochstetteri*).

2.3.2 Non-Endemic Species as an Indicator of Past Dispersal

Factors favouring the dispersal of plants between Australia and New Zealand have been assessed through an analysis of specific endemism in New Zealand (Brownsey 2001; Jordan 2001; McGlone et al. 2001). Bisexual plants, herbs, and plants with small or dust-sized propagules (Orchidaceae, ferns and fern allies, mosses and liverworts) are disproportionately non-endemic. Fern and fern ally species are 46 and 17 % endemic respectively. Plant specialists of coastal or wetland habitats (sedges, rushes and many monocot herbs), and those with hooked, barbed or sticky fruits (albeit not those with fleshy fruits, or plumed or very hairy disseminules) tend to comprise a large number of non-endemic native species. In contrast, forest plants, trees and shrubs (in particular, tall trees) tend to be highly endemic, e.g. 100 % endemism in conifer tree species.

Animal groups differ markedly in their ability to cross ocean gaps. Land and freshwater birds have relatively high endemism, and this despite being flighted: 174 of 245 prehuman species breeding in New Zealand were endemic (71.0 %), and 35 out of 110 genera (31.8 %; Holdaway et al. 2001). All three species of bat are endemic. The two mystacinids belong to a family that is restricted to New Zealand, whereas *Chalinolobus* belongs to a diverse Australian genus and is probably a more recent arrival. All native reptiles and amphibians are endemic, reflecting the difficulty that non-flighted vertebrates have in crossing oceans. Freshwater fishes are largely endemic, although a marine phase in the life cycle enhances dispersability (Waters et al. 2000; McDowall 2002).

Earthworms are 98 % endemic, with the few exceptions tolerant of seawater (Lee 1959). Molluscs are nearly completely endemic (99.6 % of 1,400 spp.; G. Barker, personal communication 2003), as are spiders (Forster and Forster 1999). Harvestmen form an endemic family. Insects are largely endemic, specific endemism varying in the range 90–100 % in most orders (Dugdale 1988; Larochelle and Larivière 2001). There are some exceptions, such as Cryptostigmata mites (82 %; Luxton 1985), dragonflies (80 %, Odonata; Rowe 1987) and thrips (37 %; Mound and Walker 1982).

2.3.3 Factors Affecting Immigration Success

The most important factor affecting immigration success is the ability of a taxon to make the transfer. More easily transported plants have propagules that are small enough to be wind-transported, or that adhere easily to birds or are well adapted to oceanic dispersal. Animals are assisted by flightedness, very small body size, or tolerance of seawater. Transport factors alone seem to have skewed the composition of the biota (Carlquist 1974): for example, New Zealand has a higher proportion of ferns and fern allies in its vascular flora (9.5 %) than do continental floras (average 7 %; Wagner 1995).

The next most important factor is the compatibility of the New Zealand physical and biotic environment with those of the source regions. New Zealand's geographic position relative to possible source areas means there are few close environmental matches. Its cool-temperate to temperate climates differentiates it from most of Australia and the subtropical islands. The deeply leached low-nutrient soils of much of Tasmania and Australia have few parallels in New Zealand. Fire is a dominant influence over most of Australia and Tasmania but was probably not at all significant in prehuman New Zealand (Ogden et al. 1998). These environmental factors probably best explain why the New Zealand biota, although largely derived from Australia, has so few characteristically "Australian" species (McGlone et al. 2001), and why rare, long-distance dispersal of cool-adapted plants from the Northern Hemisphere or southern South America, rather than the southwest Pacific, has formed the basis for much of the herbaceous flora (Winkworth et al. 2002; Rivadavia et al. 2003).

Finally, there is a cluster of factors that influence a species' ability to establish. Species with narrow specialist requirements are much less likely to successfully establish than those with broader niches. This most clearly applies to herbivores, predators and parasites with limited host preferences, and species with mutualistic relationships. The high level of endemism in invertebrates, which often have narrow feeding preferences, may partly result from this factor. A good example comes from the absence of animal groups that are specialized to feed on nectar and pollen from certain types of flowers. New Zealand has no advanced, long-tongued bees, few butterflies (no plant species have been identified as primarily adapted to butterfly pollination), no native hawkmoths, and only seven indigenous birds that visit flowers (cf. less than 1 % of indigenous angiosperms are adapted for bird pollination; Lloyd 1985). Almost certainly, this lack is a direct consequence of the improbability that a plant species and its specialized pollinator would migrate at exactly the same time.

Immigrant species that reproduce early should be favoured, whereas species with a long interval before maturity have low intrinsic population growth rates, and run the risk of not reproducing at all. The preponderance of herbs in the recently dispersed flora, and the low number of non-endemic trees and tall shrubs (Jordan 2001; McGlone et al. 2001) may in part be related to this difficulty for plants with a long juvenile phase in establishing a viable population. This should be true for animals as well.

The predominance of wetland species (birds and higher plants) in recent migration to New Zealand (McGlone et al. 2001) almost certainly results from a favourable combination of many of the above factors. Wetlands, often consisting of variable mosaics of contrasting environments governed by water depth and nutrient status, nevertheless tend to be similar over very large spatial scales. They select for stress-tolerant taxa, which typically show broad tolerance and plasticity (Santamaría 2002). Asexual reproduction and clonal spread is also a common feature of wetland plant species, facilitating establishment from a few propagules only. Finally, many wetlands are dominated by open or low vegetation and subject to constant disturbance, and are therefore highly invasible by both birds and plants. Wetland birds are often good dispersers of plants and small animals because of their wading habits, tendency to an omnivore diet, and often highly mobile lifestyles (Green et al. 2002). New Zealand has few long-established wetland bird species, and the current list is largely of recent immigrants (R. Holdaway, personal communication 2003) that probably responded to the dramatic increase in lagoons and open-water wetlands following deforestation 800–700 years ago.

2.4 Evolutionary Effects of Selective Dispersal and Island Environments

As we have seen, New Zealand's isolated oceanic setting, and cool to temperate climate in a region dominated by warm subtropical landmasses, has resulted in strong immigrant selection. Its flora and fauna are thus strongly influenced by this skewed selection from the global biota. New Zealand is therefore not a time capsule of pre-separation Gondwanan biota. Each climatic or geomorphic transition resulted in a wave of local extinctions, adaptation by surviving species, and immigration of new species suited to the new situation (Pole 1994). Strong species-selection pressure arose from the exacting requirements surrounding dispersal and establishment (Lloyd 1985; McGlone et al. 2001). Each successful establishment was followed by the conversion or radiation of the invader into typical island forms (Table 2.2).

2.5 New Zealand in an Invasive World

A vast biotic encounter is now underway between human-introduced continental invaders and our island inhabitants. The legacy of past cycles of long-distance dispersal, and the peculiarities and restricted scope of island environments have had a dramatic effect on the outcome of these exotic–indigenous encounters.

First, and most important, has been the evolutionary consequences of poor representation or absence of certain functional groups (Table 2.2). The absence of indigenous snakes, predatory terrestrial mammals, colonial bees, wasps and ants has insured a one-sided battle following human settlement between ecologically naïve islanders and sophisticated continentals. In the case of many indigenous birds, reptiles and giant invertebrates, the introduction of mammalian predators has meant rapid extinction or reduction to small, protected enclaves. Mammalian herbivore pressure has reduced many indigenous plants over wide areas, but the case for a completely novel and devastating impact is harder to argue, as indigenous avian herbivores may have had a profound effect (Atkinson and Greenwood 1989; Cooper et al. 1993; Bond et al. 2004), and there have been relatively few extinctions due to mammalian herbivore pressure. It is therefore an open question as to how resilient the native flora will prove to be.

Second, the ephemeral nature of certain habitats in New Zealand over a glacial–interglacial cycle (e.g. lowland grasslands; cold tundra; warm-temperate forest) may have limited the degree to which indigenous species could adapt (McGlone et al. 2001). Some immigrant continental species that are specialists for such habitats may thus have an advantage. For instance, a number

Table 2.2 Evolutionary effects of island environments on the biota

Driver	Effect	Consequence
Temperate, moist, highly oceanic climates		
	Moderate seasonal temperature variation	Low level of cold adaptation, little diapause in insects and few deciduous plants
	Low frequencies of fire	Lack of fire adaptations in the biota; little below-treeline grassland
	Ideal conditions for woody growth	Large areas of dense rainforest and tall scrub; very restricted grasslands
		Few annual growth forms and a limited range of lowland herbs
		Arborescence in herbaceous groups (*Hebe*, *Senecio*); giant tussocks (*Chionchloa*, *Poa*, *Aciphylla*, *Phormium*, *Astelia*)
	Moist evergreen forests; dark understories, and deep litter	Few angiosperm understory herbs; abundance of ferns, mosses and liverworts
		Abundance of forest invertebrates and high levels of speciation in forest floor invertebrates
		Survival of archaic forms
Low concentrations of plant nutrients in rocks and soils		
	Slow plant growth rates on mature soils	Dominance by slow-growing, nutrient-conserving evergreen trees and shrubs
		Few deciduous plants
		Low numbers of nitrogen-fixing trees and shrubs
Large ocean gap resulting in missing functional groups		
	Absence: mammalian herbivores	Abundance of herbivorous birds – moa, takahe, ducks, geese, rails
		Avian-specific defences common, e.g. high-strength stems, crypsis, mimicry
		Large numbers of plants sensitive to mammalian browsing
		Lack of spiny plants
	Absence: mammalian insectivores	Evolution of guilds of ground-probing carnivores (kiwi, rails, snipes, wrens)
		Proliferation of large carnivorous molluscs and centipedes

Table 2.2 (*Continued*)

Driver	Effect	Consequence
	Absence: mammalian, iguanoid and snake predators	Proliferation of large flighless or weakly flighted birds, bats and invertebrates
	Absence: aggressive colonial ants and termites	Abundant forest-floor-dwelling and litter-dwelling molluscs, insects, worms
	Absence: large, aggressive generalist carnivorous, colonial hymenoptera (wasps, hornets)	Proliferation of spiders, flies and moths
	Absence: mammalian frugivores	Many bird-dispersed fruits; few large, nutrient-rich fruits
	Absence: long-tongued bees, hawkmoths	Many flower-visiting moths, and flies; small, regular, white or green, scented promiscuous flowers; few highly specialized or coloured flowers

of cold-adapted continental trees will grow well above the New Zealand treeline, and have the potential to exclude the indigenous tussocks and low shrubs currently growing there (Wardle 1998). In lowland and montane environments, introduced rhizomatous and annual grass species have an advantage over tussock grasses in the presence of grazing (Lee 1998).

Finally, although the consequences of isolation and island environment may include some fundamental vulnerabilities, the total conversion of much of the lowland and montane New Zealand landscape into an open exotic grassland, different in basic ways to any existing in the past, has strongly affected the balance between islander and exotic. Burning and felling of dark, moist forests and their replacement with human-managed grasslands have effectively created a drier, warmer and more fire-prone prairie or steppe environment with greatly elevated soil fertility. Firing of previously extensive forest- and shrub-covered wetlands has created open-water wetlands subject to increased nutrient-rich runoff. Such environments are well suited to continental species and inimical to island specialists. The New Zealand biota, generally slow-reproducing and slow-growing, seems not to be well adjusted to the cyclical pulses that characterize such landscape. Over much of the New Zealand lowland landscapes, indigenous species are therefore reduced to a few generalists, often of historically recent Australian or southwest Pacific origin.

Acknowledgements I thank Rob Allen, Peter Bellingham, Richard Duncan and Richard Holdaway for comments on a draft of this manuscript.

References

Archer M, Hand SJ, Godthelp H (1994) Patterns in the history of Australia's mammal and inferences about palaeohabitats. In: Hill RS (ed) History of the Australian vegetation: Cretaceous to recent. Cambridge Univ Press, Cambridge, pp 80–103

Arensburger P, Buckley TR, Simon C, Moulds M, Holsinger KE (2004) Biogeography and phylogeny of the New Zealand cicada genera (Hemiptera: Cicadidae) based on nuclear and mitochondrial DNA data. J Biogeogr 31:557–569

Atkinson IAE, Greenwood RM (1989) Relationships between moas and plants. NZ J Ecol suppl 12:67–96

Barrett PJ (2003) Cooling a continent. Nature 421:221–223

Bond WJ, Lee WG, Craine JM (2004) Plant structural defences against browsing birds: a legacy of New Zealand's extinct moas. Oikos 104:500–508

Brownsey PJ (2001) New Zealand's pteridophyte flora – plants of ancient lineage but recent arrival? Brittonia 53:284–303

Carlquist SJ (1974) Island biology. Columbia Univ Press, New York

Carter RM, Naish TR (1998) Have local stages outlived their usefulness for the New Zealand Pliocene-Pleistocene? NZ J Geol Geophys 41:271–279

Chambers GK, Boon WM, Buckley TR, Hitchmough RA (2001) Using molecular methods to understand the Gondwanan affinities of the New Zealand biota: three case studies. Aust J Bot 49:377–387

Close RC, Moar NT, Tomlinson AI, Lowe AD (1978) Aerial dispersal of biological material from Australia to New Zealand. Int J Biometeorol 22:1–19

Cooper A, Cooper RA (1995) The Oligocene bottleneck and the New Zealand biota: genetic record of a past environmental crisis. Proc R Soc Lond Ser B 261:293–302

Cooper A, Atkinson IAE, Lee WG, Worthy TH (1993) Evolution of the moa and their effect on the New Zealand flora. Trends Ecol Evol 8:433–437

Cooper A, Lalueza-Fox C, Anderson S, Rambaut A, Austin J, Ward R (2001) Complete mitochondrial genome sequences of two extinct moas clarify ratite evolution. Nature 409:704–707

Cox CB (1998) From generalised tracks to ocean basins – how useful is panbiogeography? J Biogeogr 25:813–828

Craw RC, Grehan JR, Heads MJ (1999) Panbiogeography: tracking the history of life. Oxford Univ Press, New York

Daugherty CH, Gibbs GW, Hitchmough RA (1993) Mega-island or micro-continent? New Zealand and its fauna. Trends Ecol Evol 8:437–442

de Queiroz A (2005) The resurrection of oceanic dispersal in historical biogeography. Trends Ecol Evol 20(2):68–73

Dettman ME (1989) Antarctica: Cretaceous cradle of austral temperate rainforests? In: Crame JA (ed) Origins and evolution of the Antarctic biota. Geol Soc (Lond) Spec Publ 47:89–105

Dugdale JS (1988) Lepidoptera – annotated catalogue, and keys to family-group taxa. Fauna of New Zealand, 14. DSIR, Wellington

Ericson PGP, Christidis L, Irestedt M, Norman JA (2002) Systematic affinities of the lyrebirds (Passeriformes: Menura), with a novel classification of the major groups of passerine birds. Mol Phylogenet Evol 25:53–62

Falcon-Lang HJ, Cantrill DJ (2002) Terrestrial paleoecology of the Cretaceous (early Aptian) Cerro Negro Formation, South Shetlands Islands, Antarctica: a record of polar vegetation in a volcanic arc environment. Palaios 17:491–506

Fenner M, Lee WG, Wilson JB (1997) A comparative study of the distribution of genus size in twenty angiosperm floras. Biol J Linn Soc 62:225–237

Fleming CA (1979) The geological history of New Zealand and its life. Auckland Univ Press, Auckland, New Zealand

Forster RR, Forster LM (1999) Spiders of New Zealand and their world-wide kin. Univ Otago Press, Dunedin, New Zealand

Fox KJ (1978) The transoceanic migration of Lepidoptera to New Zealand – a history and a hypothesis on colonisation. NZ Entomol 6:368–380

Gleeson DM, Rowell DM, Tait NN, Briscoe DA, Higgins AV (1998) Phylogenetic relationships among *Onychophora* from Australasia inferred from the mitochondrial cytochrome oxidase subunit I gene. Mol Phylogenet Evol 10:237–248

Godley EJ (1975) Flora and vegetation. In: Kuschel G (ed) Biogeography and ecology in New Zealand. Junk, The Hague, pp 177–229

Green AJ, Figuerola J, Sánchez MI (2002) Implications of waterbird ecology for the dispersal of aquatic organisms. Acta Oecol 23:177–189

Haddrath O, Baker AJ (2001) Complete mitochondrial DNA genome sequences of extinct birds: ratite phylogenetics and the vicariance biogeography hypothesis. Proc R Soc Lond Ser B 268:939–945

Head PS, Nelson CS (1994) A high-resolution oxygen isotope record for the past 6.4 million years at DSDP Site 593, Challenger Plateau, southern Tasman Sea. In: Van der Lingen GJ, Swanson KM, Muir RJ (eds) Evolution of the Tasman Sea Basin. Proc Tasman Sea Conf 1992. Balkema, Rotterdam, pp 159–179

Herzer RH (1998) Tectonic control of terrestrial species migration to New Zealand in the early to middle Miocene. In: Cooper RA, Jones CM (eds) Geology and genes. Geol Soc NZ Misc Publ 97:35–38

Herzer RH, Chaproniere GCH, Edwards AR, Hollis CJ, Pelletier B, Raine JI, Scott GH, Stagpoole V, Strong CP, Symonds P, Wilson GJ, Zhu H (1997) Seismic stratrigraphy and structural history of the Reinga Basin and its margins, southern Norfolk Ridge system. NZ J Geol Geophys 40:425–451

Hickson RE, Slack KE, Lockhart P (2000) Phylogeny recapitulates geography, or why New Zealand has so many species of skinks. Biol J Linn Soc 70:415–433

Holdaway RN, Worthy TH, Tennyson AJD (2001) A working list of breeding bird species of the New Zealand region at first human contact. NZ J Zool 28:119–187

Jordan GJ (2001) An investigation of long-distance dispersal based on species native to both Tasmania and New Zealand. Aust J Bot 49:333–340

Kennedy EM, Spicer RA, Rees PM (2002) Quantitative palaeoclimate estimates from late Cretaceous and Palaeocene leaf floras in the northwest of the South Island, New Zealand. Palaeogeogr Palaeoclimatol Palaeoecol 184:321–345

Klimaszewski J (1997) Biodiversity of New Zealand Beetles (Insecta, Coleoptera). Mem Mus Victoria 56:659–666

Knapp M, Stockler K, Havell D, Delsuc F, Sebastiani F, Lockhart PJ (2005) Relaxed molecular clock provides evidence for long-distance dispersal of *Nothofagus* (southern beech). PloS Biol 3(1):e14

Larochelle A, Larivière M-C (2001) Carabidae (Insecta: Coleoptera): catalogue. Fauna of New Zealand, 43. Manaaki Whenua Press, Lincoln, Canterbury, New Zealand

Lee KE (1959) The earthworm fauna of New Zealand. DSIR Bull 130. Department of Scientific and Industrial Research, Wellington, New Zealand

Lee WG (1998) The vegetation of New Zealand – functional, spatial and temporal gaps. In: Lynch R (ed) Ecosystems, entomology and plants. R Soc NZ Misc Ser 48:91–101

Lee DE, Lee WG, Mortimer N (2001) Where and why have all the flowers gone? Depletion and turnover in the New Zealand Cenozoic angiosperm flora in relation to palaeogeography and climate. Aust J Bot 49:341–356

Lloyd DG (1985) Progress in understanding the natural history of New Zealand plants. NZ J Bot 23:707–722

Luxton M (1985) Cryptostigmata (Arachnida: Acari) – a concise review. Fauna of New Zealand, 7. DSIR, Wellington, New Zealand

Macphail MK (1997) Comment on M. Pole (1994): 'The New Zealand flora – entirely long-distance dispersal?' J Biogeogr 24:113–117

McDowall RM (2002) Accumulating evidence for a dispersal biogeography of southern cool temperate freshwater fishes. J Biogeogr 29:207–219

McGlone MS, Salinger MJ, Moar NT (1993) Palaeovegetation studies of New Zealand's climate since the Last Glacial Maximum. In: Wright HE, Kutzbach JE, Webb T III, Ruddiman WF, Street-Perrrott FA, Bartlein PJ (eds) Global climates since the Last Glacial Maximum. Univ Minnnesota Press, Minneapolis, pp 294–317

McGlone MS, Mildenhall DC, Pole MS (1996) History and paleoecology of New Zealand *Nothofagus* forests. In: Veblen TT, Hill RS, Read J (eds) The ecology and biogeography of *Nothofagus* forests. Yale Univ Press, New Haven, pp 83–130

McGlone MS, Duncan RP, Heenan PB (2001) Endemism, species selection and the origin and distribution of the vascular plant flora of New Zealand. J Biogeogr 28:199–216

Mildenhall DC (1980) New Zealand late Cretaceous and Cenozoic plant biogeography: a contribution. Palaeogeogr Palaeoclimatol Palaeoecol 31:197–233

Mound LA, Walker AK (1982) Terebrantia (Insecta: Thysanoptera). Fauna of New Zealand, 1. DSIR, Wellington, New Zealand

Nelson CS, Cooke PJ (2001) History of oceanic front development in the New Zealand sector of the Southern Ocean during the Cenozoic – a synthesis. NZ J Geol Geophys 44:535–553

Ogden J, Basher L, McGlone MS (1998) Fire, forest regeneration and links with early human habitation: evidence from New Zealand. Ann Bot 81:687–696

Parrish JT, Daniel IL, Kennedy EM, Spicer RA (1998) Paleoclimatic significance of mid-Cretaceous floras from the middle Clarence Valley, New Zealand. Palaios 13:149–159

Pole M (1992) Cretaceous macrofloras of Eastern Otago, New Zealand. Angiosperms. Aust J Bot 40:169–206

Pole M (1993) Keeping in touch: vegetation prehistory on both sides of the Tasman. Aust Syst Bot 6:387–397

Pole MS (1994) The New Zealand flora – entirely long-distance dispersal? J Biogeogr 21:625–635

Pole M, Douglas B (1998) A quantitative palynostratigraphy of the Miocene Manuherikia Group, New Zealand. J R Soc NZ 28:405–420

Poole I, Gottwald H (2001) Monimiaceae *sensu lato*, an element of Gondwanan polar forests: evidence from the Late Cretaceous – Early Tertiary wood flora of Antarctica. Aust Syst Bot 14:207–230

Quilty PG (1994) The background: 144 million years of Australian palaeoclimate and palaeogeography. In: Hill RS (ed) History of the Australian vegetation: Cretaceous to recent. Cambridge Univ Press, Cambridge, pp 14–43

Raine JI, Askin RA (2001) Terrestrial palynology of Cape Roberts project drillhole CRP 3, Victoria Land Basin, Antarctica. Terra Antarct 8:389–400

Raven PH (1973) Evolution of subalpine and alpine plant groups in New Zealand. NZ J Bot 11:177–200

Renner SS, Foreman DB, Murray D (2000) Timing transantarctic disjunctions in the Atherospermataceae (Laurales): evidence from coding and noncoding chloroplast sequences. Syst Biol 49:579–591

Rivadavia F, Kondo K, Kato M, Hasebe M (2003) Phylogeny of the sundews, *Drosera* (Droseraceae), based on chloroplast *rbcL* and nuclear 18S ribosomal DNA sequences. Am J Bot 90:123–130

Rowe RJ (1987) The dragonflies of New Zealand. Auckland Univ Press, Auckland, New Zealand

Santamaría L (2002) Why are most aquatic plants widely distributed? Dispersal, clonal growth and small-scale heterogeneity in a stressful environment. Acta Oecol 23:137–154

Stockler K, Daniel IL, Lockhart PJ (2002) New Zealand Kauri (*Agathis australis* (D. Don) Lindl., Araucariaceae) survives Oligocene drowning. Syst Biol 51:827–832

Sturman AP, Tyson PD, D'Abreton PC (1997) A preliminary study of the transport of air from Africa and Australia to New Zealand. J R Soc NZ 27:485–498

Swenson U, Hill RS, McLoughlin S (2001) Biogeography of *Nothofagus* supports the sequence of Gondwana break-up. Taxon 50:1025–1041

Vajda V, Raine JI, Hollis CJ (2001) Indication of global deforestation at the Cretaceous-Tertiary boundary by New Zealand fern spike. Science 294:1700–1702

Wagner WHJ (1995) Evolution of Hawaiian ferns and fern allies in relation to their conservation status. Pac Sci 49:31–41

Wagstaff SJ, Bayly MJ, Garnock-Jones PJ, Albach DC (2002) Classification, origin, and diversification of the New Zealand Hebes (Scrophulariaceae). Ann Missouri Bot Gard 89:38–63

Wardle P (1978) Origin of the New Zealand mountain flora, with special reference to trans-Tasman relationships. NZ J Bot 16:535–550

Wardle P (1998) Comparison of alpine timberlines in New Zealand and the Southern Andes. In: Lynch R (ed) Ecosystems, entomology and plants. R Soc NZ Misc Ser 48:69–90

Wardle P, Ezcurra C, Ramírez C, Wagstaff S (2001) Comparison of the flora and vegetation of the southern Andes and New Zealand. NZ J Bot 39:69–108

Waters JM, Dijkstra LH, Wallis GP (2000) Biogeography of a southern hemisphere freshwater fish: how important is marine dispersal? Mol Ecol 9:1815–1821

Watt JC (1975). The terrestrial insects. In: Kuschel G (ed) Biogeography and ecology in New Zealand. Junk, The Hague, pp 507–535

Winkworth RC, Wagstaff SJ, Glenny D, Lockhart PJ (2002) Plant dispersal N.E.W.S from New Zealand. Trends Ecol Evol 17:514–520

Worthy TH, Holdaway RN (2002) The lost world of the moa: prehistoric life of New Zealand. Canterbury Univ Press, Christchurch, New Zealand

Worthy TH, Tennyson AJD, Jones C, McNamara JA (2002) The early Miocene (15-20 Ma) Manuherikia Group reveals New Zealand's first diverse Tertiary terrestrial fauna. In: Grenfell H (ed) Abstr Vol Geological Society of New Zealand Annu Conf, 2–5 December 2002, Whangarei, Northland. Geol Soc NZ Misc Publ 112A:58–59

Zhang L-B, Renner S (2003) The deepest splits in Chloranthaceae as resolved by chloroplast sequences. Int J Plant Sci 164:S383–S392

3 Creating Gardens: The Diversity and Progression of European Plant Introductions

P.A. Williams and E.K. Cameron

3.1 Introduction

New Zealand lacked large land mammals and consequently the native flora evolved without the evolutionary pressures conducive to the development of mammal-browse-tolerant foliage and, importantly for humans, large edible fruit. Neither does the native seed plant flora include species with large edible underground food reserves, apart from bracken (*Pteridium esculentum*) and ti (*Cordyline australis*). Maori came less than 1,000 years ago from the tropics, where they did not have temperate food or agricultural plants capable of growing throughout New Zealand, although they did successfully introduce taro (*Colocasia esculenta*), kumara (*Ipomoea batatus*), yam (*Dioscorea alata*), and Pacific Island cabbage tree (*Cordyline fruticosa*) to warmer areas. Thus, to obtain plants for purely utilitarian and commercial purposes, with the notable exceptions of timber and flax (*Phormium*), European civilization was forced to import on a large scale. Like colonists everywhere, they also brought nostalgic reminders of home. Color was especially important because few New Zealand plants have conspicuous flowers, and even these are mostly white. The colonists had the advantage of being in a temperate environment, however, and in 200 years, more than 25,000 species of plants were introduced. Of these, about 2,200 now grow wild, many in restricted habitats, and others from the coast to the alpine zone. Almost half of the plant species came from Europe, this proportion increasing to three-quarters if North Africa and temperate Asia are included, and the remainder from all corners of the earth. This chapter tells the story of their arrival and escape.

Ecological Studies, Vol. 186
R.B. Allen and W.G. Lee (Eds.)
Biological Invasions in New Zealand

3.2 The Chroniclers

The documentation of this story has been sporadic. Naturalized plants were first mentioned by Banks (in Beaglehole 1962) and Darwin (1859), but the first systematic account is by J.D. Hooker (1867) who recorded 170 naturalized species. (All records have been standardized to present-day naturalized status). Many of these were from Auckland, collected by T. Kirk, who in a later account of this area (Kirk 1870) listed 257 naturalized species. Cheeseman's (1925) *Manual of the New Zealand flora* recorded 567 naturalized species, and Thomson (1922) in a study of naturalizations considered over 600 species could be regarded as "more or less" naturalized. (Here, we use the terms "fully naturalized" for species forming self-sustaining populations in the wild, and "casuals" for those that have not yet established but have been recorded occasionally. "Naturalized" we use to cover both sub-groups). Despite the publication of Allan's *Handbook of the naturalized flora of New Zealand*, the recorded number had not risen above 441 fully naturalized species, and 347 of minor or doubtful occurrence 18 years later (Allan 1940). A.J. Healy and others subsequently made many individual records; Healy and Edgar (1980) listed 168 naturalized monocotyledons (excluding grasses). With the publication of the naturalized pteridophytes, gymnosperms, and dicotyledons, a total of 1,470 species or "species equivalents" were recorded (Webb et al. 1988). The grass flora (Edgar and Connor 2000) added a further 226 species, to make naturalized 2,108 seed plants and gymnosperms (Wilton and Breitweiser 2000). Additional records since then bring the total naturalized flora of seed plants and ferns to 2,166 species, which include 23 naturalized pteridophytes. Discounting the few introduced by Maori (<10 spp. suspected), this equates to one new species every 39 days since Capt. James Cook first sighted land in 1769.

This naturalized flora added 65 families, with at least one fully naturalized species, to the total flora of New Zealand. The largest of these are Salicaceae (25 spp.), Pinaceae (18 spp.), Amaranthaceae (14 spp.), and Papavaceae (11 spp.). The largest naturalized genera containing no indigenous species are *Trifolium*, *Salix*, *Veronica*, and *Rosa*. Many new families comprise plant life-forms similar to the indigenous flora whereas others added new elements to the landscape, e.g., arum lilies (*Zantedeschia aethiopica*), and wild gingers (*Hedychium* spp.). Novelty aside, genera with native congeners have the highest naturalization rates (Duncan and Williams 2002a, b).

3.3 Plant Introductions and Their Escape

3.3.1 Early European Settlers and Farming

Plants were brought to New Zealand via many pathways, for pleasure, profit, and/or utility. The first successful European introduction was the tuberous potato (*Solanum tuberosum*), which was selected for subsistence, similarly to the kumara 800 years earlier by Maori. James Cook visited Marlborough Sounds in 1770, and introduced potatoes along with turnips (*Brassica rapa*), carrots (*Daucus carota*), and parsnips (*Pastinaca sativa*). Maori subsequently neglected the gardens, but they were familiar with root crops and the potatoes were still there on Cook's third voyage in 1777, and have remained in the area, probably continuously, ever since (Clark 1949). Wild cabbage (*Brassica oleracea*) was gathered from these gardens in 1820 (Clark 1949), and possibly represents the first cultivation escape – or perhaps this was wild onion (*Allium vineale*), reputedly wild before 1800 in the Bay of Islands from French plantings in 1772 (Thomson 1922). Vegetable gardens were well established by 1820 at various points, from Foveaux Strait in the south, to the Bay of Islands in the north. Here the missionaries, who introduced the first plough at Kerikeri in the same year, encouraged gardening. From the 1840s onward, gardening intensified at several localities throughout New Zealand, the resulting disturbance increasing the opportunities for plant naturalizations.

Potatoes were initially used to assist breaking in the ground for grains, especially on cutover forestland. Wheat became the main dietary starch source by the 1850s when the growing European population began searching for pasturage for sheep and cattle. Native plant species provided little in this respect away from the South Island uplands. In Auckland, cattle were pushed into native scrub and forest (Esler 1988). Introduced grasses and herbs were sown on land cleared from forest or fern. Browntop (*Agrostis capillaris*), danthonias (*Rytidosperma* spp., from Australia), sweet vernal (*Anthoxanthum odoratum*) and suckling clover (*Trifolium dubium*) were important. Arable land required a wider range of grasses, and by 1870 in Auckland these included perennial ryegrass (*Lolium perenne*), Italian ryegrass (*Lolium multiflorum*), bluegrass (*Poa pratensis*) and several other clover species. Herbs considered necessary for animal health, but now thought of as pasture weeds, were also sown, e.g., narrow-leaved plantain (*Plantago lanceolata*) and yarrow (*Achillea millefolium*). Scotch thistles (*Cirsium vulgare*), which arrived accidentally, were initially protected by Scottish immigrants for nostalgic reasons (Thomson 1922), soon supplanted by annoyance as their crops and pastures became infested.

These farms and habitations required protection, and exotic hedges and shelterbelts introduced by the very early colonists provide spectacular examples of invasion sources. Gorse (*Ulex europaeus*) was planted as a fence atop drainage excavations in Canterbury, from where it spread irrepressibly within

30 years. Broom (*Cytisus scoparius*) was also used as stock food. Initially less invasive than gorse, primarily because of its palatability to stock, broom is now spreading rapidly in both main islands. On the North Island, shelter was provided by several *Acacia* spp., hawthorn (*Crataegus monogyna*), and *Hakea* spp. from Australia. All these are now major weeds in parts of the country. Monkey apple (*Acmena smithii*) is used as a shelter for horticulture crops in northern parts where it is now invading intact forest. Pampas grasses (*Cortaderia* spp.) were also introduced for shelter, and cattle fodder – their legacy a major weed problem of exotic plantation forests, and for biodiversity conservation in open areas such as sand dunes.

Horticulture has been a major pathway contributing to the naturalized flora. A variety of fruits were grown by the 1840s in northern New Zealand, and the superiority of this warm-temperate region for woody plants and vines with edible fruits persists. Cape gooseberry (*Physalis peruviana*) had escaped by 1843, followed by blackberry (*Rubus fruticosus* complex), sour cherry (*Prunus cerasus*), wild peaches (*Prunus persica*), and elderberry (*Sambucus nigra*) in the 1860s (Thomson 1922). Several "edible" naturalizations have occurred since, notably species of passion vines (*Passiflora* spp.), kiwifruit (*Actinidia deliciosa*), loquat (*Eriobotrya japonica*), and even strawberry myrtle (*Ugni molinae*) on the Chatham Islands.

3.3.2 Forestry

Early settlers "mined" the native forests for timber and firewood but these regenerated too slowly for the multifarious demands made of homestead woodlots. Seeds of radiata pine (*Pinus radiata*), Scots pine (*P. sylvestris*), and macrocarpa (*Cupressus macrocarpa*) arrived in Canterbury and Auckland during the 1860s, and 47 important timber species had been introduced to Canterbury alone by 1925 (Clark 1949). Many of these are now widely naturalized.

The story of introducing conifers for forestry is highlighted in the development of Kaingaroa forest covering 188,000 ha of the central North Island, the largest man-made forest in the Southern Hemisphere. This was partly a social phenomenon; labor for planting was initially via prison reforms that sent men from unhealthy city gaols into the country to work, and later, via a steady supply of conscientious objectors to World War II (Boyd 1992). Trial plantings were made on this extensive upland volcanic plateau in 1896. Seeds were obtained from nurseries throughout the country and overseas, and planting began in 1906, mainly of Austrian pine (*Pinus nigra* subsp. *nigra*), bishop pine (*P. muricata*), Corsican pine (*P. nigra* subsp. *laricio*), and Australian gums (*Eucalyptus* spp). By 1911, 38 species had been planted (Boyd 1992). Obtaining seed for many species was difficult, but this did not apply to *P. radiata*, which soon superseded larch (*Larix decidua*) as the dominant Kaingaroa species, and thenceforth throughout the country.

Contorta pine (*Pinus contorta*) does not feature in this forest but does so at Karioi near Mt. Ruapehu, where large blocks were planted for pulp. Several pines are invasive in New Zealand, but contorta pine has the greatest ability to spread from plantations. Escapees from Karioi forest led to the largest ongoing weed control operation (apart from agricultural weeds) in New Zealand, centered on the volcanic landscapes of Tongariro National Park, adjacent tussock uplands used by the military, and the Kaweka Range. Several cold-tolerant conifer species, such as contorta pine, have large areas of herbaceous native habitat available to them because they can grow above the native tree limit. Douglas fir (*Pseudotsuga menziesii*) is the second most important forestry tree south of the Volcanic Plateau, and it too has spread. Unlike pines, however, this conifer tolerates shade, and there are concerns it may impact on *Nothofagus* forest regeneration, especially in the eastern South Island.

3.3.3 Erosion Control

The first willow introduced was the ornamental weeping willow (*Salix babylonica*), probably brought from Napoleon's grave on St. Helena to Banks Peninsula, South Island, in 1839 (van Kraayenoord et al. 1995). Many other species or cultivars were later introduced for riverbank stabilization. Crack willow (*Salix fragilis*) had naturalized along river margins and was causing flooding problems by 1880, as it has ever since (van Kraayenoord et al. 1995). Scores of other willow species and cultivars were planted widely in semi-natural wetland habitats and many have naturalized, to the detriment of the native biota.

Lowland areas were planted with many woody species to halt soil erosion. Most failed, but others, especially Tasmanian blackwood (*Acacia melanoxylon*) and false acacia (*Robinia pseudoacacia*), coral berry (*Symphoricarpus orbiculatus*), and species of *Tamarix*, persisted and spread from the 1970s onward (Sheppard 1972). Black or Lombardy poplar (*Populus nigra* cv. "Italica") survives from earlier plantings and is the most distinctive of all shelter trees, but it is less well established in the wild.

High-altitude *Alnus* and *Pinus* species from Europe and North America were hand-planted on scree slopes, and even aerially spread in an effort to control natural erosion at high altitudes. The result is a swathe of conifers in the mountains of the eastern South Island that will invade irreversibly for the foreseeable future.

Wind-blown coastal sand dunes also moved inland across farms, mostly in the western North Island following forest clearance. These were stabilized by the widespread planting of marram grass (*Ammophila arenaria*) and tree lupin (*Lupinus arboreus*), which have replaced native species over most of the sand dune country in New Zealand.

3.3.4 Horticulture

New Zealand was settled in the period of British history that coincided with the development of glasshouses, the creation of parks for the populace, the popularity of urban gardening aided by the first gardening magazines, and the invention of the Wardian Case for transporting plants. These early potted horticultural plants and those sent as seeds and bulbs, together with thousands more species later brought or sent for cultivation, constitute the main pathway for the majority of invasive species in New Zealand (Esler 1988). Most of these are now conceived as horticultural escapes, but this underestimates the early attempts at deliberately acclimatizing exotic plants directly into the landscape for aesthetic reasons. Most of these failed (Thomson 1922), but the purposeful spread of the colorful Russell lupins (*Lupinus polyphyllus*) transformed many sparsely vegetated habitats in the montane South Island, similarly to heather (*Calluna vulgaris*) that added cover and color to the Volcanic Plateau (for grouse habitat).

Flower gardens were growing in the Bay of Islands in the 1820s, and by the 1840s there were nurseries at Christchurch and Auckland. Plants initially came from England or Australia (Esler 1988), where plant nurseries were operative from the early 1800s in Sydney and the 1840s in Adelaide (Mulvaney 2001). The mid- to late 1800s was the heyday of plant collectors around the world, and New Zealand was on both the giving and receiving end of these efforts to introduce to Western Europe, North America, and by extension to a few other areas such as Australasia, as many of the world's ornamental plants as possible. In the 1860s and 1870s, Sir George Grey had an "almost endless" catalogue of plants on Kawau Island (Esler 1988), and private and public nurseries were well established in Dunedin and Christchurch. More than 200 suppliers now sell 30,000 plant taxa in New Zealand, the great majority of them exotic (Gaddum 1999).

3.3.5 Accidental Introductions

Aside from these thousands of purposeful introductions, most plant species naturalized before 1900 arrived accidentally via numerous pathways associated with goods and commodities, packing material, and ballast (Thomson 1922; Esler 1988). More than 20 European species had arrived unintentionally by the 1840s (Allan 1937), e.g., pimpernel (*Anagallis arvensis*), storksbill (*Erodium cicutarium*), alyssum (*Lobularia maritima*), broad-leaved plantain (*Plantago major*), giant buttercup (*Ranunculus acris*), and curled dock (*Rumex crispus*). As many as 104 species were reported from one load of ballast in 1896 (Esler 1988). Many species arrived as contaminants of seed for sowing, and were distributed in abundance. A single seed consignment listing

20 components actually contained 16 others, and by 1916 the Agriculture Department listed 221 exotic contaminants (Thomson 1922). Many were carried to open spaces in forest and fern on the backs of sheep imported in large numbers from Australia. Sand brome (*Bromus arenarius*) arrived probably as early as 1826 in this manner, and certainly many species of danthonia (*Rytidosperma*) and stipa (*Austrostipa* and *Nassella*) did (Allan 1937), the latter genus having originated in South America. Contaminants of potted plants had the advantage of being nurtured on their journey and distributed with the plants they hitched a ride with, e.g., oxalis (*Oxalis corniculata*). Accidental dispersal, particularly via contaminated road metal, is still an important means of species spread within New Zealand.

3.4 Regional Differences

The naturalized flora has followed the rise in total human population from 1830 onward (Fig. 3.2) but regional differences in naturalizations were evident from the outset, reflecting climate, land use and even the location of botanists. By the 1930s, 74 % of naturalized species were first recorded from the North

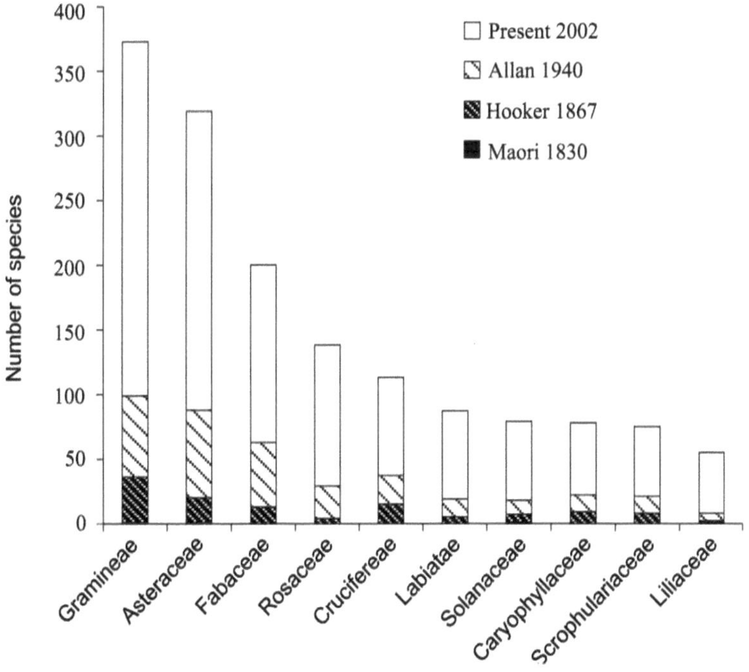

Fig. 3.1 The 10 largest families of the New Zealand naturalized flora at four time periods

Island where 65 % of the people lived (Allan 1937). Auckland was an early center of plant naturalization records, and has continued to be so – to the extent that it may be the weediest city in the world (Esler 1988). The distribution of recent naturalization records of some plants (e.g., Webb et al. 1988, 1995; Heenan et al. 1998, 1999, 2002) suggests this is closely related to the number of people in the region (Fig. 3.2). In all, 78 % of naturalization records in the last 15 years come from the North Island where the share of the population has risen to 76 %. Botanical collectors have never been evenly distributed either, which probably skews the values slightly. Certainly the moderate Auckland climate, lacking temperature extremes, and habitat diversity are important factors contributing to the city's rich naturalized flora (Esler 1988), as well as the human population. These naturalization records are mostly closely associated with either neglected places with sparse cover or mixed woody vegetation, or in and around gardens and plantations or parks (Fig. 3.3a). Many records are unspecific as to land types but land adjacent to roadsides and tracks, as well as topographical features such as banks, cliffs, gullies and riparian areas (Fig. 3.3b), are vastly over-represented as sites for naturalizations, compared with the small proportion of their aerial contribution to the landscape. Combined with long-term observations of the naturalization process (e.g., Guthrie-Smith 1953), and the factors associated with the number of weeds in forest reserves (Timmins and Williams 1991; Sullivan et al. 2005), the conclusion is inescapable that human population pressure and all it implies has been, and will continue to be, a major factor behind regional differences in plant naturalization in New Zealand.

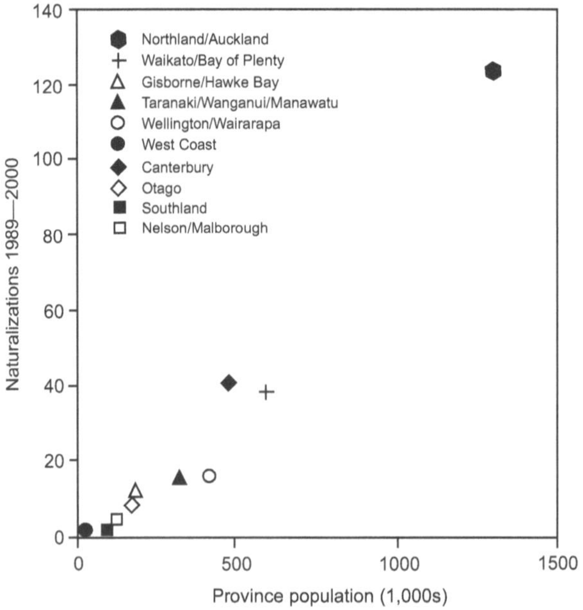

Fig. 3.2 Number of first naturalizations recorded from each of 10 provinces in New Zealand, and their human populations, for 1989–2000

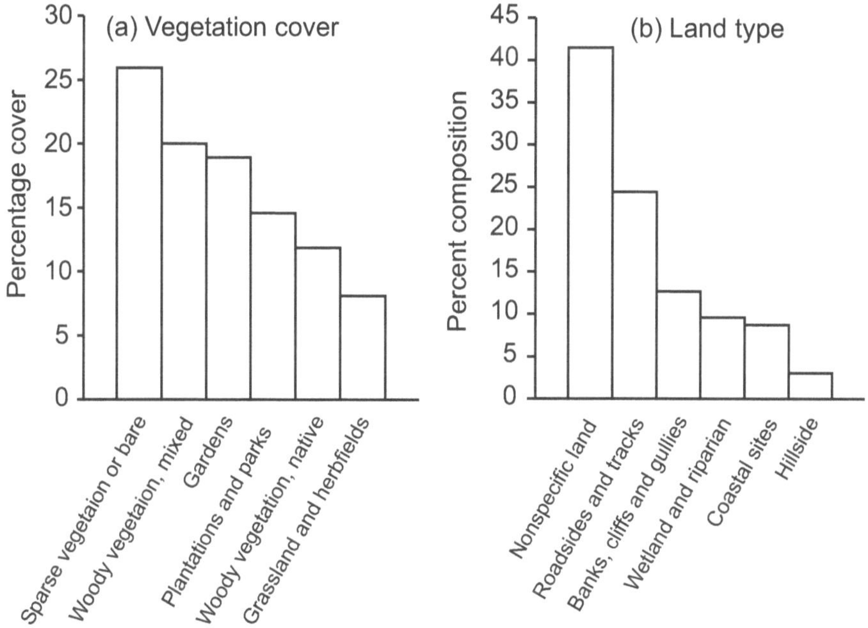

Fig. 3.3a Vegetation, and **b** land type of the localities where 267 first naturalizations were recorded in 1989–2000

3.5 Naturalized Flora

Development of the naturalized flora can be seen as occurring over four periods, after clarifying early records in the light of present-day taxonomy and greater certainty of species origins. The first period is the flora brought in by the Maori (i.e., pre-European), the second is early European to Hooker (1867), followed by middle European to Allan (1940), and finally from 1940 to 2002. The 10 families now comprising 52 % of the naturalized flora were dominant by 1867 (Fig. 3.1), but their rank order has changed, along with the pathways of introduction. Also, early escapes were mainly herbaceous, but time has allowed longer-lived woody species and climbers to escape, too (Fig. 3.4).

Not all species that appear spontaneously in the wild, nor those planted there, persist. There are numerous records of even widespread species declining for unknown reasons, e.g., water poppy (*Hydrocleys nymphoides*) and onion weed (*Asphodelus fistulosus*; Esler 1988). A larger proportion of the cultivated exotic flora fails even to reach the status of a casual naturalization, so that the naturalized flora is not simply a reflection of the total spectrum of

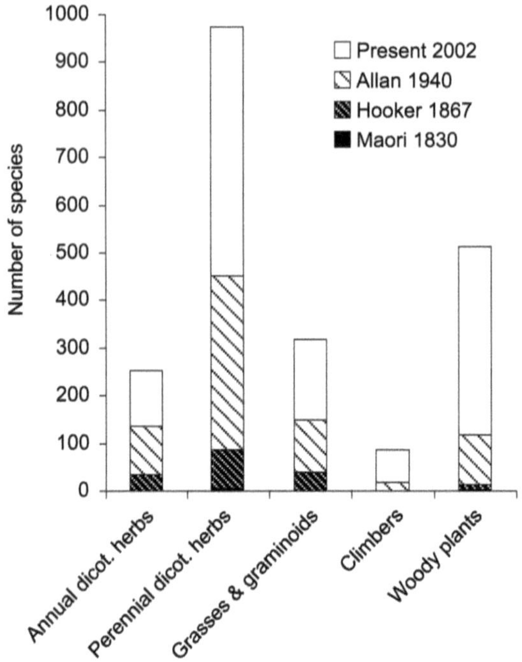

Fig. 3.4 Life-forms of the New Zealand naturalized flora at four time periods

accidental and purposeful introductions; some plant families and genera do significantly better than others. The total introduced New Zealand flora of some 28,000 species (http://nzflora.landcareresearch.co.nz) can be examined in relation to the total world flora and the naturalized flora. The 10 most species-rich families of these two groups together belong to 19 plant families (Table 3.1).

Amongst the world's most species-rich families, Melastomataceae has a low level of both introduction and naturalization in New Zealand (Table 3.1, group A), despite having many ornamental members and being arguably the worst invader of the tropical Pacific (W.R. Sykes, personal communication), e.g., miconia (*Miconia calvescens*). Rubiaceae, another large family of mainly tropical or warm-climate species that include important economic crops such as coffee, is represented in New Zealand's naturalized flora only by small herbs. Five of the world's most species-rich families are represented in New Zealand by less than 10 % of their species, but they have high rates of naturalization (Asteraceae, Fabaceae, Poaceae, Labiateae, Scrophulariaceae; Table 3.1, group D) and are major contributors to the naturalized flora (Fig. 3.1). Solanaceae, which ranks only 19th in the world, is the 7th ranked naturalized family (Fig. 3.1). These are mostly cosmopolitan families well represented amongst the world's useful plants and weeds. Only two families (Orchidaceae, Liliaceae) with high introduction rates and low naturalization rates are within

Table 3.1 Plant introductions: the 10 most species-rich angiosperm families in the world (world ranks in parentheses), together with the 10 largest families to be introduced into New Zealand[a]

Percentage of the world species introduced to New Zealand	Naturalized species as % of introductions	
	<10 %	≥10 %
≥10 %	A Orchidaceae (2) Liliaceae *s. lat.* (9) Ericaceae (15) Ranunculaceae (26) Bromeliaceae (27) Cactaceae (42)	B Euphorbiaceae (6) Brassicaceae (16) Rosaceae *s. lat.* (22) Caryophyllaceae (29) Iridaceae (39)
<10 %	C Melastomataceae (10) Rubiaceae (4)	D Asteraceae (1) Fabaceae (3) Poaceae (5) Lamiaceae (7) Scrophulariaceae (8) Solanaceae (19)

[a] The 19 families are grouped according to whether at least 10 % of the families are currently exotic in New Zealand (cultivated plus naturalized), and whether at least 10 % of these introductions are naturalized. They are divided into four groups (A–D) representing high (≥10 %) or low (<10 %) percentages of introduction and naturalization

the world's top 10 most species-rich families. The others are lesser families, although Ericaceae ranks 15th. Orchidaceae, Bromeliaceae and Cactaceae are mostly horticultural plants from climates warmer and either wetter or drier than in New Zealand, whereas Liliaceae, Ericaceae and Ranunculaceae are tropical montane or temperate families. All six families have many useful species whereas the latter three are more commonly weeds in temperate regions. Other families with both high introduction rates and high naturalization rates are mostly temperate, especially northern temperate (Brassicaceae, Rosaceae, Caryophyllaceae), widespread tropical and subtropical (Euphorbiaceae), or with a more regional distribution (Iridaceae). All are well represented in both horticultural and weed floras.

The total exotic flora is thus a very biased selection of the world's flora, consisting mostly of ornamental plants introduced for urban horticulture, and this is reflected in the composition of the resulting naturalized flora. The representativeness on a world scale of the naturalized flora is further skewed because the naturalized portion of these 19 families alone ranges from 0 % (Orchidaceae) to 49 % (Poaceae). This partly reflects climatic differences between the source countries and New Zealand, although the diversity of habitats here has enabled the development of some very distinctive communities of naturalized plants, some of which are successful in local climates similar to that of their homelands. Examples are numerous species of Crassulaceae on the volcanic cliffs of Otago, Canterbury and Rangitoto Island in Auckland, the swathes of *Hieracium* species in the South Island, the striking Chilean rhubarb (*Gunnera tinctoria*) herbfields on coastal talus in Taranaki, the nascent palm forests of Chinese windmill palm (*Trachycarpus fortunei*) and Bangalow palm (*Archontophoenix cunninghamiana*) developing in Auckland's forests, and *Banksia integrifolia* invading North Island dune lands.

3.6 Attitudes Past and Present

The attitudes of botanists and the general public toward the naturalized flora have always been ambivalent. Warnings of the potential apocalyptic impacts of the naturalized flora on indigenous plants and vegetation appeared in the earliest writings of J.D. Hooker and others, supported by the pervading belief in the superiority of the Northern Hemisphere floras held by the evolutionists in the late 1800s. Hints of greater optimism occurred almost simultaneously in the writings of T. Kirk, but the pessimists were not seriously challenged before Cockayne at the turn of the century (Cockayne 1911), followed by others (Thomson 1922; Allan 1936), emphasized the importance of habitat modification to the success of exotic species. Stable communities encompassing both native and exotic species were envisaged, to the extent that even distinguishing between the two floras was considered scientifically unjustified (Raven and Engelhorn 1971). Healy and Edgar (1980) thought "native and adventive plants had called a truce", but this optimism coincided with the period before the world at large awoke to the wider impacts of invasive species. The envisaged refuges for native plants on "bold cliffs, and extensive sand dunes" (Thomson 1922, p. 532) are among the habitats most vulnerable to invasion (Fig. 3.3b), and they contain rare species for which naturalized plants are now the greatest single threat (Reid 1998).

3.7 Responding to the Deluge

The first Act to control weeds was passed in 1854 to prevent the spread of thistles as agricultural contaminants (Thomson 1922). Many subsequent legislative moves were made to control the spread of plants, and some were prevented from entry by the treaties of Government botanists to the border authorities of the day. It was not until 1998, however, that border biosecurity legislation was passed making it illegal to import plant species not already in New Zealand that might affect the native biota. This legislation prompted a study of the weediness overseas of the exotic flora in cultivation and amongst the recent naturalizations referred to earlier, as a basis for developing effective risk-assessment systems for border control. Of the 267 first records of casual and fully naturalized species since 1989, only six were introduced accidentally, and one was for agriculture. The majority, 97 %, were introduced for urban horticulture. Of these, 156 (58.6 %) are sold in an average of 8.4 major nurseries (data from Gaddum 1999). About 60 % of these horticultural introductions, both sold and not sold, appear on an average of 7.5 weed lists overseas (data from Randall 2002). These are no less weedy than the few accidental introductions, which appear on 10.5 weed lists. Urban horticulture is responsible for the majority of new naturalizations. Well over half of these are known to be weeds in several other countries already.

The potential pool of invasive species still only in cultivation is enormous (>30,000 taxa), and moreover most naturalized species have yet to reach their present geographical limits (Fig. 3.5) – even without climate change. Those colonists who sought to transform wild New Zealand into a vast flower garden would not be disappointed if they returned, and in 100 years the colors are likely to be even more dramatic, as evidenced by the recent seaside expansions of the swathes of pink holly-leaved senecio (*Senecio glastifolius*) and yellow boneseed (*Chrysanthemoides monilifera*). In the deforested lowlands and open upland in all but well-tended places, the somber native flora has been largely overwhelmed. We now live amongst a largely untended garden of colorful exotics.

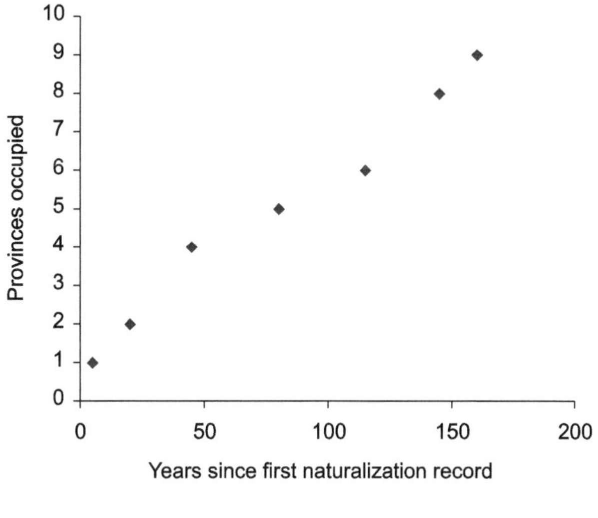

Fig. 3.5 Number of years since species first naturalized in New Zealand, and the mean number of provinces each now occupies. Data are for 1,914 species and seven periods from pre-1840 to 2000

Acknowledgements We thank Jon Sullivan and especially Bill Sykes, who has done so much to further our understanding of the cultivated flora, for many helpful comments on a draft.

References

Allan HH (1936) Indigene versus alien in the New Zealand plant world. Ecology 17:187–193

Allan HH (1937) The origin and distribution of the naturalized plants of New Zealand. Proc Linn Soc Lond 150th Session 1:25–46

Allan HH (1940) A handbook of the naturalized flora of New Zealand. DSIR Bull 83. Government Printer, Wellington, New Zealand

Beaglehole JC (1962) The endeavour journal of Joseph Banks: 1768–1771, vols 1 and 2. Public Library of New South Wales, Sydney

Boyd J (1992) Pumice and pines. GP Publications, Wellington, New Zealand

Cheeseman TF (1925) Manual of the New Zealand Flora, 2nd edn. Government Printer, Wellington, New Zealand

Clark AH (1949) The invasion of New Zealand by people, plants and animals: the South Island. Rutgers Univ Press, New Brunswick

Cockayne L (1911) Observations concerning evolution, derived from ecological studies in New Zealand. Trans NZ Inst 44:1–50

Darwin C (1859) On the origin of species by means of natural selection, or the preservation of favoured races in the struggle for life. Murray, London

Duncan RP, Williams PA (2002a) Darwin's naturalization hypothesis challenged. Nature 417:608–609

Duncan RP, Williams PA (2002b) Taxonomic patterns in the naturalisation rate of plant species in New Zealand. In: Jacob HS, Dodd J, Moore J (eds) Pap Proc 13th Australian Weeds Conf, Plant Protection Society of Western Australia, South Perth, pp 183–185

Edgar E, Connor HE (2000) Flora of New Zealand, vol V. Manaaki Whenua Press, Lincoln, New Zealand

Esler AE (1988) Naturalisation of plants in urban Auckland. DSIR, Wellington, New Zealand

Gaddum M (1999) Gaddum's plant finder 2000. New Zealand Plant Finder, Gisborne, New Zealand

Guthrie-Smith H (1953) Tutira: the story of a New Zealand sheep station, 3rd edn. William Blackwood, Edinburgh

Healy AJ, Edgar E (1980) Flora of New Zealand, vol III. Government Printer, Wellington, New Zealand

Heenan PB, Breitwieser I, Glenny DS, de Lange PJ, Brownsey PJ (1998) Checklist of dicotyledons and pteridophytes naturalized or casual in New Zealand: additional records 1994–1996. NZ J Bot 36:155–162

Heenan PB, de Lange PJ, Glenny DS, Breitwieser I, Brownsey PJ, Ogle CC (1999) Checklist of dicotyledons, gymnosperms, and pteridophytes naturalized or casual in New Zealand: additional records 1997–1998. NZ J Bot 37:629–642

Heenan PB, de Lange PJ, Cameron EK, Champion PD (2002) Checklist of dicotyledons, gymnosperms, and pteridophytes naturalized or casual in New Zealand: additional records 1999–2000. NZ J Bot 40:155–174

Hooker JD (1867) Handbook of the New Zealand flora. Reeve, London

Kirk T (1870) On the naturalized plants of New Zealand, especially with regard to those occurring in the Province of Auckland. Trans NZ Inst 2:131–146

Mulvaney M (2001) The effect of introduction pressure on the naturalisation of ornamental woody plants in south-east Australia. In: Groves RH, Panetta FD, Virtue JG (eds) Weed risk assessment. CSIRO, Melbourne, pp 186–194

Randall RP (2002) A global compendium of weeds. RG and FJ Richardson, Meredith, Victoria, Australia

Raven PH, Engelhorn T (1971) A plea for the collection of common plants. NZ J Bot 9:217–222

Reid VA (1998) The impact of weeds on threatened plants. Sci Res Internal Rep no 164. Department of Conservation, Wellington, New Zealand

Sheppard JS (1972) Evaluation of shrubs trails planted between 1956 and 1970. Ministry of Works, Wellington, New Zealand

Sullivan JJ, Timmins SM, Williams PA (2005) Movement of exotic plants into coastal native forests from gardens in northern New Zealand. NZ J Ecol 29(1):1–10

Thomson GM (1922) The naturalisation of plants and animals in New Zealand. Cambridge Univ Press, Cambridge

Timmins SM, Williams PA (1991) Weed numbers in New Zealand's forest and scrub reserves. NZ J Ecol 15:153–162

Van Kraayenoord CWS, Slui B, Knowles FB (1995) Introduced forest trees in New Zealand. Recognition, Role, and Seed Source, 15. The willows *Salix* spp. FRI Bull 124. New Zealand Forest Research Institute, Rotorua, New Zealand

Webb CJ, Sykes WR, Garnock-Jones PJ (1988) Flora of New Zealand, vol IV. DSIR Botany Division, Christchurch, New Zealand

Webb CJ, Sykes WR, Garnock-Jones PJ, Brownsey PJ (1995) Checklist of dicotyledons, gymnosperms and pteridophytes naturalized or casual in New Zealand: additional records 1988–1993. NZ J Bot 33:151–182

Wilton AD, Breitwieser I (2000) Composition of the New Zealand seed plant flora. NZ J Bot 38:537–549

4 Introduced Mammals in a New Environment

I.A.E. Atkinson

4.1 Introduction

New Zealand was originally a land without terrestrial mammals, except for three species of small bats. The Polynesian people who colonized the country in the 13th century A.D. (Anderson 1991; McFadgen et al. 1994; Higham and Hogg 1997) brought with them two mammals: the dog or kuri (*Canis familiaris*), and the Pacific rat or kiore (*Rattus exulans*). Their introduction may have been unintentional. A double-hulled canoe drawn up on the shore and partially loaded with food for a long journey would have attracted rats roaming at night. Once aboard, it is difficult to see how they could have been removed without unloading all the stores. If pigs were taken aboard, as was often the case when sailing to Pacific islands, these evidently did not survive the journey to New Zealand.

Since European settlement began in the 1790s, a further 28 species of mammal have become established, most between 1800 and 1920. At least 20 of these have had major impacts on native plants and/or animals. Apart from the Hawaiian Islands, with introductions of 21 mammal species (Tomich 1986), no other island nation has sustained this level of alien impact. This is not to deny that dogs, pigs, goats, sheep, cattle and red deer are valued as farm animals while possums, although feral, are harvested commercially for fur. Cats, dogs, horses, pigs, deer, Himalayan tahr and chamois are all valued for recreational reasons.

This chapter outlines the timing and extent of spread of each of the mammals introduced to New Zealand, considers some of their impacts on native plants and animals, and discusses reasons why these impacts have proved catastrophic for many indigenous species. A final question concerns the feasibility of restoring some mainland ecosystems to a functional state more like that of pre-human times.

Ecological Studies, Vol. 186
R.B. Allen and W.G. Lee (Eds.)
Biological Invasions in New Zealand

4.2 Mammal Introductions in New Zealand

4.2.1 Times of Introductions and Extent of Spread

Introduction times and extent of spread of each of the introduced mammals are summarized in Table 4.1. Recent dating studies of rat bones using two independent methods suggest that Pacific rats may have reached North Island and South Island (presumably with Polynesian explorers) "at least 800 years before permanent Polynesian settlement" (Holdaway 1996, 1999; Holdaway et al. 2002 c). In addition, there is evidence that the extinct and possibly flightless Finsch's duck (*Chenonetta finschi*), and the extinct and almost flightless New Zealand owlet-nightjar (*Aegotheles novaezealandiae*) began to decline before Polynesian settlement (Holdaway et al. 2002a, b). In these studies, the authors concluded that a new predator, the Pacific rat, is the most likely explanation for these declines.

This possible pre-settlement arrival time for Pacific rats in New Zealand is challenged by the dating of rat-gnawed seed cases collected from peat deposits on North Island (Wilmshurst and Higham 2004). Intact and rat-gnawed seed cases sampled from three North Island localities covered an age range from 2500 B.C. to 2000 A.D. The oldest dates measured for any rat-gnawed seed cases are consistent with rats having arrived at the time of first human settlement in New Zealand, i.e. the 13th century A.D. It is important that this kind of study be repeated in further localities, particularly on South Island, before conclusions are drawn about the timing of kiore arrival in New Zealand.

Most offshore islands are unlikely to have been visited by the transient seafarers who brought the first Pacific rats to New Zealand. Given the very poor swimming ability of this rat (Atkinson and Moller 1990), many islands escaped its introduction until after the 1250–1300 settlement period; a minority of islands escaped these rats altogether.

Dogs accompanied the Polynesian settlers and were probably used to hunt birds (Anderson 1981). They apparently remained closely associated with habitations, rather than becoming feral; their remains have not been found in deep caves acting as pitfall traps for vertebrates (Worthy and Holdaway 2002). However, packs of European dogs were roaming forests by the middle of the 19th century (Thomson 1922). Dog packs are now rarely recorded but both domestic and feral dogs continue to be predators of ground birds, including the kiwi and weka (McLennan et al. 1996; Beauchamp et al. 1998).

The frequency of mammalian introductions since European arrival can be compared between 30-year periods. Norway rats are likely to have established during the 1771–1800 period, having arrived with the first navigators and subsequently with whalers (Atkinson 1973). They had spread throughout the northern districts of North Island by the 1830s, and were common in both North Island and South Island by the 1850s (Moors 1990).

Table 4.1 Introduction times, spread as feral animals, and trophic status of terrestrial mammals in New Zealand

Species	Time or period of earliest successful introductions; extent of spread	Trophic status
Pacific rat, *Rattus exulans*	1200–1300 A.D. but possibly much earlier; widespread until 19th century; apparently now restricted to parts of Southland, Fiordland and some offshore islands	Omnivore
Dog, *Canis familiaris*	1200–1250; mainly but not wholly restricted to human settlements	Carnivore
Norway rat, *Rattus norvegicus*	1770–1800; widespread by 1820 (Atkinson 1973)	Omnivore
Cat, *Felis catus*	1770s; widespread; timing of spread unknown	Carnivore
Feral pig, *Sus scrofa*	1800s; established around most settlements by 1840 (McIlroy 1990); now widespread in forest and scrub[a]	Omnivore
Feral goat, *Capra hircus*	1800s; earlier releases probably eaten (Rudge 1990b); liberated throughout the 19th and 20th centuries; now widespread[a]	Herbivore
Feral cattle, *Bos taurus*	1819: earliest feral cattle reported; widespread by the 1860s (Taylor 1990a) but now localized in both main islands	Herbivore
Mouse, *Mus musculus*	ca. 1830 in North Island; 1850s in South Island (Taylor 1975); now very widespread	Omnivore
Feral sheep, *Ovis aries*	1840s; eight wild herds remain on the mainland, two in North Island and six in South Island (Rudge 1990a)	Herbivore
Brushtail possum, *Trichosurus vulpecula*	1858, near Riverton; 1890–1900, North and South islands (Pracy 1974); now very widespread (Cowan 2001)[a]	Herbivore/ predator
Fallow deer, *Dama dama dama*	1860–1910; 12 herds remain, six in North Island and six in South Island (Davidson and Nugent 1990)[a]	Herbivore
Red deer, *Cervus elaphus*	1861; widespread (Veblen and Stewart 1982; Challies 1990a)[a]	Herbivore
Ship rat, *Rattus rattus*	1860s in North Island, 1890s in South Island (Atkinson 1973); very widespread	Omnivore
Rabbit, *Oryctolagus cuniculus*	1870s; widespread outside forest	Herbivore
Dama wallaby, *Macropus eugenii*	1870, Kawau Island; 1912, Rotorua (Warburton and Sadleir 1990a)[a]	Herbivore

Table 4.1 (*Continued*)

Species	Time or period of earliest successful introductions; extent of spread	Trophic status
Brushtailed rock wallaby, *Petrogale penicillata*	1873; Motutapu, Rangitoto and Kawau islands (Warburton and Sadleir 1990b)[b]	Herbivore
Bennett's wallaby, *Macropus rufogriseus*	1874; South Island only (Warburton and Sadleir 1990 c)	Herbivore
Sambar deer, *Cervus u. unicolor*	1875: Bay of Plenty region, and Wanganui–Levin coastal zone and adjacent valleys (Douglas 1990a)	Herbivore
Feral horse, *Equus caballus*	1876: earliest feral horses reported (Taylor 1990b); feral horses have remained localized in both main islands	Herbivore
Ferret, *Mustela fero*	1882; widespread (Lavers and Clapperton 1990)[a]	Carnivore
Stoat, *Mustela erminea*	1885; widespread (King 1990b)	Carnivore
Weasel, *Mustela nivalis*	1885; widespread (King 1990a)	Carnivore
Hedgehog, *Erinaceus europaeus*	1885; widespread (Brockie 1990)	Carnivore
Brown hare, *Lepus europaeus*	1900s; widespread (Flux 1990)	Herbivore
Himalayan tahr, *Hemitragus jemlahicus*	1900s; South Island (Tustin 1990)[a]	Herbivore
White-tailed deer, *Odocoileus virginianus*	1905; South Island (Lake Wakatipu), Stewart Island (Davidson and Challies 1990)	Herbivore
Wapiti, *Cervus elaphus nelsoni*	1905; northern Fiordland, west of Lake Te Anau (Challies 1990b)[a]	Herbivore
Rusa deer, *Cervus timorensis*	1908; central North Island, several small herds (Douglas 1990b)[a]	Herbivore
Chamois, *Rupicapra rupicapra*	1910; South Island (Clarke 1990)[a]	Herbivore
Sika deer, *Cervus nippon*	1920s; central North Island only (Davidson 1990)[a]	Herbivore

[a] Range extensions continuing
[b] Rangitoto and Motutapu islands connected by a causeway. This population now eradicated (Sadleir and Warburton 2001)

From the time of James Cook, and particularly during the 1801–1830 period, European and American sealers, whalers and traders made gifts of pigs to Maori or used these for barter. Maori sometimes kept them in semi-feral herds from which many escaped (McIlroy 1990). Thus, it is likely that in some places feral pigs had established by the 1820s. This was probably also the case for feral goats, since Cook liberated goats in the Marlborough Sounds and in Hawke's Bay (Donne 1924; Beaglehole 1961) and further liberations were made by sealers, whalers and settlers (Rudge 1990b). Feral cattle had established in North Island by 1820 (Barton 1927) and in South Island by 1839 (Thomson 1922). Mice were first reported in the Bay of Islands, North Island, in about 1830 (Guthrie-Smith 1953) but their spread through South Island was delayed until after 1850 (Taylor 1975).

The 1831–1860 period saw the establishment of feral sheep and goats as well as cats, ship rats and possums. Feral populations of sheep were established by the 1840s (Rudge 1990a) and feral goats were present on the North and South islands by the 1850s (Thomson 1922; Rudge 1990b). Cats appear to have established in North Island in the 1830s, and in South Island in the 1850s (Fitzgerald 1990). Mistaken identifications by early observers confused the timing of ship rat arrival in New Zealand (cf. Thomson 1922) but it appears certain that this species was the last rodent to spread through the country, as was the case for most islands in the Pacific (Atkinson 1973, 1985). Their spread in North Island is most likely to have been during the 1850s, although their catastrophic effects on small birds were not noticed until the 1860s. Their spread in South Island was later – from the late 1880s to the 1890s (Atkinson 1973). The earliest possum population was established near Riverton in 1858 (Pracy 1974).

Introductions of mammals peaked during the 1861–1890 period, with introductions of fallow, red and sambar deer, rabbits, two species of wallaby, horses and further introductions of possums (Table 4.1). When rabbits were introduced in the 1870s, they caused enormous damage to pastoral farmland as well as native grasslands, thereby promoting the introduction by acclimatization societies of a succession of predators in the 1880s: ferrets, stoats and weasels. Hedgehogs were also introduced during this period.

The last mammal to become widespread was the brushtail possum, for which introductions peaked between 1890 and 1900 (Pracy 1974). They have now spread to most parts of all three main islands. Hares were not introduced until the 1900s. They were followed by Himalayan tahr and chamois in South Island, sika deer in North Island, and further introductions of red deer in both islands.

Although additional species of mammal are no longer being introduced, illegal releases of some species already present into parts of the country where they were previously absent is resulting in range extensions. Examples are pigs and goats, fallow, red, wapiti, rusa and sika deer, and tahr (Fraser et al. 2000). Ferrets have also spread more widely, particularly in Northland. Many of the Northland records are believed to be the result of escapes, or illegal releases, from fitch farms (Pierce 1996). Trading ferrets as pets is likely to be an additional factor widening their distribution (Clapperton 2001).

Introduced mammals which continue to extend their range, apparently without human assistance, include dama wallabies in the Rotorua district and possibly Bennett's wallabies in Canterbury (Sadleir and Warburton 2001), possums in south-western South Island (Cowan 2001), ferrets in Northland (Clapperton 2001) and pigs in Otago, Southland and the West Coast of South Island (McIlroy 2001). Between 1993 and 1996, 62 new herds of goats, mostly from farm escapes in various parts of the country, became nuclei for further spread (Parkes 2001). Forsyth and Tustin (2001) record one new population of tahr, south of Haast Pass, since 1990 and chamois are extending their range in Northwest Nelson, Central Otago and Fiordland (Forsyth and Clarke 2001).

4.2.2 Factors Affecting Impact of Introduced Mammals

Divergence of feeding habits between the introduced herbivorous and carnivorous mammals affects the local distribution of their impacts. Excepting possums, the herbivores are restricted to foods they can reach from the ground. However, their food preferences dictate that not all types of forest, scrub, shrubland or wetland are suitable for sustaining breeding populations. In contrast, predators such as rats and mustelids can hunt and breed throughout many different kinds of habitat. Furthermore, as a result of their climbing ability, these predators as well as possums can obtain food in a much more three-dimensional manner than most herbivores. It is doubtful whether any kind of vegetated habitat on the New Zealand mainland is free from introduced mammalian predators.

Sixteen of the 30 species of introduced mammals listed in Table 4.1 are now widespread and, of the 14 more-localized species, six are continuing to extend their ranges. Those mammals which have become widespread are the four rodent species, stoat, ferret, cat, pig, rabbit, goat, possum and red deer. Possums, for example, occupied more than 90 % of New Zealand's land area in 1990 and have continued to spread (Cowan 1990, 2001). With the possible exception of some deer species, all these animals are "generalists", in the sense that they eat a wide range of plant or animal foods and breed in a variety of habitats. The four rodents and cats, pigs, rabbits and goats have all established when introduced to islands elsewhere in the world. The ability of possums to live in a wide range of habitats is attributed (Nugent et al. 2000) to their use of high-energy foods, particularly fruit, flowers, insects and eggs, even though leaves of lower nutrient quality contribute the bulk of their diet. The success of stoats is likely to be at least partly related to their capacity to kill prey in any habitat where it is available, from sea level to altitudes well above treeline

(King 1990b). Though more restricted in habitat, ferrets prey on a wide range of invertebrate and vertebrate animals.

Reasons why some introduced mammals become widespread whereas others maintain a restricted distribution are not always clear. Competition between introduced species is almost certainly important but is difficult to measure. Competition from European rodents (including mice) is a possible explanation for the disappearance of Pacific rats (discussed above) from most of their earlier nationwide distribution (cf. Taylor 1975). Predation on Pacific rats by larger rat species, mustelids or cats may also be involved. With five species of wallaby introduced to Kawau Island, it is likely that interspecific competition for food on the island has at times been severe but four of these species survived. Failure of some deer species to spread widely in the country can possibly be attributed to competition from the more widespread red and sika deer. Davidson and Tustin (1990) suggest that competition from red deer in a habitat marginal for moose has prevented the latter from establishing in Fiordland.

The time interval since establishment of a particular mammal in part of the country provides a measure of impact duration. This interval is shortened if, as in the case of the Pacific rat, the animal has largely disappeared from the mainland. If the chronology of Holdaway et al. (2002 c) for Pacific rat presence is accepted, this rat has had a longer impact than any other introduced mammal: at least 1,400 years in North Island and 1,440 years in South Island. The difference between islands relates to the longer persistence of Pacific rats in mainland South Island; it still persists in parts of Fiordland and south Westland (Atkinson and Moller 1990; Ruscoe 2004). In contrast, Pacific rats had apparently disappeared from mainland North Island by the early 1850s (Atkinson 1973). No recent records have come to light, although very little trapping for rats has been done in remote areas.

On the main islands, it is questionable whether pest-control programmes have substantially restricted the distribution limits of any introduced mammal. Significant reductions in density and thus impact have, however, often been achieved. Such reductions are invariably temporary unless:

1. a regime of continual pest control is used to maintain mammal numbers at a low level; the most intensive application of this regime is found within the six "mainland islands" maintained by the Department of Conservation (Saunders and Norton 2001) as well as within an increasing number of privately funded "mainland islands"; and
2. specialized fencing, such as that protecting the Karori Wildlife Sanctuary, is used to completely exclude all mammals – except mice (Karori Wildlife Sanctuary Trust 1997).

Progress continues to be made, with eradications of introduced mammals from many offshore and outlying islands of particular conservation value. The most significant recent advance has been the 2001 eradication by the

Department of Conservation (confirmed in 2003) of Norway rats from sub-Antarctic Campbell Island (11,330 ha).

4.2.3 Functional Significance of the Introduced Mammals

The functional significance of an introduced mammal is related to its trophic status (herbivore, omnivore or carnivore), the kinds of foods eaten, and the characteristics of sites (including height above ground) from where food is taken. The 30 introduced species listed in Table 4.1 comprise 19 species of herbivore, six species of carnivore and five species of omnivore. Of these, brushtail possums have wide-ranging impacts because they are broad-spectrum ground feeders on seeds, seedlings and fungi as well as foliage, buds, flowers, seeds and fruits at all heights in forests. In addition, an increasing number of native animals are proving susceptible to predation by brushtail possums (see Chap. 17). In contrast, certain introduced mammals have a narrow range of impacts. For example, horses are herbivores which close-graze grassland communities.

4.3 Changes in the Herbivore–Predator Regime Since 1200 A.D.

4.3.1 Herbivores and Omnivores

Pacific rats consumed seeds and seedlings, and preyed on invertebrates and vertebrates in mainland forests, and in other kinds of community, for hundred of years before other rats established. Historical records (Meeson 1885; Rutland 1890) indicate that Pacific rats responded to heavy seedfalls of beech in South Island, at which times they reached plague numbers. Effects of this rat on native tree recruitment (see Campbell and Atkinson 1999) suggest that the composition of some forests was significantly changed long before other mammals arrived. It must be acknowledged, however, that moa (including their chicks), in addition to eating leaves, twigs and fruit, also ate seeds (Burrows et al. 1981; Burrows 1989) and almost certainly ate seedlings. They may also have been effective dispersal agents for some species. Nevertheless, it would be surprising if Pacific rats did not initiate new trends in species recruitment and, thus, forest composition. One new effect would have been their ability to climb and take fruit high up in trees where they would be competing with some native birds.

No quantitative studies appear to have been made of the effect of ship rats on the regeneration of native trees, probably because separating such effects from those of mice and larger herbivores is difficult. The plant foods of Pacific rats and ship rats are likely to overlap to a considerable extent, even though the two species co-exist in parts of Fiordland (personal observations). Following the almost complete disappearance of Pacific rats from the mainland between 1850 and 1890, it is possible that any effects they had on forest regeneration were replaced by those of ship rats as well as by more local effects stemming from Norway rats and mice. More work on the forest diets of ship rats and mice is needed to test this supposition.

The main populations of moa were probably extinct by 1400, as a result of hunting by Maori (Holdaway and Jacomb 2000; Worthy and Holdaway 2002). The loss of all moa species must have had significant effects on regeneration patterns in many kinds of forest, and probably other kinds of vegetation, too. There followed a period of some 400 years during which, apart from indigenous birds (and humans), Pacific rats would have been the most important vertebrate influence on forest regeneration (Table 4.2). From 1770 onwards, Europeans introduced a series of mammalian herbivores and omnivores, each with specific effects on indigenous forests. Beginning with the arrival of Pacific rats, changes in the introduced and endemic vertebrate fauna, interacting with site factors, have exerted a continuing influence on the country's forests for at least 750 years, and possibly much longer. Apart from the very earliest Polynesian visitors, it is doubtful whether humans ever saw "pristine forest" on mainland New Zealand.

Table 4.2 Chronological sequence of vertebrate herbivores and omnivores affecting forest regeneration in New Zealand

Time period (A.D.)	Herbivore and omnivore present during period[a]
Years B.C.–ca. 1250 A.D.	Moa + smaller leaf-eating and seed-eating native birds (see text)
ca. 1250 (or earlier)	Moa + Pacific rats
1250–1400	Transition during extermination of moa
1400–1800	Pacific rats alone
1800–1860	Norway rats + Pacific rats + pigs + mice[b]
1860–1900 (North Island)	Ship rats + pigs + (mice)[b]
1860–1890 (South Island)	Norway rats + pigs + (mice)[b]
1900 (1890)–1920	Ship rats + pigs + (mice); Norway rats localized[b]
1920–2000	Ship rats + pigs + deer + possums + (mice), combination depending on locality and time; Norway rats localized[b]

[a]Birds smaller than moa have not been included in this time sequence
[b]The distribution and abundance of mice in New Zealand forests during these periods have not been documented

In some of our oldest stands of remaining forest, particularly those dominated by kauri or podocarps, a small proportion of canopy trees are 750 or more years of age. These trees established in the presence of moa and Pacific rats. The relative densities of these very old trees may reveal something about patterns of regeneration which were possible in the presence of these two animals. A much larger number of stands have canopies formed by trees between 200 and 600 years of age. These are likely to have developed in the presence of Pacific rats alone as the most significant introduced seed consumer. Changes in climate may complicate interpretation of possible rat effects but this may be less of a problem as more detailed knowledge of past climates becomes available from physical, rather than biological measurements. The former presence of large numbers of indigenous avian seed-eaters, such as parakeets and kaka, may complicate attempts to isolate the effects of the rat. These birds, however, remained widespread until after the arrival of ship rats and stoats, so that, prior to the 19th century, they can be treated as a factor common to most forests.

From 1800 onwards, and particularly from the 1920s onwards, herbivorous and omnivorous mammals have become steadily more widespread in indigenous scrub and forest. In any particular area, unravelling the various effects of these mammal species on vegetation composition would be very difficult unless there were records of the times when each had established in that area.

4.3.2 Predators

Extinction of moa resulted in the decline of Haast's eagle (*Harpagornis moorei*), the world's largest eagle, for which Worthy and Holdaway (2002) provide compelling evidence of its ability to kill even the largest moa, now recognized as females (Bunce et al. 2003). This eagle was apparently restricted to South Island, and Worthy and Holdaway suggest that smaller prey may have included the extinct goose (*Cnemiornis calcitrans*), Finsch's duck (*Euryanus finschi*), and South Island adzebill (*Aptornis defossor*), and the extant South Island takahe (*Porphyrio hochstetteri*), weka (*Gallirallus australis*), kea (*Nestor notabilis*), kaka (*N. meridionalis*) and New Zealand pigeon. Many of these birds were hunted by Maori, and this would have further restricted food for the eagle.

Pacific rats continued to prey on indigenous invertebrates, reptiles (including tuatara) and birds on both main islands before they were replaced by new predators: Norway rats, cats, pigs, mice, ship rats, ferrets, stoats, weasels, hedgehogs and possums. This was a massive invasion of indigenous habitats by an increasingly diverse array of predatory mammals in little more than 100 years, its effects being exacerbated by continuing habitat fragmentation and loss. It had at least six major impacts on native animals:

1. increased predation on larger animals, such as kiwi and takahe, above the level already exerted by humans and their dogs;

2. increased predation on smaller animals such as large insects (e.g. wetas) and small birds (e.g. bush wrens) already preyed upon by Pacific rats and/or native birds;
3. increased predation on animals in habitats or parts of habitats little used by Pacific rats, e.g. wetlands;
4. exposure of native animals to kinds of predation not used by Pacific rats or native birds, e.g. digging and ploughing by pigs for subsoil invertebrates;
5. predation of vulnerable animals throughout the day and night by nocturnal and diurnal mammalian predators; this contrasted with Pacific rat predation, which is largely nocturnal; and
6. year-round predation of vulnerable animals by the new invaders; previously, the seasonal fluctuation in Pacific rat numbers (very few in winter) resulted in their predation being largely concentrated, if not restricted to summer months.

Ship rats, possums and stoats are apparently now the most influential mammalian predators in New Zealand forests. Innes (2001) considers "ship rats are probably the most frequent predator of eggs, chicks and sitting adults of forest birds on the New Zealand mainland". Brockie (1992), in reviewing the 24-year team study made in lowland forest of the Orongorongo Valley, near Wellington, concluded that ship rats and possums were "the most pervasive and devastating agents of change".

4.4 Response of the New Zealand Biota

The vulnerability of the New Zealand fauna to introduced mammals is underlined by an ever-increasing number of studies demonstrating significant predation on New Zealand birds. Examples are kiwi (McLennan et al. 1996), weka (Bramley 1996), kaka (Wilson et al. 1998), kereru (James and Clout 1996), banded dotterel (Rebergen et al. 1998), reptiles (Towns 1991, 1996; Towns et al. 2001) and invertebrates (e.g. C. Green in Towns et al. 1997). This vulnerability has not resulted from any lack of vertebrate predators in the pre-human fauna. The country originally supported a wider range of insectivorous birds, two species of owl and an owlet-nightjar, the flightless North Island and South Island adzebills (*Aptornis* spp.) and three species of raptor, including the giant Haast's eagle. On the ground, species of skink up to 30 cm in length, arboreal geckos, and tuatara were once widespread. All these reptiles were effective predators of small animals, and many included invertebrates in their diet. To survive in pre-human times, both native vertebrates and invertebrates must have evolved anti-predator or predator-avoidance behaviours, or other characteristics appropriate for defence against these predators. These adaptations have proved ineffective against mammalian predators introduced to New

Zealand. Remaining immobile in a nest, nesting hidden on the ground and in holes within trees or in burrows below ground, and developing postures or colours which camouflage against a predator hunting by sight or sound is of little help against an agile hunter which also has an acute sense of smell. This may be the crucial reason why New Zealand's native land animals have suffered so severely from introduced predators.

The extreme vulnerability of New Zealand birds to mammalian predators is illustrated by the kakapo (*Strigops habrotilus*). This flightless nocturnal bird is the heaviest parrot in the world, males sometimes exceeding 3.5 kg. Plumage colours are greens, yellows and browns, such that when roosting in a tree the bird is almost impossible to see with the human eye. Kakapo are lek breeders (Merton et al. 1984), mating taking place at track and bowl systems constructed by one or more males. Prolonged resonant booming by males to attract females carries for several kilometres but may also arouse the interest of mammalian predators. Once a predator finds a kakapo track, it would be easy to follow, especially as kakapo plumage is scented. After mating, the male takes no part in the protection, incubation or feeding of the young. The eggs are in the size range eaten by all three rat species (Atkinson 1985). The long incubation period (28 days) is followed by chicks remaining a further 70 days in the nest bowl during fledging and maturation (D. Eason, personal communication). During fledging, the female may leave the chicks unattended for periods of up to 4 h. Well adapted as the birds must have been to avoid avian predators, these attributes have proved quite ineffective against mammalian predators.

Many New Zealand plants such as *Asplenium bulbiferum*, *Carmichaelia* spp., supplejack (*Ripogonum scandens*), kotukutuku (*Fuchsia excorticata*), kapuka (*Griselinia littoralis*), and *Pseudopanax* spp are vulnerable to mammalian herbivores. Growth strategies by which plants may have co-existed with native herbivores, particularly moa, are discussed elsewhere in this volume (see Chap. 11).

One may ask whether the impact of introduced mammals on the New Zealand biota differs in any essential way from the impact these mammals have had on hundreds of smaller islands throughout the Pacific, Indian and Atlantic oceans. Such differences as do exist relate to the fact that most islands have received only a subset of the array of mammals which have established in New Zealand. In particular, mustelids have generally not been introduced to islands, although mongooses (Viveridae) have. As for New Zealand, the effects of all these introductions on island biotas have usually been severe and sometimes catastrophic. Examples are the spread of rats to oceanic islands (Atkinson 1985) and the effects of goats on the vegetation of many islands (Coblentz 1978). We can predict that any further introduction of a mammal to New Zealand will impact to the extent that its foods, and methods of obtaining them, differ from those of mammals already established.

4.5 Conclusions

Introduced mammalian predators have decreased the numbers of many species of New Zealand vertebrates and invertebrates, some to the point of extinction. Herbivorous and omnivorous mammals have effected major decreases in some indigenous plants but apparently have caused few extinctions. Some introduced mammals continue to spread, particularly deer from farm escapes and deliberate relocations.

Those indigenous plants and animals which have survived are to a greater or lesser extent co-existing with introduced mammals, and many other introduced animals and plants. This continuing co-existence means that indigenous species are now living in an environment where the biotic component of the selection regime is fundamentally different from that in which they evolved. The new selection regimes include new competitors, disease vectors, parasites and pollinators as well as new herbivores and predators. As a consequence, the future evolutionary pathways of the country's indigenous organisms are predicted to significantly change (Atkinson 1998, 2001).

Protecting indigenous species within fenced reserves may exclude some mammals but usually not rats, mice, introduced birds, and numerous species of introduced plants and invertebrates. With respect to species requiring protection, Walker (1992) argues that "to place equal emphasis on every species is ecologically unsound and tactically unachievable". Should we therefore concentrate on protecting a few species in each reserve and, if so, how would these be chosen? Walker considers that the best way to minimize species loss is to "maintain the integrity of ecosystem function" by identifying those species which are critical to the functioning of the system. He suggests using guild analyses to identify differing functional groups of organisms, determining the number of species within each guild, examining interactions between species within a guild and, finally, identifying those functional groups (and presumably species) which are the "major drivers of the system". Such species, whether plant or animal, would not always be charismatic.

Guild analyses have seldom been used in New Zealand. Atkinson and Millener (1991) made a preliminary attempt for native birds in prehistoric systems but they did not include other animals (e.g. bats, lizards, tuatara or invertebrates) which contribute to many functional groups. The suggestion has also been made to substitute ecologically similar and genetically related species for some extinct species in order to rebuild lost interactions or functions within a restored community (Atkinson 2001, 2002).

The crucial importance of moa as the major ground-based herbivores in the previous system is obvious. It is naïve to think that introduced deer or goats have substituted for moa as browsing equivalents in our forests, a point confirmed by the recent study of Bond et al. (2004). As discussed by Atkinson and Greenwood (1989), there are major differences in feeding ecology between moa and ungulates (cf. absence in moa of rotary mandible action, prehensile tongue and teeth). Based on present knowledge, however, restoring the functioning of a New Zealand ecosystem to a state more like that of prehistoric times is apparently not realistic. Still, is it not possible that in the future we will learn enough about moa diets to identify those groups of plants which were most affected by them? With this kind of information, we can explore the feasibility of using an extant ratite bird to partially restore the browsing influence of moa on forest understories.

Acknowledgements I am indebted to R.B. Allen, P. Cowan, T.J. Atkinson, D.J. Campbell, B.M. Fitzgerald and an anonymous referee for many helpful criticisms of earlier drafts of this chapter.

References

Anderson AJ (1981) Pre-European hunting dogs in the South Island, New Zealand. NZ J Archaeol 3:15–20

Anderson AJ (1991) The chronology of colonization in New Zealand. Antiquity 65:767–795

Atkinson IAE (1973) Spread of the ship rat (*Rattus r. rattus* L.) in New Zealand. J R Soc NZ 3:457–472

Atkinson IAE (1985) The spread of commensal species of *Rattus* to oceanic islands and their effects on island avifaunas. In: Moors PJ (ed) Conservation of island birds. ICBP Tech Publ 3:35–81

Atkinson IAE (1998) Conserving plants as evolutionary entities: successes and unanswered questions from New Zealand and elsewhere. Aliso 16:103–112

Atkinson IAE (2001) Introduced mammals and models for restoration. Biol Conserv 99:81–96

Atkinson IAE (2002) Recovery of wildlife and restoration of habitats in New Zealand. Pac Conserv Biol 8:27–35

Atkinson IAE, Greenwood RM (1989) Relationships between moas and plants. NZ J Ecol suppl 12:67–96

Atkinson IAE, Millener PR (1991) An ornithological glimpse into New Zealand's prehuman past. Acta XX Congr Int Ornithology, pp 129–192

Atkinson IAE, Moller H (1990) Kiore. In: King CM (ed) The handbook of New Zealand mammals. Oxford Univ Press, Auckland, New Zealand, pp 175–192

Barton RJ (comp) (1927) Earliest New Zealand: the journals and correspondence of the Rev John Butler. Palamontain and Petherick, Masterton, New Zealand

Beaglehole JC (ed) (1961) The Journals of Captain James Cook on his Voyages of Discovery, vol 2. The voyage of the Resolution and Discovery, 1772–1775. Cambridge Univ Press for the Hakluyt Society

Beauchamp AJ, van Berkum B, Closs MJ (1998) The decline of North Island Weka (*Gallirallus australis greyi*) at Parekura Bay, Bay of Islands. Notornis 45:31–43

Bond WJ, Lee WG, Craine JM (2004) Plant structural defences against browsing birds: a legacy of New Zealand's extinct moas. Oikos 104:500–508

Bramley GN (1996) A small predator removal experiment to protect North Island Weka (*Gallirallus australis greyi*) and the case for single-subject approaches in determining agents of decline. NZ J Ecol 20:37–43

Brockie RE (1990) European hedgehog. In: King CM (ed) The handbook of New Zealand mammals. Oxford Univ Press, Auckland, pp 99–113

Brockie B (1992) A living New Zealand forest. David Bateman, Auckland, New Zealand

Bunce M, Worthy TH, Ford T, Hoppitt W, Willerslev E, Drummond A, Cooper A (2003) Extreme reversed sexual size dimorphism in the extinct New Zealand moa *Dinornis*. Nature 425:172–175

Burrows CJ (1989) Moa browsing: evidence from the Pyramid Valley mire. NZ J Ecol suppl 12:51–56

Burrows CJ, McCulloch B, Trotter MM (1981) The diet of moas based on gizzard contents samples from Pyramid Valley, North Canterbury and Scaifes Lagoon, Lake Wanaka, Otago. Rec Cant Mus 9:309–336

Campbell DJ, Atkinson IAE (1999) Effects of kiore (*Rattus exulans* Peale) on recruitment of indigenous coastal trees on northern offshore islands of New Zealand. J R Soc NZ 29:265–290

Challies CN (1990a) Red deer. In: King CM (ed) The handbook of New Zealand mammals. Oxford Univ Press, Auckland, New Zealand, pp 436–458

Challies CN (1990b) Wapiti. In: King CM (ed) The handbook of New Zealand mammals. Oxford Univ Press, Auckland, New Zealand, pp 458–467

Clapperton BK (2001) Advances in New Zealand mammalogy 1990–2000. Feral ferret. J R Soc NZ 31:185–203

Clarke CMH (1990) Chamois. In: King CM (ed) The handbook of New Zealand mammals. Oxford Univ Press, Auckland, New Zealand, pp 380–391

Coblentz BE (1978) The effects of feral goats (*Capra hircus*) on island ecosystems. Biol Conserv 13:279–286

Cowan PE (1990) Brushtail possum. In: King CM (ed) The handbook of New Zealand mammals. Oxford Univ Press, Auckland, New Zealand, pp 68–98

Cowan PE (2001) Advances in New Zealand mammalogy 1990–2000. Brushtail possum. J R Soc NZ 31:15–29

Davidson MM (1990) Sika deer. In: King CM (ed) The handbook of New Zealand mammals. Oxford Univ Press, Auckland, New Zealand, pp 468–477

Davidson MM, Challies CN (1990) White-tailed deer. In: King CM (ed) The handbook of New Zealand mammals. Oxford Univ Press, Auckland, New Zealand, pp 506–514

Davidson MM, Nugent G (1990) Fallow deer. In: King CM (ed) The handbook of New Zealand mammals. Oxford Univ Press, Auckland, New Zealand, pp 490–506

Davidson MM, Tustin KG (1990) Moose. In: King CM (ed) The handbook of New Zealand mammals. Oxford Univ Press, Auckland, New Zealand, pp 514–520

Donne TE (1924) The game animals of New Zealand: an account of their introduction, acclimatization, and development. Murray, London

Douglas MJW (1990a) Sambar deer. In: King CM (ed) The handbook of New Zealand mammals. Oxford Univ Press, Auckland, New Zealand, pp 477–483

Douglas MJW (1990b) Rusa deer. In: King CM (ed) The handbook of New Zealand mammals. Oxford Univ Press, Auckland, New Zealand, pp 483–488

Fitzgerald BM (1990) House cat. In: King CM (ed) The handbook of New Zealand mammals. Oxford Univ Press, Auckland, New Zealand, pp 330–348

Flux JEC (1990) Brown hare. In: King CM (ed) The handbook of New Zealand mammals. Oxford Univ Press, Auckland, New Zealand, pp 161–172

Forsyth DM, Clarke CMH (2001) Advances in New Zealand mammalogy 1990–2000. Chamois. J R Soc NZ 31:243–249

Forsyth DM, Tustin KG (2001) Advances in New Zealand mammalogy 1990–2000. Himalayan tahr. J R Soc NZ 31:251–261
Fraser KW, Cone JM, Whitford EJ (2000) A revision of the established ranges and new populations of 11 introduced ungulate species in New Zealand. J R Soc NZ 30:419–437
Guthrie-Smith H (1953) Tutira: the story of a New Zealand sheep station. AH and AW Reed, Wellington, New Zealand
Higham TFG, Hogg AG (1997) Evidence for late Polynesian colonization of New Zealand. University of Waikato radiocarbon measurements. Radiocarbon 39:149–192
Holdaway RN (1996) Arrival of rats in New Zealand. Nature 384:225–226
Holdaway RN (1999) A spatio-temporal model for the invasion of the New Zealand archipelago by the Pacific rat *Rattus exulans*. J R Soc NZ 29:91–105
Holdaway RN, Jacomb C (2000) Rapid extinction of the moas (Aves: Dinornithiformes): model, test, and implications. Science 287:2250–2254
Holdaway RN, Jones MD, Beaven Athfield NR (2002a) Late Holocene extinction of Finsch's duck (*Chenonetta finschi*), an endemic, possibly flightless, New Zealand duck. J R Soc NZ 32:629–651
Holdaway RN, Jones MD, Beavan-Athfield NR (2002b) Late Holocene extinction of the New Zealand owlet-nightjar *Aegotheles novaezealandiae*. J R Soc NZ 32:653–667
Holdaway RN, Roberts RG, Beavan-Athfield NR, Olley JM, Worthy TH (2002 c) Optical dating of quartz sediments and accelerator mass spectrometry ^{14}C dating of bone gelatin and moa eggshell: a comparison of age estimates for non-archaeological deposits in New Zealand. J R Soc NZ 32:463–505
Innes J (2001) Advances in New Zealand mammalogy 1990–2000. European rats. J R Soc NZ 31:111–125
James RE, Clout MN (1996) Nesting success of New Zealand pigeons (*Hemiphaga novaeseelandiae*) in response to a rat (*Rattus rattus*) poisoning programme at Wenderholm Regional Park. NZ J Ecol 20:45–51
Karori Wildlife Sanctuary Trust Inc (1997) Karori Wildlife Sanctuary Management Plan: resource description, objectives and management policies. Wellington, New Zealand
King CM (1990a) Weasel. In: King CM (ed) The handbook of New Zealand mammals. Oxford Univ Press, Auckland, New Zealand, pp 313–320
King CM (1990b) Stoat. In: King CM (ed) The handbook of New Zealand mammals. Oxford Univ Press, Auckland, New Zealand, pp 288–312
Lavers RB, Clapperton BK (1990) Ferret. In: King CM (ed) The handbook of New Zealand mammals. Oxford Univ Press, Auckland, New Zealand, pp 320–330
McFadgen BG, Knox FB, Cole TRL (1994) Radiocarbon calibration curve variations and their implications for the interpretation of New Zealand prehistory. Radiocarbon 36:221–236
McIlroy JC (1990) Feral pig. In: King CM (ed) The handbook of New Zealand mammals. Oxford Univ Press, Auckland, New Zealand, pp 358–371
McIlroy JC (2001) Advances in New Zealand mammalogy 1990–2000. Feral pig. J R Soc NZ 31:225–231
McLennan JA, Potter MA, Robertson HA, Wake GC, Colbourne R, Dew L, Joyce L, McCann AJ, Miles J, Miller PJ, Reid J (1996) Role of predation in the decline of kiwi, *Apteryx* spp., in New Zealand. NZ J Ecol 20:27–35
Meeson J (1885) The plague of rats in Nelson and Marlborough. Trans Proc NZ Inst 17:199–207
Merton DV, Morris RB, Atkinson IAE (1984) Lek behaviour in a parrot: the Kakapo *Strigops habroptilis* of New Zealand. Ibis 126:277–283
Moors PJ (1990) Norway rat. In: King CM (ed) The handbook of New Zealand mammals. Oxford University Press, Auckland, New Zealand, pp 192–206

Nugent G, Sweetapple PJ, Coleman JD, Suisted PA (2000) Possum feeding patterns: dietary tactics of a reluctant folivore. In: Montague TL (ed) The brushtail possum: biology, impact and management of an introduced marsupial. Manaaki Whenua Press, Lincoln, New Zealand, pp 10–19

Parkes JP (2001) Advances in New Zealand mammalogy 1990–2000. Feral cattle. J R Soc NZ 31:233

Pierce RJ (1996) Impacts of ferrets on North Island brown kiwi and brown teal. Ferrets as vectors of tuberculosis and threats to conservation. R Soc Misc NZ Ser 36:47

Pracy LT (1974) Introduction and liberation of the opossum (*Trichosurus vulpecula*) into New Zealand, 2nd edn. NZ For Serv Infor Ser 45

Rebergen A, Keedwell R, Moller H, Maloney R (1998) Breeding success and predation at nests of banded dotterel (*Charadrius bicinctus*) on braided riverbeds in the central South Island, New Zealand. NZ J Ecol 22:33–41

Rudge MR (1990a) Feral sheep. In: King CM (ed) The handbook of New Zealand mammals. Oxford Univ Press, Auckland, New Zealand, pp 424–431

Rudge MR (1990b) Feral goat. In: King CM (ed) The handbook of New Zealand mammals. Oxford Univ Press, Auckland, New Zealand, pp 406–423

Ruscoe WA (2004) A new location record for kiore (*Rattus exulans*) on New Zealand's South Island. NZ J Zool 31:1–5

Rutland J (1890) On the habits of the New Zealand bush-rat (*Mus maorium*). Trans Proc NZ Inst 22:300–307

Sadleir RMF, Warburton B (2001) Advances in New Zealand mammalogy 1990–2000. Wallabies. J R Soc NZ 31:7–14

Saunders A, Norton DA (2001) Ecological restoration at mainland islands in New Zealand. Biol Conserv 99:109–119

Taylor RH (1975) What limits kiore (*Rattus exulans*) distribution in New Zealand? NZ J Zool 2:473–477

Taylor RH (1990a) Feral cattle. In: King CM (ed) The handbook of New Zealand mammals. Oxford Univ Press, Auckland, New Zealand, pp 373–379

Taylor RH (1990b) Feral horse. In: King CM (ed) The handbook of New Zealand mammals. Oxford Univ Press, Auckland, New Zealand, pp 349–357

Thomson GM (1922) The naturalisation of animals and plants in New Zealand. Cambridge Univ Press, Cambridge

Tomich PQ (1986) Mammals in Hawai'i: a synopsis and notational bibliography, 2nd edn. Bishop Museum Press, Honolulu

Towns DR (1991) Response of lizard assemblages in the Mercury Islands, New Zealand, to removal of an introduced rodent: the kiore (*Rattus exulans*). J R Soc NZ 29:119–136

Towns DR (1996) Changes in habitat use by lizards on a New Zealand island following removal of the introduced Pacific rat *Rattus exulans*. Pac Conserv Biol 2:286–292

Towns DR, Simberloff D, Atkinson IAE (1997) Restoration of New Zealand islands: redressing the effects of introduced species. Pac Conserv Biol 3:99–124

Towns DR, Daugherty CH, Cree A (2001) Raising the prospects for a forgotten fauna: a review of 10 years of conservation effort for New Zealand reptiles. Biol Conserv 99:3–16

Tustin KG (1990) Himalayan tahr. In: King CM (ed) The handbook of New Zealand mammals. Oxford Univ Press, Auckland, New Zealand, pp 392–405

Veblen TT, Stewart GH (1982) The effects of introduced wild animals on New Zealand forests. Ann Assoc Am Geogr 72:372–397

Walker BH (1992) Biodiversity and ecological redundancy. Conserv Biol 6:18–23

Warburton B, Sadleir RMF (1990a) Dama wallaby. In: King CM (ed) The handbook of New Zealand mammals. Oxford Univ Press, Auckland, New Zealand, pp 35–43

Warburton B, Sadleir RMF (1990b) Brushtailed rock wallaby. In: King CM (ed) The handbook of New Zealand mammals. Oxford Univ Press, Auckland, New Zealand, pp 58–64

Warburton B, Sadleir RMF (1990 c) Bennett's wallaby. In: King CM (ed) The handbook of New Zealand mammals. Oxford Univ Press, Auckland, New Zealand, pp 44–51

Wilmshurst JM, Higham TFG (2004) Using rat-gnawed seeds to independently date the arrival of Pacific rats and humans in New Zealand. The Holocene 14:801–806

Wilson PR, Karl BJ, Toft RJ, Beggs JR, Taylor RH (1998) The role of introduced predators and competitors in the decline of kaka (*Nestor meridionalis*) populations in New Zealand. Biol Conserv 83:175–185

Worthy TH, Holdaway RN (2002) The lost world of the moa: prehistoric life of New Zealand. Canterbury Univ Press, Christchurch, New Zealand

5 Parasites and Biological Invasions

D.M. Tompkins and R. Poulin

5.1 Introduction

Parasites are increasingly cited as a major threat to wildlife conservation (Daszak et al. 2000; Cleaveland et al. 2002), being implicated in the decline and/or extinction of many high-profile species worldwide, including the thylacine (McCallum and Dobson 1995), amphibians globally (Berger et al. 1998), UK red squirrels (Tompkins et al. 2003), and Hawaiian birds (van Riper et al. 1986, 2002). In the majority of cases, declines are ascribed to "emerging infectious diseases" (EIDs) – those disease-causing agents whose geographical range, host range, or prevalence have increased in recent years. There are several (non-exclusive) causes of disease emergence in wildlife, including changes in habitat and climate, and genomic change. However, it is the translocation of invasive species which appears to be of key importance, being implicated in more than half of all wildlife EID cases recorded (Daszak et al. 2000). A recent example is the exotic parapoxvirus introduced with grey squirrels (*Sciurus carolinensis*) into the UK at the turn of the 20th century, and which has been implicated in the nationwide decline of native red squirrels (*Sciurus vulgaris*; Tompkins et al. 2003).

In this review, we explore how parasitism and species invasions interact to influence natural communities, using New Zealand case studies complemented by global examples of potential relevance. We begin by looking at exotic parasites and how they can impact the native fauna. We then turn to exotic hosts and how, by acting as reservoirs for native parasites, they can exacerbate disease problems for native wildlife. Finally, we review the parasite threats to New Zealand land birds and freshwater fish, evaluate potential future threats to these biota, and propose management recommendations to minimize these threats.

Ecological Studies, Vol. 186
R.B. Allen and W.G. Lee (Eds.)
Biological Invasions in New Zealand

5.2 Taxonomic Affiliation of Emerging Infectious Diseases of Wildlife

Helminth and fungal parasites are believed unlikely to "emerge" as new infectious diseases of wildlife, whilst bacteria, protozoans and viruses are (Dobson and Foufopoulos 2001). However, this assumption may be premature due to a bias in the information available. Whereas microparasites can have highly visible effects, diseases caused by macroparasites are often more subtle, yet can still impact the host at the population level (Tompkins et al. 2002). Since such effects are harder to detect, many macroparasite EIDs may simply have yet to be quantified. For example, a survey of the helminth fauna of Hawaiian stream fish revealed three species of exotic parasites introduced by poeciliid fish (a nematode *Camallanus cotti*, a cestode *Bothriocephalus acheilognathi*, and a leech *Myzobdella lugubris*), occurring with higher prevalences and abundances than native parasites (Font 1998). These exotic parasites represent the most significant threat of parasite-induced disease in native stream fishes in Hawaii (Font and Tate 1994). At least one of these three species (*B. acheilognathi*) has recently emerged as a disease-causing agent in other fish species (causing bothriocephaliasis in the humpback chub *Gila cypha* in the USA; Brouder and Hoffnagle 1997).

One fungal pathogen has made a huge impact on native biodiversity globally. In 1998, the disease chytridiomycosis was first described from moribund and dead adult amphibians collected at sites of mass deaths in Australia and Panama from 1993 to 1998 (Berger et al. 1998). The causative agent was subsequently identified from captive frogs in the USA as the fungus *Batrachochytrium dendrobatidis* (Longcore et al. 1999). Since then, this pathogen has been linked to many ongoing population declines, and retrospective studies using museum specimens have found *B. dendrobatidis* infections in amphibians as early as 1974 in North America, 1978 in Australia, 1980 in South America, and 1982 in Central America. The pathogen has also been found in African specimens collected since 1998. Chytrid fungus infects the keratinized tissues of amphibians, including the epidermis of adults and the mouthparts of tadpoles, causing (in adults) abnormal posture, lethargy, and loss of righting reflex (Berger et al. 1998). Evidence suggests that its emergence has been driven by the anthropogenic introduction of the pathogen to new regions and host species ("pathogen pollution"; Daszak et al. 2000). For example, comparisons of ssrRNA sequences indicate that the New Zealand chytrid strain is almost identical to that initially identified in the USA but varies by up to 3 % from the Australian strain (Waldman et al. 2001). This suggests that New Zealand chytrids are recent arrivals, most likely from America via the international pet trade.

5.3 Determinants of Establishment Success by Introduced and Invading Parasites

A survey of the parasite communities of three species of exotic freshwater poeciliid fishes in Queensland, Australia, revealed that these introduced hosts harbour more species-poor parasite infracommunities and regional faunas than 15 ecologically similar native species (Dove 2000). This is because, while some parasites have been picked up by the exotic hosts from the native fish fauna, many traditional parasites have been lost on introduction to the new geographic region. Other surveys indicate that such losses are the general pattern being documented also in introduced molluscs, crustaceans, birds, mammals, amphibians and reptiles (Freeland 1993; Torchin et al. 2003). In addition to the number of parasite species found in exotic host populations being, on average, only half that found in native populations, introduced hosts also tend to have a lower prevalence of infection. For example, brushtail possums (*Trichosurus vulpecula*) in New Zealand have a very limited endoparasite fauna, compared with Australia (Obendorf et al. 1998; Stankiewicz et al. 1998), and the possum-specific parasites present are patchily distributed nationally (Cowan et al. 2000).

There are several reasons for this loss of parasite species. Primarily, when a host population is subsampled in such a way, the individuals involved are unlikely to support the entire parasite fauna because of the characteristic aggregation of parasites among hosts (Shaw and Dobson 1995). An extreme form of bottleneck occurs when eggs or juvenile hosts are introduced to a new environment, resulting in the loss of all parasites dependent on the adult form. Exotic salmonids in New Zealand are generally not host to any of their traditional parasites, since they were introduced as either fry or eggs (Dix 1968; Margolis and Boyce 1990). The three exceptions to this are the protozoan *Myxobolus cerebralis*, which causes "whirling disease" and is believed to have been introduced in infected bait (Boustead 1993; Champion et al. 2002), the protozoan *Ichthyophthirius multifilis*, a cosmopolitan pathogen which causes "whitespot" and probably spread from native fish (Hine et al. 2000), and the cestode *Ligula intestinalis*, which appears not to persist but is introduced occasionally by infected birds (the definitive hosts) from Australia (Weekes and Penlington 1986).

Even if parasites are introduced with their hosts, they may still be lost. Small numbers of introduced parasites may disappear due to either stochastic effects, or the new host population initially being of insufficient size or density (below the persistence "threshold"; Swinton et al. 2002). The directly transmitted fish louse *Argulus*, a well-known and widespread parasite of freshwater fish, does not occur in New Zealand. However, when imports of exotic goldfish (*Carassius auratus*) were more frequent (now severely restricted to limit disease entry), *Argulus* did arrive from time to time but apparently never became established in wild fish populations (Pilgrim 1967). Some parasites may be unable to persist in the new environment, due to either inappropriate abiotic conditions, the absence of other obligate components of the parasites' life cycle (such as suitable intermediate hosts or vectors), or even out-competition by a native parasite (Bauer 1991).

5.4 Exotic Parasite Impact on Native Fauna

Introduced parasites can directly cause the extinction of native host populations (Cleaveland et al. 2002). Since infectious agents tend to have threshold levels of host population size below which they are unable to persist ("fade-out"; Swinton et al. 2002), it is those parasites with alternative "reservoirs" available which pose the greatest threat. Indeed, most local extinctions and population crashes in threatened wildlife due to disease have been ascribed to the "spillover" of infection from alternative hosts (Daszak et al. 2000). For example, the decline of the native grey partridge (*Perdix perdix*) in the UK is partly due to the cross-species transmission of the exotic nematode *Heterakis gallinarum* from its introduced traditional host, the ring-necked pheasant (*Phasianus colchicus*; Tompkins et al. 2000). Certain native hosts may also act as "source" populations for the transmission of exotic infectious agents to "sink" populations of other native species. For example, infection trials have demonstrated that whereas chytridiomycosis is fatal for certain amphibians, others (such as the bullfrog *Rana catesbeiana*) can be infected with no apparent harmful effects (Daszak et al. 2003). Introduced parasites can also indirectly cause the extinction of native host populations by suppressing their size or resilience (Cleaveland et al. 2002).

The anthropogenic introduction of exotic bird pathogens into Hawaii is perhaps the most cited example of how exotic microparasites can play a key role in the mass decline and extinction of native fauna. It is often overlooked, however, that it was the introduction of an exotic vector, the mosquito *Culex quinquefasciatus*, which is thought to have in fact enabled the introduced pathogens to reach epizootic proportions (van Riper et al. 2002). Both avian pox (*Poxvirus avium*) and avian malaria (*Plasmodium relictum*) have likely been repeatedly introduced to the Hawaiian Islands since their discovery by Captain Cook in 1778, with the importation of domestic avian stock, but it would not have been until the introduction of their principal vector in 1827 that transmission levels were sufficient for establishment to eventually occur. Avian pox is subsequently implicated in the numerous extinctions of endemic birds occurring during the late 1800s (van Riper et al. 2002), and avian malaria in the extinctions of the early 1900s (van Riper et al. 1986). Furthermore, the distribution of *C. quinquefasciatus*, due to its role as the primary vector of these introduced pathogens, is currently a major factor limiting the distribution of many of the remaining native bird populations across the Hawaiian Islands. This is a clear illustration of how the introduction of exotic vectors (and intermediate hosts) can often trigger greater effects in native communities than those stemming from pathogen introduction, and their subsequent distribution can determine any ongoing effects.

Specialist parasites are less successful than generalists in invading new environments, since the former are less likely to encounter new hosts which they can infect, reducing their potential impact on the native fauna. The implications of a broad host range are seen in examples such as avian malaria and avian pox in Hawaii, where two introduced pathogens with very broad ranges have affected the vast majority of the native avifauna. Likewise, the recent impact of West Nile Virus in North America can be linked to its broad host specificity – since its introduction from the Middle East in 1999, the virus had reached 42 of the 48 southern states in the USA (as well as four of Canada's 10 provinces) by the end of 2002, being found in over 70 bird species and more than 40 mosquito species (Enserink 2002). The impact of exotic infectious agents is also more severe when native species related to introduced hosts are present in the new environment. This is presumably because relatedness facilitates host switching, yet the new hosts lack the adaptive defences against the parasite of the traditional host and, thus, can suffer serious pathogenic effects (Bauer 1991).

5.5 Introduced Species as New Hosts for Native Parasites

An alternative disease impact of biological invasions on native fauna, somewhat overlooked in the literature, is when introduced species act as new hosts for native parasites from which potentially deleterious "spillback" of infection to native hosts could occur (Daszak et al. 2000). Whilst many parasites are lost from hosts upon introduction to a new environment, such hosts tend to acquire generalist parasites from the native fauna. For example, a total of 31 parasite species have been recorded from introduced populations of the mallard duck (*Anas platyrhynchos*), 21 of which do not occur in its native range and have evidently been acquired from other hosts worldwide (Torchin et al. 2003). Likewise, of 28 helminth parasites recorded from introduced populations of the cane toad (*Bufo marinus*), 20 do not occur in its native range (Barton 1997). Such acquisition of native parasites is also clear in the Hawaiian and Queensland freshwater fish parasite surveys discussed above. Thus, epizootics of previously undocumented parasites should not, de facto, be considered introduced species simply because they are associated with the introduction of exotic hosts. Rather, they may represent cases where formerly rare local parasites are able to colonize the introduced hosts, the likelihood of which would probably increase if the region invaded contains phylogenetically and ecologically similar native host species (Torchin et al. 2003).

Native parasites can show a relatively low degree of fitness when infecting introduced hosts. Trejo (1992) demonstrated that the fitness of the acanthocephalan parasite *Pomphorhynchus patagonicus* when infecting introduced rainbow trout (*Oncorhynchus mykiss*) in Argentinian waters is far lower than when infecting the traditional host *Patagonina hatchery* (Patagonian pejerrey). Under such circumstances, there would be limited threat to native hosts. In other cases, however, introduced species have proven competent hosts for native parasites. Rauque et al. (2003) demonstrated that the fitness of the parasitic ancanthocephalan *Acanthocephalus tumescens* infecting freshwater fish (again in Argentina) is greater in introduced rainbow trout than in many of the parasite's native hosts. Here, the spillback of infection from introduced hosts may prove deleterious to susceptible native species. Such effects can potentially occur whenever introduced host populations act as sources, rather than sinks, of infective stages. In their study, Rauque et al. (2003) concluded that the rainbow trout now harbours a large proportion of the adult component population of *A. tumescens*, effects on native hosts due to raised parasitemia being easily conceivable. Since the acquisition of native generalist parasites by exotic hosts tends to be the rule, rather than the exception, spillback effects have the potential to be more common in the wild than are EIDs caused by introduced parasites.

5.6 Context Dependence of Exotic Parasite Impact

Multiple factors, both biotic and abiotic, are known to interact with parasite emergence, invasion, and impact on native fauna. On this basis, observed disease effects are often "context dependent" (Blaustein and Kiesecker 2002). Climate change is thought to play a major role in the emergence of wildlife diseases by altering the geographic distribution of intermediate hosts and vectors (Harvell et al. 1999). For example, in Hawaii, the decline in endemic forest birds caused by avian malaria and avian pox follows an elevation gradient whereby pathogen transmission is most intense in mid-elevation forests where the mosquito vectors and endemic birds have the greatest overlap (van Riper et al. 1986, 2002). Thus, high-elevation forests currently function as refugia for several of these bird species. Landscape analysis of these refugia, however, shows that climate change is likely to drive many of the remaining endemics extinct, mainly through warmer temperatures increasing the abundance of vectors in the high-elevation forests (Benning et al. 2002). In addition to biting flies, there are also concerns that climate change will alter the distribution of many tick species (Sutherst 2001), acting as vectors for a variety of pathogens of concern to wildlife health (Randolph et al. 2002). Habitat disturbance can also lead to, and exacerbate the emergence and spread of infectious diseases, via alterations in intermediate host and vector distributions and dynamics (Patz et al. 2000). On certain Hawaiian islands, the abundance of mosquitoes is strongly increased by the damage caused by feral pigs increasing the number of ephemeral water pools available in which the larvae of these insects develop. Under such circumstances, the best approach available for managing the impact of malaria and pox in the avifauna is to reduce vector abundance by controlling pig numbers.

Factors which increase host susceptibility to infection can make them more prone to introduced parasites. For example, atmospheric pollutants are believed to interact with chytridiomycosis in its impact on amphibian populations (Pounds 2001; Blaustein and Kiesecker 2002). Similarly, experiments have documented how carp (*Cyprinus carpio*) exposed to cadmium are more susceptible to the protozoan parasite *Ichthyophthirius multifilis* (Combes 2001). Pollution may also increase parasite incidence via effects on intermediate hosts (Lafferty 1997). For example, eutrophication of freshwater can increase the incidence of "swimmers itch" in lakes, caused by the trematode *Trichobilharzia*, by improving conditions for intermediate snail hosts (Beer and German 1993). In New Zealand, disturbance and pollution interacting with macroparasites may increase the susceptibility of waterfowl to infection by microparasites such as avian paramyxoviruses (causative agents of Newcastle disease), influenza, and botulism (Alley 2002).

Stress may also influence host susceptibility, via suppressive effects on immunocompetence (Buchanan 2000). The causes of stress to wildlife are multiple, including habitat alteration, degradation and reduction, anthropogenic translocation, and artificial rearing (Dobson and Foufopoulos 2001). Pathogen emergence fostered by both the increased potential for pathogen transmission and the increased susceptibility of fish hosts in crowded hatchery conditions, although well recognized by aquaculturists, is an acute problem (Kennedy 1994). Such fish-breeding activities are believed to have driven the global emergence of the protozoan *Myxobolus cerebralis* in wild fish populations (Dobson and Foufopoulos 2001). Habitat changes can also increase pathogen impacts if such changes alter or disrupt social systems. For example, both free-living and captive hihi (*Notiomystis cincta*) in New Zealand are believed to be rendered susceptible to infection to the opportunistic fungal pathogen *Aspergillus* by immunosuppression caused by intense social interaction during the breeding season (Alley et al. 1999; Alley 2002). Down-regulation of host immunocompetence can also occur when endangered species suffer reduced genetic diversity as a result of population bottlenecks (Altizer et al. 2003). Such effects may be at least partly responsible for the high incidence of avian pox (*Poxvirus avium*) clinical symptoms in black robin (*Petroica traversi*) populations in New Zealand (Tisdall and Merton 1988).

5.7 Disease Threat of Biological Invasions to New Zealand's Avifauna

In 1918 it was postulated that it "may be possible to connect the comparatively rapid disappearance of New Zealand native birds with the introduction of exotic protozoa through the medium of imported fauna" (Doré 1918). Myers (1923) added to this by stating "the hypothesis of an introduced avian disease which might conceivably commit the same havoc among the indigenous birds as measles among aboriginal races of man supplies perhaps the only theory which can even partially explain the wholesale disappearance of certain species from untouched areas either before weasels or stoats were introduced or before they or any other cause of sufficient magnitude had conceivably reached the area in question". Although exotic birds in New Zealand have, as one would expect, apparently lost some parasites on introduction (e.g. the absence of *Haemoproteus* blood parasites in silvereyes (*Zosterops lateralis*) and rock pigeons (*Columba livia*); Laird 1950), there are undoubtedly some which successfully established. In his summation of the available evidence, Laird (1950) stated "a very full investigation of the haematozoa of New Zealand birds, both indigenous and exotic, will have to be made before any decision can be reached concerning the effect of these parasites on the native avian fauna". However, 50 years on, our position regarding the blood parasites of New Zealand birds remains relatively unchanged, with surveys of sufficient size having been conducted on only a few host species (McKenna 1998). Although such surveys would be of limited use in discerning past population and species extinctions, they may inform on cases where populations are currently being impacted by the pathogenic effects of such parasites (Atkinson and van Riper 1991), or indicate regions where such pathogens do not currently occur. The surveys conducted thus far, however, do show that *Plasmodium relictum* occurs in both native and introduced bird species in New Zealand (Table 5.1). As discussed above, such sharing of parasites commonly has deleterious effects for the more susceptible hosts.

The impact of blood parasites on New Zealand birds is likely to increase in the near future, due to the ongoing introduction and range expansion of exotic mosquitoes, primary vectors of such parasites (Atkinson and van Riper 1991). Four exotic species of mosquito have established in New Zealand to date, three of which (*Culex quinquefasciatus*, *Aedes australis* and *A. notoscriptus*) are considered to have been introduced in the 19th and early 20th centuries, and one of which (*A. camptorhynchus*) has only recently established (Laird 1995). Of key concern is *C. quinquefasciatus*, the primary vector of *Plasmodium relictum* in Hawaii. Although it appears that mosquitoes native to New Zealand can also vector *P. relictum*, since avian malaria infections have been recorded in birds outside the ranges of the introduced mosquitoes

Table 5.1 Parasitic protozoans and helminths known to infect both native and introduced birds in New Zealand (data obtained from McKenna 1998; it is unclear whether any of the parasites listed are introduced)

Parasite	Native hosts	Exotic hosts
Protozoa		
Plasmodium relictum[a]	Dotterel Yellow-eyed penguin Fiordland crested penguin	Song thrush Blackbird House sparrow
Trematoda		
Dendritobilharzia pulverulenta	Grey duck New Zealand shoveler New Zealand scaup Paradise shelduck	Mallard duck Black swan Mute swan
Echinoparyphium recurvatum	Australasian bittern Paradise shelduck New Zealand shoveler New Zealand scaup	Black swan Canada goose Domestic goose
Echinostoma revolutum	Paradise shelduck New Zealand shoveler	Canada goose Domestic goose Rock pigeon
Notocotylus gippyensis	Paradise shelduck Grey duck New Zealand shoveler	Canada goose Domestic goose
Notocotylus tadornae	Paradise shelduck	Domestic goose
Psilochasmus oxyurus	New Zealand shoveler New Zealand scaup	Black swan
Trichobilharzia sp.[b]	Grey duck	Mallard duck
Nematoda		
Echinuria uncinata	Grey duck	Mallard duck
Heterakis gallinarum	Morepork	Ring-necked pheasant Brown quail Domestic fowl Domestic turkey
Porrocaecum ensicaudatum	Kiwi	Blackbird Song thrush
Syngamus trachea	Silvereye Kea	Ring-necked pheasant Chukor partridge Domestic fowl Blackbird Starling
Trichostrongylus tenuis	Paradise shelduck	Brown quail Peafowl

[a] *Plasmodium* sp. (most likely *relictum*) has also been recorded from native grey duck, New Zealand pipit, New Zealand dotterel, and exotic skylark

[b] Identified as a single species unique to New Zealand by Davis (2000)

(Holder et al. 1999), the increasing distribution of *C. quinquefasciatus* is likely to enhance the incidence of avian malaria in three distinct ways.

1. Parasitemia may increase in those species already infected, such as the endangered and endemic Fiordland crested (*Eudyptes pachyrhynchus*) and yellow-eyed (*Megadyptes antipodes*) penguins (Duignan 2001), with an increasing impact on the health of individuals and populations (Graczyk et al. 1995; Reed 1997; Alley 2001).
2. *P. relictum* may increase its host range in the areas in which it is already present. For example, whereas black swans (*Cygnus atratus*) in Australia have tested 50 % positive for *Plasmodium*, the parasite has not been found in black swans in New Zealand, now a common and widespread gamebird since its introduction during the 19th century (Laird 1950). However, increasing vector densities may allow avian malaria to sustain infections in black swans in this country as well, which would then become potential reservoirs of infection to other species.
3. *P. relictum* may increase its geographic range in New Zealand, potentially reaching offshore islands where its impact may be particularly severe due to a lack of prior exposure to blood parasites (McCallum and Dobson 1995). This is likely, since the indigenous portion of the New Zealand avifauna which has been examined for hematozoa appears exceptionally free of these parasites, and no indigenous species of hematozoa have yet been described from New Zealand birds (Laird 1950; McKenna 1998).

Increases in the distribution and/or abundance of other mosquitoes in New Zealand may also lead to increases in the prevalence of other infections such as avian pox (Holder et al. 1999). Furthermore, the threat of mosquitoes carrying exotic parasites being either transported or blown to New Zealand is ever present (Laird et al. 1994). The spread of such infectious agents, and incursions by new mosquito species, could be exacerbated by even moderate increases in average annual temperatures due to global warming (Liehne 1988). Similar considerations also apply to the introduction and spread of exotic pathogens by ticks (Sutherst 2001; Heath 2002; Randolph et al. 2002).

There may be other parasite impacts on native species, caused by biological invasions and which are going unnoticed. This is illustrated in Table 5.1, which lists all known instances of native and introduced birds in New Zealand sharing the same protozoan or helminth parasite species. Thus, the presence in New Zealand of alternative hosts for these parasites could potentially be raising parasitemia in native populations. For example, both introduced gallinaceous birds (pheasant, partridge, quail, fowl, turkey) and waterfowl may be exacerbating disease problems for native species through a range of shared parasites.

5.8 Disease Threat of Biological Invasions to New Zealand Freshwater Fish

Few parasites have been associated with the introduction of exotic freshwater fish to New Zealand (although some have come in by other routes – see above). Perhaps the likeliest candidates are the three monogeneans *Dactylogyrus ctenopharyngodonis*, *Gyrodactylus ctenopharyngodontis*, and *Gyrodactylus* sp. Monogeneans transmit horizontally and directly, fish to fish, making them one of the most common groups to be introduced with movements of host fish. Since they are the only three monogenean parasites reported in New Zealand, and two of them have been found infecting only introduced grass carp (*Ctenopharyngodon idella*), they have almost certainly been introduced into New Zealand on exotic cyprinid hosts (Hine et al. 2000).

As discussed above, the lack of traditional parasites on exotic salmonids in New Zealand is primarily due to their introduction as either fry or eggs, excluding parasite species dependent on adult hosts (Dix 1968; Margolis and Boyce 1990). These hosts have evidently acquired new parasite assemblages from the native fish fauna (McDowall 2000; Poulin and Mouillot 2003). Of particular interest is the copepod *Paenodes nemaformis* which, although occurring only in New Zealand, has been recorded infecting only introduced chinook salmon (*Oncorhynchus tshawytscha*) and brown trout (*Salmo trutta*). One suggestion is that this copepod was once a parasite of the now extinct grayling *Prototroctes oxyrhynchus* (McDowall 1990). The only documented harmful effect of parasites associated with introduced salmonids is that of the protozoan causing "whirling disease" (and then, only to the salmonids themselves). However, since the parasites acquired by salmonids in New Zealand waters have (as theory would predict) tended to be highly generalist, there is potential for these new hosts to be impacting native fish via the spillback of infection. This is illustrated in Table 5.2, which lists all known instances of native and introduced freshwater fishes in New Zealand sharing the same protozoan, helminth, or copepod parasite species. The native fishes at greatest risk from such effects are likely to be the galaxiids, including the koaro (*Galaxias brevipinnis*) and the inanga (*G. maculatus*), which share many parasites with the salmonids, and are the most endangered of the indigenous freshwater species (Champion et al. 2002). Such considerations may help explain why salmonids have displaced galaxiids from certain freshwater bodies in New Zealand but not from others (Townsend 2003).

Introduced non-salmonids, such as the cyprinid grass carp (*Ctenopharyngodon idella*) and goldfish (*Carassius auratus*), are also host to parasites shared with native species (Table 5.2). Thus, they also have the potential to drive deleterious disease effects in native fish populations. There are also concerns that the further spread of introduced cyprinid fish within New Zealand

Table 5.2 Parasitic protozoans, helminths, and copepods known to infect both native and introduced freshwater fish in New Zealand (data obtained from Hine et al. 2000; only the cestode *Ligula intestinalis* has been introduced)

Parasite	Native hosts	Exotic hosts
Protozoa		
Ichthyophthirius multifiliis	Short-finned eel Long-finned eel Common bully Koaro	Grass carp Rainbow trout Chinook salmon
Digenea		
Coitocaecum parvum	Short-finned eel Long-finned eel Koaro Inanga Upland bully Common bully Giant bully Blue-gilled bully Red-finned bully Smelt	Goldfish Brown trout
Telogaster opisthorchis	Short-finned eel Long-finned eel Koaro Dwarf galaxias Banded kokopu Inanga Short-jawed kokopu Common river galaxias Upland bully Common bully Blue-gilled bully Red-finned bully Brown mudfish Smelt	Brown trout
Tubulovesicula angusticauda	Short-finned eel Long-finned eel	Chinook salmon
Cestoda		
Ligula intestinalis	Common bully	Rainbow trout
Nematoda		
Eustrongylides ignotus	Short-finned eel Long-finned eel Inanga Upland bully Giant bully	Rainbow salmon Sockeye salmon Brown trout Tench

Table 5.2 (*Continued*)

Parasite	Native hosts	Exotic hosts
Hedruris spinigera	Short-finned eel Long-finned eel Inanga Cran's bully Brown mudfish Smelt	Chinook salmon Red-finned perch Brown trout
Copepoda		
Abergasilus amplexus	Short-finned eel Long-finned eel Inanga Smelt	Goldfish Red-finned perch

may lead to the persistence and range expansion of the introduced cestode *Ligula intestinalis*, for which cyprinids tend to be more typical hosts (McDowall 1990). Potential consequences of this would be the introduction of this generalist parasite to many more native fishes, and an increased impact on the native common bully (*Gobiomorphus cotidianus*), from which infection has already been recorded (Weekes and Penlington 1986). In a similar vein, there is the constant threat that the New Zealand fish fauna, which has largely escaped the disease and parasite problems which have plagued some fish species in other countries, would be highly vulnerable if such organisms were introduced (Champion et al. 2002). As with the threats to the endemic avifauna, habitat degradation and climate change could potentially increase the threat of successful introduction and range expansion of such parasites.

5.9 Conclusions

The points highlighted above, regarding the parasite threat from biological introductions, are also applicable to diseases of other biota (notably the chyrid fungus threat to New Zealand's endemic frogs). A common thread is that anthropogenic factors are almost always involved, either as direct causes of parasite emergence or as interacting factors. One relevant example is how the poeciliid fish species introduced to Hawai'i to prey on mosquito larvae, with the goal of reducing the incidence of the introduced mosquito-borne pathogens *Plasmodium relictum* and *Poxvirus avium* in native birds, are now themselves implicated as a major threat to the native Hawaiian freshwater fish fauna via the parasites which were introduced with them (Font and Tate

1994). This illustrates how human introduction of animals to new geographic regions is almost always accompanied by unforeseen deleterious effects on native communities. One would thus hope that the utmost is done to prevent such introductions (either accidental or deliberate) in the future. Recommendations specific to preventing the accidental introduction of exotic pathogens (or vectors) include strict border control biosecurity regimes, and controlling human traffic to remote regions and populations of endangered species (Jackson et al. 2000; Duignan 2001; Alley 2002). In addition, separation should be maintained between captive or domestic and wild animal populations at all time, since domestics commonly act as sources of either novel parasite introduction or increased levels of infection by endemic parasites (Cleaveland et al. 2002). Furthermore, contact between closely related species (e.g. in hatcheries where multiple fish species are reared) should also be minimized, since such contact increases the potential for the cross-species transfer of infectious agents (Dobson and Foufopoulos 2001).

Safeguards should be put in place whenever it is necessary to move animals among wild populations, or from captivity into the wild (as is common in endangered species management), since disease problems can also occur under such circumstances. For example, the mass mortalities of the sardine *Sardinops sagax* which occurred throughout its range in Australia during the 1990s were caused by an exotic pathogen introduced with imported fish of the same species (Gaughan 2002). Thus, as stated by many authors (e.g. Jakob-Hoff 2001; Alley 2002), health screening prior to any translocation of animals between populations should be considered a necessity. Also, three important factors to consider when translocating animals into new areas are (1) whether hosts (of the same or other species) already present in the area may be deleteriously affected by parasites which may be introduced, (2) whether parasites present in the native fauna in the area may deleteriously affect the introduced animals, and (3) whether the introduced animals may act as a new biological reservoir for native parasites, from which infection may spill back into existing wild populations.

No matter how successful management regimes are, the increasing level of global movement means that there will undoubtedly always be instances when exotic pathogens or vectors are introduced or invade new geographic regions, host species, or host populations (Daszak et al. 2003). Quantifying the probability of such events is the realm of risk assessment (e.g. Jakob-Hoff 2001). However, certain steps can be taken to ensure that native wildlife populations are more resistant to such incursions if and when they do occur. For example, since habitat reduction and alteration can increase both parasite transmission between host individuals and host susceptibility to infection, management to minimize such effects should be a top priority. Furthermore, although habitat fragmentation may be beneficial to native species faced with exotic parasites (if it acts to limit the spread of infection), it is also more likely to exacerbate disease problems through effects such as crowding and habitat degradation (McCallum and Dobson 2002). Likewise, pollution has the potential to interact deleteriously with introduced parasites. Thus, unless there is a good reason not to, both the fragmentation and the pollution of native habitat should be minimised to reduce potential host susceptibility to infection, safeguarding against the eventuality that exotic parasites or vectors invade or are inadvertently introduced into new regions.

This chapter has highlighted two broad gaps in our knowledge relevant to the disease effects of biological invasions in New Zealand. First, we still do not have a complete picture of the existing parasite fauna of many native species. Second, there has been no real consideration of the role which already introduced hosts may be playing as biological reservoirs for native parasites. As is often the case in the study of wildlife disease, more fundamental research is required to address these gaps.

References

Alley MR (2001) Avian malaria in a yellow-eyed penguin. Kokako 8(1):12–13

Alley MR (2002) Avian wildlife diseases in New Zealand: current issues and achievements. NZ Vet J 50:S118–S120

Alley MR, Castro I, Hunter JEB (1999) Aspergillosis in hihi (*Notiomystis cincta*) on Mokoia Island. NZ Vet J 47:88–91

Altizer S, Harvell D, Friedle E (2003) Rapid evolutionary dynamics and disease threats to biodiversity. Trends Ecol Evol 18:589–596

Atkinson CT, van Riper C (1991) Pathogenicity and epizootiology of avian haematozoa: Plasmodium, Leucocytozoan, and Haemoproteus. In: Loye JE, Zuk M (eds) Bird-parasite interactions: ecology, evolution, and behaviour. Oxford Univ Press, Oxford, pp 19–48

Barton DP (1997) Introduced animals and their parasites: the cane toad, *Bufo marinus*, in Australia. Aust J Ecol 22:316–324

Bauer ON (1991) Spread of parasites and diseases of aquatic organisms by acclimatization: a short review. J Fish Biol 39:679–686

Beer SA, German SM (1993) Ecological prerequisites of worsening of the cercariosis situation in cities of Russia (Moscow region as an example). Parazitologiya (St Petersburg) 27:441–449

Benning TL, LaPointe D, Atkinson CT, Vitousek PM (2002) Interactions of climate change with biological invasions and land use in the Hawaiian Islands: modeling the fate of endemic birds using a geographic information system. Proc Natl Acad Sci USA 99:14246–14249

Berger L, Speare R, Daszak P, Green DE, Cunningham AA, Goggin CL, Slocombe R, Ragan MA, Hyatt AD, McDonald KR, Hines HB, Lips KR, Marantelli G, Parkes H (1998) Chytridiomycosis causes amphibian mortality associated with population declines in the rain forests of Australia and Central America. Proc Natl Acad Sci USA 95:9031–9036

Blaustein AR, Kiesecker JM (2002) Complexity in conservation: lessons from the global decline of amphibian populations. Ecol Lett 5:597–608

Boustead NC (1993) Detection and New Zealand distribution of *Myxobolus cerebralis*, the cause of whirling disease of salmonids. NZ J Mar Freshwater Res 27:431–436

Brouder MJ, Hoffnagle TL (1997) Distribution and prevalence of the Asian fish tapeworm, *Bothriocephalus acheilognathi*, in the Colorado River and tributaries, Grand Canyon, Arizona, including two new host records. J Helminthol Soc Wash 64:219–226
Buchanan KL (2000) Stress and the evolution of condition-dependent signals. Trends Ecol Evol 15:156–160
Champion P, Clayton J, Rowe D (2002) Lake managers handbook: alien invaders. Ministry for the Environment, Wellington, New Zealand
Cleaveland S, Hess GR, Dobson AP, Laurenson MK, McCallum HI, Roberts MG, Woodroffe R (2002) The role of pathogens in biological conservation. In: Hudson PJ, Rizzoli A, Grenfell BT, Heesterbeek H, Dobson AP (eds) The ecology of wildlife diseases. Oxford Univ Press, Oxford, pp 139–150
Combes C (2001) Parasitism: the ecology and evolution of intimate interactions. Univ Chicago Press, Chicago
Cowan PE, Clark JM, Heath DD, Stankiewicz M, Meers J (2000) Predators, parasites, and diseases of possums. In: Montague TL (ed) The brushtail possum: biology, impact and management of an introduced marsupial. Manaaki Whenua Press, Lincoln, New Zealand, pp 82–91
Daszak P, Cunningham AA, Hyatt AD (2000) Emerging infectious diseases of wildlife – threats to biodiversity and human health. Science 287:443–449
Daszak P, Cunningham AA, Hyatt AD (2003) Infectious disease and amphibian population declines. Divers Distrib 9:141–150
Davis NE (2000) Cercarial dermatitis and the possibility of biological control in Lake Wanaka, New Zealand. PhD Thesis, University of Otago, Dunedin, New Zealand
Dix TG (1968) Helminth parasites of brown trout (*Salmo trutta* L.) in Canterbury, New Zealand. NZ J Mar Freshwater Res 2:363–374
Dobson AP, Foufopoulos J (2001) Emerging infectious pathogens of wildlife. Philos Trans R Soc Lond Ser B 356:1001–1012
Doré AB (1918) Rat trypanosomes in New Zealand. NZ J Sci Technol 1:200
Dove ADM (2000) Richness patterns in the parasite communities of exotic poeciliid fishes. Parasitology 120:609–623
Duignan PJ (2001) Diseases of penguins. Surveillance 28(4):5–11
Enserink M (2002) West Nile's surprisingly swift continental sweep. Science 297:1988–1989
Font WF (1998) Parasites in paradise: patterns of helminth distribution in Hawaiian stream fishes. J Helminthol 72:307–311
Font WF, Tate DC (1994) Helminth parasites of native Hawaiian freshwater fishes: an example of extreme ecological isolation. J Parasitol 80:682–688
Freeland WJ (1993) Parasites, pathogens and the impacts of introduced organisms on the balance of nature in Australia. In: Moritz C, Kikkawa J (eds) Conservation biology in Australia and Oceania. Surrey Beatty, Chipping Norton, New South Wales, pp 171–180
Gaughan DJ (2002) Disease-translocation across geographic boundaries must be recognized as a risk even in the absence of disease identification: the case with Australian Sardinops. Rev Fish Biol Fisher 11:113–123
Graczyk TK, Cockrem JF, Cranfield MR, Darby JT, Moore P (1995) Avian malaria seroprevalence in wild New Zealand penguins. Parasite 2:401–405
Harvell CD, Kim K, Burkhdder JM, Colwell RR, Epstein PR, Grimes DJ, Hofmann EE, Lipp EK, Osterhaus ADME, Overstreet RM, Porter JW, Smith GW, Vasta GR (1999) Review: marine ecology – emerging marine diseases – climate links and anthropogenic factors. Science 285:1505–1510
Heath ACG (2002) Recently introduced exotic animals and their parasites: what risk to New Zealand's biosecurity? Surveillance 29(4):15–17

Hine PM, Jones JB, Diggles BK (2000) A checklist of the parasites of New Zealand fishes, including previously unpublished records. NIWA Tech Rep 75. Wellington, New Zealand

Holder P, Brown G, Bullians M (1999) The mosquitoes of New Zealand and their animal disease significance. Surveillance 26(4):12–15

Jackson R, Morris RS, Boardman W (2000) Development of a method for evaluating the risk to New Zealand's indigenous fauna from the introduction of exotic diseases and pests - including a case study on native parrots. Science for Conservation 138. Department of Conservation, Wellington, New Zealand

Jakob-Hoff R (2001) Disease risk assessment for translocation of kaki (black stilt), *Himantopus novaezelandiae*, from captivity to the wild. DOC Science Internal Ser 16. Department of Conservation, Wellington, New Zealand

Kennedy CR (1994) The ecology of introductions. In: Pike AW, Lewis JW (eds) Parasitic diseases of fish. Samara, Cardigan, UK, pp 189–208

Lafferty KD (1997) Environmental parasitology: what can parasites tell us about human impacts on the environment? Parasitol Today 13:251–255

Laird M (1950) Some parasites of New Zealand birds. Victoria Univ Coll Zool Publ 5:1–20

Laird M (1995) Background and findings of the 1993–94 New Zealand mosquito survey. NZ Entomol 18:77–90

Laird M, Calder L, Thornton RC, Syme R, Holder PW, Mogi M (1994) Japanese *Aedes albopictus* among four mosquito species reaching New Zealand in used tires. J Am Mosquito Control Assoc 10:14–23

Liehne PFS (1988) Climatic influences on mosquito-borne diseases in Australia. In: Pearman GI (ed) Greenhouse: planning for climate change. CSIRO, Australia, pp 624–637

Longcore JE, Pessier AP, Nichols DK (1999) *Batrachochytrium dendrobatidis* gen. et sp. nov., a chytrid pathogenic to amphibians. Mycologia 91:219–227

Margolis L, Boyce NP (1990) Helminth parasites from North Pacific anadromous chinook salmon, *Oncorhynchus tshawytscha*, established in New Zealand. J Parasitol 76:133–135

McCallum H, Dobson A (1995) Detecting disease and parasite threats to endangered species and ecosystems. Trends Ecol Evol 10:190–194

McDowall RM (1990) New Zealand freshwater fishes: a natural history and guide, rev edn. Heinemann Reed, Auckland, New Zealand

McDowall RM (2000) Biogeography of the southern cool-temperate galaxioid fishes: evidence from metazoan macroparasite faunas. J Biogeogr 27:1221–1229

McKenna PB (1998) Checklist of helminth and protozoan parasites of birds in New Zealand. Surveillance 25:S3–S12

Myers JG (1923) The present position of the endemic birds of New Zealand. NZ J Sci Technol 6(2):65–99

Obendorf D, Spratt D, Beveridge I, Presidente P, Coman B (1998) Parasites and diseases in Australian brushtail possums, *Trichosurus* spp. Biological control of possums. R Soc NZ Misc Ser 45:6–9

Patz JA, Graczyk TK, Geller N, Vittor, AY (2000) Effects of environmental change on emerging parasitic diseases. Int J Parasitol 30:1395–1405

Pilgrim RLC (1967) *Argulus japonicus* Thiele (Crustacea: Branchiura) – a new record for New Zealand. NZ J Mar Freshwater Res 1:395–398

Poulin R, Mouillot D (2003) Host introductions and the geography of parasite taxonomic diversity. J Biogeogr 30:837–845

Pounds JA (2001) Climate and amphibian declines. Nature 410:639–640

Randolph SE, Chemini C, Furlanello C, Genchi C, Hails RS, Hudson PJ, Jones LD, Medley G, Norman RA, Rizzoli AP, Smith G, Woolhouse MEJ (2002) The ecology of tick-borne infections in wildlife reservoirs. In: Hudson PJ, Rizzoli A, Grenfell BT, Heesterbeek H,

Dobson AP (eds) The ecology of wildlife diseases. Oxford Univ Press, Oxford, pp 119–138

Rauque CA, Viozzi GP, Semenas LG (2003) Component population study of *Acanthocephalus tumescens* (Acanthocephala) in fishes from Lake Moreno, Argentina. Folia Parasitol 50:72–78

Reed CEM (1997) Avian malaria in New Zealand dotterel. Kokako 4:3

Shaw DJ, Dobson AP (1995) Patterns of macroparasite abundance and aggregation in wildlife populations: a quantitative review. Parasitology suppl 111:S111–S133

Stankiewicz M, Cowan PE, Heath DD (1998) Endoparasites of possums from selected areas of North Island, New Zealand. NZ J Zool 25:91–97

Sutherst RW (2001) The vulnerability of animal and human health to parasites under global change. Int J Parasitol 31:933–948

Swinton J, Woolhouse MEJ, Begon ME, Dobson AP, Ferroglio E, Grenfell BT, Guberti V, Hails RS, Heesterbeek JAP, Lauazza A, Roberts MG, White PJ, Wilson K (2002) Microparasite transmission and persistence. In: Hudson PJ, Rizzoli A, Grenfell BT, Heesterbeek H, Dobson AP (eds) The ecology of wildlife diseases. Oxford Univ Press, Oxford, pp 83–101

Tisdall DJ, Merton D (1988) Chatham Islands black robin-disease. Surveillance 15:15–16

Tompkins DM, Greenman JV, Robertson PA, Hudson PJ (2000) The role of shared parasites in the exclusion of wildlife hosts: *Heterakis gallinarum* in the ring-necked pheasant and the grey partridge. J Anim Ecol 69:829–840

Tompkins DM, Dobson AP, Arneberg P, Begon ME, Cattadori IM, Greenman JV, Heesterbeek JAP, Hudson PJ, Newborn D, Pugliese A, Rizzoli AP, Rosá R, Wilson K (2002) Parasites and host population dynamics. In: Hudson PJ, Rizzoli A, Grenfell BT, Heesterbeek H, Dobson AP (eds) The ecology of wildlife diseases. Oxford Univ Press, Oxford, pp 45–62

Tompkins DM, White AR, Boots M (2003) Ecological replacement of native red squirrels by invasive greys driven by disease. Ecol Lett 6:189–196

Torchin ME, Lafferty KD, Dobson AP, McKenzie VJ, Kuris AM (2003) Introduced species and their missing parasites. Nature 421:628–630

Townsend CR (2003) Individual, population, community, and ecosystem consequences of a fish invader in New Zealand streams. Conserv Biol 17(1):38–47

Trejo A (1992) A comparative study of the host-parasite relationship of *Pomphorhynchus patagonicus* (Acanthocephala) in two species of fish from Lake Rosario (Chubut, Argentina). J Parasitol 78:711–715

Van Riper C III, van Riper SG, Goff ML, Laird M (1986) The epizootiology and ecological significance of malaria in Hawaiian land birds. Ecol Monogr 56:327–344

Van Riper C, van Riper SG, Hansen WR (2002) Epizootiology and effect of avian pox on Hawaiian forest birds. The Auk 119:929–942

Waldman B, van de Wolfshaar KE, Klena JD, Andjic V, Bishop PJ, de B Norman RJ (2001) Chytridiomycosis in New Zealand frogs. Surveillance 28(3):9–11

Weekes PJ, Penlington B (1986) First records of *Ligula intestinalis* (Cestoda) in rainbow trout, *Salmo gairdneri*, and common bully, *Gobiomorphus cotidanus*, in New Zealand. J Fish Biol 28:183–190

6 Export of Plant and Animal Species from an Insular Biota

G.W. Yeates and P.A. Williams

6.1 Introduction

The deliberate and accidental establishment of exotic species in New Zealand is widely known. However, increasingly New Zealand native species are being introduced in other temperate regions of the world, and a significant proportion has become invasive. This chapter describes the establishment of native New Zealand plant and animal species elsewhere, and discusses the (varied) reasons for their success as exotic species. Charles Darwin (1859) wrote ... "we may doubt, if all the productions of New Zealand were set free in Great Britain, whether any considerable number would be enabled to seize on places now occupied by our native plants and animals". The assumption that insular biotas are inherently poor competitors continues, but the increasing spread of island species suggests otherwise. Because trade from New Zealand has only occurred over the past 150 years, opportunities for the native species to establish in other biogeographic regions are relatively recent. We adopt a case study approach to indicate the range of New Zealand species that have recently surfaced as invasive species in other countries.

6.2 Plant Emigrants

The major driver for the export of New Zealand native plants is the horticultural industry where the focus has been on interesting evergreen forms. The main New Zealand plant species considered invasive overseas are described, and a complete list presented in Table 6.1.

Ecological Studies, Vol. 186
R.B. Allen and W.G. Lee (Eds.)
Biological Invasions in New Zealand

Table 6.1 New Zealand plant species (in some cases also native to Australia) that are recorded as having naturalized or are listed as weeds in other countries[a]

Species	Family	Region of naturalization[b]	Dispersal method
Acaena novae-zelandiae	Rosaceae	Eur., N. Am.	Contaminant
Cardamine corymbosa	Brassicaceae	Aus.	Contaminant
Carex albula	Cyperaceae	Aus.	Horticulture
C. flagellifera	Cyperaceae	Aus.	Horticulture
C. testaceae	Cyperaceae	Aus.	Horticulture
Coprosma repens	Rubiaceae	Aus., Eur., N. Am., S. Afr.	Horticulture
C. robusta	Rubiaceae	Aus., Eur., N. Am.	Horticulture
Cordyline australis	Agavaceae	Pac.	Horticulture
Cortaderia richardii	Poacaeae	Aus., Eur., N. Am.	Horticulture
Corynocarpus laevigatus	Corynocarpaceae	N. Am., Aus., SEA, Pac.	Horticulture
Crassula helmsii	Crassulaceae	Eur.	Horticulture
C. sieberiana	Crassulaceae	N. Am.	?
Epilobium billardierianum	Onagraceae	Aus., Eur., N. Am.	Horticulture
E. brunnescens	Onagraceae	Eur.	Horticulture
E. komorovianum	Onagraceae	Eur.	Horticulture
E. rotundifolium	Onagraceae	Aus.	Horticulture
Hebe barkeri	Scrophulariaceae	Eur.	Horticulture
H. brachysiphon	Scrophulariaceae	Eur.	Horticulture
H. dieffenbachii	Scrophulariaceae	Eur.	Horticulture
H. elliptica	Scrophulariaceae	Aus.	Horticulture
H. parviflora	Scrophulariaceae	Aus.	Horticulture
H. salicifolia	Scrophulariaceae	Eur.	Horticulture
H. speciosa	Scrophulariaceae	Aus., N. Am.	Horticulture
Hoheria populnea	Malvaceae	N. Am.	Horticulture
Hydrocotyle tripartita	Apiaceae	Aus.	?
Kunzea ericoides	Myrtaceae	Pac.	Horticulture
Leptospermum scoparium	Myrtaceae	Pac.	Horticulture
Metrosideros excelsa	Myrtaceae	Aus., S. Afr.	Horticulture
Muehlenbeckia complexa	Polygonaceae	Aus.	?
M. ephedroides	Polygonaceae	Aus.	?
Myoporum laetum	Myoporaceae	Aus., N. Am.	Horticulture
Myriophyllum propinquum	Haloragaceae	Aus.	?
Olearia avicenniaefolia	Asteraceae	Eur.	Horticulture
O. traversii	Asteraceae	Eur.	Horticulture
Phormium cookianum	Agavaceae	Eur.	Horticulture
P. tenax	Agavaceae	Aus., N. Am., Pac.	Horticulture
Pittosporum crassifolium	Pittosporaceae	Aus.	Horticulture
P. eugenioides	Pittosporaceae	Aus.	Horticulture
Solanum aviculare	Solanaceae	Aus., Carib., N. Am.	Horticulture

[a] Data from Randall (2002)

[b] Regions denoted by: Aus., Australia; Carib., Caribbean; Eur., Europe; N. Am., North America; Pac., Pacific; South Africa, S. Afr.; SEA, South East Asia

Phormium tenax

A tall (2–3 m), closely tufted, perennial monocotyledon, *Phormiun tenax* ("New Zealand flax", Agavaceae), is perhaps the best-known invasive New Zealand plant species. Introduced to St. Helena in the South Atlantic for a fiber industry, it was grown on 1,200 ha before the industry collapsed in the 1960s. During this time, flax spread to modified woodlands and upland areas where it has since become a serious pest (Cronk 1989). Flax also grows on islands in the Tristan da Cunha group in the South Atlantic, having been used for traditional thatched houses (Wace and Holdgate 1976); it is currently a problem on Nightingale and Inaccessible islands (Roux et al. 1992). It was introduced to Hawai'i in the 18th century for fiber (Harris 1998), and has since escaped into gullies in mesic areas below 300 m a.s.l., e.g., on Moloka'i, and the northern Hamakua coastline (Smith 1998). In all these island groups, it forms dense thickets, similar to those of its native range, which exclude other species. Since the formation of the Diana's Peak National Park in St. Helena, there has been a control program by the Agricultural and Natural Resources Department to clear flax and replace it with a native species (R. Cairns-Wick, personal communication). *P. tenax* is also used horticulturally in the UK and USA (California). In both countries, it is infected with the native mealy bug *Balanococcus diminutus* (see Sect. 6.3.6).

Crassula helmsii

Crassula helmsii is a fine-stemmed, succulent, perennial herb ("New Zealand pigmyweed" or "Australian swamp stonecrop") that has invaded a wide range of static water habitats in the British Isles, where it changes wetland habitats and out-competes native species (Leach and Dawson 1999). Initially sold in 1927, the first naturalized collection was made in 1957. Since then, it has spread widely, partly as a result of its increasing horticultural popularity; one recent estimate suggests 10,000 sites have been colonized. *C. helmsii* grows mainly as a low turf, as it does in New Zealand, but stems may reach 1.3 m in length in open water, which is much longer than in its native range. It has yet to make the transition from static water bodies to flowing water, as in its native range (Centre for Ecology and Hydrology 2001). Despite the highly invasive nature of *C. helmsii*, it has no official weed status and little attempt is made to control it (Williamson 2002).

Metrosideros excelsa

Metrosideros excelsa, the "New Zealand Christmas tree", named after its show of red flowers in December, has become a major weed of the floristically rich fynbos (fine bush, in Afrikaans) in the Western Cape, South Africa. It had been there since the 1940s, and by the 1980s was starting to spread and form dense thickets on sandy, nutrient-poor soils. After measuring the potential propagule pressure from established trees, Rejmánek et al. (2004) showed "wet" areas were more invasible by *M. excelsa* than "dry" areas. Efforts are underway to control its spread with the help of volunteer labor (Richardson and Rejmánek 1999). Interestingly, the exotic species that suppress *M. excelsa* in New Zealand are absent from the Western Cape. In New Zealand, the introduced brushtail possum (*Trichosurus vulpecula*) causes heavy damage on adult trees, and invasive herbs, particularly subtropical C4 grasses such as kikuyu grass (*Pennisetum clandestinum*), prevent seedling establishment. *M. excelsa* is used horticulturally in Australia and the USA (California), but has yet to become naturalized.

Coprosma spp.

Several native *Coprosma* species have been popular amenity plantings in Australia and South Africa. *Coprosma repens*, a coastal small tree in New Zealand, is a serious weed in south-east Australia and Tasmania where it invades vegetation types ranging from warm temperate rainforest to dry coastal vegetation (Blood 2001). It has also started to invade coastal areas of the Western Cape, South Africa. Seed dispersal seems to be via stormwater drains; unlike the situation in New Zealand and Australia, the local frugivorous birds are not acting as dispersal agents. It is not yet subject to control (D.M. Richardson, personal communication). *C. robusta* is naturalized in higher-rainfall areas of Victoria and Tasmania, but is not yet as widespread as *C. repens* (Blood 2001).

Carex spp.

Carex comans, *C. flagellifera*, and *C. testacea* are all naturalized in Australia where they were introduced for horticulture and are still widely sold. They have been introduced to Tasmania, and planted for roadside soil binding and beautification but have spread to adjoining pastures and woodlands where they impact on pasture production, and were proclaimed noxious in 1989 (Parsons and Cuthbertson 2001).

Corynocarpus laevigatus

Corynocarpus laevigatus ("New Zealand laurel") is a tall tree with a long history of cultivation on the Hawaiian Island group, including being broadcast from the air for reforestation purposes (Harris 1998). It has been further spread by birds, and now forms dense thickets in mesic habitats at altitudes of 700–1,500 m a.s.l. (Smith 1998). These threaten the heau (*Exocarpus luteolus*), one of Hawaii's most endangered plant species (Harris 1998).

Kunzea ericoides and *Leptospermum scoparium*

These two shrubs or small trees were also introduced to the Hawaiian Island group about 80 years ago, and they have established on a few islands at mesic sites between 300 and 700 m a.s.l. (Smith 1998). They form dense thickets on ridge tops that have been bared by overgrazing.

6.3 Animal Emigrants

Flatworms

The "New Zealand flatworm" (*Arthurdendyus triangulatus*) stimulated interest in the spread and distribution of terrestrial planarians when it was reported to be responsible for reducing lumbricid earthworm populations below detectable levels in Northern Ireland (Blackshaw 1990). Similar adverse effects on Lumbricidae have since been confirmed in the Faroe Islands, and Scotland (Mather and Christensen 1992; Boag et al. 1999). *A. triangulatus* has the potential to become established in western continental Europe, stimulating increasing interest in the biology of, and quarantine risk posed by, this and other terrestrial planarians (Alford et al. 1998; Boag and Yeates 2001). There is concern that reduction in earthworm populations could adversely affect both soil conditions and wildlife dependent on native earthworms.

The verified European distribution of *A. triangulatus* in the mid-1960s was a single site in each of Northern Ireland, England and Scotland; there are, however, anecdotal reports of its presence in Northern Ireland and Scotland in the 1950s (B. Boag, personal communication). In Scotland, it was initially confined to central areas from where it spread principally north and west during the 1970s and 1980s, to reach the extremities of mainland Scotland by 1990 (Boag et al. 1997). Its spread from botanic gardens, garden centers and nurseries, to home gardens, and finally to farmland suggests that it was dispersed in association with container-grown plants from grower to wholesaler and eventually to retail purchasers. The early records of *A. triangulatus* from botanic gardens in Britain indicate it may have arrived with imported plant material. *A. triangulatus* has been recorded from a nursery on Banks Peninsula that had at one time exported plants to Britain (Alford et al. 1998), and so there is an association between the plant trade and spread of *A. triangulatus* on both local and global scales. DNA evidence suggests *A. triangulatus* has been spread passively within both New Zealand and Northern Ireland (Dynes et al. 2001). Phylogenetic analysis supports the hypothesis of multiple introductions to the British Isles from New Zealand. Within the UK, prohibition on release of *A. triangulatus* has been made under the Wildlife and Countryside Act 1981.

In contrast to the UK, in New Zealand the native *A. triangulatus* and introduced Lumbricidae apparently coexist. Under New Zealand conditions, terrestrial planarians are generally not found in the extensive agricultural lands where Lumbricidae are abundant; they are rare in open fields that are hot and dry in summer. Terrestrial planarians are, however, common in older-style nurseries where plant pots sit on damp soil, and under debris around farm buildings and paddock boundaries. Populations of Lumbricidae in a long-term zero-till maize (*Zea mays*) paddock were greatly reduced, the maize debris providing refugia for planarians and, in the absence of cultivation-induced mortality, they multiplied sufficiently to reduce the Lumbricidae (Yeates et al. 1999).

The New Zealand species *A. australis* was found in Scotland in 1997 (Jones 1998). In addition to *A. triangulatus* and *A. australis*, two further terrestrial planarians introduced to the British Isles probably have New Zealand origins. The large, earthworm-eating *A. albidus* was first found in Midlothian, Scotland, in 1996. It is assumed to be a native of (but is not yet recorded in) New Zealand (Jones and Gerard 1999). The origins of the "Australian flatworm" (*Australoplana sanguinea alba*), which is established in England and apparently feeds on lumbricid earthworms (Santoro and Jones 2001), are uncertain but it may also be native to New Zealand. New Zealand and Australia have several species of terrestrial planarians in common, and their individual origins are uncertain (Boag and Yeates 2001). Outside Australasia, *Australoplana sanguinea alba* was first detected in the Isles of Scilly, and has spread to many sites in England and Ireland, with records from Scotland and Wales.

Antiponemertes pantini

The establishment of *Antiponemertes pantini* in the Scilly Isles provided its first record outside New Zealand (Moore et al. 2001). Terrestrial nemerteans are poorly known, but their increasing spread is attributed to international movement of plants, and from garden center movements within countries (Moore et al. 2001).

Potamopyrgus antipodarum

This small, freshwater snail was apparently transported from New Zealand to Tasmania in drinking water supplies on ships (Ponder 1988), and became established in streams around Hobart between 1829–1830 and the 1860s; it has since spread in south-eastern Australia. It probably entered Europe in the same way, suddenly appearing in the mid-19th century. The first record was from the Thames Estuary, and it has since spread through much of Europe, at first in brackish water and later in freshwater (Winterbourn 1972). In New Zealand, diploid sexual and triploid clonal individuals can frequently be found in mixed populations (Negovetic and Jokela 2001), whereas European populations are mostly, if not entirely, parthenogenetic (Wallace 1992). Using DNA fingerprinting, Jacobsen et al. (1996) concluded that *P. antipodarum* successfully invaded Europe by the proliferation of very few clones. In the USA, Lake Ontario (NY) is inhabited by the European freshwater clone, whereas the clone in four disparate areas in the western United States matches an Australian genotype (Dybdahl 2002). Monoclonal samples from Japan and Tasmania were similar to North Island populations. In New Zealand, snails may be sterilized when parasitized by trematode larvae, the prevalence of infection being related to the frequency of males (Lively 2001). There appear to be no reports of adverse effects of *P. antipodarum* colonization on European habitats, and Eno et al. (1997, p. 121) describe its effects on the environment as "unknown other than it eats water cress but that is not a concern as this snail is so small".

The first record of *P. antipodarum* in North America was in 1987, from the Snake River system in Idaho (Zaranko et al. 1997). It may have been first introduced via a fish farm, or ballast water from European boats (Gangloff 1998). *P. antipodarum* has been reported from the Great Lakes and several American states, where it is known as the "New Zealand mud snail". It is considered harmful, due to its potential to be a bio-fouler and to compete for food and space occupied by native snails (OTA 1993; Richards et al. 2001; USGS 2002). However, recent work suggests that snail grazing may stimulate algal growth, and that *P. antipodarum* may coexist with the endemic snail *Pyrgulopsis robusta* (Riley et al. 2002). North American *P. antipodarum* populations are predominately females, and no trematode infections have been recorded (Gangloff 1998). In at least Wyoming, there are regulations prohibiting its importation (Wyoming Game and Fish Commission 1998).

Achaearanea veruculata

The "cobweb spider", described from Auckland, New Zealand, is known from several collections on Tresco, Scilly Isles. It has recently been reported from Belgium (van Keer 1993). In both New Zealand and Australia, *A. veruculata* has potential to control spider mites and leafroller caterpillars (Wilson et al. 1998; HortResearch 1999).

Stick Insects

The sexually reproducing "smooth stick insect" (*Clitarchus hookeri*) is known from several collections from a garden in the Scilly Isles (Uvarov 1950; Haes and Harding 1997). *Acanthoxyla prasina geisovii*, the parthenogenetic "prickly stick insect", is known from the Scilly Isles, Cornwall and Devon, mainly from gardens. The earliest records are from 1907 and 1908 (Uvarov 1944; Haes and Harding 1997), and Uvarov (1944) considered that its appearance in the British Isles may be a sequel to the importation of a large number of live plants into the Scilly Isles in 1907 and 1908. *Acanthoxyla prasina inermis*, the parthenogenetic "unarmed stick insect", was first recognized in Britain in 1987, having previously been confused with *C. hookeri*. It has apparently been present in Cornwall since the 1920s, and there was a further introduction in the 1970s (Haes and Harding 1997). It is also established in the Kenmare River area of south-west Ireland (Haes and Harding 1997). The stick insects *Acanthoxyla* spp. and *C. hookeri* have a similar mode of colonization to *Achaearanea veruculata* (Merrett and Rowe 1961).

Balanococcus diminutus

This native species is a pest of ornamental *Phormium* (Agavaceae), not only in New Zealand, but also in South Australia, New South Wales, Europe and North America (Cox 1987; CSIRO 2002). It was described from New Zealand flax in Italy in 1918. It was first reported in the USA in 1906, and until the 1960s was localized in the San Francisco Bay area but became more widespread with the popularity of New Zealand flax in the 1980s (San Marcos Growers 2002). *B. diminutus* is considered to have been introduced into the UK on imported plants in the 1970s (Royal Horticultural Society 2002).

Ambeodontus tristis

The two-toothed longhorn beetle occurs throughout New Zealand wherever suitable tree hosts are found (Hosking 1978). The larvae bore into dead softwoods, and into living trees where wounds give access to heartwood. It is established in Australia (Tasmania), where it has been found in sawn timber imported from New Zealand. It is one of 15 wood-destroying insects claimed as susceptible to borate timber preservative (Advance Guard 2002).

Weevils

The wood-boring weevil, *Euophryum confine*, occurs throughout Britain and mainland Europe. In Britain, it is a significant factor in wood decay, both on domestic premises and in the wild (Thompson 1989). It was first recorded in England in 1940 (Green and Pitman 2003). In Britain, populations in wooden cellars and damp ground floors are usually in conjunction with wet rot fungi. *Euophryum rufum* is a similar wood-boring weevil to *E. confine*, and was also introduced into Britain from New Zealand (Thompson 1989), the first record being in 1934. Often it is not differentiated from *E. confine* in building surveys (Green and Pitman 2003).

Feather Wing Beetles

Ptinella cavelli and *P. taylori* are New Zealand species that were apparently introduced into the British Isles through the timber trade, and live under the bark of dead trees (Johnson 1982). *P. errabunda* Johnson was described from England, and is well established there but probably originated from New Zealand; it has also spread to Europe (Vorst 1993).

Rove Beetles

Teropalpus unicolor Sharp is well established on sandy beaches along the south and south-west coasts of Britain where it is associated with tidal detritus. It had colonized Britain by the beginning of the 20th century, and also occurs in Australia, South Africa and on the Pacific coast of North America (Hammond 2000). Although there are unresolved taxonomic problems, two species of *Carpelimus* from New Zealand are recent colonists in Britain. One, the smaller of the two species, had become established in England by 1968, and has since spread widely in Britain and is also found elsewhere in Europe. The second (apparently *C. zealandicus*) had become established in southern

England by 1997 (D. Lott, personal communication; P. Hammond, personal communication). No dispersal mechanisms have been postulated.

Dryadaula pactolia

This fungus/wood-eating moth has been established in parts of Britain since 1911. Several records are associated with cellars and warehouses (Heath and Emmet 1985). The larvae are recorded from the cellar fungus (*Coniophora puteana*). In New Zealand, the species is usually found indoors.

Planotortix notophaea

This leafroller was originally described from a specimen intercepted in 1921 on shoots of *Acacia decurrens* sent from Epping, New South Wales. *P. notophaea* is a New Zealand species, and in the 1920s there was an established population in a nursery that specialized in New Zealand plants in New South Wales (Dugdale 1966).

6.4 Conclusions

Very few native New Zealand plants have been economically useful and widely planted overseas (Harris and Heenan 1992) – among the invasive species, only flax was mass planted as a crop. Consequently, their economic impact as invaders has also been negligible on a global scale (Pimentel 2002). However, at least 39 New Zealand plant species in 22 genera have naturalized (i.e., formed self-maintaining populations in the wild) or become weeds in other regions, with impacts on a local scale (Table 6.1). Most of these 39 species are found in southern Australia or Europe (mainly southern England), and a few occur in North America (mainly California), South East Asia, the Pacific (mainly Hawai'i), and coastal South Africa. Most were introduced for horticulture, and even those regarded as contaminants have been spread via potted plants. This includes the many species of *Epilobium* that are more of a nuisance in gardens than serious invasive environmental weeds. *Acaena novae-zelandiae* ("bidibid") is widespread and was probably distributed by several vectors, including exported wool (e.g., Gynn and Richards 1985).

Those New Zealand plant species that have become invasive overseas derive mainly from non-forest habitats: open lands (*Acaena novae-zelandiae*, *Cortaderia richardii*, many species of *Epilobium*); wetlands (*Carex* spp., *Crassula* spp., *Phormium tenax*); exposed coastal habitats (*Coprosma repens*, *Hebe* spp., *Metrosideros excelsa*, *Muehlenbeckia complexa*, *Myoporum laetum*,

Olearia avicenniaefolia, *O. traversii*, *Phormium colensoi*); and early-successional vegetation (*Coprosma robusta*, *Hebe* spp., *Leptospermum scoparium*, *Kunzea ericoides*, *Pittosporum* spp., *Solanum aviculare*). Their native habitats are characterized by stress and disturbance, and in general, the habitats they have invaded offshore have similar characteristics. There are similarities with New Zealand climates in some regions, particularly for southern England and southern Australia. Other areas where New Zealand plants have invaded tend to be hotter (southern Africa, California and Hawai'i).

Mostly the invasive migrants produce copious quantities of seed that are either wind blown or dispersed by birds. Vegetative reproduction has not been generally conducive to the spread of New Zealand species overseas, although *Crassula helmsii* is a spectacular exception. Countries wishing to import New Zealand species for horticulture should avoid species with these characteristics and from the above habitats. Purposely planting exotic species into the natural environment frequently results in invasions requiring official control, illustrated here by *Carex* and *Phormium*. This practice, too, should be avoided.

The 39 plant species naturalized overseas represent 2.0% of the New Zealand native angiosperm flora. That this is of the same order of magnitude as the percentage of the world's flora that has become naturalized in New Zealand (2,000 species) casts doubt on the assumption that island species are generally poor competitors. Exotic plants invading New Zealand show lags in their spread (Chap. 3). Assuming New Zealand species behave the same overseas, then both the number and extent of New Zealand species can be expected to increase when barriers to invasion are removed, e.g., when southern African birds learn to eat *Coprosma repens* berries.

In contrast to plants, the animals known to have been exported from New Zealand largely represent accidental transfers. *Potamopyrgus antipodarum* could be regarded as a forerunner of the late 20th century concern with ballast-water contaminants. *Balanococcus diminutus* has been translocated as an obligate associate of its host plant, *Phormium tenax*. Terraria (or Wardian cases; Ward 1842) played a part in some animal translocations but they were more commonly associated with the export of New Zealand plants. Timber has also been important in animal transfer. The contrast between deliberate and accidental transfers is highlighted by the formal taxonomic description of *Arthurdendyus albidus* from Scotland, of *Ptinella errabunda* from England, and of *B. diminutus* from Italy, rather than from New Zealand. Similarly, the terrestrial amphipod *Arcitalitrus dorrieni* was described from the Scilly Isles, rather than from Australia (O'Hanlon and Bolger 1997). The cluster of Australasian invertebrates established in the Scilly Isles is remarkable, and raises the issue of how many establishment events elsewhere remain undetected.

Potamopyrgus antipodarum apparently reached England from New Zealand via Australia. Subsequently, it has moved to continental Europe, Japan and North America. The continuing new records of exotic terrestrial planarians from the British Isles also indicate that phytosanitary regulations are at least partly ineffective (Boag and Yeates 2001); perhaps they serve to at least reduce the flood that could otherwise accompany increasing volumes of international trade.

For several New Zealand animal species that are today regarded as problem organisms overseas, there appears to have been a variable lag between successful export and pest status. *P. antipodarum* was recorded as causing blockage in water supplies almost a century after its arrival in Australia (Ponder 1988), but concern about its effects in North American waterways was almost immediate. The terrestrial planarian *Arthurdendyus triangulatus* was apparently present in both Scotland and the Faroe Islands for some 25 years before adverse effects were observed. These two species appear to be utilizing resources not used by native species of the respective groups. Climatic conditions have been used to explain actual or potential distributions of both species (Ponder 1988; Boag et al. 1995). In the case of *P. antipodarum*, a lack of parasitism may be a contributing factor. If the delay in onset of parasitism of the millipede *Ommatoiulus moreletii* by a native rhabditid nematode in Australia (McKillup et al. 1988) is a realistic indication, natural development of parasitic control is likely to be a long-term phenomenon.

Sixteen species of stick insects are recognized in New Zealand, 15 of them being sexually reproducing whereas the parthenogenetic genus *Acanthoxyla* has a single species with seven subspecies (Salmon 1991). Although more than 10 exotic species of *Phasmida* occur in continental Europe, only three taxa have established as breeding outdoor populations in the British Isles. All three are from New Zealand; two other species are recorded from the British Isles following deliberate releases (Haes and Harding 1997).

The horticultural use of *Phormium*, *Metrosideros* and other New Zealand plants internationally, and the accumulating records of terrestrial planarians from New Zealand in the British Isles indicate that species from this insular biota continue to successfully challenge offshore ecosystems. The propensity for New Zealand plants to establish in, and exploit, disturbed habitats and the ability of New Zealand animals to exploit available resources in areas with suitable climates not only provide opportunities for further study, but should also alert ecologists to the potential for island biota to become invasive in any part of the world.

Acknowledgements John Dugdale, Peter Hammond and Derek Lott increased our awareness of New Zealand insects in the British Isles.

References

Advance Guard (2002) Advance Guard Borate Protected Wood. http://www.advance-guard.com/preservative.htm (25 September 2002)

Alford DV, Boag B, Johns PM, Yeates GW (1998) Report on the OECD workshop on terrestrial flatworms. Pedobiologia 42:385–388

Blackshaw RP (1990) Studies on *Artioposthia triangulata* (Dendy) (Tricladida: Terricola), a predator of earthworms. Ann Appl Biol 116:169–176

Blood K (2001) Environmental weeds: a field guide for SE Australia. CRC Weed Management Systems, CH Jerram, Melbourne, Australia

Boag B, Yeates GW (2001) The potential impact of the New Zealand flatworm, a predator of earthworms, in Western Europe. Ecol Appl 11:1276–1286

Boag B, Evans KA, Neilson R, Yeates GW, Johns PM, Mather JG, Christensen OM, Jones HD (1995) The potential spread of terrestrial planarians *Artioposthia triangulata* and *Australoplana sanguinea* var. *alba* to continental Europe. Ann Appl Biol 127:385–390

Boag B, Jones HD, Neilson R (1997) The spread of the New Zealand flatworm (*Artioposthia triangulata*) within Great Britain. Eur J Soil Biol 33:53–56

Boag B, Jones HD, Neilson R, Santoro G (1999) Spatial distribution and relationship between the New Zealand flatworm *Arthurdendyus triangulatus* and earthworms in a grass field in Scotland. Pedobiologia 43:340–344

Centre for Ecology and Hydrology (2001) Botany research: invasive species - *Crassula helmsii*. http://dorset.ceh.ac.uk (23 August 2002)

Cox JM (1987) Pseudococcidae (Insecta: Hemiptera). Fauna of New Zealand, vol 11. DSIR, Wellington, New Zealand

Cronk QCB (1989) The past and present vegetation of St Helena. J Biogeogr 16:47–64

CSIRO (2002) New Zealand flax mealybug *Balanocccus diminutus* (Leonardi). http://www.ento.csiro.au/aicn/name_c/a_2366.htm (4 July 2003)

Darwin C (1859) On the origin of species by means of natural selection, or the preservation of favoured races in the struggle for life. Murray, London

Dugdale JS (1966) A new genus for the New Zealand "Elusive Tortrix" (Lepidoptera: Tortricidae: Tortricinae). NZ J Sci 9:391–398

Dybdahl MF (2002) Multiple worldwide invasion pathways of a freshwater snail: the revenge of New Zealand. In: Abstr Vol 87th ESA Annu Meet, Tucson, AR, pp 119–120

Dynes C, Fleming CC, Murchie AK (2001) Genetic variation in native and introduced populations of the 'New Zealand flatworm', *Arthurdendyus triangulatus*. Ann Appl Biol 139:165–174

Eno NC, Clark RA, Sanderson WG (eds) (1997) Non-native marine species in British waters: a review and directory. Joint Nature Conservation Committee, Peterborough, UK. http://www.jncc.gov.uk/marine (8 July 2002)

Gangloff MM (1998) The New Zealand mud snail in Western North America. Aquat Nuis Spec Dig 2:25, 28–30

Green M, Pitman AJ (2003) Effects of temperature and humidity on flight initiation and mortality of *Euophryum confine*. NZ J Zool 30:11–15

Gynn EG, Richards AJ (1985) Biological Flora of the British Isles, no 161. *Acaena novae-zelandiae* T. Kirk. J Ecol 73:1055–1063

Haes ECM, Harding PT (1997) Atlas of grasshoppers, crickets and allied insects in Britain and Ireland. The Stationery Office, London

Hammond PM (2000) Coastal Staphylinidae (rove beetles) in the British Isles, with special reference to saltmarshes. In: Sherwood BR, Gardiner BG, Harris T (eds) British saltmarshes. Linnean Society of London, London, pp 1–100

Harris G (1998) Invasive New Zealand weeds: our native plant invaders. CalEPPC News 6(4):8–9

Harris W, Heenan PB (1992) Domestication of the New Zealand flora – an alternative view. NZ J Crop Hort Sci 20:257–271

Heath J, Emmet AM (1985) The Moths and Butterflies of Great Britain and Ireland, vol 2. Cossidae to Heliodinidae. Harley Books, Colchester, UK

HortResearch (1999) Brownheaded leafroller – natural enemies and diseases. HortNET – an information centre for New Zealand's plant-based industries, BugKEY, The Horticulture and Food Research Institute of New Zealand. http://www.hortnet.co.nz/key/keys/info/enemies/bhl-enem.htm (23 August 2002)

Hosking GP (1978) *Ambeodontus tristis* (Fabricius) (Insecta: Coleoptera: Cerambycidae). Two-toothed longhorn. NZ Forest Service, Forest and Timber Insects in New Zealand, no 26

Jacobsen R, Forbes VE, Skovgaard O (1996) Genetic population structure of the prosobranch snail *Potamopyrgus antipodarum* (Gray) in Denmark using PCR-RAPD fingerprints. Proc R Soc Lond B 263:1065–1070

Johnson C (1982) An introduction to the Ptiliidae (Coleoptera) of New Zealand. NZ J Zool 9:333–376

Jones HD (1998) The African and European land planarian faunas, with an identification guide for field workers in Europe. Pedobiologia 42:477–489

Jones HD, Gerard BM (1999) A new genus and species of terrestrial planarian (Platyhelminthes; Tricladida; Terricola) from Scotland, and emendation of the genus *Artioposthia*. J Nat Hist (Lond) 33:387–394

Leach L, Dawson H (1999) *Crassula helmsii* in the British Isles – an unwelcome invader. Br Wildl 10:234–239

Lively CM (2001) Trematode infection and the distribution and dynamics of parthenogenetic snail populations. Parasitology suppl 123:S19–S26

Mather JG, Christensen OM (1992) The exotic land planarian *Artioposthia triangulata* in the Faroe Islands: colonisation and habitats. Frodskaparrit 40:49–60

McKillop SC, Allen PG, Skewes MA (1988) The natural decline of an introduced species following its initial increase in abundance; an explanation for *Ommatoiulus moreletii* in Australia. Oecologia 77:339–342

Merrett P, Rowe JJ (1961) A New Zealand spider, *Achaearanea veruculata* (Urquhart), established in Scilly, and new records of other species. Annu Mag Nat Hist Ser 13 4:89–96

Moore J, Gibson R, Jones HD (2001) Terrestrial nemerteans thirty years on. Hydrobiologia 456:1–6

Negovetic S, Jokela J (2001) Life-history variation, phenotypic plasticity, and subpopulation structure in a freshwater snail. Ecology 82:2805–2815

O'Hanlon RP, Bolger T (1997) Biomass, growth, and secondary production of *Arcitalitrus dorrieni* (Crustacea: Amphipoda: Talitridae) at two sites in Co. Galway, Ireland. J Zool (Lond) 241:409–428

OTA (1993) Harmful non-indigenous species in the United States. US Congress, Office of Technology Assessment OTA-F-565, US Government Printing Office, Washington, DC

Parsons WT, Cuthbertson EG (2001) Noxious weeds of Australia, 2nd edn. CSIRO, Canberra

Pimentel D (ed) (2002) Biological invasions: economic and environmental costs of alien plant, animal, and microbe species. CRC Press, Boca Raton, FL

Ponder WF (1988) *Potamopyrgus antipodarum* – a molluscan coloniser of Europe and Australia. J Molluscan Stud 54:271–285

Randall RP (2002) A global compendium of weeds. RG and FJ Richardson, Meredith, Victoria, Australia

Rejmánek M, Richardson DM, Pysek P (2004) Plant invasions and invasibility of plant communities. In: Van der Maarel E (ed) Vegetation ecology. Blackwell, Oxford, pp 332–355

Richards DC, Cazier LD, Lester GT (2001) Spatial distribution of three snail species, including the invader *Potamopyrgus antipodarum*, in a freshwater spring. West Nth Am Nat 61:375–380

Richardson DM, Rejmánek M (1999) *Metrosideros excelsa* takes off in the fynbos. Veld Flora 85:14–16

Riley LA, Dybdahl MF, Hall RO (2002) Interactions between invasive and endemic freshwater snails: algal stimulation and relaxed competition. In: Abstr Vol 87th ESA Annu Meet, Tucson, AR, abstr no 247

Roux JP, Ryan PG, Milton SJ, Moloney CL (1992) Vegetation and checklist of Inaccessible Island, central South Atlantic Ocean, with notes on Nightingale Island. Bothalia 22:93–109

Royal Horticultural Society (2004) Phormium mealybug (*Balanococcus diminutus*). http://www.rhs.org.uk/advice (8 July 2002)

Salmon JT (1991) The stick insects of New Zealand. Reed, Auckland, New Zealand

San Marcos Growers (2002) New Zealand flax mealybug. http://www.smgrowers.com/info/flaxnzmealybug.asp (8 July 2002)

Santoro G, Jones HD (2001) Comparison of the earthworm population of a garden infested with the Australian land flatworm (*Australoplana sanguinea alba*) with that of a non-infested garden. Pedobiologia 45:313–328

Smith CW (1998) Hawaiian alien plant studies. University of Hawai'i at Monoa, Botany Department, http://www.botany.hawaii.edu/faculty/cw_smith/aliens.htm (23 August 2002)

Thompson RT (1989) A preliminary study of the weevil genus *Euophryum* Broun (Coleoptera: Curculionidae: Cossoninae). NZ J Zool 16:65–79

USGS (2002) Nonindigenous species information bulletin: New Zealand mudsnail, *Potamopyrgus antipodarum* (J. E. Gray, 1853) (Mollusca: Hydrobiidae). http://cars.er.usgs.gov/mudsnail5.pdf (23 August 2002)

Uvarov BP (1944) A New Zealand phasmid (Orthoptera) established in the British Isles. Proc R Entomol Soc Lond B 13:94–96

Uvarov BP (1950) A second New Zealand stick-insect (Phasmatodea) established in the British Isles. Proc R Entomol Soc Lond B 19:174–175

Van Keer J (1993) Mededeling: *Achaearanea veruculata* (URQUHART, 1885): nieuw voor België of toevallige gast? Niwsber Belg Arachnol Ver 8:21

Vorst O (1993) *Ptinella errabunda* nieuw voor de Nederlandse fauna (Coleoptera: Ptiliidae). Entomol Ber 53(12):174–175

Wace NM, Holdgate MW (1976) Man and nature in the Tristan da Cunha islands. IUCN Monogr 6:1–114

Wallace C (1992) Parthenogenesis, sex and chromosomes in *Potamopyrgus*. J Molluscan Stud 58:93–107

Ward NB (1842) On the growth of plants in closely glazed cases. Van Voorst, London

Williamson M (2002) Alien plants in the British Isles. In: Pimentel D (ed) Biological invasions: economic and environmental costs of alien plant, animal, and microbe species. CRC Press, Boca Raton, FL, pp 91–112

Wilson LJ, Bauer LR, Lally DA (1998) Effect of early season insecticide use on predators and outbreaks of spider mites (Acari: Tetranychidae) in cotton. Bull Entomol Res 88:477–488

Winterbourn MJ (1972) Morphological variation of *Potamopyrgus jenkinsi* (Smith) from England and a comparison with the New Zealand species, *Potamopyrgus antipodarum* (Gray). Proc Malacol Soc Lond 40:133–145

Wyoming Game and Fish Commission (1998) Game and Fish Department Regulations, chap 10. Regulation for importation, possession, confinement, transportation, sale and disposition of live wildlife. http://gf.state.wy.us (8 July 2002)

Yeates GW, Ross CW, Shepherd TG (1999) Populations of terrestrial planarians affected by crop management: implications for long-term management. Pedobiologia 43:360–363

Zaranko DT, Farara DG, Thompson FG (1997) Another exotic mollusc in the Laurentian Great Lakes: the New Zealand native *Potamopyrgus antipodarum* (Gray 1843) (Gastropoda, Hydrobiidae). Can J Fish Aquat Sci 54:809–814

B Controls on Invasion Success

7 Genetics of Invasive Species in New Zealand

D. Gleeson, H. Harman and T. Armstrong

7.1 Introduction

The genetic architecture of invasive species is now considered to be an important factor governing the long-term success of colonization in a new environment, as this is critical for the subsequent ability of invasives to respond to natural selection (Lee 2002). A clearer understanding of the genetics of invasive species has the potential to provide insights into the mechanisms of invasions as well as determining the genetic changes associated with the process of invasions. Genetic information may also be used to determine factors such as the origin and relative success of invasive species, and provide potential for investigating novel methods of control.

New Zealand offers opportunities to investigate the genetic processes underlying invasive species. Mainly this is due to a significant number of invasive species in New Zealand being deliberately introduced and, as a consequence, having well-documented introduction histories that enable post-colonization genetic investigations to be carried out. In addition, New Zealand is a heterogeneous landscape that possesses many natural barriers along with varying climatic conditions, facilitating multiple founder events, as well as providing colonizers with a range of selection pressures. Given these factors, relatively few studies have taken advantage of the opportunities presented by such introductions. This chapter reviews the research that has attempted to address the genetics of alien species in New Zealand, and suggests potential future research directions that can take advantage of these systems.

7.2 Genetics of Colonization

The principles of colonization genetics have been well documented through a range of publications that have investigated these in relation to classical models of founder events and allopatric speciation (Mayr 1963; Baker and Steb-

Ecological Studies, Vol. 186
R.B. Allen and W.G. Lee (Eds.)
Biological Invasions in New Zealand

bins 1965; Carson 1971; Barton and Charlesworth 1984). Generally, invasive alien species experience the same effects as do natural colonizers, in that there is an initial establishment phase followed by rapid growth and subsequent invasion. The genetic consequence of such founder events has been one of the central tenets of allopatric speciation (Barton and Charlesworth 1984). The predicted genetic outcomes following a dramatic reduction in population size associated with founder events are a reduction in genetic variability, and, if species experience several independent founder events, this can result in genetic divergence among founding populations. Random genetic drift (Wright 1931) is the chance process most often advocated as the cause of the observed genetic change.

The extent of genetic change following founder events is predicted to be highly contingent on the size of the initial founding population. If the founding population is large, then there may be no substantial loss of genetic variability (i.e., allelic diversity) in the newly established range, as there is a high probability of all the genetic variants in the parental population being represented in the founder population (Kambhampati et al. 1990). Alternatively, if the initial founding population is small and subsequent establishment in the new range results in even smaller isolated subpopulations, then random genetic drift will play an increasingly important role in the loss or restructuring of variation. Thus, the effects of founding events and subsequent genetic drift are often difficult to distinguish (Barton and Charlesworth 1984).

The most recognized genetic consequence of colonization is the loss of genetic variability, due to founding individuals representing only a subset of the species' variability. Recessive alleles have a higher probability of becoming homozygous and being exposed to the effects of selection, with many alleles potentially becoming disadvantageous due to the new imbalance in the epigenotype or change in selective value (Mayr 1963). From the perspective of colonizing ability, genetic changes can either facilitate this or have a negative effect. An extreme example of where a reduction in genetic diversity post-colonization has resulted in significant changes in behavior in an invasive species is the Argentine ant, *Linepithema humile* (Tsutsui et al. 2000, 2001; Tsutsui and Case 2001). This species becomes a successful invader due to the loss of both self-recognition and aggressive territoriality, enabling the formation of "supercolonies" extending up to thousands of kilometers. However, there have been few studies that have addressed the importance of genetics in the establishment and early success of invasive species (Roderick and Howarth 1999). Rather, most studies have used genetic markers to investigate the origins and population genetic structure of invasive species.

7.3 New Zealand Examples

The most studied systems within New Zealand in regard to genetics of species arriving after European colonization have been a result of either the deliberate introductions of a desired species (e.g., birds, biocontrol agents) or the accidental arrival of an economically important pest species (insects, weeds). These studies have primarily investigated the genetic changes that have occurred after colonization, or used genetic data to retrace the history of an invasion to determine the source population.

7.3.1 Bird Species

The many passerine birds that were introduced to New Zealand from England in the 19th century provide opportunity for evolutionary biologists to investigate population differentiation within a microevolutionary time frame. Chapter 9 addresses the ecological factors that have enabled some species of introduced birds to successfully establish in New Zealand where others have not. We will review what is known regarding the levels of genetic differentiation and founder effect following their introduction.

Genetic studies that have investigated genetic differentiation and rates of evolution in introduced bird species include those of the house sparrow, *Passer domesticus* (Parkin and Cole 1985), the common myna, *Acridotheres tristis* (Baker and Moeed 1987), the starling, *Sturnus vulgaris* (Ross 1983), and the chaffinch, *Fringilla coelebs* (Baker 1992). All of these employed isozyme electrophoresis to compare population genetic structure and levels of genetic variability between introduced populations and those from the source of introduction. Bird introductions into New Zealand are relatively well documented, with historical records of dates, numbers and origins of those individuals involved in the founding process, enabling clear comparisons to be made. Each of these species showed high rates of genetic differentiation in the introduced populations, which can be attributed to genetic shifts promoted by bottlenecks and random drift. This has occurred to such a degree in mynas, for example, that after a relatively short evolutionary period of 100–130 years, these genetic shifts are equivalent to differences observed between subspecies of birds (Baker and Moeed 1987). Some losses of genetic variability were observed in terms of reduction of alleles per locus, but mostly these alleles were rare in the native population initially. Heterozygosity levels and proportion of polymorphic loci remained stable between introduced and native source populations.

Each of these studied bird species has been successful at establishing in New Zealand, and has not been compromised by the reduction in genetic variation relative to the source population. As a factor that may limit the potential for success in a new habitat, genetics is difficult to quantify, as data are unavailable from unsuccessful introduction attempts. However, Duncan et al. (1999; and see Chap. 9) have shown that successful introductions and subsequent range size of birds in New Zealand are positively related to the number of individuals introduced. This could be partly attributed to the likelihood of larger introductions providing the required genetic architecture for adaptation to the new environment, rather than resulting in a dramatic loss of genetic variability.

7.3.2 Biocontrol Agents

Biological control practitioners have long been concerned with how best to select and release biocontrol agents. Maximizing both the establishment of biocontrol agents and the suppression of target species are critical factors in using biological control as a tool for pest management. Genetic information can help in improving the probability of selecting appropriate biotypes with desirable attributes that are able to adapt to introduction into a new environment. Retrospective analysis of two biocontrol agents in New Zealand has attempted to determine genetic factors that may have contributed to their success of establishment.

Argentine Stem Weevil Parasitoid

One concern of biological control practitioners is whether some ecotypes of an agent will establish more readily or be more effective in suppressing the target species. In New Zealand, the strategic use of an existing program for the biological control of Argentine stem weevil, *Listronotus bonariensis* (Kuschel), has allowed scientists to investigate the relative success in establishment of various ecotypes of the Argentine stem weevil parasitoid, *Microctonus hyperodae* (Winder et al. 1997; Iline et al. 2002).

Parasitoids were collected from different locations in South America across a diverse range of geographic and climatic zones. Collections of eight "ecotypes" were then established, and over 240 lines generated (Goldson et al. 1990). Equal numbers of each ecotype were reared and released in order to maximize the genetic diversity of *M. hyperodae*, and thus facilitate establishment and success in New Zealand. It was hypothesized that the populations most suited to the conditions prevailing at each release site would out-compete those that were less "pre-adapted", and eventually become dominant. The use of genetic data subsequent to the release could provide insight into the importance of biocontrol agent "pre-adaptation" and the nature of any inter-ecotypic competition.

Studies in the laboratory have shown that the ecotypes varied significantly in aspects such as fecundity, photoperiodic response, and morphometrics. A morphometric method showed that parasitoids from east of the Andes had out-competed those from west of the Andes throughout New Zealand during 1992–1995 (Phillips et al. 1994). Efforts to develop more precise methods of discriminating between the populations using genetic data have been unable to differentiate between *M. hyperodae* ecotypes (Winder et al. 1997). Recent developments of tandem repeat primers, and use of the polymerase chain reaction (PCR) to amplify anonymous DNA regions have now enabled differentiation of the populations into two genetically distinct types that can be classified as originating from either east or west of the Andes. A preliminary survey of allozyme variation has also shown the same trans-Andes pattern of subdivision between east and western populations, although this is not reflected by inter-population variations in diapause behavior or egg load (Iline et al. 2002).

Broom Twig Miner

Broom twig miner, *Leucoptera spartifoliella* Hübner (Lepidoptera: Lyonetiidae), an accidentally introduced species, shows considerable potential as a biocontrol agent for broom, *Cytisus scoparius* (L.) Link (Syrett and Harman 1995). Since its arrival, this species has successfully colonized and dispersed throughout the country. Harman (2003) has compared the genetic variability of the broom twig miner from populations in New Zealand with that of populations from the native range in Western Europe. The aim was to determine the geographic origin of the broom twig miner in New Zealand, and to investigate post-colonization genetic changes using mitochondrial DNA (mtDNA) and amplified fragment length polymorphisms (AFLPs).

The genetic variability, measured by AFLP analysis, indicated that populations from the native range of broom twig miner showed little differentiation, which implies widespread gene flow. Although no matches for the New Zealand mtDNA were found in any of the other populations surveyed, the information does not support the hypothesis of Syrett (1990) that broom twig miner arrived in New Zealand on ornamental cultivars of broom imported from North America. A single founder event of small propagule size is suggested by the mtDNA data. Although both AFLP and mtDNA data show some loss of genetic diversity in New Zealand populations, compared with those from the native range, the AFLP data show that this could largely be attributed to the loss of less frequent alleles. In general, the New Zealand populations still show considerable genetic diversity. This is probably a result of broom twig miner populations having the capacity to grow steadily, as indicated by estimates of per capita rate of increase (r) in New Zealand conditions (Harman 2003). Therefore, any bottlenecks that might occur through colonization events are likely to be short, with little impact on genetic variation.

7.3.3 Pest Species

Determining the genetic variation of an introduced pest species can enable post-colonization population genetic structure to be characterized, and the origin of the source population to be identified. This information can be used to design realistic and effective integrated control programs, e.g., ability to source appropriate biocontrol agents, and to identify the presence of insecticide resistance genes.

Australian Sheep Blowfly

One of the most comprehensive studies investigating the post-colonization genetics of a pest species in New Zealand has been carried out on *Lucilia cuprina* (Wiedemann), the Australian sheep blowfly. This species is the principal fly involved in flystrike of sheep in Australia and South Africa (Mackerras and Fuller 1937). In Australia, the first record of strike on the mainland was from Queensland in 1883, and by 1915 flystrike had become a major problem of the sheep industry. Collection material in Australia contains no *L. cuprina* specimens dated earlier than 1912, other than one specimen assumed to have been collected in southwestern Australia between 1864 and 1867 (Norris 1990). The origin of *L. cuprina* in Australia is hypothesized to be South Africa by Norris (1990), with the subspecies *L. cuprina cuprina* likely to have been introduced repeatedly into eastern Australia from a variety of Pacific regions. Therefore, the origin of *L. cuprina* in Australia is speculative, with only the estimated time of introduction relatively accurate, being based on museum collection material.

Flystrike in New Zealand had been apparent since the 1870s, but only became serious enough to warrant attention in the 1920s when the actual species were characterized and control measures put into place. From 1982 onward, there was a change in prevalence and severity of flystrike, one contributing factor being the arrival of *L. cuprina*, first detected in 1988. Subsequent investigation of preserved material found *L. cuprina* present since the early 1980s, when collections were first made. Hence, it is possible that the presence of *L. cuprina* in New Zealand could have gone unnoticed for at least a decade. Following the arrival of *L. cuprina*, the prevalence of this species in flystrike samples progressed from 20 to over 60 % representation in a period of 2 years (Heath et al. 1991). It is likely then that this species was initially present in relatively low numbers, the first surveys showing that only 1.8 % of flies trapped were this species (Dymock and Forgie 1993).

Initial interest in the genetics of *L. cuprina* was whether this arrival in New Zealand also introduced resistance to organophosphates (OPs). The genetic basis of OP resistance in Australian populations of *L. cuprina* has been well documented and attributed to mutations in the esterase gene *aE7* (Parker et al. 1991; Newcomb et al. 1997). One mutation encodes a phosphatase with enhanced ability to hydrolase diethyl OPs over dimethyl OPs, whereas another mutation has the opposite ability. Dose-mortality data together with biochemical assays were carried out on New Zealand populations, and showed that resistance to the diethyl OP diazinon was present at a very high frequency (Gleeson et al. 1994). Subsequent molecular data have been collected that revealed two diethyl-resistance mutations associated with haplotypes present in extant flies from New Zealand (Newcomb et al. 2005), along with a haplotype possessing the amino acid substitution conferring resistance to dimethyl OPs, found in a pinned museum specimen collected from Wellsford, New Zealand in 1989 (C. Yong, D. Gleeson, R. Newcomb, unpublished data).

Further genetic studies of *L. cuprina* in New Zealand were carried out using allozyme electrophoresis and mtDNA data. The aims were to determine the genetic effects of colonization in terms of genetic variability and differentiation, and to attempt to identify the geographic origin of *L. cuprina* in New Zealand. The allozyme study revealed high levels of genetic variability (allelic diversity and differentiation) in some New Zealand and Australian populations (Gleeson 1995). This was attributed to an initially large founding population, seasonal fluctuations in population size further facilitating the effects of genetic drift. Although the isozyme results were able to determine population structure, the results were unable to accurately assess the level of relatedness between geographic populations in order to investigate potential origin or to estimate the time of colonization. Source was investigated using mtDNA sequence divergence through restriction fragment length polymorphisms (RFLPs) and cytochrome oxidase subunit I (COI) sequence data (Gleeson and Sarre 1997). Potential colonization time of *L. cuprina* in New Zealand was estimated by screening populations for evidence of the fitness modifier gene, which enables flies that are resistant to diazinon to be equal in fitness to those that are susceptible, in the absence of the insecticide (McKenzie and Clarke 1988). This gene was found to be predominant in Australian populations in the late 1970s, and is closely linked to the white-eye (*w*) locus on chromosome III (McKenzie and Game 1987; McKenzie and Clarke 1988). Evidence for multiple introductions was apparent from the RFLP and DNA sequence data (Gleeson and Sarre 1997), with several haplotypes existing in New Zealand. Measures of fluctuating asymmetry gave evidence for a fitness modifier present in New Zealand populations, and a crossing experiment showed the New Zealand fitness modifier locus is closely linked with *w* (D. Gleeson, unpublished data).

This *L. cuprina* study showed the importance of multiple datasets in understanding the genetics of colonization. It is clear from the syntheses of insecticide resistance, allelic diversity, mtDNA, and fitness modifier genetic data that, rather than a single introduction with limited numbers of colonizers, either large or multiple introductions from a predominantly Australian source occurred.

Rose-Grain and Green Spruce Aphids

The rose-grain aphid (*Metopolophium dirhodum*) and the green spruce aphid (*Elatobium abietum*) were both accidentally introduced into New Zealand. The former colonizes barley predominantly, whereas the latter feeds on spruce (*Picea* spp.). Investigation of the levels of genetic diversity in both species in New Zealand has been carried out and comparisons made with populations from the UK (Nicol et al. 1997, 1998). The rose-grain aphid is native to the UK, whereas the green spruce aphid is an introduced pest that probably arrived in the UK before 1500 A.D., along with Norway spruce (*P. abies*).

The rose-grain aphid is a recent introduction into New Zealand, having been first recorded in the North Island in 1981 and in Canterbury (South Island) in 1982. In Europe, this species causes direct damage to cereal crops, and damage through virus transmission. In Europe, *M. dirhodum* overwinters as eggs on Rosaceae or as parthenogenic females on Graminae. In New Zealand, however, only nonsexual forms have been recorded and the aphid overwinters parthenogenetically on cereals. Genetic variation was investigated at both a microgeographic (<1 km) and a macrogeographic scale, comparing *M. dirhodum* in New Zealand with that in a country where they are native (Scotland), using randomly amplified polymorphic DNA (RAPD)-PCR analysis. There was a significant temporal effect on the ratio of genotypes in populations collected in the field, with no significant spatial aggregation of genotypes. All populations from New Zealand and Scotland showed significant genetic diversity, with Scottish populations showing more diversity than there is in New Zealand. Given the relatively high levels of genetic diversity in New Zealand, despite the populations being entirely parthenogenic, it is likely that there have been multiple introductions from other locations within the native range of the rose-grain aphid (Nicol et al. 1997).

Genetic diversity of the green spruce aphid was assessed using RAPD-PCR analysis to evaluate this method for determining the origin and number of introductions of this aphid into New Zealand (Nicol et al. 1998). It also provided an opportunity to investigate an invading species in two countries with contrasting population dynamics. Samples were collected from four locations in the UK that were up to 240 km apart, and from up to six locations in New Zealand that were up to 1,200 km apart. A high degree of genotypic diversity was found in the UK populations, in comparison with New Zealand. This is presumed to be due to a very limited founder population, a short period of establishment, continued isolation, and lack of sexual reproduction in New Zealand, in contrast to the UK.

Argentine Stem Weevil

Native to South America, the introduced pasture pest Argentine stem weevil (*Listronotus bonariensis*) is thought to have been accidentally introduced into New Zealand and Australia through grasses or hay used as stockfeed on trading ships. This species was first reported in New Zealand in 1927, with a maximum of 200–250 generations elapsing since its establishment.

Geographic populations of *L. bonariensis* were analyzed using RAPD-PCR in an attempt to determine the geographic origin of the pest (Lenney-Williams et al. 1994). Morphologically indistinguishable individuals were collected from nine South American, five New Zealand, and one Australian population. The results indicated that the sampled New Zealand populations originated from the east coast of South America. This initial screen of the genetic makeup of these populations has provided useful information to assist in identifying appropriate source populations of biocontrol agents.

White-Fringed Weevil

White-fringed weevil, *Naupactus leucoloma* Boheman, was first found in New Zealand in 1945 (Todd 1968), with the origin being hypothesized as North America. Although this species arrived in New Zealand 50–60 years ago, the fact that it is parthenogenic outside of its native range (Lanteri and Normack 1995), coupled with a long generation time, would result in limited opportunities for natural genetic variation to arise outside of major mutation or recombination events. Hardwick et al. (1997) investigated the level of RAPD-PCR variation in New Zealand populations, compared with that in Victoria, Australia. In this relatively small study, three different genetic profiles were revealed from two New Zealand populations, with two profiles being revealed in a single paddock. It is possible that this variation has arisen after colonization, or that there had been multiple introductions of this pest species, although there is currently no knowledge of the extent of genetic variation occurring in other native or introduced populations of *N. leucoloma*.

7.3.4 Invasive Plants

Few studies have sought to determine the genetic characteristics of successful invasive plants, although they are likely to be subject to genetic constraints similar to those of other colonizing species (Barrett 1992). It is generally accepted that the founder effect and subsequent genetic drift in small, isolated populations result in lower genetic diversity and increased spatial genetic differentiation in adventive populations, compared with native populations

(Schierenbeck et al. 1995). The impacts of these processes on plant fitness and invasive potential are unclear, although population genetics theory predicts that founding populations may be at a disadvantage relative to native taxa. This conclusion is based on expectations of decreased fitness resulting from (1) inbreeding depression in small founding populations, and (2) genetic drift resulting in the loss of adaptive genetic diversity (Young et al. 1996).

Paradoxically, despite these potential genetic constraints, a large number of plant species have become extremely successful invaders, out-competing locally adapted native species over wide areas and across diverse ecosystems. In New Zealand, the ratio of naturalized alien plant species to native species is approximately 1:1, which is reported as one of the highest ratios of any country or region (Williams and Lee 2001). Evidently, a narrow population genetic focus alone is inadequate to explain the contribution of genetic factors to the apparent success of invasive plants.

Recent studies point to the importance of plant reproductive strategies and genomic factors in enabling species to overcome the genetic constraints associated with founder processes, and in facilitating the successful colonization of new habitats. Many plant species exhibit variable levels of sexual and vegetative reproduction, possess polyploid genomes, and are able to hybridize with close relatives. These attributes can potentially increase levels of heterozygosity in colonizing populations, guard against the negative consequences of inbreeding depression, and facilitate rapid adaptation to new conditions through hybridization and genomic reorganization (Barrett 1992; Arnold 1997; Levin 2000).

Hawkweed

Pilosella officinarum F.W. Schultz & Sch. Bip. (usually known in New Zealand as *Hieracium pilosella* L.), mouse-ear hawkweed, is native to Britain and Europe and has successfully colonized New Zealand after being accidentally introduced in the late 19th century as a contaminant of grass seed. Since its introduction, the species has spread over a wide geographic area and forms a dominant component of the vegetation across a diverse range of habitats. *P. officinarum* reproduces both vegetatively through ramet fragmentation and also through the production of sexual and apomictic (maternally derived) seed (Chapman et al. 2000).

Despite the expectation of reduced genetic diversity arising from bottleneck events associated with its introduction, Chapman et al. (2000) found that levels and patterns of genetic diversity in New Zealand populations of *P. officinarum* were similar to those reported for closely related facultatively apomictic species in their native ranges (Widén et al. 1994). New Zealand populations of *P. officinarum* differed only in that they exhibited slightly increased genetic differentiation, possibly reflecting the effect of local genetic drift in founder populations at the periphery of an expanding range. *P. officinarum* is known to have a flexible and complex breeding system, involving predominant apomixis and low levels of sexual reproduction among clones possessing various levels of polyploidy (Jenkins and Jong 1997). It has been suggested that the levels of clonal diversity present in the species reflect latent variation that arose through sexual reproduction and hybridization in ancestral populations, and that has since become fixed through polyploidy and apomixis (Chapman et al. 2000). Occasional events of sexual reproduction between polyploid clones are capable of generating a range of new recombinant genotypes, exposing this latent genetic variation to natural selection in new environmental contexts. Such a strategy may result in the species having an increased potential to colonize new habitats, evolve rapidly and persist. The patterns of genetic variation present in New Zealand populations of *P. officinarum* suggest that such processes may be at least partially responsible for its success in colonizing a wide range of new habitats.

7.3.5 Hybridization and Invasive Potential

Genetic studies (Warwick et al. 1989; Rieseberg et al. 1995; O'Hanlon et al. 1999) have increasingly highlighted the importance of hybridization and subsequent genomic reorganization to the success of plant invasions worldwide. The potential for hybridization between different species or races of invasive plants that may have been geographically or ecologically isolated in their native ranges is greatly increased when such taxa are brought together under disturbed or novel ecological contexts (Warwick et al. 1989; Arnold 1997). Hybridization can generate a new gene pool with greatly increased genetic diversity and heterozygosity relative to levels present in the parental taxa. This process can facilitate rapid adaptation by exposing a wide range of novel recombinant genotypes to natural selection, many of which may be better adapted to the new conditions than are the parental species, and may therefore have increased invasive ability. There is great potential for this phenomenon to be examined in New Zealand, with so many invasive genera present comprising a number of potentially interfertile entities (e.g., in the genera *Carduus*, *Hieracium*, *Brassica*, *Festuca*, *Passiflora*). The importance of understanding hybridization within the context of invasive potential is highlighted by the development of genetically manipulated crops, some of which may be capable of interbreeding with invasive species already established in the environment. Genetically engineered traits such as *Bt* expression may confer considerable selective advantages to weedy relatives.

7.4 Potential Genetic Impacts on Indigenous Biota

The "invasion" of alien genes into indigenous species represents a potentially serious and intractable problem that has so far received little scientific attention in New Zealand. In plant species, one particular genetic consequence of plant invasion is genetic assimilation through the introgression and permanent incorporation of genes from adventives into the gene pools of native species through hybridization. The possibility of hybridization between indigenous and naturalized species has been studied infrequently in New Zealand, the only examples being in *Acaena* (Rosaceae; Webb et al. 1988), *Carpobrotus* (Aizoaceae; Chinnock 1972), and *Epilobium* (Onagraceae; Raven and Raven 1976). Spontaneous natural hybrids have been reported between the naturalized *Epilobium ciliatum* and two indigenous species, *E. brunnescens* subsp. *brunnescens* and *E. melanocaulon* (Raven and Raven 1976). In *Epilobium*, a number of fertile hybrids between indigenous and naturalized species have been generated experimentally (Brockie 1966), highlighting the potential for hybridization to occur between indigenous and naturalized species in the wild. Experimental studies also suggest that hybridization may occur naturally between the invasive *Calystegia silvatica* and the indigenous *C. sepium* (Convolvulaceae; Ogden 1978), and also between naturalized *Polygonum persicaria* and the indigenous *P. salicifolium* (Polygonaceae; Webb et al. 1988). Despite these experimental findings, and the fact that some of the most aggressive naturalized species have closely related indigenous congeners, including *Myosotis* (Boraginaceae), *Ranunculus* (Ranunculaceae), and *Senecio* (Asteraceae), there are few documented accounts of natural hybridization between indigenous and naturalized species in New Zealand. However, hybridization might be difficult to detect without the use of molecular markers, as hybridizing species commonly possess characters that are morphologically similar, difficult to interpret, and could be masked by environmental factors that influence growth habit. For example, observations that cultivated Tasmanian *Myoporum insulare* R. Br. is hybridizing extensively with the New Zealand native *Myoporum laetum* Forster. (ngiao) in the Auckland region (NZ Department of Conservation, unpublished data) require genetic analysis to verify the identity of putative hybrids and the local extent of introgression. There is clearly a wider need to investigate the possibility of hybridization using molecular approaches to better understand and predict the genetic consequences of plant invasions for the New Zealand biota.

7.5 Conclusions

Genetic effects following the colonization and invasion of New Zealand have been studied in relatively few cases for a variety of reasons. The outcomes of these studies have shown that there are ranges of genetic effects that do occur after colonization, although it is often difficult to attribute genetics as a factor in the success or failure of invasive species. However, specific genotypes can facilitate success in some cases, e.g., sheep blowfly and insecticide resistance gene. It is clear that further invasions are likely, and that many risks are posed to the indigenous biota, in particular through hybridization with exotics. The extent to which this is currently occurring is little known and presents opportunities for future research, particularly when considering potential impacts of releasing genetically modified plants into the environment.

Another area where genetic research can assist in the study of biological invasions is in a better understanding of the invasive process and adaptive genetic variance that underlies the success of invasive species. Recent research suggests that only a small number of genes along with epistatic genetic variance can have a major impact on colonizing ability. Advancements in the field of genomics are now providing the tools necessary to investigate the genomic characteristics of successful invaders.

It has also been suggested that genetics should have a larger role in the development of policy to manage and control invasive species through a better understanding of the risks that particular genotypes pose (Allendorf and Lundquist 2003). At present, focus on invasive species in New Zealand is at the taxon level, whereas in the future, genetic risk associated with invasives is likely to have greater emphasis.

References

Allendorf FW, Lundquist LL (2003) Introduction: population biology, evolution, and control of invasive species. Conserv Biol 17:24–30

Arnold ML (1997) Natural hybridization and evolution. Oxford Univ Press, New York

Baker AJ (1992) Genetic and morphometric divergence in ancestral European and descendent New Zealand populations of chaffinches (*Fringilla coelebs*). Evolution 46:1784–1800

Baker AJ, Moeed A (1987) Rapid genetic differentiation and founder effect in colonizing populations of common mynas (*Acridotheres tristis*). Evolution 41:525–538

Baker HG, Stebbins GL (eds) (1965) The genetics of colonizing species. Academic Press, New York

Barrett SCH (1992) Genetics of weed invasions. In: Jain SK, Botsford LW (eds) Applied population biology. Kluwer, Dordrecht, The Netherlands, pp 91–119

Barton NH, Charlesworth B (1984) Genetic revolutions, founder effects, and speciation. Annu Rev Ecol Evol Syst 15:133–164

Brockie WB (1966) Artificial hybridization of New Zealand species and varieties of *Epilobium*. NZ J Bot 4:366–391

Carson HL (1971) Speciation and the founder principle. Univ Missouri Stadler Symp 3:51–70

Chapman HM, Parh D, Oraguzie N (2000) Genetic structure and colonizing success of a clonal, weedy species, *Pilosella officinarum* (Asteraceae). Heredity 84:401–409

Chinnock RJ (1972) Natural hybrids between Disphyma and Carpobrotus (Aizoaceae) in New Zealand. NZ J Bot 10:615–626

Duncan RP, Blackburn TM, Veltman CJ (1999) Determinants of geographical range sizes: a test using introduced New Zealand birds. J Anim Ecol 68:963–975

Dymock JJ, Forgie SA (1993) Habitat preferences and carcase colonization by sheep blowflies in the northern North Island of New Zealand. Med Vet Entomol 7:155–160

Gleeson DM (1995) The effects on genetic variability following a recent colonization event: the Australian sheep blowfly, *Lucilia cuprina* arrives in New Zealand. Mol Ecol 4:699–707

Gleeson DM, Sarre S (1997) Mitochondrial DNA variability and geographic origin of the sheep blowfly, *Lucilia cuprina* (Diptera: Calliphoridae), in New Zealand. Bull Entomol Res 87:265–272

Gleeson DM, Barry SC, Heath ACG (1994) Insecticide resistance status of *Lucilia cuprina* in New Zealand using biochemical and toxicological techniques. Vet Parasitol 53:301–308

Goldson SL, McNeill MR, Stufkens MW, Proffitt JR, Pottinger RP, Farrell JA (1990) Importation and quarantine of *Microctonus hyperodae*, a South American parasitoid of Argentine stem weevil. In: Proc 43rd New Zealand Weed and Pest Control Conf, New Zealand Plant Protection Society. Manaaki Whenua Press, Lincoln, New Zealand, pp 334–338

Hardwick S, Armstrong KF, Wratten SD, Prestidge RA (1997) Genetic comparison of whitefringed weevil species and populations by RAPD-PCR. In: Proc 50th New Zealand Plant Protection Conf, New Zealand Plant Protection Society. Manaaki Whenua Press, Lincoln, New Zealand, pp 327–332

Harman HM (2003) Establishment of a beneficial invader, broom twig miner (*Leucoptera spartifoliella* Hübner), on broom (*Cytisus scoparius* (L.) Link) in New Zealand. PhD Thesis, Lincoln University, Canterbury, New Zealand

Heath ACG, Bishop DM, Cole DJW, Dymock JJ (1991). Exotic blowflies in New Zealand. In: Abstr Proc New Zealand Society for Parasitology. NZ J Zool 18:85

Iline II, Phillips CB, Goldson SL, Chapman HM (2002) Allozyme variation between South American geographic populations of *Microctonus hyperodae*. NZ Plant Protect 55:267–271

Jenkins TA, Jong K (1997) Significance of polyploid variation in New Zealand *Pilosella* and *Hieracium* (Asteraceae). Bot J Scot 49:75–88

Kambhampati S, Black WC, Rai KS, Sprenger D (1990) Temporal variation in genetic structure of a colonising species: *Aedes albopictus* in the United States. Heredity 64:281–287

Lanteri AA, Normark BB (1995) Parthenogenesis in the tribe Naupactini (Coleoptera: Curculionidae). Ann Entomol Soc Am 88:722–731

Lee CE (2002) Evolutionary genetics of invasive species. Trends Ecol Evol 17:386–391

Lenney Williams C, Goldson SL, Baird DB, Bullock DW (1994) Geographical origin of an introduced insect pest, *Listronotus bonariensis* (Kuschel), determined by RAPD analysis. Heredity 72:412–419

Levin DA (2000) The origin, expansion and demise of plant species. Oxford Univ Press, New York

Mackerras IM, Fuller M (1937) A survey of the Australian sheep blowflies. J Council Sci Industr Res Aust 10:261–270

Mayr E (1963) Animal species and evolution. Harvard Univ Press, Cambridge, MA

McKenzie JA, Clarke GM (1988) Diazinon resistance, fluctuating asymmetry and fitness in the Australian sheep blowfly, *Lucilia cuprina*. Genetics 120:213–220

McKenzie JA, Game AY (1987) Diazinon resistance in *Lucilia cuprina*; mapping of a fitness modifier. Heredity 59:371–381

Newcomb RD, Campbell PM, Ollis DL, Cheah E, Russell RJ, Oakeshott JG (1997) A single amino acid substitution converts a carboxylesterase to an organophosphorus hydrolase and confers insecticide resistance on a blowfly. Proc Natl Acad Sci USA 94:7464–7468

Newcomb R, Gleeson D, Russell R, Oakeshott R (2005) Multiple point mutations and gene duplications conferring organophosphorus insecticide resistance have been selected at the Rop-1 locus of the sheep blowfly, *Lucilia cuprina*. J Mol Evol 60:207–220

Nicol D, Armstrong KF, Wratten SD, Cameron CM, Frampton C, Fenton B (1997) Genetic variation in an introduced aphid pest (*Metopolophium dirhodum*) in New Zealand and relation to individuals from Europe. Mol Ecol 6:255–265

Nicol D, Armstrong KF, Wratten SD, Walsh PJ, Straw NA, Cameron CM, Lahmann C, Frampton CM (1998) Genetic diversity of an introduced pest, the green spruce aphid *Elatobium abietinum* (Hemiptera: Aphididae) in New Zealand and the United Kingdom. Bull Entomol Res 88:537–543

Norris KR (1990) Evidence for the multiple exotic origin of Australian populations of the sheep blowfly, *Lucilia cuprina* (Wiedemann) (Diptera: Calliphoridae). Aust J Zool 38:635–648

Ogden J (1978) Variation in *Calystegia* R.Br. (Convolvulaceae) in New Zealand. NZ J Bot 16:123–140

O'Hanlon PC, Peakall R, Briese DT (1999) Amplified fragment length polymorphism (AFLP) reveals introgression in weedy *Onopordum* thistles: hybridization and invasion. Mol Ecol 8:1239–1246

Parker AG, Russell RJ, Delves AC, Oakeshott JG (1991) Biochemistry and physiology of esterases in organophosphate-susceptible and resistant strains of the Australian sheep blowfly, *Lucilia cuprina*. Pestic Biochem Physiol 41:305–318

Parkin DT, Cole SR (1985) Genetic differentiation and rates of evolution in some introduced populations of the House Sparrow, *Passer domesticus* in Australia and New Zealand. Heredity 54:15–23

Phillips CB, Baird DB, Goldson SL (1994) The South American origins of New Zealand *Microctonus hyperodae* parasitoids as indicated by morphometric analysis. In: Proc 47th New Zealand Plant Protection Conf, New Zealand Plant Protection Society. Manaaki Whenua Press, Lincoln, New Zealand, pp 220–226

Raven PH, Raven TE (1976) The genus *Epilobium* (Onagraceae) in Australasia: a systematic and evolutionary study. DSIR Bull 216. Christchurch, New Zealand

Rieseberg LH, van Fossen C, Desrochers AM (1995) Hybrid speciation accompanied by genomic reorganization in wild sunflowers. Nature 375:313–316

Roderick GK, Howarth FG (1999) Invasion genetics: natural colonizations, non-indigenous pests, and classical biological control. In: Yano E, Matsuo K, Shiyomi M, Andow D (eds) Biological invasions of ecosystem by pests and beneficial organisms. NIAES Ser 3. National Institute of Agro-Environmental Science, Tsukuba, Japan

Ross HA (1983) Genetic differentiation of starling (*Sturnus vulgaris*: Aves) populations in New Zealand and Great Britain. J Zool Lond 201:351–362

Schierenbeck KA, Hamrick JL, Mack RN (1995) Comparison of allozyme variability in an introduced species of *Lonicera*. Heredity 75:1–9

Syrett P (1990) The impact of the self-introduced twigmining moth *Leucoptera spartifoliella* on broom (*Cytisus scoparius*) in New Zealand. In: Proc 7th Int Symp Insect-Plant Relationships, Budapest, Hungary, 1989, Akadémiae Kiadó, Budapest, pp 527–540

Syrett P, Harman HM (1995) *Leucoptera spartifoliella* Hübner as a biological control agent for broom in New Zealand. Plant Protect Q 10:75–78

Todd DH (1968) Laboratory evaluation of insecticides against white-ringed weevil (*Graphognathus leucoloma* (Boheman) (Coleoptera: Curculionidae)) larvae. NZ J Agric Res 11:903–910

Tsutsui ND, Case TJ (2001) Population genetics and colony structure of the Argentine ant (*Linepithema humile*) in its native and introduced ranges. Evolution 55:976–985

Tsutsui ND, Suarez AV, Holway DA, Case TJ (2000) Reduced genetic variation and the success of an invasive species. Proc Natl Acad Sci USA 97:5948–5953

Tsutsui ND, Suarez AV, Holway DA, Case TJ (2001) Relationships among native and introduced populations of the Argentine ant (*Linepithema humile*) and the source of introduced populations. Mol Ecol 10:2151–2161

Warwick SI, Bain JF, Wheatcroft R, Thompson BK (1989) Hybridization and introgression in *Carduus nutans* and *C. acanthoides* reexamined. Syst Bot 14:476–494

Webb CJ, Sykes WR, Garnock-Jones PJ (eds) (1988) Flora of New Zealand, vol IV. Botany Division, DSIR, Christchurch, New Zealand

Widén B, Cronberg N, Widén M (1994) Genotypic diversity, molecular markers and spatial distribution of genets in clonal plants, a literature survey. Folia Geobot Phytotax Praha 29:245–263

Williams PA, Lee WG (2001) Why screen for weediness? NZ Garden J 4:17–21

Winder LM, Goldson SL, Lenney Williams C (1997) Genetic variation between two *Microctonus hyperodae* populations imported for control of Argentine stem weevil. In: Proc 50th New Zealand Plant Protection Conf 1997, New Zealand Plant Protection Society. Manaaki Whenua Press, Lincoln, New Zealand, pp 333–337

Wright S (1931) Evolution in Mendelian populations. Genetics 16:97–159

Young A, Boyle T, Brown T (1996) The population genetic consequences of habitat fragmentation for plants. Trends Ecol Evol 11:413–418

8 Are the Marine Biotas of Island Ecosystems More Vulnerable to Invasion?

G.J. Inglis, B.J. Hayden and W.A. Nelson

8.1 Introduction

New Zealand's native marine ecosystems are facing challenges from more than 130 nonindigenous species (NIS) which have become established in New Zealand's coastal waters in the past 200 years (Cranfield et al. 1998). The Asian date mussel *Musculista senhousia*, the smooth cord grass *Spartina alterniflora*, and the Japanese kelp *Undaria pinnatifida* have substantially modified habitats, changing their suitability for native species. Others, such as the Pacific oyster (*Crassostrea gigas*) and Pacific salmon (*Oncorhynchus tshawytscha*), now form the basis of valued industries and are actively cultured. Most NIS (~96 %) have arrived accidentally, on the hulls or in the ballast of ocean-going ships (Cranfield et al. 1998). A dependence on maritime trade, together with changes in the diversity of our major trading partners and global increases in the number, size and speed of ocean-going vessels, means that the influx of foreign species is likely to continue in the future (Ruiz et al. 2000).

The susceptibility of island biotas to invaders has been a central theme in invasion biology, with the changes wrought to New Zealand's terrestrial and freshwater ecosystems featuring prominently in most international reviews of the subject (e.g. Elton 1958; Simberloff 1995; Williamson 1996). Despite differences between marine and terrestrial ecosystems, Ruiz et al. (1997) predicted that the effects of marine invaders would also be greatest on small islands which have large numbers of endemic marine species. The long ocean distances between New Zealand and continental neighbours mean that its marine plants and animals have evolved in relative isolation from other coastal assemblages and, in many ways, are as distinctive as its terrestrial biota (Dell 1968; Towns and Ballantine 1993). In this chapter, we review current ideas about the susceptibility of island biotas to invasion, and discuss their application to the marine environments of New Zealand and other oceanic islands.

Ecological Studies, Vol. 186
R.B. Allen and W.G. Lee (Eds.)
Biological Invasions in New Zealand

8.2 Island Bioinvasion Theory

The success of NIS on islands has been attributed to a variety of evolutionary and ecological causes. Central to most explanations has been the generally smaller pool of native species and, consequently, limited competition for relatively abundant resources (Elton 1958). Both the "maladaptation" and the "empty niche" hypotheses have since been challenged. Simberloff (1995) noted that island species are not always poorer competitors than their mainland counterparts and that resources on islands were typically never underutilized. The success of island invaders, therefore, is not necessarily because there are fewer native species present on islands, but because entire ways of living (i.e. functional groups) may be absent, allowing particular types of invaders to flourish (Simberloff 1995).

Contemporary treatments of "niche opportunity" define the conditions which promote invasions as a combination of available resources (such as food and space), lack of natural enemies, and suitable environmental conditions (Williamson 1996; Shea and Chesson 2002). A community's susceptibility to invasion is not a static or permanent attribute (Davis et al. 2000), resulting in a dynamic definition of an "available niche". Frequent inoculations of a particular species ("propagule pressure") enhance the chance that its arrival will coincide with suitable conditions for establishment. Species which have a strong association with human transport pathways will generally have greater odds of success (Wonham et al. 2000). Similarly, species with physiological attributes or life-history characteristics which allow establishment under a greater range of conditions, such as broad habitat tolerances and high reproductive potential, should be more successful invaders than less adaptable species.

8.3 Are Marine Biotas of Islands More Invaded?

8.3.1 Number of Nonindigenous Species in Various Regions

Carlton (1987) analysed available data on the distribution of NIS in the Pacific and noted that Hawai'i, the Pacific coast of North America, and Australasia (including New Zealand) appeared to be major recipients of introduced species, whereas the Pacific coast of Asia and the south-western Pacific were major source regions. Like many other countries, New Zealand has only recently begun to document the extent of invasions in its coastal environments. Inventories of marine NIS, similar to that compiled for New Zealand by Cranfield et al. (1998), have been developed for (at least) parts of Australia, the USA, UK, Mediterranean Sea, North Sea, Black Sea, and Baltic Sea (see Table 8.1). In

Table 8.1 Numbers of non-indigenous marine species reported from coastal regions

Oceans and seas	Fishes	Molluscs	Barnacles	Macroalgae	Sponges	Decapods	Bryozoans	Total
New Zealand[a]	3	22	18	21	9	6	19	130
Hawai'ian Islands[b]	20	53	4	24	23	16	16	343
Guam[c]	6	14	2	?	5	5	3	85
United Kingdom[d]	1	13	3	15	1	3	3	79
North Sea[e]	0	11	4	20	0	4	2	80
BalticSea[f]	29	12	1	7	?	6	1	101
Black Sea[g]	7	7	2	2	0	4	0	53
Mediterranean Sea[h]	55	75	?	61	?	33	?	>300
Bays and harbours								
Pearl Harbour (Hawai'i)[i]	4	15	4	1	5	5	8	69
Kane'ohe Bay (Hawai'i)[j]	12	23	4	6	11	4	12	82
Apra Harbour (Guam)[k]	?	?	?	?	?	?	?	46
Port Phillip Bay (Australia)[l]	4	7	1	16	6	2	23	99
San Francisco Bay (USA)[m]	6	27	2	6	5	7	11	212
Venice Lagoon (Italy)[n]	?	7	?	8	?	3	2	30

[a] Cranfield et al. (1998)
[b] Eldredge and Carlton (2002)
[c] Paulay et al. (2002)
[d] Minchin and Eno (2002)
[e] Reise et al. (1999, 2002)
[f] Leppäkoski et al. (2002)
[g] Gomoiu et al. (2002)
[h] Galil (2000)
[i] Coles et al. (1999a)
[j] Coles et al. (2002)
[k] Coles and Eldredge (2002)
[l] Hewitt et al. (1999)
[m] Cohen and Carlton (1995)
[n] Occhipinti Ambrogi (2000)

most cases, these encompass both historical (since biological records were first kept) and more recent introductions.

The number of NIS known to have established in New Zealand waters is markedly higher than that recognized from the North Sea, Black Sea or UK (Table 8.1). Similarly, recent studies in the Hawai'ian Islands have recorded almost as many nonindigenous marine species as have been recorded from the whole of mainland USA (Table 8.1; Eldredge and Carlton 2002). Around 298 nonindigenous invertebrates and algae, and 100 species of fish have so far been reported from the coastal waters of the continental USA (Ruiz et al. 2000). A reliable comparison of the rates of invasion in the marine biotas of island and mainland areas is not yet possible, however, because the methods used to compile lists of NIS, the geographic scales of consideration, and the taxonomic criteria used vary widely among the published studies (Ruiz et al. 2000). In some cases (e.g. Cranfield et al. 1998; Reise et al. 2002), the lists were compiled solely from published records and museum specimens. Others were developed using a combination of historical records and targeted field studies (e.g. Coles et al. 1999a, b).

In Australia, standardized sampling techniques have been used to determine the extent of invasions in major shipping ports throughout the country. Results reveal a trend towards greater numbers of nonindigenous species in southern temperate environments than in tropical ports (Hewitt 2002). The southern marine ecosystems of Australia are relatively isolated from other temperate environments, and contain a large proportion of endemic species. However, the southern ports of Australia also have a longer history of settlement than does tropical Australia, and the latitudinal trend may simply reflect a longer history and quantity of supply of NIS to southern Australia, rather than any inherent difference in susceptibility to invasion (Hewitt 2002).

8.3.2 Problem of Cryptogenic Species

"Cryptogenic" species (i.e. species which are not demonstrably introduced or indigenous) can comprise up to 30% of some coastal marine assemblages (Ruiz et al. 2000). Their prevalence creates an uncertain regulatory environment for managers of marine pests, since eradication and control measures are usually not sanctioned against native species. This was highlighted recently by two pest alerts in New Zealand in which decisions about remedial action revolved around the geographic origins of the species.

Chaetopterus sp.: a Pesky Worm

In 1997, commercial scallop fishers in north-eastern New Zealand reported catching large numbers of a parchment tubeworm which was clogging dredges and reducing catch efficiency. The worm, a species of polychaete in the genus *Chaetopterus*, has since been found along the eastern coast of northern New Zealand. Its tubes formed dense aggregations (>120 individuals per 0.02 m^2) up to 2 m in diameter and 25 cm deep, and it colonized a wide variety of habitats including rocky reef and soft sediment substrates from the shallow intertidal to depths>60 m (Tricklebank et al. 2001). Museum collections suggest that isolated individuals may have been present in New Zealand during the 1960s, but these specimens were poorly preserved and not formally identified at the time. The complex and confusing taxonomy of this group (most species are inadequately described) and the poor condition of museum specimens has meant that it has not been possible to determine whether the species is indigenous to New Zealand or is introduced. A similar, unidentified species of *Chaetopterus* has become abundant in Kane'ohe Bay and Pearl Harbour, Hawai'i, since the 1970s where it is also regarded as cryptogenic (DeFelice et al. 2001).

Didemnum vexillum sp. nov.: a Worrying Ascidian

In 2001, the Harbourmaster of a small coastal harbour in north-eastern New Zealand reported an unusual growth which blanketed wharf piles and some boats in the harbour. Subsequent surveys showed it to be a previously undescribed colonial ascidian in the genus *Didemnum* (Kott 2002). The species was subsequently recorded in Tauranga, and on a barge in Picton Harbour which had been relocated from Tauranga. Because of its habit of overgrowing other fouling species, discovery of the ascidian in the Marlborough Sounds raised alarm bells with the Greenshell™ mussel industry. There is no evidence that *Didemnun vexillum* sp. nov. has been introduced to New Zealand and, because it has not been described elsewhere, it is currently assumed to be a native species which underwent a sudden, unusual bloom in abundance (Kott 2002).

8.4 Are the Marine Biotas of Islands Less Diverse?

Elton (1958) considered that islands contained a smaller pool of species than was present in comparable areas of continental landmasses, and therefore provided greater opportunity for invaders. The relationship between species richness and invasibility has received little consideration in marine environments. A recent experimental study (Stachowicz et al. 2002) showed that invaders of space-limited marine invertebrate communities tend to be more successful when the native assemblage has low species richness. At the small scale at which the experiments were done (10x10 cm tiles), the mechanism for this success appeared to be the greater temporal stability in resource (open space) utilization by diverse assemblages, thereby providing less opportunity for an invader to establish. Whether similar mechanisms apply in different types of marine assemblages or at the larger, biogeographic scales relevant to island bioinvasion theory is still unclear. However, patterns of species distribution and diversity in the oceans are complex and still poorly understood (Warwick 1996), in part because the description of biogeographic patterns is contingent upon the state of taxonomy and systematics – both of which are patchy for marine biotas (Gordon 2001). Also, because most biological surveys of marine environments commenced centuries after ships began traversing the globe, many species which are currently considered "cosmopolitan" or even "native" could well have been spread throughout the world before any records were made (Ruiz et al. 2000).

Biodiversity gradients for coastal marine species tend to occur as a mosaic at a wide range of spatial scales, and reflect the geological history of the shoreline as well as contemporary and historical biogeographic barriers to dispersal (Myers 1997; Hooper et al. 2002; Roberts et al. 2002). This can mean that, for particular groups of organisms, variation in species richness can be greater within a continent than between mainland and island biotas. Some islands also have particularly rich marine floras and faunas. For example, southern Australia and New Zealand have rich marine macroalgal floras and bryozoan assemblages. Hawai'i and French Polynesia have particularly rich crab faunas. Conversely, estuaries of continental north-eastern America, and the Baltic, Black and eastern Mediterranean seas, which have relatively large numbers of NIS, are known to have quite depauperate native biotas (see Cohen and Carlton 1998; Galil 2000). Currently, there is little direct evidence that marine biotas of oceanic islands are, on average, less speciose than comparable areas of continental shoreline.

8.5 Insularity and Niche Opportunity

The unique evolutionary history of New Zealand's coastal marine biota has resulted in both missing and functionally distinctive groups of species. Compared with other temperate coastlines, for example, New Zealand has a relatively depauperate crab fauna (Dell 1968), with around 63 species of crab (McLay 1988) – about two-thirds as many as South Australia (93), and less than one-fifth the number of species found in Japan (>337). Most genera (82 %) are represented by only a single species (typical of isolated biotas), with a large percentage of endemic species (53 %). Dell (1968) suggested that many of the niches which are filled by crabs elsewhere appear to be occupied by other organisms or are filled by only a single species in New Zealand.

The recent establishment in Auckland of the large (max. carapace width~110 mm) estuarine swimming crab *Charybdis japonica*, a native of northern Asia, supports the view that New Zealand niches are underrepresented for some groups (Webber 2001). Large portunid (swimming) crabs are important predators of benthic invertebrates in most temperate and subtropical estuarine environments (e.g. *Portunus pelagicus* in temperate Australia; *Callinectes sapidus* on the Atlantic coast of North and South America; various species of *Charybdis* and *Thalamita* in northern Asia). There was no comparable species assemblage in New Zealand estuaries, where native swimming crabs are typically much smaller (e.g. *Nectocarcinus antarcticus*, max. carapace width~87 mm; *Liocarcinus corrugatus*, max. width~26 mm) or are uncommon in estuarine habitats (e.g. *Ovalipes catharus*; McLay 1988). If *C. japonica* becomes more widespread and abundant, then it could have large impacts on native estuarine invertebrate assemblages.

Similarly, the composition of New Zealand's algal assemblage is quite different from those of temperate environments in the Northern Hemisphere. For example, New Zealand has many indigenous species of large brown algae (Phaeophyta) in the order Fucales (31 species from 10 indigenous genera), but comparatively few kelps from the order Laminariales (six species from three indigenous genera). Members of the Fucales are distinctive in that the mature plants (sporophytes) develop directly from settled gametes after fertilization, whereas laminarian kelps have two alternating life-history phases – the large kelp stage, which is the sporophyte generation, and a microscopic gametophyte generation, which functions somewhat like the seed bank of terrestrial angiosperms. Most Fucalean species are perennial, occupy the coastal shallow subtidal zone, and many are fertile in winter. Comparatively few species occur in intertidal habitats in New Zealand. All three of the New Zealand laminarian genera also occur in upper subtidal habitats, and none are intertidal in distribution.

The invasive Japanese kelp *Undaria pinnatifida*, which appeared in New Zealand in 1987 (Hay 1990), is a laminarian alga. Its gametophyte stage is able to tolerate relatively high temperatures, allowing it to penetrate higher on the shoreline than many native fucalean or laminarian species (Dieck 1993). *Undaria* has spread to shorelines throughout southern and eastern New Zealand where it reaches maximal abundance in low intertidal areas and shallow subtidal habitats to around 3 m, extending to depths of 18 m. *Undaria* grows on both sheltered and open coastlines, in clear seas or in highly turbid estuarine waters, on natural substrates or in a wide range of artificial habitats including ropes, buoys, boat hulls, and concrete retaining walls (Hay and Villouta 1993). None of the indigenous laminarian kelps or fucalean species is able to occupy a similarly broad range of substrates or conditions.

Although niche opportunity may explain the success of *Undaria* and *Charybdis* in New Zealand, there are other plausible explanations. *Undaria* has also established adventive populations in regions with diverse laminarian floras, such as on the coasts of Argentina, California, France (both Atlantic and Mediterranean coasts), Spain, and Italy. Its success worldwide may reflect both its high reproductive output and broad habitat tolerances, features which have been selected for during decades of mariculture in Korea, Japan, and China (Hay and Villouta 1993).

Nonindigenous swimming crabs have also invaded elsewhere. *Charybdis helleri*, a close relative of *C. japonica*, recently established populations in Colombia, Cuba, Venezuela, Florida and Brazil, regions which have rich native faunas of swimming crabs (Dineen et al. 2001). Along the Brazilian coast where *C. helleri* has spread, there are up to 20 species of portunid crab, at least four of which occupy similar habitats (Mantelatto and Garcia 2001). The success of *C. helleri* in the Western Atlantic has been attributed to several life-history traits, including a relatively long larval life, short generation time, and the ability of females to store sperm and produce multiple broods from a single insemination. The expanding global range of *C. helle*ri may reflect the tendency for pre-moult and moulting crabs to seek shelter in the crevices of ocean-going ships. In 2000, a single specimen of *C. helleri* was recovered from the sea chest of a fishing vessel which was hauled up for maintenance in Nelson, in New Zealand's South Island (Dodgshun and Coutts 2002).

Marine invaders exhibit considerable variability in resource use (Grosholz and Ruiz 1996). Some, like *Undaria* and *Charybdis*, use distinctly different resources from native species in their introduced range (Lohrer et al. 2000; Byers 2002), others compete directly with native counterparts (Byers and Goldwasser 2001), while still others use different resources in different parts of their introduced range (Grosholz and Ruiz 1996). It is unlikely, therefore, that "empty niche" concepts alone will explain geographic variation in the success of marine invaders.

8.6 Are Marine Species of Islands Poor Competitors?

If island species were poorer competitors than continental species, we should expect that mainland species fail less often when introduced to islands than do island species when introduced to continental areas (Simberloff 1995). Though there are few data to test this hypothesis directly, it is notable that some of the most successful marine introductions have been of island (Japanese) species (e.g. oysters and algae) which were transported to the Americas, Europe and Australia for mariculture. A least 45 NIS have established permanent populations as a result of introductions for fisheries or aquaculture on the mainland coast of the USA (Ruiz et al. 2000). In contrast, only five species appear to have persisted in Hawai'i, although at least 18 marine species have been introduced deliberately for fisheries (Coles et al. 1999a, b; Eldredge and Carlton 2002). Similarly, there have been concerted attempts over the last 50 years to introduce a range of aquatic plants and animals to the Pacific

Islands for the purposes of aquaculture or fisheries (Eldredge 1994). More than 100 different species, including at least 38 marine organisms, have been deliberately transported to the islands (excluding Hawai'i), and most have failed. At least 50 % of the marine species either did not establish wild populations or their status in the introduced environment is unknown (45 %), presumably indicating that they are (at best) uncommon.

In New Zealand, deliberate introductions of marine species have also met with limited success. During the late 1800s, attempts were made to establish wild populations of the European herring (*Clupea harengus*), turbot (*Scophthalmus maximus*), lobster (*Homarus gammarus*), European cancer crab (*Cancer pagurus*), Pacific salmon, and three species of cord grass (*Spartina*). Only the Pacific salmon, and cord grass gained a lasting foothold (Cranfield et al. 1998). Many of these attempts appear to have failed because the original stock suffered high mortality during shipment, and relatively few individuals were actually released into the wild. Nevertheless, between 1908 and 1914, around $1x10^6$ lobster (*H. gammarus*) larvae and $12x10^6$ crab (*C. pagurus*) larvae were liberated from hatchery facilities in Dunedin (Thomson 1922). Neither established successfully in New Zealand waters.

No endemic marine species from Hawai'i are known to have established in other parts of the world (Carlton 1987), but a number of endemic New Zealand marine species have been successfully transported elsewhere. During the 1920s, crates of New Zealand oysters (*Tiostrea chilensis*) were shipped alive submerged in water to Tasmania, Australia (Cranfield et al. 1998). At least nine other native New Zealand species (*Astrostole scabra*, *Patiriella regularis*, *Neilo australis*, *Venerupis* (*Paphirus*) *largillierti*, *Maoricolpus roseus*, *Cancer novaezelandiae*, *Halicarcinus innominatus*, *Petrolisthes elongatus*, *Chiton* (*Amaurochiton*) *glaucas*) are thought to have accompanied these shipments and subsequently became established there. However, most have not spread far from the point of introduction, with one notable exception, the New Zealand screw shell, *Maoricolpus roseus*, which occurs in vast beds in northern Bass Strait, south-eastern Australia. This area contains among the most diverse native marine soft-sediment assemblages anywhere in the world (Gray et al. 1998). *Maoricolpus* has also spread to the coasts of eastern Tasmania, Victoria and New South Wales. Other New Zealand species have been recorded elsewhere in Australia (e.g. green mussel, *Perna canaliculus* – Tasmania and South Australia; flat oyster, *Tiostrea chilensis* – Victoria; isopod, *Eurylana arcuata* – New South Wales and South Australia; fish, *Fosterygion varium* – Tasmania; fish, *F. lapillum* – Victoria), the USA (e.g. isopod, *Sphaeroma quoyana*; sea slug, *Philine auriformis*), the UK (barnacle, *Elminius modestus*; flat oyster, *Tiostrea chilensis*), and Europe (macroalga, *Asparagopsis armata*, also a native of Western Australia; amphipod, *Corophium sextonae*; Gosliner 1995; Cranfield et al. 1998; Lockett and Gomon 2001; Reise et al. 2002). The relatively high number of native New Zealand species which have been successfully exported (mostly unintentionally) to other regions of the world is disproportionate to the country's size and importance as a shipping nation, and argues against the idea that island species are necessarily poor competitors.

8.7 Do NIS Have Greater Impacts in Island Environments?

Native species with restricted geographic ranges (such as those occurring on islands) are thought to be more prone to extinction simply because an invader can affect a larger proportion of the total population (Parker et al. 1999). Intuitively, this should also hold true for marine populations, despite the high dispersal capabilities commonly assumed for marine species. Indeed, recent ecological, biogeographic and genetic studies have challenged conventional ideas about the "openness" of marine populations by showing that population subdivision, narrow endemism, and small range sizes are more common than previously suspected (Myers 1997; Benzie 1998; Hooper et al. 2002). In New Zealand, up to 90 % of the known species of marine molluscs, 66 % of shallow water amphipods, 95 % of sponges, 60 % of crabs, 60 % of bryozoans, 35 % of macroalgae and 28 % of reef fishes are endemic (Towns and Ballantine 1993; Myers 1997; Francis and Nelson 2003). The geographic range sizes of many of these species are extremely small. For example, several of the small islands within the New Zealand archipelago, such as the Three Kings (484 ha), Chatham (97,244 ha), Snares (333 ha) and Antipodes (2,095 ha) islands, contain species endemic to individual islands (Nelson 1994). Other isolated island groups such as the Marquesas, Mauritius, Easter Island, Society Islands, St. Helena, Ascension Island and the Galápagos Islands also support many marine species whose distribution is restricted to the island chain (Norse 1993; Roberts et al. 2002). Analogously, movement of tectonic plates and dynamic discontinuities in ocean currents have, over evolutionary time, created sharp barriers to dispersal which limit the range of marine species on continental coastlines. We predict that the impacts of invasions will be most severe in regions where many species have relatively restricted ranges, and that these will not necessarily be confined to oceanic islands. Importantly, the review by Roberts et al. (2002) covered only a few marine groups, these being relatively large and well studied. We expect that other marine groups, many of which have poorer dispersal capabilities than those examined by Roberts et al. (2002), will also contain high numbers of species with very restricted ranges.

Marine invaders are capable of a variety of ecological effects, the most cited of which tend to be competition for resources, habitat alteration, and trophic interactions (predation and herbivory) with native species (Parker et al. 1999; Ruiz et al. 1999). Much information on impacts, however, is anecdotal and relatively few studies have measured the three fundamental dimensions which determine the total impact of an invader: its range, abundance, and per capita impact in the introduced range (Parker et al. 1999). A comparison of the vulnerability of island and continental biotas also requires (at least) contrasting the per capita effects of invaders in each setting independently of the invader's range and abundance. There are few examples of such studies, in part because they are methodologically challenging. Creese et al. (1997) and Crooks (1998) determined the effects of habitat alteration by dense mats of the exotic bivalve *Musculista senhousia* in Auckland, New Zealand, and San Diego, USA respectively. In the New Zealand study, species richness and total abundance of macroinvertebrates were generally lower inside the mats than in adjacent control areas (Creese et al. 1997). Suspension-feeding bivalves were most affected, with an 87 % reduction in average densities inside the mats at one site. In contrast, Crooks (1998) found that macrofaunal densities and species richness were typically greater inside than outside the mats. Although the contrasting impacts found in these two studies could be interpreted as evidence for a regional difference in the effects of *M. senhousia*, a more parsimonious explanation is that they represent different points on a spectrum of possible outcomes in each country. The composition of marine soft-sediment assemblages varies widely in space and time, resulting in localized and idiosyncratic effects of an invader. Greater spatial replication of the observations could have revealed a range of effects of *M. senhousia* mats in each country. Indeed, of the two sites sampled by Creese et al. (1997), one exhibited a deleterious effect of the mats on macrofauna, whereas no significant changes in invertebrate richness or abundance were found at the other. Demonstrating regional differences in the per capita effects of an invader requires tests which show variation between island and continental ecosystems is greater than that within each ecosystem.

8.8 Is Propagule Pressure to Islands Greater than to Mainland Areas?

Many successful marine invaders have strong associations with human transport vectors, and have repeatedly been transported by ships around the globe. Their success outside their native range may be related as much to the frequency with which they are transported ("propagule pressure") as to the susceptibility of recipient ecological assemblages ("niche availability"). Although small island states such as New Zealand, Hawai'i and the Marianas are heavily dependent on maritime trade (Nawadra and Gilbert 2002), they receive comparatively little of the world's shipping traffic. The majority of international shipping occurs within and between continental Europe, North America and East Asia. The International Maritime Organization estimates that around $3x10^9$–$5x10^9$ m^3 of ballast water is transferred internationally each year (2000–2005) from more than 45,000 commercial cargo vessels (Carlton 2001;

GloBallast 2005). Each cubic metre of ballast water can contain up to 1,000 marine organisms from as many as 16 taxa (Smith et al. 1999). Compliance records kept by the New Zealand Ministry of Fisheries show that at least 2.8×10^6 m^3 of ballast water was discharged in New Zealand in 1999 from ca. 2,000 international vessel movements (Inglis 2001). This is twice the volume discharged into the Port of Melbourne, Australia, each year (1.2×10^6 m^3; Walters 1996), but about half the amount discharged within the State of Victoria, Australia (5.8×10^6 m^3; Walters 1996), and less than one-fifth of the volume of ballast discharged into the port of New Orleans, USA, annually ($\sim 13.5 \times 10^6$ m^3; Smith et al. 1999).

Ballast water discharged into New Zealand ports is currently sourced from more than 23 different countries, but the majority comes from the temperate northwest Pacific (Japan, China, Hong Kong, Taiwan, South Korea, 71 %) and Australia (29 %; Inglis 2001). Most of the recent arrivals of NIS to New Zealand also come from these two regions. Since 1960, at least three species of crab, three macroalgae, five species of mollusc, one bryozoan, and one species of fish have arrived in New Zealand from the north-western Pacific (Cranfield et al. 1998). In Hawai'i, the main source regions for NIS appear to be the western and central Indo-Pacific (Eldredge and Carlton 2002), reflecting wartime shipping during the 1940s. Most ballast water entering Hawai'i is now sourced from the west coast of North and Central America, but large numbers of fishing boats, which do not discharge ballast, still enter the islands from Asia (Godwin and Eldredge 2001).

8.9 Conclusions

Nonindigenous species invasions are transforming coastal marine ecosystems throughout the world. While the initial results are alarming, the current status of marine taxonomy, systematics and biogeography do not allow the true extent of the problem, or regional variations in the prevalence and impacts of introduced marine species, to be determined clearly. We have attempted to show why conventional explanations for the susceptibility of island biotas to invasions are not currently amenable to tests in marine environments and, indeed, may not be applicable. The case studies which we have reviewed make it clear that marine invaders can have severe impacts on the ecology of both island and continental biotas. We found no evidence that native marine biotas of islands are any more or less susceptible to invasion, or that they are more severely affected by them, than is the case for continental biotas. Many of these hypotheses await more rigorous treatment as more data become available.

As in other ecosystems, we expect the effects of marine invaders to be of greatest consequence for those native species which have highly restricted distributions and limited capacity to expand their range, and for local populations which are not replenished by recruitment from outside infested areas.

Successful establishment by an NIS outside its natural range is a highly probabilistic outcome which depends upon the coincidence between delivery of the species to the new location and suitable conditions for establishment, including the absence of enemies and the availability of resources. Both the supply characteristics (i.e. propagule pressure) and opportunity for establishment (i.e. niche opportunity) are likely to be highly variable in space and time in marine systems. Although there are few data on rates of transport, the geographic origins of invaders in different parts of the world reflect the predominant shipping routes into each region (Carlton 1996). Species with habitat preferences (e.g. Wonham et al. 2000), environmental tolerances, or life-history strategies which predispose them to transport by human vectors are likely to reach new locations more often than are other organisms.

The composition of marine assemblages and ecological interactions within them are strongly influenced by the rate of supply of new recruits and the frequency of resource-releasing disturbances (Underwood and Fairweather 1989). We believe, therefore, that the most profitable studies of invasion success will be those which vary supply characteristics in combination with resource availability. By shifting the theoretical emphasis away from considering invasion resistance as a static function of species diversity to a more dynamic consideration of resource utilization and availability, we should gain a better understanding of how the frequency of invasion is influenced by temporal and spatial variability in the host environment.

References

Benzie JAH (1998) Genetic structure of marine organisms and SE Asian biogeography. In: Hall R, Holloway JD (eds) Biogeography and geological evolution of SE Asia. Backhuys, Leiden, The Netherlands, pp 197–209

Byers JE (2002) Physical habitat attribute mediates biotic resistance to non-indigenous species invasion. Oecologia 130:146–156

Byers JE, Goldwasser L (2001) Exposing the mechanism and timing of impact of non-indigenous species on native species. Ecology 82:1330–1343

Carlton JT (1987) Patterns of transoceanic marine biological invasions in the Pacific Ocean. Bull Mar Sci 41:452–465

Carlton JT (1996) Pattern, process, and prediction in marine invasion ecology. Biol Conserv 78:97–106

Carlton JT (2001) Introduced species in U.S. coastal waters: environmental impacts and management priorities. Pew Oceans Commission, Arlington, Virginia

Cohen AN, Carlton JT (1995) Nonindigenous aquatic species in a United States estuary: a case study of the biological invasions of the San Francisco Bay and Delta. NTIS Rep no PB96-166525 to the US Fish and Wildlife Service, Washington, DC

Cohen AN, Carlton JT (1998) Accelerating invasion rate in a highly invaded estuary. Science 279:555–558

Coles SL, Eldredge LG (2002) Nonindigenous species introductions on coral reefs: a need for information. Pac Sci 56:191–209

Coles SL, DeFelice RC, Eldredge LG, Carlton JT (1999a) Historical and recent introductions of non-indigenous marine species into Pearl Harbour, Oahu, Hawai'ian Islands. Mar Biol 135:147–158

Coles SL, DeFelice RC, Eldredge LG, Godwin LS, Pyle RL, Suzumoto A (1999b) Non-indigenous marine species introductions in the harbors of the south and west shores of Oahu, Hawai'i. Bishop Museum Tech Rep no 15. Bishop Museum, Honolulu, Hawai'i

Coles SL, DeFelice RC, Eldredge LG (2002) Nonindigenous species in Kane'ohe Bay, O'ahu, Hawai'i. Bishop Museum Tech Rep no 24. Bishop Museum, Honolulu, Hawai'i

Cranfield H J, Gordon DP, Willan RC, Marshall BA, Battershill CN, Francis MP, Nelson WA, Glasby CJ, Read GB (1998) Adventive marine species in New Zealand. NIWA Tech Rep 34. Wellington, New Zealand

Creese R, Hooker S, de Luca S, Wharton Y (1997) Ecology and environmental impact of *Musculista senhousia* (Mollusca: Bivalvia: Mytilidae) in Tamaki Estuary, Auckland. NZ J Mar Freshwater Res 31:225–236

Crooks JA (1998) Habitat alteration and community-level effects of an exotic mussel, *Musculista senhousia*. Mar Ecol Prog Ser 162:137–152

Davis MA, Grime JP, Thompson K (2000) Fluctuating resources in plant communities: a general theory of invasibility. J Ecol 88:528–534

DeFelice RC, Eldredge LG, Carlton JT (2001) Nonindigenous invertebrates. In: Eldredge LG, Smith C (coordinators) Guidebook to the introduced marine species in Hawaiian waters. Bishop Museum Tech Rep 21. Bishop Museum and University of Hawai'i, Honolulu, Hawai'i

Dell RK (1968) Composition and distribution of the New Zealand Brachyuran fauna. Trans R Soc NZ (Zool) 10:225–240

Dieck TI (1993) Temperature tolerance and survival in darkness of kelp gametophytes (Laminariales, Phaeophyta) – ecological and biogeographical implications. Mar Ecol Prog Ser 100:253–264

Dineen JF, Clark PF, Hines AH, Reed SA, Walton HP (2001) Life history, larval description, and natural history of *Charybdis hellerii* (Decapoda, Brachyura, Portunidae), an invasive crab in the western Atlantic. J Crustacean Biol 21:774–805

Dodgshun T, Coutts A (2002) Ships' sea chests: a "side door" for marine pests? Cawthron Institute, Nelson, New Zealand, http://www.cawthron.org.nz/Assets/seachest.pdf

Eldredge LG (1994) Perspectives in Aquatic Exotic Species Management in the Pacific Islands, vol I. Introduction of commercially significant aquatic organisms to the Pacific islands. South Pacific Commission, Noumea, New Caledonia

Eldredge LG, Carlton JT (2002) Hawai'ian marine bioinvasions: a preliminary assessment. Pac Sci 56:211–212

Elton CS (1958) The ecology of invasions by animals and plants. Methuen, London

Francis M, Nelson WA (2003) Biogeography. In: Andrew NL, Francis M (eds) The living reef: the ecology of New Zealand's rocky reefs. Craig Potton, Nelson, pp 32–37

Galil BS (2000) A sea under siege – alien species in the Mediterranean. Biol Invasions 2:177–186

GloBallast (2005) The problem. Global Ballast Water Management Programme, 2000–2005. International Maritime Organization, London, UK, http://globallast.imo.org/index.asp?page=problem.htm&menu=true

Godwin LS, Eldredge LG (2001) South Oahu Marine Invasions Shipping Study (SOMISS). Bishop Museum Tech Rep no 20. Report to the State of Hawai'i Department of Land and Natural Resources, Division of Aquatic Resources

Gomoiu M-T, Alexandrov B, Shadrin N, Zaitsev Y (2002) The Black Sea – a recipient, donor and transit area for alien species. In: Leppäkoski E, Gollasch S, Olenin S (eds) Invasive aquatic species of Europe – distribution, impacts and management. Kluwer, Dordrecht, The Netherlands, pp 341–350

Gordon DP (2001) Marine biodiversity. Alpha 108. The Royal Society of New Zealand, Wellington, New Zealand

Gosliner TM (1995) Introduction and spread of *Philine auriformis* (Gastropoda: Opisthobranchia) from New Zealand to San Francisco Bay and Bodega Harbor. Mar Biol (Berl) 122:249–255

Gray JS, Poore GCB, Ugland KI, Wilson RS, Olsgard F, Johannessen Ø (1998) Coastal and deep-sea benthic diversities compared. Mar Ecol Prog Ser 159:97–103

Grosholz ED, Ruiz GM (1996) Predicting the impact of introduced marine species: lessons from the multiple invasions of the European green crab *Carcinus maenas*. Biol Conserv 78:59–66

Hay CH (1990) The dispersal of sporophytes of *Undaria pinnatifida* by coastal shipping in New Zealand, and implications for further dispersal of *Undaria* in France. Br Phycol J 25:301–313

Hay CH, Villouta E (1993) Seasonality of the adventive Asian kelp *Undaria pinnatifida* in New Zealand. Bot Mar 36:461–476

Hewitt CL (2002) Distribution and biodiversity of Australian tropical marine bioinvasions. Pac Sci 56:213–222

Hewitt CL, Campbell ML, Thresher RE, Martin RB (eds) (1999) Marine biological invasions of Port Philip Bay, Victoria. Tech Rep no 20. Centre for Research on Introduced Marine Pests, Hobart, Tasmania

Hooper JNA, Kennedy JA, Quinn RJ (2002) Biodiversity 'hotspots', patterns of richness and endemism, and taxonomic affinities of tropical Australian sponges (Porifera). Biodivers Conserv 11:851–885

Inglis GJ (2001) Criteria for selecting New Zealand ports and other points of entry that have a high risk of invasion by new exotic marine organisms. Report for Ministry of Fisheries Research Project 2BS2000/04. National Institute of Water and Atmospheric Research, Wellington, New Zealand

Kott P (2002) A complex didemnid ascidian from Whangamata, New Zealand. J Mar Biol Assoc UK 82:625–628

Leppäkoski E, Gollasch S, Gruszka P, Ojaveer H, Olenin S, Panov V (2002) The Baltic – a sea of invaders. Can J Fish Aquat Sci 59:1175–1188

Lockett MM, Gomon MF (2001) Ship mediated fish invasions in Australia: two new introductions and a consideration of two previous invasions. Biol Invasions 3:187–192

Lohrer AM, Whitlatch RB, Wada K, Fukui Y (2000) Home and away: comparisons of resource use by a marine species in native and invaded habitats. Biol Invasions 2:41–57

Mantelatto FLM, Garcia RB (2001) Biological aspects of the nonindigenous portunid crab *Charybdis hellerii* in the western tropical South Atlantic. Bull Mar Sci 68:469–477

McLay CL (1988) Brachyura and crab-like Anomura of New Zealand. University of Auckland, Auckland, New Zealand, Leigh Lab Bull 22:1–463

Minchin D, Eno NC (2002) Exotics of coastal and inland waters of Ireland and Britain. In: Leppäkoski E, Gollasch S, Olenin S (eds) Invasive aquatic species of Europe – distribution, impacts and management. Kluwer, Dordrecht, The Netherlands, pp 267–275

Myers AA (1997) Biogeographic barriers and the development of marine biodiversity. Estuarine Coastal Shelf Sci 44:241–248

Nawadra S, Gilbert T (2002) Risk of marine spills in the Pacific Islands region and its evolving response arrangements. In: Proc PACPOL SPILLCON 2002, www.spillcon.com/2002

Nelson WA (1994) Distribution of macroalgae in New Zealand – an archipelago in space and time. Bot Mar 37:221–233

Norse EA (1993) Global marine biological diversity: a strategy for building conservation into decision making. Island Press, Washington, DC

Occhipinti Ambrogi A (2000) Biotic invasions in a Mediterranean lagoon. Biol Invasions 2:165–176

Parker IM, Simberloff D, Lonsdale WM, Goodell K, Wonham M, Kareiva PM, Williamson MH, von Holle B, Moyle PB, Byers JE, Goldwasser L (1999) Impact: toward a framework for understanding the ecological effects of invaders. Biol Invasions 1:3–19

Paulay G, Kirkendale L, Lambert G, Meyer C (2002) Anthropogenic biotic interchange in a coral reef ecosystem: a case study from Guam. Pac Sci 56:403–422

Reise K, Gollasch S, Wolff WJ (1999) Introduced marine species of the North Sea coasts. Helgol Meeresunters 52:219–234

Reise K, Gollasch S, Wolff WJ (2002) Introduced marine species of the North Sea coasts. In: Leppäkoski E, Gollasch S, Olenin S (eds) Invasive aquatic species of Europe – distribution, impacts and management. Kluwer, Dordrecht, The Netherlands, pp 260–266

Roberts CM, McClean CJ, Veron JEN, Hawkins JP, Allen GR, McAllister DE, Mittermeier CG, Schueler FW, Spalding M, Wells F, Vynne C, Werner TB (2002) Marine biodiversity hotspots and conservation priorities for tropical reefs. Science 295:1280–1284

Ruiz GM, Carlton JT, Grosholz ED, Hines AH (1997) Global invasions of marine and estuarine habitats by non-indigenous species: mechanisms, extent, and consequences. Am Zool 37:621–632

Ruiz GM, Fofonoff P, Hines AH, Grosholz ED (1999) Non-indigenous species as stressors in estuarine and marine communities: assessing invasion impacts and interactions. Limnol Oceanogr 44:950–972

Ruiz GM, Fofonoff PW, Carlton JT, Wonham MJ, Hines AH (2000) Invasion of coastal marine communities in North America: apparent patterns, processes, and biases. Annu Rev Ecol Syst 31:481–531

Shea K, Chesson P (2002) Community ecology theory as a framework for biological invasions. Trends Ecol Evol 17:170–176

Simberloff D (1995) Why do introduced species appear to devastate islands more than mainland areas? Pac Sci 49:87–97

Smith LD, Wonham MJ, McCann LD, Ruiz GM, Hines AH, Carlton JT (1999) Invasion pressure to a ballast-flooded estuary and an assessment of inoculant survival. Biol Invasions 1:67–87

Stachowicz JJ, Fried H, Osman RW, Whitlatch RB (2002) Biodiversity, invasion resistance, and marine ecosystem function: reconciling pattern and process. Ecology 83:2575–2590

Thomson GM (1922) The naturalisation of animals and plants in New Zealand. Cambridge Univ Press, Cambridge

Towns DR, Ballantine WJ (1993) Conservation and restoration of New Zealand island ecosystems. Trends Ecol Evol 8:452–457

Tricklebank KA, Nesbit G, Read G, Acosta H (2001) The distribution of the parchment tubeworm *Chaetopterus* off the coast of northeastern New Zealand. Final Research Report to the Ministry of Fisheries. Auckland Uniservices, Auckland, New Zealand

Underwood AJ, Fairweather PG (1989) Supply-side ecology and benthic marine assemblages. Trends Ecol Evol 4:16–20

Walters S (1996) Ballast water, hull fouling and exotic marine organism introductions via ships – a Victorian study. Publication 494. Environment Protection Authority (Victoria), Melbourne, VIC, Australia

Warwick RM (1996) Marine biodiversity: a selection of papers presented at the conference "Marine Biodiversity: Causes and Consequences", York, UK, 30 Aug–2 Sept 1994. J Exp Mar Biol Ecol 202:IX–X
Webber R (2001) Space invaders: crabs that turn up in New Zealand unannounced. Seafood NZ (Nov):80–84
Williamson M (1996) Biological Invasions. Chapman and Hall, London
Wonham MJ, Carlton JT, Ruiz GM, Smith LD (2000) Fish and ships: relating dispersal frequency to success in biological invasions. Mar Biol 136:1111–1121

9 Factors Affecting the Release, Establishment and Spread of Introduced Birds in New Zealand

R.P. Duncan, T.M. Blackburn and P. Cassey

> *"it was not good work bringing those birds out here; they eat all the potatoes and the oats; they are not good birds to bring out ... were there not plenty of good birds in New Zealand that eat no man's food?"*
> Te Whiti of Parihaka in 1883

9.1 Introduction

Attempts to introduce non-native birds to New Zealand began with the arrival of the first Europeans in 1773, when Captain James Cook liberated geese (*Anser anser*) in Dusky Sound. Not until the mid-1800s, however, did bird introductions begin in earnest with the setting up of acclimatization societies in New Zealand whose major aim was "the introduction, acclimatization and domestication of all animals, birds, fishes, and plants, whether useful or ornamental" (McDowall 1994). The acclimatization societies kept records of the birds they introduced and the outcome of those introductions which, along with information from other sources, was compiled into a book by George Thomson in 1922. Thomson's book, along with other sources, provide a unique record of the history and outcome of bird introductions to New Zealand. Because this record appears reasonably complete and frequently includes details such as the location of individual introductions and the numbers of birds released, these data have been used extensively to address questions about the factors determining whether introductions succeed to establish wild populations (see references in Table 9.1). Given the difficulties of conducting experiments at an appropriate scale to address such questions, these detailed records provide a rare opportunity to gain insight into the mechanisms underlying the successful establishment and spread (i.e. invasion) of species into a novel environment.

Ecological Studies, Vol. 186
R.B. Allen and W.G. Lee (Eds.)
Biological Invasions in New Zealand

Table 9.1 Studies which have used the record of New Zealand bird introductions to quantitatively address questions concerning factors affecting establishment success or spread following introduction[a]

Reference	Taxonomic group of interest	Species used in analysis (*n*)
Veltman et al. (1996)	All species	79
Duncan (1997)	Passeriformes	42
Green (1997)	All species	47
Sorci et al. (1998)	All species	79
Duncan et al. (1999)	All established species	34
Legendre et al. (1999)	Passeriformes	31
Sol and Lefebvre (2000)	All species	39
Blackburn et al. (2001)	Established passeriformes	12
Cassey (2001)	All species	118
	All established species	31
Moulton et al. (2001)	Galliformes	20
Duncan and Blackburn (2002)	Galliformes	20
Duncan and Forsyth (2005)	Passeriformes	42

[a]We identified 120 bird species which have been introduced to New Zealand, of which 34 succeeded in establishing. For various reasons, most earlier studies considered only a subset of these species

In this chapter, we review what is known about factors explaining the release, successful establishment and spread of introduced bird species in New Zealand. As a framework for our review, we recognize four stages through which a species must progress in order to successfully invade a new location (Williamson 1996; Kolar and Lodge 2001; Duncan et al. 2003). First, the species must be transported to its new location. Second, the species must be released (or escape) into the environment. Third, the species must establish a self-sustaining wild population. Fourth, to become widespread, that population has to increase in size and spread from its point of establishment. In this chapter, we consider variables likely to explain patterns in the outcome of bird introductions through the last three stages in this invasion process (release, establishment and spread). We do not consider the first stage in the process, and therefore exclude five species transported to New Zealand by acclimatization societies, but which records indicate were not released (*Anas falcata*, *Anser cygnoides*, *Chrysolophus pictus*, *Lagopus mutus*, *Lophura nycthemera*, Thomson 1922), or species which have been transported to New Zealand only to be kept as cage birds. While we will draw extensively on the results of previous studies, most have focused on only a subset of the 120 bird species introduced to New Zealand (Table 9.1).

9.2 Data Sources and Analysis

To identify variables which might explain the patterns of release, establishment and spread, we compiled a list of the 120 bird species which have been recorded as introduced (i.e. transported and released) to New Zealand, based primarily on the information in Thomson (1922), supplemented with data from other sources (Hutton 1871; Buller 1888; Drummond 1907; Williams 1950, 1969; Westerskov 1953; Lamb 1964; Wellwood 1968; Heather and Robertson 1996). Unless historical records stated otherwise, we assumed that all birds introduced for acclimatization purposes were released. Our list of introduced birds differs from previous lists in several ways. First, as far as we can tell, this list is comprehensive and includes more species than have been identified in previous lists (Table 9.1). Second, our list excludes several species which appear in previous lists but for which we could find no evidence of release. For example, Veltman et al. (1996) and Sorci et al. (1998) list the nightingale (*Luscinia megarhynchos*) as having failed to establish following introduction to New Zealand. Thomson (1922) states that all of the nightingales transported to New Zealand died on the journey out, except one which died soon after arriving. We could find no other record confirming the release of this species. We excluded *Pycnonotus cafer*, which established a population in Auckland but which was subsequently eradicated (Turbott 1990).

We obtained an estimate of introduction effort (the total number of individuals released in New Zealand) for 112 of the 120 introduced species. These data varied in quality – for many species, the number of individuals released was recorded whereas for others, only a qualitative estimate was available for some locations (e.g. a few, several). We used this information to classify species into one of three broad effort categories (a total of ≤20, 21–100 or >100 individuals released). These categories result in a loss of information for some species but have the advantage that, given the data available, we could classify all species into categories. In addition, for all species we obtained the following variables: native latitudinal range (degrees), body mass (g), clutch size, incubation period (days), migratory status (migratory, partially migratory, or sedentary) and plumage dichromatism (monochromatic or dichromatic, the latter if the sexes are recognizable on the basis of plumage differences). Relative brain size (the residual deviation for a species from an interspecific regression of brain mass on body mass) was available for 46 species (Sol and Lefebvre 2000; Sol et al. 2002), and year of first release for 113 species.

Our analyses focus on which of the above variables can explain patterns in the release and establishment of species; variables which account for the pattern of spread of introduced birds in New Zealand have previously been identified (Duncan et al. 1999; Cassey 2001). We used generalized linear models (GLMs), specifying a logit link function and binomial error distribution, to model establishment success or failure (a binary response variable) as a function of the explanatory variables listed above. Closely related species tend to share traits in common, some of which may affect the probability of establishment. If this is the case, then establishment outcomes may be phylogenetically correlated: closely related species are more likely to share the same outcome than are less closely related species, violating the assumption that each observation represents an independent data point. To control for possible non-independence due to phylogeny, we also modelled establishment success or failure using generalized estimating equations (GEEs), an extension of GLMs, to include situations where errors are likely to be correlated (Diggle et al. 1994). We measured phylogenetic correlation for each species pair as the distance from the last common ancestor to the root of the phylogenetic tree divided by the distance from the root to the tip of the tree (i.e. that proportion of the total evolutionary history of the species pair which is shared in common; Martins and Hansen 1997; Garland and Ives 2000). Distance was measured as the number of nodes separating species on a complete phylogeny of the world's birds. This phylogeny was derived from Sibley and Ahlquist (1990), supplemented with information in Sibley and Monroe (1990). GLMs and GEEs were fitted using PROC GENMOD in SAS (1990), with the matrix of between-species phylogenetic distances included as a user-specified correlation matrix when fitting GEEs.

9.3 Factors Affecting Release

Birds which have been released in new locations around the world are not a random subset of the world's bird species (Lockwood 1999; Lockwood et al. 2000; Blackburn and Duncan 2001b). Rather, there is a strong bias towards species in certain taxonomic groups (reflecting the "types" of birds which humans desired for introduction) and species from certain geographic regions (often reflecting the origin of the settlers who transported birds to new locations). Of the 25 families of birds introduced to New Zealand, two families (Anatidae, ducks; Phasianidae, gamebirds) are significantly over-represented relative to their availability in the global avifauna (Table 9.2). Over one-third of the species introduced are in these two families. Species in these two families are also significantly over-represented in bird introductions globally (Lockwood 1999; Blackburn and Duncan 2001b), reflecting repeated attempts by European settlers to establish waterfowl and gamebird populations for hunting in the areas they colonized.

Over half of the bird species introduced to New Zealand originate from Palaearctic and Australasian geographic regions (Table 9.3). This mirrors the pattern for global bird introductions (Blackburn and Duncan 2001b), where settlers at new locations tended to introduce birds from their home region (in this case, western Europe) and birds found in nearby regions (in this case, Australasia), places with which trade and transport would have been particularly important and frequent.

Nevertheless, only a small proportion of the bird species native to a particular geographic region were transported and released in new locations. The majority of early settlers to New Zealand came from Britain, yet of the 214

Table 9.2 Species introduced and successfully established in New Zealand, by family

Family	Species introduced (*n*)	Species successfully established (*n*)	Species in family[a] (*n*)	*P*[b]
Anatidae	23	4	148	*<0.001*
Phasianidae	21	5	176	*<0.001*
Casuariidae	2		4	0.003
Columbidae	10	3	313	0.005
Odontophoridae	3	1	31	0.007
Pteroclidae	2	–	16	0.016
Passeridae	10	2	388	0.021
Numididae	1	1	6	0.079
Charadriidae	3	–	88	0.094
Fringillidae	17	7	995	0.106
Turnicidae	1		17	0.192
Tytonidae	1	–	17	0.193
Megapodiidae	1	–	19	0.204
Maluridae	1	–	26	0.278
Strigidae	3	1	156	0.297
Alaudidae	2	1	91	0.312
Cracidae	1	–	50	0.471
Halcyonidae	1	1	61	0.530
Paridae	1	–	65	0.556
Sturnidae	2	2	148	0.572
Meliphagidae	2	–	182	0.664
Psittacidae	3	2	360	0.834
Muscicapidae	3	2	452	0.919
Corvidae	4	2	650	0.966
Sylviidae	2	–	560	0.997

[a] Total number of species in each introduced family follows Sibley and Monroe (1990)

[b] The probability of observing as many or more introductions from that family, given the total number of species in the family and the number of species introduced to New Zealand. Italics indicate probabilities which are significantly (α=0.05) lower than expected once a sequential Bonferroni correction for multiple tests has been applied. *P* values were calculated by selecting 120 species at random, and without replacement, from the total global avifauna (9,702 species) and summing the number of these randomly chosen species in each family. This simulation process was repeated 4,000 times, and *P* was calculated as the proportion of simulations in which the simulated number of introduced species in the family exceeded the observed number

Table 9.3 Species introduced and successfully established in New Zealand, by biogeographic region of origin

Biogeographic region of origin[a]	Species introduced (*n*)	Species successfully established (*n*)
Palaearctic	41	14
Australasia	30	6
Nearctic	14	3
South-east Asia	12	4
Holarctic	8	2
Afrotropics	6	2
Multi-regional	6	3
Central/South America	2	0
Pacific	1	0

[a] Region definitions follow Blackburn and Duncan (2001b)

breeding bird species in Britain, only 37 were released in New Zealand. These 37 species are a non-random subset of the British birds, being strongly biased towards species which are widespread, abundant and resident in Britain (Blackburn and Duncan 2001b). Hence, British settlers caught, transported and released birds which were common and readily available in their homeland.

Overall, a strong set of historical and cultural factors appear to have influenced the birds released in New Zealand, with European settlers desiring certain types of bird species for introduction. It seems likely that similar patterns in selectivity will hold for other groups of non-native species purposefully introduced to New Zealand.

9.4 Factors Affecting Establishment

In all, 34 of the 120 bird species in our data set successfully established a self-sustaining wild population following release in New Zealand. We consider four categories of variables which might explain why these 34 species succeeded whereas the remainder failed. These are categories which have dominated previous attempts to explain establishment success in many taxa and regions, and not only for New Zealand birds.

9.4.1 Introduction Effort

Small populations are particularly vulnerable to extinction due to demographic and environmental stochasticity, with the probability of extinction declining as population size increases (Richter-Dyn and Goel 1972). For introduced species, we would therefore expect that the more individuals liberated during a release event, the lower the chance of stochastic population extinction, and hence the greater the chance of successful population establishment. Likewise, there is a greater chance of at least one population establishing if a species is subject to a greater number of release events of a given size (Crawley 1989). Several studies show that for birds introduced to New Zealand, both the total number of birds released and the number of separate release events can serve as particularly strong predictors of establishment success (Veltman et al. 1996; Duncan 1997; Green 1997; see Table 9.4). Most species introduced to New Zealand were released in low numbers, which is unsurprising considering the large distances the birds had to travel by ship. In many cases, the introduction of species was poorly planned, and it was not unheard of for entire shipments to die during transport (Thomson 1922). Fewer than 20 individuals were released for 62 of the 112 species for which we obtained data on introduction effort, of which only one species established (cirl bunting, *Emberiza cirlus*). Of the 23 species with 21–100 individuals released, seven established, and of the 27 species with more than 100 individuals released, 20 established. This result for New Zealand birds highlights a consistent pattern: introduction effort is a key factor explaining the outcome of historical introductions in many taxa (Newsome and Noble 1986; Pimm 1991; Hopper and Roush 1993; Duncan et al. 2001; Forsyth and Duncan 2001). The consistency of this relationship implies that it is likely to be a general rule for bird introductions, although the effort required to establish a species may vary between taxa and regions.

Given the importance of introduction effort in explaining establishment success, what factors might determine how many individuals of each species were released? At least for those species introduced to New Zealand from Britain, species which were more abundant in Britain were released in New Zealand more often and in greater numbers (Blackburn and Duncan 2001b). These common British birds were also more likely to establish in New Zealand, at least in part because the more common species tended to be released in greater numbers. Again, the reasons for this are likely to be because common British species were familiar and desired for introduction, and were more easily caught and transported in greater numbers.

In addition, however, species may differ in their ability to survive long-distance transport in ships or in their condition upon arrival. Cassey (2001) found that the greater the distance between the source and the location of introduction for birds, the lower the probability of establishment, and suggested that birds with a greater distance to travel may have arrived in worse condition. For birds introduced to New Zealand, we lack sufficient data on the numbers which died in transit to test the hypothesis that species differed in their ability to tolerate long-distance travel, but differential survival in transport may have systematically biased the number of individuals of each species available for release.

Table 9.4 Variables included to explain establishment success for birds introduced to New Zealand[a]

Variable	Category	n[b]	Across-species[c]		Phylogeny controlled[d]	
			Estimate	P	Estimate	P
Introduction effort						
No. individuals	≤20	112	–5.16	<0.001	–3.31	<0.001
Released	21–100		–1.88	0.003	–1.90	<0.001
	>100		0			
Environmental match						
Latitudinal overlap	No	120	–1.44	0.062	–1.32	0.025
	Yes		0			
Log_{10}latitudinal range		120	2.76	0.010	1.68	<0.001
Life history traits						
Log_{10}body mass		120	–0.03	0.911	NC[e]	
Log_{10}clutch size		120	–0.86	0.320	–2.61	0.118
Log_{10}incubation		120	–1.60	0.220	–2.41	0.254
Plumage	No	120	0.38	0.355	0.02	0.907
Dichromatism	Yes		0			
Migration	Migratory	120	–1.79	0.094	–0.76	0.078
	Partial		0.68	0.124	0.46	0.049
	Sedentary		0		0	
Relative brain size		46	0.01	0.934	0.28	0.503
Time of introduction						
Year of 1st release		113	-0.02	0.152	-0.01	0.014

[a] The response variable is binary (a species established in New Zealand or not)
[b] Number of species included in a model
[c] Results of fitting a generalized linear model, specifying a logit link function and binomial error distribution, with the explanatory variable in the left-hand column
[d] Results of fitting a generalized estimating equation model, with the same link and error functions and a working correlation matrix specifying the degree of phylogenetic relatedness among species, measured as the number of nodes along the longest branch separating species on a phylogenetic tree containing all of the world's bird species. Negative parameter estimates further from zero indicate a lower probability of establishment
[e] Model failed to converge

9.4.2 Environmental Match

We would predict that species introduced to a new environment which closely matches that in their natural range should have a greater probability of establishing than species introduced into a very different environment (Mack 1996; Williamson and Fitter 1996). When environmental features such as climate and habitat are closely matched, introduced species are more likely to posses the necessary pre-adaptations for survival and reproduction in the new environment (Daehler and Strong 1993). This hypothesis is not easy to test because it is difficult to quantify both a species' environmental niche and its match to the location of introduction, although the few attempts to do so suggest this is an important factor. For bird species introduced to Australia, those with a closer match between the climate in their natural range and the climate in Australia were significantly more likely to establish (Duncan et al. 2001). For global bird introductions, species introduced to latitudes closer to the midpoint of their natural range and species introduced to locations in their native biogeographic region were also more likely to establish (Blackburn and Duncan 2001a), perhaps because locations at similar latitudes within the same biogeographic region are more likely to share similar climate and habitat than locations elsewhere.

For birds introduced to New Zealand, we assessed whether the natural latitudinal range of each species included the latitudinal range of the three main islands of New Zealand (North, South and Stewart islands; 34–47°S), without reference to hemisphere. Species whose natural range includes latitudes between 34 and 47°S were significantly more likely to establish following introduction to New Zealand than species whose natural range falls outside these latitudes, although the relationship is weak (Table 9.4). Assuming that latitude acts as a surrogate for some broad features of climate, this provides limited support for the hypothesis that a closer environmental match increases the chance of establishment (see also Green 1997).

In a variety of taxa, species with larger geographic ranges are consistently found to have a greater probability of establishing following introduction (Moulton and Pimm 1986; Rejmánek 1996; Reichard and Hamilton 1997; Blackburn and Duncan 2001a). Species with larger geographic ranges have more chance of being captured and transported in greater numbers, which could account, at least in part, for their higher probability of establishment (Forcella and Wood 1984; Forcella et al. 1986; Goodwin et al. 1999; see above). In addition, however, species occupying larger geographic ranges are more likely to be pre-adapted to a wider range of environments (Stevens 1989), and are therefore more likely to encounter suitable conditions when released at a new location than species occupying smaller ranges. For example, Cassey (2002) found that the probability of successful establishment was greater for species which occupied a greater number of habitat types. For bird species introduced to New Zealand, natural latitudinal range is a highly significant predictor of establishment success (Table 9.4).

9.4.3 Life History Traits

Species differ in many characteristics which could influence their probability of establishing a population following release in a novel environment. For example, characteristics which reduce vulnerability to extinction due to demographic and environmental stochasticity may increase establishment success (Pimm 1991). Thus, species with rapid rates of population increase, or long-lived species which are relatively unaffected by environmental fluctuations, may have a greater probability of establishing.

Results from studies which have examined life history traits among introduced birds suggest that species are more likely to establish if they are sedentary (Veltman et al. 1996; Sol and Lefebvre 2000; Cassey 2002), have larger body masses (Green 1997; Duncan et al. 2001), larger clutch sizes (Green 1997), larger relative brain size (Sol and Lefebvre 2000; Sol et al. 2002, 2005) and are habitat generalists (Cassey et al. 2004). Sexually dichromatic species are less likely than monochromatic species to have established (McLain et al. 1995, 1999; Green 1997; Cassey 2002), even when the number of individuals introduced is statistically controlled (Sorci et al. 1998; but see Cassey et al. 2004). Legendre et al. (1999) found that monogamous mating led to a higher extinction risk, and that demographic uncertainty imposed a greater probability of introduction failure on populations of short-lived species.

Of the traits we examined, only migratory tendency explained significant variation in establishment success for birds introduced to New Zealand (Table 9.4). A number of additional traits, which have been shown in previous studies to significantly predict establishment success in birds introduced to New Zealand, were not significant predictors in this study. The most likely reason for this inconsistency is that the lists of species used in previous studies were incomplete (see Table 9.1). Overall, life history and demographic traits were weak predictors of establishment success. It appears that, in New Zealand, these characteristics have had much less effect on establishment probability than introduction effort and the degree to which the new environment is suitable for the introduced species. Indeed, when we fitted a multivariate model and eliminated non-significant predictor variables by backward selection, only introduction effort and latitudinal range were retained in the model as significant predictors of establishment.

9.4.4 Competition

Several studies have suggested that interspecific competition is an important determinant of establishment success for birds introduced to islands (e.g. Moulton and Pimm 1983, 1986, 1987; Moulton 1993; Lockwood and Moulton 1994; Brooke et al. 1995; Moulton et al. 1996). Two patterns in introduction data have been used to infer the importance of competition. First, bird species introduced later tend to have a lower rate of establishment because later introductions face a higher level of competition from already established species. Second, interspecific competition should be greater between morphologically similar species, and hence establishment success should be higher for species introduced to islands where no morphologically similar species are present.

For passerine birds introduced to New Zealand, there is a significant order of introduction effect: later introductions are significantly less likely to establish (Duncan 1997). This pattern holds for all bird species introduced to New Zealand, but is much weaker and significant only after controlling for phylogeny (Table 9.4). For passerine birds, this order of introduction effect is confounded with introduction effort because less effort was expended in later introductions (Duncan 1997; this is true also for the entire data set). The median year of introduction for releases involving ≤20 individuals is 1871, for 21–100 individuals it is 1868, and for >100 individuals it is 1864 (Kruskal-Wallis χ^2=15.0, P<0.001). Hence, the lower establishment rate of later introductions could be due to the diminished effort put into introducing these species. For passerine birds, however, variation in introduction effort fails to explain all of the order of introduction effect, because introduction effort remains a significant predictor of establishment success once effort is controlled for (Duncan 1997). Nevertheless, this residual effect could be due to factors other than competition which vary throughout the period of introductions. In particular, establishment success may be influenced by level of predation (e.g. Lovegrove 1996), which almost certainly increased during the period of bird introductions to New Zealand as introduced predators established and spread throughout the country (King 1990; Holdaway 1999).

For a subset of the gamebirds introduced to New Zealand, there is a significant pattern of morphological overdispersion: successfully established species are less similar to each other than would be expected if introductions had succeeded or failed at random (Moulton et al. 2001). Moulton et al. (2001) attributed this pattern to competition between introduced gamebird species. However, a subsequent reassessment of the data on gamebird introductions showed that the observed pattern of morphological overdispersion could not be due to interspecific competition because gamebird species were rarely introduced to the same location at the same time (Duncan and Blackburn 2002). This implies that factors other than competition can generate a pattern of morphological overdispersion among successfully introduced birds (see also Simberloff and Boecklen 1991).

To untangle the importance of competition from that of other confounding factors, Duncan and Forsyth (2005) investigated how three factors (viz. introduction effort, the abundance of a competitor, and the strength of competition) interact to determine the probability that an introduced species will establish in the presence of a competitor, using a simple mathematical model. They showed that these factors interact such that the importance of competition varies depending on the relative abundances of the introduced and competitor species, and on introduction effort. Competition should be of much less importance in affecting establishment when a species is introduced in small numbers (because stochastic processes dominate) or when the resident competitor occurs at low abundance. The effects of competition should be most apparent when both the introduced and competitor species occur at high abundance. These model predictions were strongly supported by the data for passerine introductions to New Zealand, implying that competition was an important factor structuring this assemblage, having accounted for the timing and effort put into introductions (Duncan and Forsyth 2005).

9.5 Factors Affecting Spread

Once a species has established a viable population in a new environment, that population may persist only around the site of release, or it may expand and spread more widely. Characteristics which have been suggested to influence the spread of introduced species include the time available for population expansion to occur, the population growth rate, the extent of suitable environment, and dispersal tendency or ability (Duncan et al. 1999).

Time since introduction is not a significant predictor of range size in birds successfully introduced to New Zealand (Duncan et al. 1999). Birds have the potential to disperse rapidly and over large distances. Indeed, following introduction, the spread of most introduced birds in New Zealand appears to have been rapid, such that by the early 1900s most species were occupying close to their present range sizes (Thomson 1926; Heather and Robertson 1996). The same may not be true for other taxa whose dispersal capabilities are lower. Among the birds established in New Zealand, species which are partially migratory in their native range tend to have larger range sizes in New Zealand, which may reflect higher rates of dispersal among partial migrants (Duncan et al. 1999).

There is also some evidence that the geographic range size of introduced species in New Zealand is greater when the species uses habitats which are themselves more widespread (Duncan et al. 1999). More specifically, species which readily use extensive human-modified habitats, predominantly farmland, attain the highest range sizes in New Zealand. In addition, the proportion of non-native bird species in a habitat may increase with increasing habitat modification (McLay 1974; Diamond and Veitch 1981; Case 1996). Cassey (2001) found that of the ten species with the greatest geographic distributions across mainland New Zealand, only two, *Gerygone igata* and *Rhipidura fuliginosa*, were native before the arrival of humans. Two other widespread native species, *Zosterops lateralis* and *Circus approximans*, naturally colonized New Zealand from Australia only after the widespread loss of forest cover and habitat modification which accompanied human arrival. The remaining six widespread species were all introduced during European settlement.

Both Duncan et al. (1999) and Cassey (2001) reported that the introduced species which have established and maintained larger geographic ranges in New Zealand are those which possess life history traits associated with high population growth rates. These are characteristically small-bodied, rapidly maturing species, with high fecundity. The same is also true for bird species introduced to Australia (Duncan et al. 2001). For British birds introduced to New Zealand, these are also characteristics associated with larger native geographic ranges (Blackburn et al. 1996; Duncan et al. 1999).

One unexpected finding is that the range size of introduced bird species in New Zealand is positively related to the number of individuals introduced (Duncan et al. 1999). One explanation for this finding is that species with large founding populations initially captured a greater proportion of any shared resources from species with smaller founding populations. This initial advantage could have been reinforced – those species initially able to capture a greater share of resources would have had a faster population growth and rate of spread, allowing them to dominate resource utilization at newly colonized sites as their ranges expanded. In addition, the species with larger founding populations were released at more sites throughout the country (Veltman et al. 1996; Duncan 1997), giving them a further head-start in range expansion and "first come, first served" resource utilization. This effect should be most pronounced between closely related species because these are more likely to compete for similar resources, which is indeed the case (Duncan et al. 1999). This explanation implies that competition plays a role in limiting range sizes, along with affecting establishment success (Duncan and Forsyth 2005). Indeed, we might expect competitive interactions to be more apparent among established populations which have reached sufficiently high density (Forsyth and Hickling 1998). Nevertheless, other explanations may underlie the relationship between range size and introduction effort; this relationship is not evident among introduced birds in Australia (Duncan et al. 2001).

9.6 Conclusions

The record of bird introductions into New Zealand is remarkably complete and detailed, and thus provides a unique opportunity to identify factors responsible for success at different stages in the invasion process. Analysis of this record shows that historical and cultural factors dominate in determining the probability that a bird species was released and the probability that it established a wild population in New Zealand. Species from certain families were preferentially selected for release (reflecting the types of birds most desired for introduction); these species tended to originate in the Palaearctic (where most settlers to New Zealand in the 19th century originated) or in nearby Australasia, and to be widespread, abundant species at their source locations. People put more effort into introducing these widespread, abundant species and consequently they had a higher rate of successful establishment and, at least in New Zealand, became more widespread. Species also appeared more likely to succeed in establishing and spreading if they had a large natural geographic range size, and if the environment in their native range matched that in New Zealand. Overall, life history and demographic traits were relatively unimportant in explaining patterns of release and establishment, although the widespread introduced bird species tend to be those with small body mass and high reproductive potential.

The establishment of exotic bird species may currently pose less of an ecological and economic threat to New Zealand than the establishment of species in other groups, particularly plants and invertebrates. This does not imply, however, that understanding the outcome of bird introductions is less of a priority than unravelling the factors underlying invasion success in other taxa. The advantage with birds is that the record of introductions allows us to answer key questions which we presently cannot address for other taxa. Moreover, there is evidence that the general factors identified as affecting invasion success in birds are similar to those which influence invasion success in other taxa (see Kolar and Lodge 2001), making bird introductions to New Zealand a particularly useful model system.

Acknowledgements P.C. acknowledges the assistance of the French Ministry of Education and Research, Action Concertée Incitative "Jeunes Chercheurs 2000", awarded to the group "Eco-Evolution Mathématique", and the French Ministry for the Environment, Action Concertée Incitative "Invasions Biologiques".

References

Blackburn TM, Duncan RP (2001a) Determinants of establishment success in introduced birds. Nature 414:195–197

Blackburn TM, Duncan RP (2001b) Establishment patterns of exotic birds are constrained by non-random patterns in introduction. J Biogeogr 28:927–939

Blackburn TM, Lawton JH, Gregory RD (1996) Relationships between abundances and life histories of British birds. J Anim Ecol 65:52–62

Blackburn TM, Gaston KJ, Duncan RP (2001) Population density and geographical range size in the introduced and native passerine faunas of New Zealand. Divers Distrib 7:209–221

Brooke RK, Lockwood JL, Moulton MP (1995) Patterns of success in passeriform bird introductions on Saint Helena. Oecologia 103:337–342

Buller WL Sir (1888) A history of the birds of New Zealand, 2nd edn. Sir WL Buller, London

Case TJ (1996) Global patterns in the establishment and distribution of exotic birds. Biol Conserv 78:69–96

Cassey P (2001) Determining variation in the success of New Zealand land birds. Global Ecol Biogeogr 10:161–172

Cassey P (2002) Life history and ecology influences establishment success of introduced land birds. Biol J Linn Soc 76:465–480

Cassey P, Blackburn TM, Sol D, Duncan RP, Lockwood JL (2004) Global patterns of introduction effort and establishment success in birds. Proc R Soc Lond B 271:S405–S408

Crawley MJ (1989) Chance and timing in biological invasions. In: Drake JA, Mooney HA, di Castri F, Groves RH, Kruger FJ, Rejmánek M, Williamson M (eds) Scope 37 biological invasions: a global perspective. Wiley, Chichester, pp 407–435

Daehler CC, Strong DR Jr (1993) Prediction and biological invasions. Trends Ecol Evol 8:380

Diamond JM, Veitch CR (1981) Extinctions and introductions in the New Zealand avifauna: cause and effect? Science 211:499–501

Diggle PJ, Liang K-Y, Zeger SL (1994) Analysis of longitudinal data. Clarendon Press, Oxford

Drummond J (1907) On introduced birds. Trans Proc NZ Inst 39:227–252

Duncan RP (1997) The role of competition and introduction effort in the success of passeriform birds introduced to New Zealand. Am Nat 149:903–915

Duncan RP, Blackburn TM (2002) Morphological over-dispersion in game birds (Aves: Galliformes) successfully introduced to New Zealand was not caused by interspecific competition. Evol Ecol Res 4:551–561

Duncan RP, Forsyth DM (2005) Competition and the assembly of introduced bird communities. In: Cadotte MW, McMahon SM, Fukami T (eds) Conceptual ecology and invasions biology. Springer, Berlin Heidelberg New York, pp 415–431

Duncan RP, Blackburn TM, Veltman CJ (1999) Determinants of geographical range sizes: a test using introduced New Zealand birds. J Anim Ecol 68:963–975

Duncan RP, Bomford M, Forsyth DM, Conibear L (2001) High predictability in introduction outcomes and the geographical range size of introduced Australian birds: a role for climate. J Anim Ecol 70:621–632

Duncan RP, Blackburn TM, Sol D (2003) The ecology of bird introductions. Annu Rev Ecol Evol Syst 34:71–98

Forcella F, Wood JT (1984) Colonization potentials of alien weeds are related to their 'native' distributions: implications for plant quarantine. J Aust Inst Agric Sci 50:35–41

Forcella F, Wood JT, Dillon SP (1986) Characteristics distinguishing invasive weeds within *Echium* (Bugloss). Weed Res 26:351–364
Forsyth DM, Duncan RP (2001) Propagule size and the relative success of exotic ungulate and bird introductions in New Zealand. Am Nat 157:583–595
Forsyth DM, Hickling GJ (1998) Increasing Himalayan tahr and decreasing chamois densities in the eastern Southern Alps, New Zealand: evidence for interspecific competition. Oecologia 113:377–382
Garland T Jr, Ives AR (2000) Using the past to predict the present: confidence intervals for regression equations in phylogenetic comparative methods. Am Nat 155:346–364
Goodwin BJ, McAllister AJ, Fahrig L (1999) Predicting invasiveness of plant species based on biological information. Conserv Biol 13:422–426
Green RE (1997) The influence of numbers released on the outcome of attempts to introduce exotic bird species to New Zealand. J Anim Ecol 66:25–35
Heather BD, Robertson HA (1996) The field guide to the birds of New Zealand. Viking, Auckland, New Zealand
Holdaway RN (1999) Introduced predators and avifaunal extinction in New Zealand. In: McPhee RDE (ed) Extinctions in near time: causes, contexts and consequences. Kluwer/Plenum, New York, pp 189–238
Hopper KR, Roush RT (1993) Mate finding, dispersal, number released, and the success of biological control introductions. Ecol Entomol 18:321–331
Hutton FW (1871) Catalogue of the birds of New Zealand. Hughes, Wellington, New Zealand
King CM (ed) (1990) The handbook of New Zealand mammals. Oxford Univ Press, Auckland, New Zealand
Kolar CS, Lodge DM (2001) Progress in invasion biology: predicting invaders. Trends Ecol Evol 16:199–204
Lamb RC (1964) Birds, beasts and fishes: the first hundred years of the North Canterbury Acclimatisation Society. The Society, Christchurch
Legendre S, Clobert J, Møller AP, Sorci G (1999) Demographic stochasticity and social mating system in the process of extinction of small populations: the case of passerines introduced to New Zealand. Am Nat 153:449–463
Lockwood JL (1999) Using taxonomy to predict success among introduced avifauna: relative importance of transport and establishment. Conserv Biol 13:560–567
Lockwood JL, Moulton MP (1994) Ecomorphological pattern in Bermuda birds: the influence of competition and implications for nature preserves. Evol Ecol 8:53–60
Lockwood JL, Brooks TM, McKinney ML (2000) Taxonomic homogenization of the global avifauna. Anim Conserv 3:27–35
Lovegrove TG (1996) Island releases of saddlebacks *Philesturnus carunculatus* in New Zealand. Biol Conserv 77:151–157
Mack RN (1996) Predicting the identity and fate of plant invaders: emergent and emerging properties. Biol Conserv 78:107–121
Martins EP, Hansen TF (1997) Phylogenies and the comparative method: a general approach to incorporating phylogenetic information into the analysis of interspecific data. Am Nat 149:646–667
McDowall RM (1994) Gamekeepers for the nation: the story of New Zealand's acclimatization societies, 1861–1990. Canterbury Univ Press, Christchurch
McLain DK, Moulton MP, Redfearn TP (1995) Sexual selection and the risk of extinction of introduced birds on oceanic islands. Oikos 74:27–34
McLain DK, Moulton MP, Sanderson JG (1999) Sexual selection and extinction: the fate of plumage-dimorphic and plumage-monomorphic birds introduced onto islands. Evol Ecol Res 1:549–565

McLay CL (1974) The species diversity of New Zealand forest birds: some possible consequences of the modification of beech forests. NZ J Zool 1:179–196
Moulton MP (1993) The all-or-none pattern in introduced Hawaiian passeriforms: the role of competition sustained. Am Nat 141:105–119
Moulton MP, Pimm SL (1983) The introduced Hawaiian avifauna: biogeographic evidence for competition. Am Nat 121:669–690
Moulton MP, Pimm SL (1986) Species introductions to Hawaii. In: Mooney HA, Drake JA (eds) Ecology of biological invasions of North America and Hawaii. Ecological Studies 58. Springer, Berlin Heidelberg New York, pp 231–249
Moulton MP, Pimm SL (1987) Morphological assortment in introduced Hawaiian passerines. Evol Ecol 1:113–124
Moulton MP, Sanderson JG, Simberloff D (1996) Passeriform introductions to the Mascarenes (Indian Ocean): an assessment of the role of competition. Écologie 27:143–152
Moulton MP, Sanderson JG, Labisky RF (2001) Patterns of success in game bird (Aves: Galliformes) introductions to the Hawaiian Islands and New Zealand. Evol Ecol Res 3:507–519
Newsome AE, Noble IR (1986) Ecological and physiological characters of invading species. In: Groves RH, Burdon JJ (eds) Ecology of biological invasions: an Australian perspective. Australian Academy of Science, Canberra, pp 1–20
Pimm SL (1991) The balance of nature? Ecological issues in the conservation of species and communities. Chicago Univ Press, Chicago
Reichard SH, Hamilton CW (1997) Predicting invasions of woody plants introduced into North America. Conserv Biol 11:193–203
Rejmánek M (1996) A theory of seed plant invasiveness: the first sketch. Biol Conserv 78:171–181
Richter-Dyn N, Goel NS (1972) On the extinction of a colonizing species. Theor Popul Biol 3:406–433
SAS Institute (1990) SAS/STAT user's guide. Version 6. SAS Institute, Cary, NC
Sibley CG, Ahlquist JE (1990) Phylogeny and classification of birds: a study in molecular evolution. Yale Univ Press, New Haven, CT
Sibley CG, Monroe BL Jr (1990) Distribution and taxonomy of birds of the world. Yale Univ Press, New Haven, CT
Simberloff D, Boecklen W (1991) Patterns of extinction in the introduced Hawaiian avifauna: a reexamination of the role of competition. Am Nat 138:300–327
Sol D, Lefebvre L (2000) Behavioural flexibility predicts invasion success in birds introduced to New Zealand. Oikos 90:599–605
Sol D, Timmermans S, Lefebvre L (2002) Behavioural flexibility and invasion success in birds. Anim Behav 63:495–502
Sol D, Duncan RP, Blackburn TM, Cassey P, Lefebvre L (2005) Big brains, enhanced cognition, and response of birds to novel environments. Proc Natl Acad Sci USA 102:5460–5465
Sorci G, Møller AP, Clobert J (1998) Plumage dichromatism of birds predicts introduction success in New Zealand. J Anim Ecol 67:263–269
Stevens GC (1989) The latitudinal gradient in geographical range: how so many species coexist in the tropics. Am Nat 133:240–256
Thomson GM (1922) The naturalization of animals and plants in New Zealand. Cambridge Univ Press, Cambridge
Thomson GM (1926) Wild life in New Zealand, part II. Introduced birds and fishes. Government Printer, Wellington, New Zealand

Turbott EG (convener) (1990) Checklist of the birds of New Zealand and the Ross Dependency, Antarctica, 3rd edn. Random Century in association with the Ornithological Society of New Zealand Inc, Auckland
Veltman CJ, Nee S, Crawley MJ (1996) Correlates of introduction success in exotic New Zealand birds. Am Nat 147:542–557
Wellwood JM (ed) (1968) Hawke's Bay Acclimatisation Society centenary 1868–1968. Hawke's Bay Acclimatisation Society, Hastings, New Zealand
Westerskov KE (1953) Introduction into New Zealand of the Australian blue wren in 1923. Notornis 5:106–107
Williams GR (1950) Chukar in New Zealand. NZ Sci Rev 8:2–6
Williams GR (1969) Introduced birds. In: Knox GA (ed) The natural history of Canterbury. AH & AW Reed, Wellington, pp 435–451
Williamson M (1996) Biological invasions. Chapman and Hall, London
Williamson MH, Fitter A (1996) The characters of successful invaders. Biol Conserv 78:163–170

10 Biological Control Agents: Can They Tell Us Anything About the Establishment of Unwanted Alien Species?

S.V. Fowler, H.M. Harman, R. Norris and D. Ward

10.1 Introduction

Biological control introductions aim to establish a new beneficial alien species in a country, and are now sanctioned only after careful safety testing and risk analysis. Biological control programs have formed a major part of the management of invertebrate pests and weeds in New Zealand (Cameron et al. 1993; see Chap. 23). In many cases, however, release strategies for biocontrol agents remain based on untested assumptions and intuition, both in New Zealand and overseas (Memmott et al. 1998). In this chapter, we review recent research efforts focused on improving the rates of establishment of biocontrol agents, increasing the efficiency of biocontrol releases, and understanding the factors that influence the establishment and extinction of small populations (Memmott et al. 1996, 1998; Grevstad 1999a; Syrett et al. 2000a). Establishment is a process common to all invasions, whether they are deliberate releases of biocontrol agents or accidental arrivals of potential pests. We ask whether information gained from attempts to understand, and improve on the process of establishing biocontrol agents could offer insights into factors affecting the establishment of accidentally arriving, unwanted alien species in New Zealand. Where possible, we have taken examples from New Zealand biological control programs, but have included studies and observations from such programs elsewhere to illustrate the potential to use these systems for understanding invasion processes.

Ecological Studies, Vol. 186
R.B. Allen and W.G. Lee (Eds.)
Biological Invasions in New Zealand

10.2 Establishment Rates for Biological Control Agents

For New Zealand, Cameron et al. (1993) report that the rates of establishment of species released as biocontrol agents were 31 % for releases targeting insect pests, and 41 % for releases of weed biocontrol agents. However, data from 2000 show that establishment success of agent species released against weeds in New Zealand had reached 76 % (Table 10.1). This rapid increase in the rate of successful establishment of biocontrol agents released against weeds has been attributed to the enhanced effort put into technical information transfer programs, and to the extensive support of release programs by regional councils, the Department of Conservation, and non-government bodies in New Zealand (Fowler et al. 2000). A relationship between increased effort put into releasing a species and increased establishment success has been shown for other groups, e.g., passeriform birds (Duncan 1997) and ungulates (Forsyth and Duncan 2001) in New Zealand, and plants in South Africa (Rouget and Richardson 2003).

10.3 Release Size

Establishment can be hampered by factors that put small populations at particular risk of extinction. These include demographic stochasticity, environmental variability, and specific issues affecting small populations, known as Allee effects, that include reduced ability to find mates when population density is low (Grevstad 1999a; McEvoy et al. 2000). A higher probability of establishment is often correlated with both larger release sizes and greater frequency of releases (Grevstad 1999a). However, the ability to make both larger releases and more of them can be constrained by difficulties in the collection,

Table 10.1 Biological control agent species released against weeds in New Zealand that have established or failed to establish[a]

	Species established (*n*)	Species failed to establish (*n*)	Establishment rate (%)
Monophagous species	16	5	76
Oligophagous species	6	2	75
Total	22	7	76

[a] Data extracted from Fowler et al. (2000). Feeding types defined as in Hoare (2001): monophagous species restrict their feeding to a single genus, oligophagous species to several genera within a plant family

rearing and handling of agents, which along with time and finances, often result in a limited number of agents available for release. A single large release might improve the chance of successful establishment, but the alternative strategy of many small releases has the advantage of spreading the risk of local extinction. Simulation models (Grevstad 1999a) suggest that demographic stochasticity alone is not likely to be important in the establishment of biocontrol agents, whereas Allee effects and environmental variability are crucial. In a variable environment, a large number of very small releases will maximize the chance of overall establishment, because environmental variability reduces the likelihood of establishment over the entire range of colony sizes (Grevstad 1999a). In contrast, when an Allee effect is present (in a constant environment), a single large release is optimal because the presence of an Allee effect results in a population establishment threshold. For colony sizes below the threshold, a population will become extinct, whereas those above the threshold will establish (Grevstad 1999a).

Memmott et al. (1998) carried out a manipulative field experiment in New Zealand that investigated the relationship between release size and the probability of establishment of gorse thrips, *Sericothrips staphylinus*, a biocontrol agent for gorse *Ulex europaeus*. Replicated releases of 10, 30, 90, 270, and 810 gorse thrips were made onto isolated gorse bushes, and the thrips sampled 1 year after release. A higher proportion of the small releases became extinct during this time period: thrips were recovered from 100 % of the releases of 270 and 810 thrips, but from only 33 % of releases of 10, 30, and 90 thrips. The demonstrated relationship between the probability of establishment and the release size was used to determine an optimum release strategy for gorse thrips. Calculations that took into consideration the number of releases that could be made for a fixed number of available insects suggested that the optimum release size for gorse thrips in New Zealand might be fewer than 100 (Memmott et al. 1998). This contrasted with the then current strategy of 1,000 gorse thrips per release in New Zealand. A similar trade-off between increasing release size and the numbers of releases made occurred with the parasitoid *Aphytis melinus*, where a tenfold increase in the numbers released only doubled the probability of establishment (Campbell 1976).

In a longer-term experiment in New Zealand, different release sizes (2, 4, 10, 30, 90, and 270) of the psyllid *Arytainilla spartiophila* to control Scotch broom, *Cytisus scoparius*, were made and monitored for 5 years (Memmott et al. 2005). Local extinction was greatest in the first year (compared with later years), and although the probability of extinction in the first year was related to release size, several releases of only one pair of psyllids did result in established populations. Moreover, the persistence of populations after the first year and population growth rates were independent of release size (Memmott et al. 2005).

Similarly, Grevstad (1999b) examined the effect of release size on establishment of two chrysomelid beetles, *Galerucella calmariensis* and *G. pusilla*, for control of purple loosestrife, *Lythrum salicaria*, in New York State, USA. Releases of 20, 60, 180, or 540 beetles were monitored at 36 sites over a 3-year period. Although the probability of establishment increased with release size, subsequent experiments also showed that a single gravid female (one of the 20 individuals released) could establish a population (Grevstad 1999b). Thus, releases of even very small numbers can result in establishment, albeit with a higher probability of early extinction, compared with releases of larger numbers of individuals. It is difficult to know whether this is a general pattern, and consequently current release strategies for insects for biological control of weeds in New Zealand tend to use a range of release sizes. The aim is to maximize efficiency, while reducing the risk of complete failure in case there is a currently unknown threshold release size below which establishment is very unlikely. As a result, a data set of release size and establishment success across a range of species is slowly being accumulated.

10.4 Climate and Phenology

New Zealand has a wide range of climates in close spatial proximity, which can make exact climate matching with other locations problematic (Cameron et al. 1993). Clearly, the term "climate" encompasses many variables, and it is difficult to match the parameters that affect populations of any particular species. Despite this, biocontrol agents are usually sought from areas of their native range that have broadly similar climates to those common in New Zealand. This approach is thought to lessen the chance that an agent will fail to establish for climatic reasons alone (Cameron et al. 1993).

In New Zealand, populations of the gorse spider mite, *Tetranychus lintearius*, collected from England to control gorse, had an establishment rate of 22 % in areas characterized by high rainfall and relatively warm winters (the upper North Island and on the West Coast of the South Island), which was much lower than the 87 % establishment rate elsewhere in New Zealand (Hill et al. 1991). Additional collections of gorse spider mite were made from coastal Portugal and northwestern Spain – climates that matched more closely the areas in New Zealand where the English mites had shown poor establishment rates (Hill et al. 1993). Experimental releases of the new strains indicated that the Portuguese and Spanish strains grew better, with colonies up to 30 times larger than those of English mites in the upper North Island and West Coast (Hill et al. 1993).

Small, vulnerable founder populations of biocontrol agents may also be lost due to chance events, such as adverse climate conditions (Lawton 1990). For example, Hill et al. (1993) investigated the effect of rainfall on the establishment of gorse spider mites, *T. lintearius*. When this mite was introduced to New Zealand to control gorse, these authors found that the mite established poorly in areas of heavy rainfall. Hill et al. (1993) constructed wooden frames with clear plastic roofs over individual gorse bushes to deflect direct rainfall, and found that protected colonies survived significantly better and were up to 10 times larger than the previously unprotected colonies, confirming the negative influence of heavy rainfall on mite establishment.

Efforts to establish an English gorse seed weevil, *Exapion ulicis*, failed in Hawai'i, despite the success of weevils from the same source in New Zealand (Markin and Yoshioka 1990). Subsequent efforts to release New Zealand-collected specimens in Hawai'i also failed, with the conclusion that English material was not suited to the Hawaiian climate. Gorse seed weevils collected from southern France were then released in Hawai'i, and became established. Additional observation on the flowering phenology of gorse showed large differences between Hawai'i and New Zealand and England (Markin and Yoshioka 1990). Furthermore, the English strain of weevil has only one generation per year, and is inactive over winter, whereas in Hawai'i the weevils reportedly have >1 generation per year and are active all year round – a characteristic of the weevils collected from southern France (Markin and Yoshioka 1990).

Climate matching helps to ensure that the life history of the biocontrol agent is synchronized with the phenology of the target host (Cameron et al. 1993). This synchrony is essential if a biocontrol agent is to achieve its maximum damage, and for it to be available to attack the target over its full period of susceptibility (Lawton 1990). For example, Harman (1999) illustrated that variability in the timing of the reproductive stage of Scotch broom, *C. scoparius*, may influence the establishment of the broom seed beetle, *Bruchidius villosus*, in New Zealand. The phenology of broom varies considerably across different sites in any one season; however, where broom seed beetle has established, it has synchronized its life cycle with the reproductive stages of the host. Further research showed the timing of the reproductive maturity of the beetle can be manipulated under artificial conditions, thus providing an opportunity to maximize the efficiency of the release strategy (Harman 1999). Poor synchronization of life stages also appears to be a major factor in the failure of a seed fly *Botanophila jacobaeae* to establish in parts of New Zealand for the control of ragwort, *Senecio jacobaea* (Dymock 1988). As a result, 80–90 % of ragwort seeds escape predation (Dymock 1987).

The most obvious situation in New Zealand biological control programs where phenology is important is when agents from the Northern Hemisphere are introduced. Rephasing Northern Hemisphere populations to the Southern Hemisphere seasons is often a challenge for biological control programs, and can lead to less than ideal timings for releases, or even to the rejection of agents where rephasing is particularly difficult. For example, the broom seed weevil, *Exapion fuscirostre*, was rejected in favor of the broom seed beetle, *Bruchidius villosus*, in part because of its obligatory winter diapause. In contrast, the broom seed beetle becomes sexually mature in response to feeding on pollen of its host plants (Harman 1999).

Climate matching thus appears to be important for establishment success of biocontrol agents, but has seldom been tested rigorously, largely because of a lack of resources to undertake scientific studies in operational biological control programs.

10.5 Genetic Issues

Mackauer (1976) and Hopper et al. (1993) have reviewed the genetic issues concerning the rearing and release of insects for biological control. Genetic issues, which are acute in small populations, include the loss of alleles through genetic drift; inbreeding depression, especially where the mating of siblings increases the likelihood of homozygous deleterious recessive alleles; and the selection for laboratory-adapted populations, particularly when colonies are maintained for too long in artificial conditions. Genetic diversity of collected agents may be adversely affected by rearing methods, but this is usually only a problem if many generations are reared before release. Many genetic issues can also be minimized by making large releases of agents. A population will be affected in the long term by the level of genetic variability or the potential to adapt to change – in general, the greater the diversity of the initial release, the greater the likelihood of being able to adapt to a particular environment (Hopper et al. 1993; Williams et al. 2000).

The significance of inbreeding depression in field populations has been contentious, but in a landmark study, Nieminen et al. (2001) carried out a series of laboratory and field studies showing that inbreeding depression did increase the extinction risk in small populations of the Glanville fritillary butterfly, *Melitaea cinxia*. In laboratory studies, female butterflies were mated either to brothers (inbred lines) or to males from the other population (crossbred lines). After the mated females had laid eggs on potted host plants of *Plantago lanceolata*, the eggs were incubated under standard conditions and the egg-hatching rate recorded. Nieminen et al. (2001) found that the egg-hatching rate was significantly lower in egg batches laid by inbred females, compared with crossbred females. In complementary field studies, experimental populations consisting of inbred and crossbred lines were randomly assigned to 12 patches of dry meadow habitat. Adult butterflies were captured and marked, and their breeding and overwintering success was determined. All six crossbred populations, but only two inbred populations, produced offspring that survived until late summer (Nieminen et al. 2001).

Intraspecific genetic variation may also be important. For example, Phillips et al. (1997) released eight South American populations of the parasitoid *Microctonus hyperodae* in New Zealand to control the Argentine stem weevil, *Listronotus bonariensis*. Using morphometric analysis of adult parasitoids of known South American origins to determine which populations had established successfully in New Zealand, Phillips et al. (1997) showed significantly more parasitoids derived from east of the Andes were prevalent at the release sites. This suggested that for *M. hyperodae*, intraspecific variation has probably been important in their establishment success in New Zealand. Populations of *M. hyperodae* from east of the Andes had greater fecundity, and there was a greater likelihood that these populations had co-evolved more recently with *L. bonariensis* populations that are in New Zealand (Phillips et al. 1997). If they had released only *M. hyperodae* from west of the Andes, then the evidence suggests that the introduction and biocontrol program would have failed.

Overall, the significance of genetic issues in releasing biocontrol agents remains unresolved. Crucially, we have no evidence that releases that fail to establish, or do well, do so because of genetic factors such as inbreeding depression or lack of adaptability. This is an area where experimental research during the release and establishment of biocontrol agents could offer considerable insights.

10.6 Natural Enemies

Natural enemies have been implicated in the establishment failure of approximately half of published weed biological control programs (Goeden and Louda 1976). However, experimental evidence for the impact of natural enemies upon the failure of biocontrol agents in the field remains limited, because of a lack of detailed monitoring, particularly of agents that fail to establish or impact on their target pests.

Kluge (1991) proposed that the failure of an arctiid moth, *Pareuchaetes pseudoinsulata*, to become established in South Africa to control triffid weed, *Chromolaena odarata*, was because of excessive ant and chrysopid predation of their eggs. In Australia, Briese (1986) carried out a manipulative field experiment designed to identify and quantify factors affecting the establishment of the geometrid moth *Anaitis efformata* on St. John's wort, *Hypericum perforatum*. Neonate larvae were placed on St. John's wort in the field, and natural enemy levels were manipulated. Natural enemies were capable of suppressing the populations of the moth, and Briese (1986) predicted that these natural predators could prevent a population of *A. efformata* from reaching levels that contribute significantly to the control of St. John's wort.

In New Zealand, several biocontrol agents are thought to be adversely affected by predation. For example, gorse spider mite *T. lintearius* populations are regulated by *Stethorus bifidus*, a predatory coccinellid beetle, and by a predatory mite, *Phytoseiulus persimilis* (Hill et al. 2000). A possible, but unconfirmed, reason for the failure of the heather beetle *Lochmaea suturalis* to establish at 13 of 14 release sites for control of heather *Calluna vulgaris* (L.)

Hull is attack by natural enemies, which can severely affect European populations of *L. suturalis* (Waloff 1987; Syrett et al. 2000b; Landcare Research, unpublished data).

Natural enemies are thought to play an important role in preventing establishment of some biocontrol agents, but again we have a paucity of supportive data and experimental evidence.

10.7 Other Factors

Crawley (1989) examines a wide range of factors leading to the success or failure of weed biological control. A number of additional factors to those considered above appear to be important in the establishment of biocontrol agents, although the reasons for biocontrol failures are complex and interacting. The analysis suggested that smaller-sized insects are significantly more likely to become established, probably because body size is associated with a high intrinsic rate of increase, an important demographic characteristic (Crawley 1989). There also appear to be taxa that are poor invaders and have a high failure rate in establishing populations as biocontrol agents. For example, Lepidoptera are amongst the worst performers, possibly because their eggs may be particularly vulnerable to generalist predators (Crawley 1986; Kluge 1991). In contrast, Diptera are relatively successful at establishment, possibly because they are mainly internal feeders that can escape generalist predation, and although they may be heavily parasitized in their native range, can be freed from this as biocontrol agents (Crawley 1989). Biological control introductions use a limited range of taxa, mostly insect parasitoids, predators or herbivores, and in a few cases plant diseases. Species from these groups can also pose risks as accidental alien invaders, but generalizing about establishment success to unrelated taxa of invaders may be difficult. Even where invading taxa are reasonably closely related, their different functional ecology – for example, with social insects such as ants and wasps – may make generalizations difficult.

The way biocontrol agents are handled during transportation to their release sites could have important consequences for their establishment success. Hight et al. (1995) found that although mortality of the leaf-feeding beetles *Gallerucella calmariensis* and *G. pusilla* was low during shipment from Europe to America, 10 % of the beetles died during quarantine in the week following shipment. An assessment of the "hardiness" of a species, and of the life stages most at risk during transportation, could aid in an overall assessment of the probability of establishment for invasive species. There is scope for using biocontrol agents to investigate survival of exotic organisms along different pathways of entry to a new environment.

Host specificity may be significant because species that can exploit a range of resources might be expected to establish more easily. All biological control agents are carefully selected to be sufficiently host specific to minimize the risk of unwanted impacts on non-target species. Does this limit the value of trying to generalize to all potential invasive species? Although no polyphagous species have ever been introduced as biocontrol agents against weeds, a range of monophagous and oligophagous herbivorous species have been used (Table 10.1; Fowler et al. 2000). Table 10.1 shows that establishment rates of monophagous (76 %) and oligophagous (75 %) biocontrol agents released against weeds in New Zealand are very similar. We can also ask whether more polyphagous species invade New Zealand compared with more-specific feeders, and what are the relative impacts of polyphages compared with more-specialist feeders? In one insect group, the Lepidoptera, enough is known about the alien species that have established in New Zealand to address these questions. Table 10.2 shows that significantly fewer generalist lepidopteran species established in New Zealand (up to 1998), compared with more-specific feeders (monophages and oligophages; $P<0.01$). However, although only 22 % of the total were polyphagous species, a higher portion (64 %) of these were regarded as pests, compared with the monophagous (20 %) or oligophagous (27 %) species ($P<0.05$). Generalist insect herbivores therefore do not seem to either establish more easily, or arrive more often, than more-specialist species. If they do become established, however, they are more likely to be pests. Generalizations from studying herbivorous biocontrol agents, regarding factors affecting establishment, appear to be acceptable, but caution would have to be used if interest was centered only on species that become pests.

Table 10.2 Herbivorous alien lepidopteran species established in New Zealand (excluding deliberately released biocontrol agents) up to 1998[a]

	Pest species (%)[b]	Non-pest species	Totals (%)[c]
Monophagous species	5 (25)	16	20 (31)
Oligophagous species	11 (37)	22	30 (47)
Polyphagous species	10 (71)	5	14 (22)

[a] Data and feeding type classifications taken from Hoare (2001): monophagous species restrict their larval feeding to a single plant genus, oligophagous species to a plant family, whereas polyphagous species feed on more than one plant family

[b] Percentage of the total number of species falling into each feeding type. Species were classified as pests if they were reported as major, minor, occasional or potential pests in Scott (1984) and Hoare (2001)

[c] Percentages of pest species are shown for each feeding type. Chi-squared analysis was used to compare total monophages and oligophages with total polyphages ($P<0.001$, $\chi^2=22.04$, d.f.=1), and feeding type with pest status ($P<0.05$, $\chi^2=7.35$, d.f.=2)

10.8 Conclusions

After a long history of mostly ad hoc approaches to release and establishment in biological control, there have been recent research efforts aimed at increasing our understanding of the factors that are likely to affect establishment success rates. Although biocontrol agents are only a small subset of the range of alien species that can invade a country such as New Zealand, we suggest that some generalizations may apply to all invasions of alien species. The apparent benefit from increasing the effort put into biological control releases (e.g., making many repeat releases, increasing the size of releases, careful climate matching, avoiding adverse weather conditions, and natural enemies) does suggest that establishment is a major barrier for many species in a new environment. From a biosecurity perspective, this is encouraging as it suggests that a lot of incursions probably fail to establish. This is not meant to encourage complacency, but rather to discourage a tendency to despair that, as we can never completely seal our border, eventually everything harmful that can arrive will arrive.

Certainly, the deliberate invasions that biological control programs create offer one of the very few opportunities for controlled experimental studies of the invasion process. We suggest that a small amount of experimental ecology built into the early stages of biocontrol release programs could make an appreciable difference not only to our understanding of establishment success of beneficial agents for biological control, but also to understanding the establishment of harmful invasive species (Memmott et al. 1998). We also agree with Crawley (1989) that biocontrol releases represent some of the largest field experiments carried out in ecology, although they have seldom had control treatments or been adequately randomized.

Acknowledgements Rob Allen and two anonymous referees made helpful comments on earlier versions of the manuscript. Guy Forrester assisted with statistical analysis. Funding was provided by the New Zealand Foundation for Research, Science and Technology (contract C09X0210).

References

Briese DT (1986) Factors affecting the establishment and survival of *Anaitis efformata* (Lepidoptera: Geometridae) introduced into Australia for the biological control of St. John's Wort, *Hypericum perforatum*. II. Field trials. J Appl Ecol 23:821–839

Cameron PJ, Hill RL, Bain J, Thomas WP (1993) Analysis of importations for biological control of insect pests and weeds in New Zealand. Biocontrol Sci Technol 3:387–404

Campbell MM (1976) Colonisation of *Aphytis melinus* DeBach (Hymenoptera: Aphelinidae) in *Aonidiella aurantii* (Mask.) (Hemiptera, Coccidae) on citrus in South Australia. Bull Entomol Res 65:659–668

Crawley MJ (1986) The population biology of invaders. Philos Trans R Soc Lond Ser B 314:711–731

Crawley MJ (1989) The successes and failures of weed biocontrol using insects. Biocontrol News Info 10:213–223

Duncan RP (1997) The role of competition and introduction effort in the success of passeriform birds introduced to New Zealand. Am Nat 149:903–915

Dymock JJ (1987) Population changes of the seedfly, *Pegohylemyia jacobaeae* (Diptera: Anthomyiidae) introduced for biological control of ragwort. NZ J Zool 14:337–342

Dymock JJ (1988) Pre-oviposition period of the ragwort seedfly, *Pegohylemyia jacobaeae*, introduced for the biological control of ragwort, *Senecio jacobaea*. NZ J Zool 15:507–511

Forsyth DM, Duncan RP (2001) Propagule size and the relative success of exotic ungulate and bird introductions to New Zealand. Am Nat 157:583–595

Fowler SV, Syrett P, Hill RL (2000) Success and safety in the biological control of environmental weeds in New Zealand. Austral Ecol 25:553–562

Goeden RD, Louda SM (1976) Biotic interference with insects imported for weed control. Annu Rev Entomol 21:325–342

Grevstad FS (1999a) Factors influencing the chance of population establishment: implications for release strategies in biocontrol. Ecol Appl 9:1439–1447

Grevstad FS (1999b) Experimental invasions using biological control introductions: the influence of release size on the chance of population establishment. Biol Invas 1:313–323

Harman HM (1999) The effect of variability in the phenology of the reproductive stages of Scotch broom (*Cytisus scoparius*) on the synchronization of the life stages of broom seed beetle (*Bruchidius villosus*) in New Zealand. Biol Control 15:228–234

Hight SD, Blossey B, Laing J, Declerck-Floate D (1995) Establishment of insect biological control agents from Europe against *Lythrum salicaria* in North America. Environ Entomol 24:967–977

Hill RL, Grindell JM, Winks CJ, Sheat JJ, Hayes LM (1991) Establishment of gorse spider mite as a control agent for gorse. In: Proc 44th New Zealand Weed and Pest Control Conf, New Zealand Plant Protection Society. Manaaki Whenua Press, Lincoln, pp 31–34

Hill RL, Gourlay AH, Winks CJ (1993) Choosing gorse spider mite strains to improve establishment in different climates. In: Prestidge RA (ed) Proc 6th Australasian Conf Grassland Invertebrate Ecology, Ruakura Agricultural Research Centre, Hamilton, New Zealand, pp 377–383

Hill RL, Gourlay AH, Fowler SV (2000) The biological control programme against gorse in New Zealand. In: Spencer NR (ed) Proc 10th Int Symp Biological Control of Weeds, Montana State University, Bozeman, Montana, pp 909–917

Hoare RJB (2001) Adventive species of Lepidoptera recorded for the first time in New Zealand since 1988. NZ Entomol 24:23–47

Hopper KR, Roush RT, Powell W (1993) Management of genetics of biological-control introductions. Annu Rev Entomol 38:27–51

Kluge RL (1991) Biological control of triffid weed, *Chromolaena odarata* (Asteraceae), in South Africa. Agric Ecosyst Environ 37:193–197

Lawton JH (1990) Biological control of plants: a review of generalisations, rules and principles using insects as agents. In: Bassett C, Whitehouse LJ, Zabkiewicz JA (eds) Alternatives to the chemical control of weeds. FRI Bull 155. Forest Research Institute, Rotorua, New Zealand, pp 3–7

Mackauer M (1976) Genetic problems in the production of biological control agents. Annu Rev Entomol 21:369-385

Markin GP, Yoshioka ER (1990) Present status of biological control of the weed gorse (*Ulex europaeus* L.) in Hawaii. In: Delfosse ES (ed) Proc 7th Int Symp Biological Control of Weeds, Instituto Sperimentale per la Patalogia Vegetale, Rome, Italy, pp 357–362

McEvoy P, Grevstad FS, Schooler S, Schat M, Coombs EM (2000) Weed biocontrol as an invasion process. In: Spencer NR (ed) Proc 10th Int Symp Biological Control of Weeds, Montana State University, Bozeman, Montana, pp 596

Memmott J, Fowler SV, Harman HM, Hayes LM (1996) How best to release a biological control agent. In: Moran VC, Hoffman JH (eds) Proc 9th Int Symp Biological Control of Weeds, University of Cape Town, Stellenbosch, South Africa, pp 291–296

Memmott J, Fowler SV, Hill RL (1998) The effect of release size on the probability of establishment of biological control agents: gorse thrips (*Sericothrips staphylinus*) released against gorse (*Ulex europaeus*) in New Zealand. Biocontrol Sci Technol 8:103–115

Memmott J, Craze PG, Harman HM, Syrett P, Fowler SV (2005) The effect of propagule size on the invasion of an alien insect. J Anim Ecol 74:50–62

Nieminen M, Singer MC, Fortelius W, Schöps K, Hanski I (2001) Experimental confirmation that inbreeding depression increases extinction risk in butterfly populations. Am Nat 157:237–244

Phillips CB, Baird DB, Goldson SL (1997) South American origins of *Microctonus hyperodae* Loan (Hymenoptera: Braconidae) established in New Zealand as defined by morphometric analysis. Biocontrol Sci Technol 7:247–258

Rouget M, Richardson DM (2003) Inferring process from pattern in plant invasions: a semimechanistic model incorporating propagule pressure and environmental factors. Am Nat 162:713–724

Scott RR (1984) New Zealand pest and beneficial insects. Lincoln University College of Agriculture, Lincoln, New Zealand

Syrett P, Briese DT, Hoffmann JH (2000a) Success in biological control of terrestrial weeds by arthropods. In: Gurr G, Wratten S (eds) Biological control: measures of success. Kluwer, Dordrecht, pp 189–230

Syrett P, Smith LA, Bourner TC, Fowler SV, Wilcox A (2000b) A European pest to control a New Zealand weed: investigating the safety of heather beetle, *Lochmaea suturalis* (Coleoptera: Chrysomelidae) for biological control of heather, *Calluna vulgaris*. Bull Entomol Res 90:169–178

Waloff N (1987) Observations on the heather beetle *Lochmaea suturalis* (Thomson) (Coleoptera: Chrysomelidae) and its parasitoids. J Nat Hist 21:545–556

Williams PA, Nicol E, Newfield M (2000) Assessing the risk to indigenous New Zealand biota from new exotic plant taxa and genetic material. Sci Conserv 143. Department of Conservation, Wellington, New Zealand

11 The Context of Plant Invasions in New Zealand: Evolutionary History and Novel Niches

J.M. Craine, W.G. Lee and S. Walker

"The annals of Tutira can be read in its weeds. Each phase ... has been marked by the arrival and establishment of aliens particularly fitted for the particular condition"
Herbert Guthrie-Smith in 1921

11.1 Introduction

With the large environmental and economic costs of species invasions, predicting the invasion potential of non-native species and managing them post-establishment is an important component of New Zealand's biosecurity. Current efforts to assess the invasion potential of plant species utilize phenomenological approaches where the invasion probabilities of species are identified based on their invasion history in other regions, or climatic matching of the distribution of the non-native species with local environments (Pheloung et al. 1999). These phenomenological approaches are easily applied to numerous species, and may aid in recognizing new and latent invaders, but they are likely to be a coarse filter that is not calibrated to the uniqueness of New Zealand's habitats and flora. Consequently, the weed potential of many species will be overestimated, whereas others that may aggressively spread will be overlooked. Phenomenological approaches are likely to fall short in identifying invasive species, are also unable to help manage existing non-native species' expansions, and can not identify vulnerable communities.

In contrast to phenomenological approaches, a mechanistic, predictive invasion theory identifies the processes by which one species invades a habitat and/or the factors that allow an assemblage to resist invasion (Wiser et al. 1998; Kolar and Lodge 2001). Long recognized as a major challenge for ecologists (Elton 1958), a mechanistic invasion theory requires knowledge of the potentially numerous plant traits that are critical for plants to both establish

Ecological Studies, Vol. 186
R.B. Allen and W.G. Lee (Eds.)
Biological Invasions in New Zealand

and dominate (Mooney and Drake 1986; Williamson 1999; Olden and Poff 2003). The mechanisms of establishment also need to be coupled with the mechanisms by which assemblages resist invasion (Lodge 1993).

Our knowledge of the relationship between the structuring of assemblages and the traits of individuals is still rudimentary, but generalizations are emerging along stress and disturbance gradients (Reich et al. 2003; Verheyen et al. 2004). The success of a plant species invading a habitat depends not only on the traits of the invasive species and the abiotic characteristics of the habitat, but also on the spectrum of traits represented in the habitat's biotic assemblage (Robinson et al. 1995; Hector et al. 2001; Miller et al. 2002). Plant invaders appear to be less successful where similar functional groups are present in the resident flora (Fargione et al. 2003), and when unutilized resource levels remain low (Davis and Pelsor 2001). Operationally, there are far too many plant species, traits, and environmental factors to reasonably incorporate into a mechanistic framework, and our knowledge of critical plant traits determining a species' stress and disturbance tolerance is always likely to be limited. Yet, understanding and predicting invasions can be made more efficient if phenomenological approaches are adjusted by mechanistic theory of invasions.

The habitats of New Zealand can be considered a collection of niches that differ in their similarity to niches elsewhere in the world. We consider a niche to be described by a series of environmental axes that define the way in which plants acquire, allocate, and maintain resources to be successful. Important factors that determine the niche include resource supplies, disturbance regime, biotic interactions, and abiotic stresses. Niches differ not only along these major axes, but also in their spatiotemporal extent and consistency over evolutionary timescales.

For a given set of niches that make up a habitat, the functional trait diversity of the extant assemblage is driven by evolutionary-scale patterns of species radiation, immigration and extirpation, and is constrained by the size and diversity of the regional species pool and characteristics of the dominant abiotic and biotic drivers. Applying theories of island biogeography (MacArthur and Wilson 1967), immigration increases functional diversity as propagules with novel functional characteristics come into contact with appropriate habitats. Extinction rates depend on the spatiotemporal extent and continuity of the habitat, the composition of the resident biota, the availability of refugia, and the temporal pattern of major environmental changes. The expansion of functional diversity reflects the spatiotemporal extent of a habitat and the trait diversity found in the regional species pool.

Understanding the evolutionary and ecological history of New Zealand plant assemblages can assist in identifying which assemblages are more likely to be invaded, and suggesting some approaches to managing established non-native plants. Although biogeographic processes may limit the size and variation of the regional species pool, the spatiotemporal extent and continuity of ecosystems plays a large role in shaping the level of native functional diversity in an assemblage. When environmental conditions associated with a current habitat have been spatiotemporally rare and/or discontinuous, the assemblage that currently occupies the habitat is less likely to contain species or traits that will be competitively successful in the habitat, and therefore resist invaders. In novel habitats, the appropriate native functional diversity is likely to be lower than for donor assemblages, and the native assemblage therefore highly vulnerable to invasion. We propose that the large proportion of invasions in New Zealand are dependent on the presence of these novel niches, whereas core niches, which have been more continuously present for a longer period and/or over a larger spatial extent, are less likely to be invaded. The key to applying this approach to New Zealand begins with identifying the core and novel niches of the country.

11.2 New Zealand's Core and Novel Niches

Novel niches for a region can be defined a priori though an understanding of the region's evolutionary-scale history. Novel niches are more recent in origin, less extensive in space, and more discontinuous in space or time. Core niches, in contrast, are defined by having greater spatiotemporal extent and/or continuity through time. For simplicity, we refer to niche novelty categorically (i.e., core or novel), rather than trying to express the degree of novelty associated with niches. In core habitats, native diversity of functional traits in relation to stress and disturbance gradients should be higher and better adapted than is the case for native functional diversity in novel niches, especially when compared with similar niches in other parts of the world where the niche has been less extensive over space and time.

For most of the past 70 million years, New Zealand has been dominated by warm, wet, low-fertility habitats that lacked regular fire (Wardle 1991). The vegetation was largely forest, with a distinctive fauna of avian herbivores, lizards, and large insects. Novel niches have been produced by recent (i.e., less than 2.5 million years) geological and climatic changes as well as human alteration of habitats over the last millennia. Glacial activity, tectonism, a climate that has oscillated in temperature, herbivorous mammals, fire, and soil disturbance are the major factors that have created novel niches in the latter part of the Tertiary. Each has reduced the relative abundance of core niches, making a larger proportion of New Zealand more likely to be invaded by nonnative species.

11.2.1 High Nutrient Supply

Throughout most of the Cenozoic, strongly weathered lateritic and infertile volcanic soils predominated, perhaps with local nutrient enrichment around coastal bird colonies. Relatively high phosphorus availability across much of the landscape was created only during the Pleistocene by sediment redistribution associated with mountain erosion and glacial activity (O'Connor 1980). European agricultural practices have also involved phosphate and nitrate fertilizer addition, especially in the lowlands, together with the widespread use of pasture legumes. The long-term absence of nutrient-rich soils in New Zealand has restricted the development of functional diversity for high-nutrient-supply environments, and limited the evolutionary capacity of plants to efficiently utilize or compete with introduced plants on resource-rich soils (Lloyd et al. 2002).

Native grasslands in New Zealand occupied cold or wet, low-nutrient-supply ecosystems and few of the grasses were well suited for growth at high nutrient supply. Craine and Lee (2003) analyzed the functional traits of roots and leaves of grasses from 30 grasslands across a broad altitudinal gradient on South Island. When compared at a common altitude, native New Zealand grasses have traits that are associated with low nutrient supply, whereas non-native species have traits associated with high nutrient supply. This reflects both the lack of native species that are adapted to high nutrient supplies, and that the non-native species that have been introduced are the ones that perform well in high-nutrient-supply sites. The agronomic species Europeans brought to New Zealand were either largely selected for their ability to grow in pastures that were fertilized and/or they were grown in association with non-native nitrogen-fixing *Trifolium* species. This is evident from their higher nutrient concentrations, superior morphology for light competition, and low structural investment into leaves and roots that confers short longevity, a trait associated with high nutrient supply. Over the past 100 years, areas where fertilization or clover growth has ceased have often reverted to native grasses and other low-N plants such as bracken fern.

Adaptations to low soil fertility also extend to native woody species. Few New Zealand trees have short leaf longevity and high rates of photosynthesis, traits associated with high nutrient supply, and deciduous species are sparse (McGlone et al. 2004). Nitrogen-fixing plants, which are dependent on high phosphorus levels, are uncommon in the native flora, being largely limited to a few shrubs (*Coriaria*, *Carmichaelia*, *Discaria*). Compared with the widespread introduced N_2-fixing woody species (*Cytisus* and *Ulex*), the natives are slow growing and far less competitive on the nutrient-rich soils associated with agricultural development and in natural alluvial systems.

11.2.2 Low and High Temperature

Compared to Northern Hemisphere trees, indigenous tree species are remarkably cold-sensitive; no species tolerates temperatures below -30 °C (Wardle 1991). This can be linked to the oceanic influences that ameliorated climatic extremes during successive Pleistocene glaciations. Geographic isolation and the limited regional species pool created by warm-temperate/subtropical antecedents also contributed to the weak frost tolerance amongst native plants. Extensive terrain above the upper limit of trees became available during the early Pleistocene when New Zealand was well isolated from cold-adapted woody species. The youthfulness of high-altitude cold habitats and the fluctuations in temperature have left open a zone of alpine grasslands above New Zealand's treeline that, based on climate data, would be occupied by trees in other parts of the world (Wardle 1985). Not surprisingly, Northern Hemisphere high-altitude trees such as *Pinus contorta* are quickly spreading from erosion control plantings throughout this zone, and displacing the native mixed shrubland and grassland communities.

Cold temperatures began relatively recently in the Pliocene in New Zealand, and extirpated at least 36 plant genera with strong subtropical affinities, which are now mostly confined to New Caledonia and Australia (Lee et al. 2001). The impact on the vegetation of cooling climates and of the extreme climate cycles during the Pleistocene was exacerbated by the absence of refugia at lower latitudes. These extinctions, and New Zealand's relative isolation, created a flora depauperate in warm-temperate and subtropical plants. Subsequent reintroduction by humans of many of the lost flora (e.g., *Eucalyptus*, *Acacia*, and many Proteaceae), and their demonstrable invasive ability in warmer climates, suggests numerous vacant niches in our current warm-temperate/subtropical communities.

11.2.3 Low Precipitation

Sub-humid environments characterized New Zealand for most of the Cenozoic, and strong continental climates probably developed only with the origin of the Southern Alps, which created summer-dry climates in eastern regions. Though there are areas where annual precipitation has averaged less than 500 mm in historical times, they are rarely consistently seasonally summer-dry and can receive ecologically significant rainfall at any time. Worldwide, annuals are a notable feature of these dryland habitats but are rare in the New Zealand flora, which has less than five summer-drought-adapted annual species (Rogers et al. 2002). Present in the native flora are xeromorphic features such as deep tap roots or leaves that roll during water stress and/or have waxy cuticles. Although these may have evolved in response to drying of the climate, they are likely to be primary adaptations related to nutrient-poor ecosystems during the Cenozoic. Adaptations to low water availability may be similar in coexisting native and non-native shrubs in a semi-arid environment (Kissel et al. 1987); the range of functional types present is markedly less amongst the natives. Continental climates in New Zealand have experienced probably the greatest invasion of non-native plant species of any biome. Over the last 150 years, a great range of annual, biennial, and perennial plants, grasses and forbs from California and the Mediterranean have invaded low-rainfall areas (Walker and Lee 2000).

11.2.4 Mammalian Herbivory

New Zealand is highly unusual in having had a rich fauna of avian megaherbivores, with no terrestrial herbivorous mammals. Bird browsing was likely a strong selective force in the past, and a major component of many core niches in New Zealand. For example, a high proportion (~20 %) of native plant species in New Zealand are represented in ephemeral turf communities that occupy less than 1 % of the current landscape (Johnson and Rogers 2003). The turf species (<3 cm tall) are well adapted to seasonally dry wetlands, and the removal of aboveground biomass and local nutrient input that is likely to have occurred when numerous birds browsed in these habitats. Observations indicate that these functionally diverse communities maintain their native dominance under strong avian herbivory, nowadays largely by introduced waterfowl, and under these conditions are remarkably resistant to invasion by introduced plants.

In addition to turf communities that may have been browsed by birds, there are also a large number of medium-fertility woody plant species that are well defended against browsing birds, within 3 m of the ground (Bond et al. 2004). Adaptations to the distinctive modes of feeding by the large dinornid ratites are widespread amongst woody plants (Bond et al. 2004), and include strong narrow stems, small widely spaced leaves, and complex three-dimensional branching patterns. However, this architectural syndrome fails against mammals that can shear and manipulate woody shoots. The importation of diverse mammals by Europeans in the early 19th century found an ecologically naïve flora with few defenses against the feeding strategy of herbivorous mammals. Extensive grasslands comprised of tall (*Chionochloa*) and short (*Festuca*, *Poa*) caespitose species were similarly poorly adapted to foraging by grazing mammals, with elevated and exposed meristems, slow growth rates, and episodic flowering. For example, recovery rates in *Chionochloa pallens*, a fast-growing tall tussock species associated with nutrient-rich soils, are exceedingly slow, taking at least two decades to restore biomass removed after one clipping of lamina material (Lee et al. 2000). Low foliar nutrient concentrations may have initially deterred avian herbivores from eating these plants; however, cattle, sheep and rabbits rapidly consumed juvenile plants and those on more fertile soils, and overstocking quickly depleted most montane grasslands.

Multiple mammalian herbivores are now virtually ubiquitous on the mainland, and are strongly filtering plant assemblages. In this selection process, non-native species with a long evolutionary history of mammalian herbivory are increasingly displacing native species – most dramatically in non-forest communities with the highest loadings of mammals. The reduction of palatable and susceptible native species increased resource availability for species that are better adapted to grazing and browsing, due either to chemical defenses (*Hieracium*, *Anthoxanthum*) or physical defenses (spines in *Ulex europeaus*, *Hakea* species), or having a grazing-tolerant growth form (*Agrostis*).

11.2.5 Fire

Natural ignition events have a long history in New Zealand, associated mainly with volcanic eruptions and lightning strikes, but these have never been sufficiently frequent or widespread in a dissected landscape to result in strong selection for fire-adapted traits in the flora. For example, serotiny is limited to populations of the myrtaceous shrub *Leptospermum* and is extremely low, and most native woody and herbaceous species are extremely fire-sensitive. Burning by Maori led to an expansion in range of some native species such as bracken, *Leptospermum*, and *Chionochloa rigida* grasslands (McGlone 2001), but during the pre-European era there was a low diversity of non-native plants available to exploit the reduction in forest. European burning, however, enhanced the spread of an imported pool of fire-adapted non-natives with serotinous or heavily protected capsules (e.g., *Hakea*) and/or large seed banks (e.g., *Ulex europaeus*, *Calluna vulgaris*). Fire also enhances the opportunity for the expansion of numerous ruderal non-native species with prolific seed production (e.g., *Melilotus* spp., *Medicago* spp.).

11.2.6 Soil Disturbance

Soil disturbance in New Zealand is largely created by windthrow, landslides, and a range of erosional and depositional processes associated with colluvial and alluvial systems. Native ruderal plants are rare, and are easily outcompeted by a large number of non-native herbaceous and woody plant species from regions of the world with a long history of human disturbance, and coexistence with burrowing mammals. Native burrowing animals in New Zealand soils are limited to earthworms and although some species reach over 1,000 mm in length, these are generally subsoil specialists (Lee 1959). Terrestrial birds and sea birds such as petrels may have been significant soil-forming and burrowing agents before their widespread demise over a century ago on the mainland by introduced mammalian predators, but evidence for their role in enhancing ecosystem processes remains unclear. Annual and biennial life-forms are well represented amongst non-native plant species, together with fecund, small-seeded, wind-dispersed perennials that effectively colonize disturbed areas created by large-mammal grazing, and a range of human activities associated with settlement and agriculture that maintain open ground.

11.2.7 Interactions Among Factors

Human activities and the template of novel niches that existed when humans arrived overlap to create niches that were even more divergent from what was present over evolutionary timescales in New Zealand. For example, nutrient-rich sites on valley floors that were previously burnt are heavily browsed or grazed by mammalian herbivores and represent a combination of novel niches for native plants, and would be highly susceptible sites for invasion because each factor has generally been rare or absent in New Zealand until recently. This is confirmed for grasslands associated with alluvial systems in eastern South Island, where non-native species richness increases with grazing pressure, flooding frequency, and soil fertility (Walker and Lee 2000).

The large-scale fertilization of tussock grasslands was coupled with a rise in the abundance of European pasture grasses. Native grasses were unable to successfully compete against non-native grasses and legumes under the dual conditions of high nutrient availability and mammalian grazing. The native grasses are unlikely to have experienced equivalent herbivory pressures from the avian fauna because these grasses characteristically inhabited low-nutrient-supply habitats, and the nutritional quality of the grass would seldom be above the minimum level required to support herbivores. Consequently, it is likely that ratite browsers would rarely have removed leaf biomass from these grasses. Native shrubs do invade these systems, but they are either spinescent (*Discaria*), unpalatable (*Ozothamnus*, *Leptospermum*), or prostrate (*Coprosma petriei*).

Phosphorus enrichment in non-forest communities favors herbivore-resistant N_2-fixing plants. The few native N_2-fixers are mainly shrubs, but many of these appear to have little defense against mammalian herbivores and are generally rare. Only when mammals are locally removed, and nutrient levels maintained, do native N_2-fixing woody species increase greatly in abundance (Walker et al. 2003). It is the combination of mammalian grazing and phosphorus addition that consolidates the presence of non-native woody and herbaceous N_2-fixers in many montane grasslands. In the absence of mammals, native shrubs would be much more abundant and more likely to have resisted invasion by shrubs such as gorse and broom.

11.3 Core and Novel Niches: a Predictive, Useful Framework?

Current efforts to address invasive species can be sharpened by understanding the distinctive ecological and evolutionary history of New Zealand. We suggest that the niches that have been present in New Zealand with the least disruption, for the longest time, and over the greatest space should have relatively high relevant functional diversity, and therefore be the least likely to be invaded by plants not native to New Zealand. By implication, identification of factors that create novel niches should be useful not only in predicting invasions, but also in managing plants that have already invaded New Zealand.

The novel-niche theory predicts that the native assemblages of recent niches are more likely to be invaded than those in core or ancient niches. For example, low-fertility, warm, sub-humid, forested sites that lack a recent history of herbivorous mammals and frequent fire should have a higher proportion of native species as structural dominants, and fewer non-native species than those habitats characterized with novel niches. There have been few studies in New Zealand to directly test novel-niche theory, but plot data are gradually becoming available across the country to assess invasions in different environments (Wiser et al. 2001). One problem is that very few unmodified core habitats remain anywhere on the mainland, although the success of invasive species into intact, warm-temperate, infertile, forest systems appears to be relatively low compared with other habitats in New Zealand.

Anthropogenic niches can be managed to decrease the impact of invaders by reducing the impact of the novel factors associated with human activity. Dominance of native plant species is more probable when fire can be eliminated, mammals removed, and exogenous nutrient supply reduced. Introducing surrogate species, especially birds, for native species that have gone extinct should also help reestablish the core-niche factors, and increase community resistance to invasion. Few floras worldwide have evolved solely in the presence of avian herbivores, and it is likely that many introduced plant species would be at a disadvantage under this type of selection pressure. Some anthropogenic disturbances that mimic long-term natural disturbances may also facilitate the success of native plants and decrease invisibility. For example, actions such as mowing and maintaining grazing by small mammals may foster native turf plants while decreasing invasive, sward-forming introduced grasses.

Novel niches that are independent of humans will be the most difficult to maintain free of non-native species. This will be especially true when other areas with similar niches outside of New Zealand have had a greater evolutionary extent, and have greater native diversity than the New Zealand assemblages. For example, even in the absence of mammalian herbivores, alpine and southern beech forests have low native diversity relative to similar habitats in other parts of the world. It is probable that these habitats have not been invaded yet due to low propagule pressure, not because of high long-term resistance to invasion. If these habitats are also subjected to additional novel-niche factors, such as increased fertility that comes with fertilization or nitrogen deposition, we hypothesize that invasions would become ever more likely at the same propagule pressure.

A complete theory of plant invasions will be a union of the mechanisms of invasion within the constraints imposed by evolution and dispersal associated with varying environmental conditions. A predictive theory will include plant traits that are associated with resource acquisition and maintenance, but will extend to other factors that are important in a plant's life cycle: predator escape, propagule availability, dispersers and pollinators abundance, stresses, and disturbances. These mechanistic, important details about the invasion process need to be contextualized to the evolutionary history of the niche as well as the native assemblage and non-native species. These broader analyses are helpful in structuring our questions regarding invaders, and determining where to focus scarce human resources. Operationally, incorporating our understanding of the history of habitats and their assemblages in New Zealand with mechanistic understanding of the invasion processes is the best way to proceed to manage invasions in the country.

References

Bond WJ, Lee WG, Craine JM (2004) Plant structural defences against browsing birds: a legacy of New Zealand's extinct moas. Oikos 104:500–508

Craine JM, Lee WG (2003) Covariation in leaf and root traits for native and non-native grasses during an altitudinal gradient in New Zealand. Oecologia 134:471–478

Davis MA, Pelsor M (2001) Experimental support for a resource-based mechanistic model of invasibility. Ecol Lett 4:421–428

Elton CS (1958) The ecology of invasions by animals and plants. Methuen, London

Fargione J, Brown CS, Tilman D (2003) Community assembly and invasion: An experimental test of neutral versus niche processes. Proc Natl Acad Sci USA 100:8916–8920

Guthrie-Smith H (1921) Tutira: the story of a New Zealand sheep station. Blackwood, London

Hector A, Dobson K, Minns A, Bazeley-White E, Lawton JH (2001) Community diversity and invasion resistance: an experimental test in a grassland ecosystem and a review of comparable studies. Ecol Res 16:819–831

Johnson PN, Rogers GM (2003). Ephemeral wetlands and their turfs in New Zealand. Department of Conservation, Wellington, New Zealand

Kissel RM, Wilson JB, Bannister P, Mark AF (1987) Water relations of some native and exotic shrubs of New Zealand. New Phytol 107:29–37

Kolar CS, Lodge DM (2001) Progress in invasion biology: predicting invaders. Trends Ecol Evol 16:199–204

Lee KE (1959) The earthworm fauna of New Zealand. DSIR Bull 130. Wellington, New Zealand

Lee WG, Fenner M, Loughnan A, Lloyd KM (2000) Long-term effects of defoliation: incomplete recovery of a New Zealand alpine tussock grass, *Chionochloa pallens*, after 20 years. J Appl Ecol 37:348–355

Lee DE, Lee WG, Mortimer N (2001) Where and why have all the flowers gone? Depletion and turnover in the New Zealand Cenozoic angiosperm flora in relation to palaeogeography and climate. Aust J Bot 49:341–356

Lloyd KM, Lee WG, Wilson JB (2002) Competitive abilities of rare and common plants: comparisons using *Acaena* (Rosaceae) and *Chionochloa* (Poaceae) from New Zealand. Conserv Biol 16:975–985

Lodge DM (1993) Biological invasions: lessons for ecology. Trends Ecol Evol 8:133–137

MacArthur RH, Wilson EO (1967) The theory of island biogeography. Princeton Univ Press, Princeton, NJ

McGlone MS (2001) The origin of the indigenous grasslands of southeastern South Island in relation to pre-human woody ecosystems. NZ J Ecol 25(1):1–15

McGlone MS, Dungan RJ, Hall GMJ, Allen RB (2004) Winter leaf loss in the New Zealand woody flora. NZ J Bot 42:1–19

Miller TE, Kneitel JM, Burns JH (2002) Effect of community structure on invasion success and rate. Ecology 83:898–905

Mooney HA, Drake JA (1986) Ecology of biological invasions of North America and Hawaii. Ecological Studies Analysis and Synthesis, vol 58. Springer, Berlin Heidelberg New York

O'Connor KF (1980) The use of mountains: a review of New Zealand experience. In: Anderson AG (ed) The land our future: essays on land use and conservation in New Zealand in honour of Kenneth Cumberland. Longman Paul/New Zealand Geographical Society, Auckland, New Zealand, pp 193–222

Olden JD, Poff NL (2003) Toward a mechanistic understanding and prediction of biotic homogenization. Am Nat 162:442–460

Pheloung PC, Williams PA, Halloy SR (1999) A weed risk assessment model for use as a biosecurity tool evaluating plant introductions. J Environ Manage 57:239–252

Reich PB, Buschena C, Tjoelker MG, Wrage K, Knops J, Tilman D, Machado JL (2003) Variation in growth rate and ecophysiology among 34 grassland and savanna species under contrasting N supply: a test of functional group differences. New Phytol 157:617–631

Robinson GR, Quinn JF, Stanton ML (1995) Invasibility of experimental habitat islands in a California winter annual grassland. Ecology 76:786–794

Rogers G, Walker S, Tubbs M, Henderson J (2002) Ecology and conservation status of three "spring annual" herbs in dryland ecosystems of New Zealand. NZ J Bot 40:649–669

Verheyen K, Vellend M, van Calster H, Peterken G, Hermy M (2004) Metapopulation dynamics in changing landscapes: a new spatially realistic model for forest plants. Ecology 85:3302–3312

Walker S, Lee WG (2000) Alluvial grasslands in south-eastern New Zealand: vegetation patterns, long-term and post-pastoral change. J R Soc NZ 30:69–103

Walker S, Wilson JB, Lee WG (2003) Recovery of short tussock and woody species guilds in ungrazed *Festuca novae-zelandiae* short tussock grassland with fertiliser or irrigation. NZ J Ecol 27:179–189

Wardle P (1985) New Zealand timberlines. 3. A synthesis. NZ J Bot 23:263–271

Wardle P (1991) Vegetation of New Zealand. Cambridge Univ Press, Cambridge

Williamson M (1999) Invasions. Ecography 22:5–12

Wiser SK, Allen RB, Clinton PW, Platt KH (1998) Community structure and forest invasion by an exotic herb over 23 years. Ecology 79:2071–2081

Wiser SK, Bellingham PJ, Burrows LE (2001) Managing biodiversity information: development of New Zealand's National Vegetation Survey databank. NZ J Ecol 25(2):1–17

12 Controls on the Population Dynamics of Invading Mammals

D.M. FORSYTH

12.1 Introduction

This chapter outlines the development of concepts (models) of controls on the population dynamics of invading large mammalian herbivores in New Zealand. The primary aim is to reconcile models with available data. The focus is on models appropriate to the New Zealand context. These models are all based on the assumption that, in the absence of predators, food determines the abundance of large herbivores. A secondary aim is to suggest directions for further work.

12.2 Models of Invading Large Herbivore–Plant Dynamics

12.2.1 Eruptive

Holloway (1950) gave the first description of a sequence of concomitant changes in the density (number per unit area) of invading large mammalian herbivores (red deer *Cervus elaphus scoticus*) and their food supply (forests in western Southland) in New Zealand. Holloway proposed that the forest initially provides a maximum amount of browse, such that the herbivore population increases most rapidly. The deer feed selectively, such that preferred plants are removed first. The population increases and eventually "overtakes ... available food supply". Animals begin to starve and population density declines. The deer population eventually achieves a static ("equilibrium") density dependent upon the renewal rate of food plants. Holloway recognized that changes in the abundance of the food supply were not uniform, but rather varied according to the palatability of the plant species to deer. He postulated

Ecological Studies, Vol. 186
R.B. Allen and W.G. Lee (Eds.)
Biological Invasions in New Zealand

that the most palatable species could become locally extinct, and that unpalatable species would increase in abundance.

Subsequent workers (Howard 1964a, b; Riney 1964; Caughley 1970; Challies 1985, 1990) have elaborated upon this model by subdividing it into four stages, but their stage definitions differ. The stages of Howard (1964a, b) and Riney (1964) were based upon attributes of the herbivore and the food supply. Caughley (1970) noted that these were "not congruent", and based his "hypothetical" stages only on attributes of the herbivore's population (Fig. 12.1). The "initial increase" represents the period between establishment and the initial population peak. The "initial stabilization" lasts from the population peak until the start of the decline. The "decline" continues until the population's rate of increase changes from negative to zero. The "postdecline" is characterized by small fluctuations around an equilibrium density. Riney (1964) suggested that the postdecline density is lower than the initial peak because the availability of forage has been modified.

Because herbivore density is postulated to undergo a large increase followed by a decline, this conceptual framework has become known as the eruptive (or irruptive) model. Caughley's (1970) major insight was realizing that changes in rate of increase (from positive to zero to negative to zero) defined the eruptive model. Challies (1985, 1990) confused the issue by attempting to link stages in the eruptive model with the food supply. For example, the zero rate of increase postulated for the initial stabilization was "when deer are temporarily in balance with their habitat" (Challies 1990, p. 448).

The eruptive model has become the paradigm for understanding the population dynamics of introduced ungulates in New Zealand. For example, Challies (1990, p. 448) claimed "many of the local red deer populations have passed through a classic population irruption". The eruptive model has also been an influential concept in North America (review in McCullough 1997).

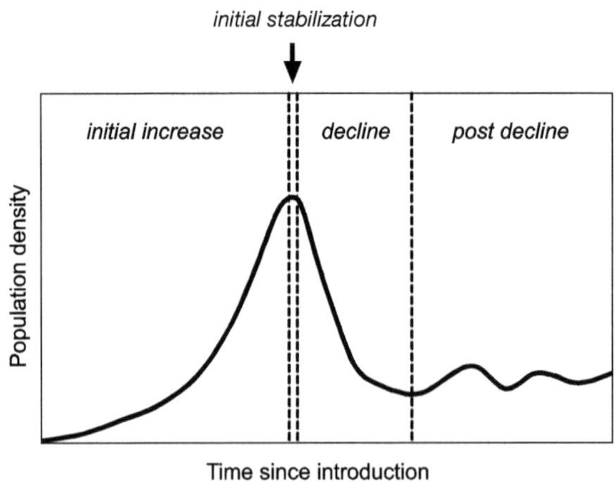

Fig. 12.1 Representation of the eruptive model showing the four stages defined by Caughley (1970)

Forsyth and Caley (2006) recently developed single-species mathematical models that can generate the eruptive dynamics described above.

12.2.2 Logistic

The simplest class of herbivore population model is the θ-logistic (Gilpin and Ayala 1973), which can be written as:

$$N_{t+1} = N_t \exp\left[r_m\left(1-\left(\frac{N_t}{K}\right)^{\theta}\right)\right] \quad (12.1)$$

where N is the population size in year t, r_m the intrinsic rate of increase, K the carrying capacity, and θ represents the form of density dependence (Fig. 12.2a, c). A number of variants on this model are available (see Barlow

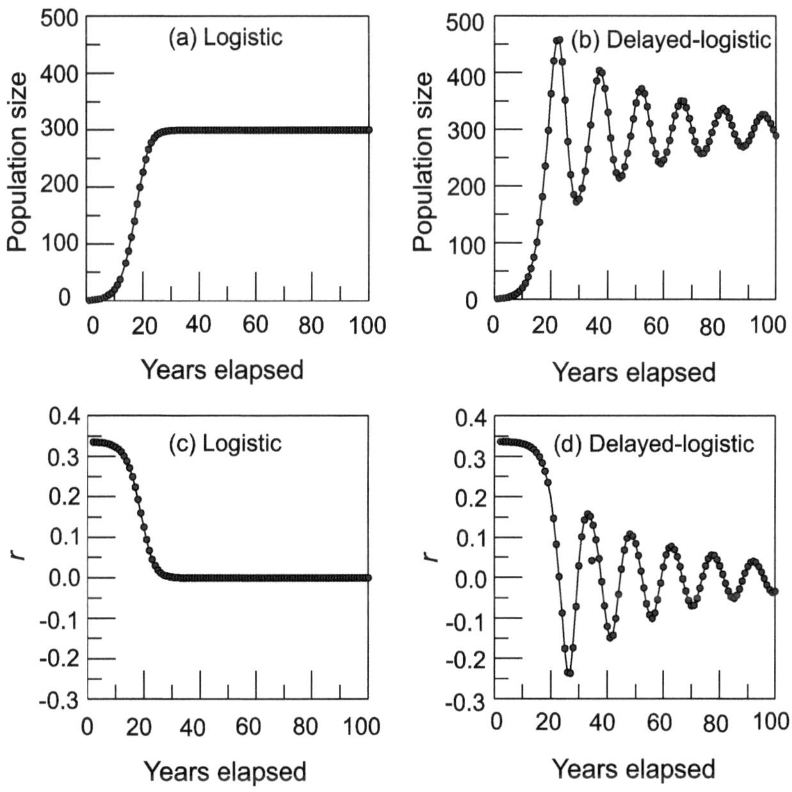

Fig. 12.2a–d Comparison of the θ-logistic (θ=1) and delayed-logistic models (T=3). r_m and K were arbitrarily set at 0.4 and 300, respectively, for both models. **a** and **b** show changes in herbivore population size with time; **c** and **d** show changes in the annual exponential rate of increase (r) with time

and Clout 1983). The intrinsic rate of increase is the maximum increase in population size possible from t to $t+1$ (review in Sibly and Hone 2002).

In the θ-logistic, the population's exponential rate of increase (r) is greatest at lowest densities. However, unlike the eruptive model, the θ-logistic population does not undergo any oscillation, instead increasing to equilibrium. The logistic model assumes that the rate of renewal of the food supply is constant, and not influenced by the number of animals using it or by its initial biomass (Caughley 1976).

May and Oster (1976) showed how the θ-logistic model can generate complicated dynamics, ranging from a stable cycle to chaos, when $r_m>2$ (see also May 1976). Hence, it can be difficult to distinguish data generated by these deterministic models from stochastic variation and sampling and/or measurement error.

The θ-logistic model has been widely used to model changes in the abundance of possums (*Trichosurus vulpecula*) in New Zealand (e.g., Barlow and Clout 1983; Barlow 1991) because of "its simplicity and tractability" (Barlow and Clout 1983, p. 256). Barlow and Clout (1983) argued that the two most-likely forms of population regulation in possums – competition for refuges and competition for food – both give growth curves peaked to the right (i.e., $\theta>1$).

Hutchinson (1948) modified the logistic model to include the assumption of a delay between changes in resource availability and changes in the rate of increase. The delayed-logistic,

$$N_{t+1} = N_t \exp\left[r_m\left(1-\frac{N_{t-T}}{K}\right)\right] \tag{12.2}$$

has the population changing as a function of *lagged* density (T). Although the delayed-logistic model still assumes that the rate of food renewal is constant and independent of N, it may be useful because the lag time could (fortuitously) describe the generation time of vegetation (May 1973). The delayed-logistic therefore appears to better represent the dynamics of the eruptive model than does the θ-logistic (Fig. 12.2b, d).

12.2.3 Interactive

Caughley (1976) considered that the logistic models described above did not adequately represent the food dynamics of the system. He therefore modified the Lotka-Volterra equations into an *interactive* model that related the dynamics of the herbivore to the dynamics of the food resource. Based on May (1973), Caughley's interactive model included one equation for the change in vegetation biomass (V) as a consequence of what is present, potential maximum (K), minus the amount removed by herbivores:

$$V_{t+1} = V_t + a_1 V_t \left(1 - \frac{V_t}{K}\right) - c_1 N_t \left(1 - e^{-d_1 V_t}\right) \tag{12.3a}$$

The first two components are described in the θ-logistic model, with a_1 the maximum rate at which vegetation biomass increases. The last component describes how much a single herbivore would eat given that biomass was present (i.e., the functional response), multiplied by the number of herbivores present (N). d_1 is the rate at which intake changes as V increases, and c_1 the rate at which vegetation growth is slowed by animals eating it. The change in the number of herbivores was also modeled using the functional response:

$$N_{t+1} = N_t + N_t \left[-a_2 + c_2 \left(1 - e^{-d_2 V_t}\right)\right] \tag{12.3b}$$

where a_2 is the rate at which the herbivore population will decline when V=0, c_2 the rate at which the population increases when animals can eat unconstrained, and d_2 determines the rate at which c_2 declines as vegetation biomass is consumed.

With this model, Caughley (1976) captured what he considered to be the dynamics of the eruptive model for a hypothetical introduction of white-tailed deer into a mosaic of grassland and forest (Fig. 12.3). He used parame-

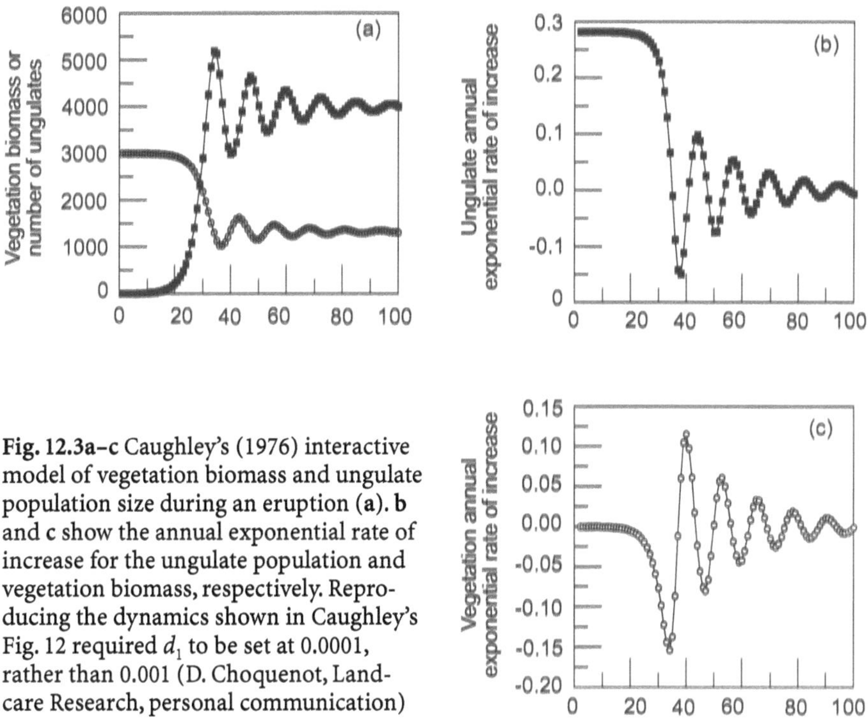

Fig. 12.3a–c Caughley's (1976) interactive model of vegetation biomass and ungulate population size during an eruption (**a**). **b** and **c** show the annual exponential rate of increase for the ungulate population and vegetation biomass, respectively. Reproducing the dynamics shown in Caughley's Fig. 12 required d_1 to be set at 0.0001, rather than 0.001 (D. Choquenot, Landcare Research, personal communication)

ter values that "might be expected", and considered that his *interactive* model was a more informative simulation of ungulate population dynamics than either the logistic or delayed-logistic models. Owen-Smith (2002) elaborated Caughley's (1976) model to include density-dependent interference in herbivore feeding, and several mechanisms of vegetation change.

12.3 Reconciling Models with Data

12.3.1 Ungulates in New Zealand

The deterministic models above assume that, following arrival into suitable habitat, populations will grow following the same trajectory. However, there is a strong theoretical basis for expecting that the number of individuals involved ("propagule size") will be an important determinant of subsequent dynamics. The likelihood of a propagule becoming extinct is inversely related to propagule size because of the combined effects of demographic and environmental stochasticity, and random catastrophes (Dennis 2002). Of the 14 ungulate (order: Artiodactyla) species released into New Zealand, 11 were judged by Forsyth and Duncan (2001) to have established. Across species, there was a threshold introduction size of about six individuals, below which ungulate releases in New Zealand were likely to fail, and above which they were likely to succeed (Forsyth and Duncan 2001). A similar relationship was found for the 16 introductions of six ungulate species into Victoria, Australia, with introductions involving four or fewer individuals going extinct, whereas those introductions involving seven or more individuals established (Forsyth et al. 2004). Hence, there appears to be a strong stochastic component to the establishment of introduced mammal populations.

Caughley (1970) tested the eruptive model using Himalayan tahr (*Hemitragus jemlahicus*), a mountain ungulate that had established near Mt. Cook and subsequently expanded its range north and south along the Southern Alps. Caughley assumed that the four stages of the eruption (Fig. 12.1) would be represented by tahr at different locations along the gradient of colonization, with those in the most recently colonized areas being in the initial increase stage, and those in the longest-colonized areas being in the postdecline stage. His sampling sites were hypothesized to represent the four stages of the eruptive model. Tahr were shot, sexed and aged. Life-tables for female tahr supported parts of the eruptive model, the rate of increase being high (r=0.13) at the site predicted to be in the initial increase stage, and close to zero (r=0.01) at the site predicted to be in the initial stabilization stage. However, because data were not available for the decline stage, these results also support a θ-logistic model. Only long-term monitoring of density can discriminate between the postdecline stage of the eruptive model and K in the logistic model. Caughley observed that rate of increase was determined mainly by variation in survival rates, rather than by reproductive rates.

As well as noting the relationship between changes in r and stages of the eruptive model, Caughley (1970) predicted a relationship between r (a parameter of the population) and body condition (a parameter of the individual). Adult female kidney fat index (an index of body condition) of tahr supported these predictions – highest during the initial increase stage (when r is largest), lowest during the decline ($r<0$), but slightly higher in the postdecline ($r=0$), although sample sizes were small (≤ 27) for the latter two stages. These predictions were supported by analysis of kidney fat data from a much larger sample ($n=4{,}789$) of tahr shot during May–September 1972–1976, and with many "replicates" of colonization history (Forsyth et al. 2005).

Challies (1978, 1985, 1990) further explored the link between time since establishment, demography and body condition. He estimated carcass parameters from hunted populations of red deer established for different periods of time. The carcass weight of both 2-year-old males and females was highest in the most recently established population (sampled ca. 10 years after colonization), but declined thereafter. A similar pattern was observed for the proportion of lactating 2- and 3-year-old females (Challies 1978, 1985, 1990). Following the advent of intensive commercial hunting, which reduced deer densities >90 % in 15 years (Challies 1977), carcass weight subsequently increased, presumably due to increased per capita food availability.

12.3.2 Possums in New Zealand

Patterns in annual kill-trapping of possums in a southern North Island catchment during 1945–1989 were reported as being consistent with "irruptive fluctuation" (Thomas et al. 1993). Possums colonized the catchment in ca. 1920, and kill-trapping was conducted along the same lines in 1945, 1951, and annually from 1964 onward. There appeared to be a large decrease in the abundance of possums from 1945 to 1965, and a further decline in 1977. Possums subsequently increased, such that by 1989 the numbers trapped were comparable with those reported for the period 1965–1976. Although the eruptive model was originally developed for "large herbivores", possums have many of the demographic characteristics of ungulates (Thomas et al. 1993). Hence, the eruptive model could reasonably be expected to apply to possums, too.

A mark-recapture study of possums on a 15-ha grid in mixed lowland podocarp-broadleaved forest in the Orongorongo Valley, North Island, during 1967–1997 provides strong evidence for an "equilibrium" (Efford 2000). The population has fluctuated between 5.5 and 13.3 possums ha^{-1} for 30 years, and shows a return tendency (sensu Murdoch 1994). The Orongorongo possum population was more stable (i.e., the fluctuations around a long-term mean were lower) than most other mammals that have been studied (Efford 2000). However, the equilibrium is consistent with both the "postdecline" of the eruptive model and *K* in the logistic models.

12.3.3 International Studies

Two long-term studies of the abundance of invading large herbivores are summarized in Fig. 12.4. Sæther et al. (2002) reported annual counts of a population of ibex (*Capra ibex*) derived from seven animals introduced into the Swiss National Park in 1920. The population increased to a peak of 280 animals in 1963, then declined to a low of 128 in 1970, before increasing again (Fig. 12.4a, d). Scheffer (1951) reported the number of reindeer (*Rangifer tarandus*) counted annually on St. Paul and St. George islands, Pribilof Group, USA, following their introduction in 1911 (Fig. 12.4b, c, e, f). Both populations

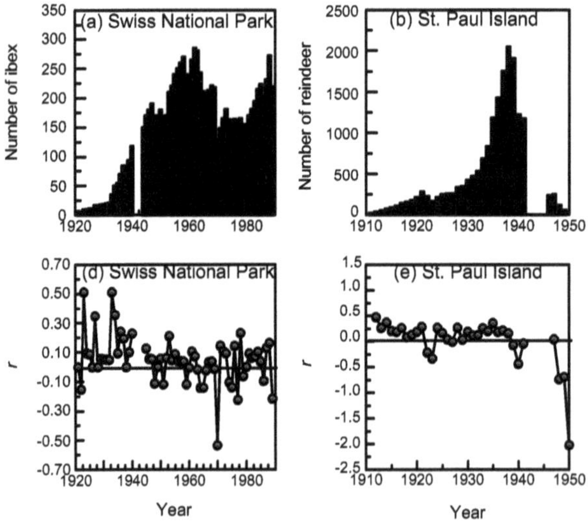

Fig. 12.4a Number of ibex counted annually in the Swiss National Park following their introduction in 1920. Note that no counts were conducted during 1941–1943. **b, c** Number of reindeer counted annually on St. Paul Island and St. George Island, Pribilof Group, USA, following their introduction in 1911 (data from Scheffer 1951). No counts were conducted during 1942–1945 on St. Paul, and during 1942–1944 and in 1950 on St. George. On St. Paul, harvests were nil or <9 %, except in 1941 and 1942 when 21 and 27 % of animals were killed, respectively. On St. George, harvests were negligible, except during 1918–1926, when 10–22 % of the annual count were killed (see Scheffer 1951). **d–f** Respective annual exponential rates of increase (*r*) of ibex and reindeer calculated from those data

increased substantially before declining to very low numbers. The population on St. George appeared to fluctuate around some postdecline equilibrium predicted by the eruptive model, although it was close to extinction in 1949 (eight animals). Insufficient data prevent assessment of the postdecline stage on St. Paul. Forsyth and Caley (2006) showed that the dynamics of these three populations, plus three of four other populations either introduced to new range or released from harvesting, were best described by eruptive models (of which the delayed logistic was one model): only one of the seven populations did not display the predicted eruptive dynamic (i.e., was best described by the θ-logistic model).

To my knowledge, only one other study has attempted to compare which of the θ- and delayed-logistic models best describes the dynamics of invading mammals. Maas (1997) experimentally reduced the abundance of a population of feral goats (*Capra hircus*) in the semi-arid rangelands of New South Wales, Australia, and compared its subsequent dynamics (abundance estimated quarterly) with a nearby population that was not manipulated. Both the θ- and delayed-logistic models fitted the data poorly. This may have been because the density of goats was not sufficiently reduced to detect either form of logistic growth with the monitoring used, even if logistic growth was the true model (Maas 1997). Maas concluded "the dynamics could not be represented by single species models ... probably due to the stochastic environment in which they lived".

12.4 Environmental Variation and the Dynamics of Invading Mammals

Sæther (1997) concluded that stable equilibria in large ungulate populations and their environment were unlikely because of the effects of both environmental variation and population density on demography. Environmental variation can vary the food supply available to herbivores (Caughley et al. 1987), and McLeod (1997) argued that the notion of a relatively constant carrying capacity (K) is inappropriate in variable environments. Bayliss and Choquenot (2002) illustrate how even small stochastic variation in K (5 % year-to-year) can mask θ-logistic growth. These studies suggest that variation in food supply caused by the environment independent of the herbivore will be an important determinant of the dynamics of invading herbivores. Such variation, at least in some circumstances, may be extreme (Caughley et al. 1987).

If it is assumed that the abundance of large herbivores, in the absence of predators, is determined by food availability, then two sources of variation in food availability and quality could affect the dynamics of animal populations

(Andrewartha and Birch 1954). *Intrinsic* variation arises from the effect of animals consuming their own food resources; *extrinsic* variation arises from the effects of factors independent of the animals' foraging. Choquenot (1998) evaluated the relative influence of intrinsic and extrinsic variation in food availability on the rate of increase of feral pig (*Sus scrofa*) populations in New South Wales, Australia. In a large-scale field experiment, neither pasture biomass nor r declined with increasing pig density, suggesting that food availability was dominated by extrinsic factors. Modeling indicated that such a result was likely because pigs, relative to other herbivores present, were inefficient grazers and occurred at low densities.

A continuum of disturbance exists, from frequent but small-scale gap formation, through to infrequent but large-scale catastrophes such as fires, floods, hurricanes, and earthquakes. In Westland, New Zealand, there is evidence of three massive earthquakes within the last 650 years, with a large proportion of the forest cover comprising a cohort of trees established after the most recent disturbance (Wells et al. 2001).

One difference between grasslands and forests is that in the former ungulates have virtually constant access to all of their food supply, whereas in forests these animals can access forage only within the browse tier (e.g., assumed to be 2 m for red deer; Nugent et al. 2001a). The amount of food available in a forest is dependent on tree demographics, which can be unrelated to deer (see Coomes et al. 2003). Ungulates can therefore remove only a small proportion of foliage from most canopy trees, and are unlikely to affect the mortality of mature trees. However, there may be a lack of regeneration, because saplings and seedlings are killed faster than they are replaced. Litterfall is an important component of the diet of deer in some New Zealand forests (Nugent et al. 2001a).

Choquenot (1998) proposed a continuum between grazing systems dominated by intrinsic and extrinsic sources of variation in food availability. Increasing environmental stochasticity and declining efficiency of the vegetation–herbivore feedback loop means that food availability would be of an increasingly extrinsic origin. Feral pigs in the rangelands of New South Wales are at the extrinsic end of the continuum (Choquenot 1998). New Zealand forests, in which the *supply* of a major food of deer (litterfall) is independent of the density of deer, will be at the extrinsic end of the continuum (D. Choquenot, Landcare Research, personal communication). However, litterfall could also be a consequence of deer historically altering canopy composition. Decreasing environmental stochasticity and increasing efficiency of the vegetation–herbivore feedback means that food shortages would be increasingly intrinsic in origin.

Temperature-driven synchrony in fruiting by masting species may link simultaneous fluctuations in animal condition and breeding in New Zealand (see Chap. 16). At four long-term study sites spanning 120 km in lower North Island, possum breeding was much reduced in one year (1996) at all sites

(28 % of adult females bred, compared to >80 % in other years; Ramsey et al. 2002). The probability of adult females breeding declined as body condition fell below average. Hinau (*Elaeocarpus dentatus*) is a masting tree species (Schauber et al. 2002), and possums eat its fruit. An index of hinau fruiting plus possum population density in the preceding year were the most important predictors of both body condition and breeding rate (Ramsey et al. 2002). Data from two of the sites (ca. 50 km apart) suggested a delayed density-dependent effect on breeding, through competition for highly nutritious components of the diet reducing body condition.

12.5 Harvesting and the Dynamics of Invading Mammals

Harvesting can be an important driver of the dynamics of some populations. Forsyth et al. (2003) used a θ-logistic model to investigate how the form and size of the harvest relative to recruitment can regulate a population already at *K*. Three outcomes are possible for a harvested population: eradication, regulation at low density, and regulation at high density. There are well-documented examples of feral goats being eradicated from offshore islands (e.g., Raoul Island; Parkes 1984). Red deer appear to be regulated at low density by helicopter-based commercial hunting, at least in areas with a large proportion of open habitat (e.g., Fiordland; Nugent et al. 1987). Helicopter-based commercial hunting reduced the density of Himalayan tahr in Carneys Creek by >90 % (Tustin and Challies 1978), enabling recreational hunters to subsequently regulate this population at low density (Forsyth 1999).

12.6 Interspecific Competition and the Dynamics of Invading Mammals

Smaller ungulates may have the potential to out-compete larger species by depressing vegetation biomass below that required to meet the greater absolute requirements of the latter (review in Owen-Smith 2002). The decline and possible extinction of moose (*Alces alces*) in Fiordland, New Zealand, has been attributed to the reduced availability of food caused by the arrival and subsequent eruption of the smaller red deer (Nugent et al. 2001b). Sika deer (*Cervus nippon*) have apparently out-competed the larger red deer in central North Island (Davidson and Fraser 1991). Red deer have failed to expand their range into the Greenstone and Caples valleys (Otago), which were first colonized by fallow deer (*Dama dama*). Red deer also have a limited distribution on Stewart Island, where the smaller white-tailed deer is widespread and abundant (Challies 1990). However, a counter-example is the observation that increasing densities of Himalayan tahr are associated with declining densities of the smaller chamois (*Rupicapra rupicapra*) in the eastern Southern Alps (Forsyth and Hickling 1998). Both species overlap in diet (Parkes and Thomson 1995) and habitat selection (Forsyth 2000), so scramble competition for food may be important. An alternative mechanism is that chamois avoid the more social tahr (Forsyth and Clarke 2001).

12.7 Synthesis and Directions for Further Work

How can the sources of variation described above be incorporated into models of the dynamics of invading mammals? Some interactive models can capture components of variation affecting the food supply (review in Bayliss and Choquenot 2002), and may be appropriate for systems in which the herbivore has access to all of a homogenous food resource (e.g., feral pigs and pasture; Choquenot 1998). However, when dealing with multi-species forest communities, where turnover may take hundreds of years, these models appear simplistic.

Mechanistic and spatial simulation models appear to have the greatest potential to extrapolate from short-term and small-scale field measurements to long-term and large-scale dynamics of mixed forests, and to incorporate disturbances (i.e., model the food supply). One of these models (SORTIE: Pacala et al. 1996) is being parameterized for the forests of eastern Fiordland (Coomes et al. 2003). The challenge is to link the foraging and energetics of the herbivore (Moen et al. 1997) to the recruitment, growth and mortality of the forest species at the appropriate scales. The energetic intake of the herbivore should predict age-specific changes in body mass and body condition (see Illius and Gordon 1999), the determinants of survival, and reproduction. In an important recent advance, Owen-Smith (2002) describes methods for scaling ungulate foraging across spatial and temporal scales, and for linking intake to nutritional gain and then individual growth, storage, reproduction, and survival. Since the latter two life-history variables are determinants of *r* (Sibly and Hone 2002), the dynamics of both the invading herbivore and vegetation could be simulated under various scenarios. Owen-Smith (2002) proposes that the apparent complexity of ungulates foraging on potentially many species can be reduced to a manageable level by classifying plant species into classes based on the likelihood that they will be consumed when encountered by a foraging herbivore (see also Forsyth et al. 2002).

This approach could therefore capture the dynamics described for both the herbivore (changes in: herbivore foraging behavior, body mass and condition, age-specific fecundity and survival, *r*, and abundance) and plants (changes in: growth, survival, demographics, dispersal, recruitment, and abundance) during the invasion of New Zealand habitats by large herbivores.

Such models could be modified to include the effects of harvesting (predation) risk and interspecific competition outlined above (see Owen-Smith 2002). Parallel to model development should be the collection of independent data by which to test and refine the model.

References

Andrewartha HG, Birch LC (1954) The distribution and abundance of animals. Univ Chicago Press, Chicago

Barlow ND (1991) Control of endemic bovine Tb in New Zealand possum populations: results from a simple model. J Appl Ecol 28:794–809

Barlow ND, Clout MN (1983) A comparison of 3-parameter, single-species population models, in relation to the management of brushtail possums in New Zealand. Oecologia 60:250–258

Bayliss P, Choquenot D (2002) The numerical response: rate of increase and food limitation in herbivores and predators. Philos Trans R Soc Lond B 57:1233–1248

Caughley G (1970) Eruption of ungulate populations, with emphasis on Himalayan thar in New Zealand. Ecology 51:53–72

Caughley G (1976) Wildlife management and the dynamics of ungulate populations. In: Coaker TH (ed) Applied Biology, vol 1. Academic Press, London, pp 183–246

Caughley G, Shepherd N, Short J (eds) (1987) Kangaroos: their ecology and management in the sheep rangelands of Australia. Cambridge Univ Press, Cambridge

Challies CN (1977) Effects of commercial hunting on red deer densities in the Arawata Valley, South Westland, 1972-76. NZ J For Sci 7:263–273

Challies CN (1978) Assessment of the physical well-being of red deer (*Cervus elaphus* L.) populations in South Westland, New Zealand. PhD Thesis, University of Canterbury, Christchurch

Challies CN (1985) Establishment, control, and commercial exploitation of wild deer in New Zealand. In: Fennessy PF, Drew KR (eds) Biology of deer production. R Soc NZ Bull 22:23–36

Challies CN (1990) Red Deer. In: King CM (ed) The handbook of New Zealand mammals. Oxford Univ Press, Auckland, pp 436–458

Choquenot D (1998) Testing the relative influence of intrinsic and extrinsic variation in food availability on feral pig populations in Australia's rangelands. J Anim Ecol 67:887–907

Coomes DA, Allen RB, Forsyth DM, Lee WG (2003) Factors preventing the recovery of New Zealand forests following control of invasive deer. Conserv Biol 17:450–459

Davidson MM, Fraser KW (1991) Official hunting patterns, and trends in the proportions of sika (*Cervus nippon*) and red deer (*C. elaphus scoticus*) in the Kaweka Range, New Zealand, 1958–1988. NZ J Ecol 15:31–40

Dennis B (2002) Allee effects in stochastic populations. Oikos 96:389–401

Efford M (2000) Possum density, population structure, and dynamics. In: Montague TL (ed) The brushtail possum: biology, impact and management of an introduced marsupial. Manaaki Whenua Press, Lincoln, Canterbury, pp 47–61

Forsyth DM (1999) Long-term harvesting and male migration in a New Zealand population of Himalayan tahr *Hemitragus jemlahicus*. J Appl Ecol 36:351–362

Forsyth DM (2000) Habitat selection and coexistence of the Alpine chamois (*Rupicapra rupicapra*) and Himalayan tahr (*Hemitragus jemlahicus*) in the eastern Southern Alps, New Zealand. J Zool (Lond) 252:215–225

Forsyth DM, Caley P (2006) Testing the irruptive paradigm of large-herbivore dynamics. Ecology 87:2997–303
Forsyth DM, Clarke CMH (2001) Advances in New Zealand mammalogy 1990–2000: chamois. J R Soc NZ 31:243–249
Forsyth DM, Duncan RP (2001) Propagule size and the relative success of exotic ungulate and bird introductions to New Zealand. Am Nat 157:583–595
Forsyth DM, Hickling GJ (1998) Increasing Himalayan tahr and decreasing chamois densities in the eastern Southern Alps, New Zealand: evidence for interspecific competition. Oecologia 113:377–382
Forsyth DM, Coomes DA, Nugent G, Hall GMJ (2002) Diet and diet preferences of introduced ungulates (Order: Artiodactyla) in New Zealand. NZ J Zool 29:323–343
Forsyth DM, Hone J, Parkes JP, Reid G, Stronge D (2003) Feral goat control in Egmont National Park, New Zealand, and the implications for eradication. Wildl Res 30:437–450
Forsyth DM, Duncan RP, Bomford M, Moore G (2004) Climatic suitability, life-history traits, introduction effort and the establishment and spread of introduced mammals in Australia. Conserv Biol 18:557–569
Forsyth DM, Duncan RP, Tustin KG, Gaillard J-M (2005) A substantial energetic cost to male reproduction in a sexually dimorphic ungulate. Ecology 86:2154–2163
Gilpin ME, Ayala FJ (1973) Global models of growth and competition. Proc Natl Acad Sci USA 70:3590–3593
Holloway JT (1950) Deer and the forests of western Southland. NZ J For 6:123–137
Howard WE (1964a) Introduced browsing mammals and habitat stability in New Zealand. J Wildl Manage 28:421–429
Howard WE (1964b) Modification of New Zealand's flora by introduced mammals. Proc NZ Ecol Soc 11:59–62
Hutchinson GE (1948) Circular causal systems in ecology. Ann NY Acad Sci 50:221–246
Illius AW, Gordon IJ (1999) Scaling up from functional response to numerical response in vertebrate herbivores. In: Olff H, Brown VK, Drent RH (eds) Herbivores: between plants and predators. Blackwell, Oxford, pp 397–425
Maas S (1997) Population dynamics and control of feral goats in a semi-arid environment. Master Appl Sci Thesis, University of Canberra, Australia
May RM (1973) Stability and complexity in model ecosystems. Princeton Univ Press, Princeton
May RM (1976) Models for single populations. In: May RM (ed) Theoretical ecology: principles and applications. Blackwell, Oxford, pp 4–25
May RM, Oster GF (1976) Bifurcations and dynamic complexity in simple ecological models. Am Nat 110:573–599
McCullough DR (1997) Irruptive behavior in ungulates. In: McShea WJ, Underwood HB, Rappole JH (eds) The science of overabundance: deer ecology and population management. Smithsonian Institution Press, Washington, DC, pp 69–98
McLeod SR (1997) Is the concept of carrying capacity useful in variable environments? Oikos 79:529–542
Moen R, Pastor J, Cohen Y (1997) A spatially explicit model of moose foraging and energetics. Ecology 78:505–521
Murdoch WW (1994) Population regulation in theory and practice. Ecology 75:271–287
Nugent G, Parkes JP, Tustin KG (1987) Changes in the density and distribution of red deer and wapiti in northern Fiordland. NZ J Ecol 10:11–21
Nugent G, Fraser W, Sweetapple P (2001a) Top down or bottom up? Comparing the impacts of introduced arboreal possums and "terrestrial" ruminants on native forests in New Zealand. Biol Conserv 99:65–79

Nugent G, Fraser KW, Asher GW, Tustin KG (2001b) Advances in New Zealand mammalogy 1990–2000: deer. J R Soc NZ 31:263–298

Owen-Smith RN (2002) Adaptive herbivore ecology: from resources to populations in variable environments. Cambridge Univ Press, New York

Pacala SW, Canham CD, Saponara J, Silander JA Jr, Kobe RK, Ribbens E (1996) Forest models defined by field measurements: estimation, error analysis and dynamics. Ecol Monogr 66:1–43

Parkes JP (1984) Feral goats on Raoul Island. I. Effect of control methods on their density, distribution, and productivity. NZ J Ecol 7:85–94

Parkes JP, Thomson C (1995) Management of thar, part II. Diet of thar, chamois and possums. Science for Conservation 7, Department of Conservation, Wellington, New Zealand

Ramsey D, Efford M, Cowan P, Coleman J (2002) Factors influencing annual variation in breeding by common brushtail possums (*Trichosurus vulpecula*) in New Zealand. Wildl Res 29:39–50

Riney T (1964) The impact of introductions of large herbivores on the tropical environment. IUCN Publ New Ser 4:261–273

Sæther B-E (1997) Environmental stochasticity and population dynamics of large herbivores: a search for mechanisms. Trends Ecol Evol 12:143–149

Sæther B-E, Engen S, Filli F, Aanes R, Schröder W, Andersen R (2002) Stochastic population dynamics of an introduced Swiss population of the ibex. Ecology 83:3457–3465

Schauber EM, Kelly D, Turchin P, Simon C, Lee WG, Allen RB, Payton IJ, Wilson PR, Cowan PE, Brockie RE (2002) Masting by eighteen New Zealand plant species: the role of temperature as a synchronizing cue. Ecology 83:1214–1225

Scheffer VB (1951) The rise and fall of a reindeer herd. Sci Monthly 73:356–362

Sibly RM, Hone J (2002) Population growth rate and its determinants: an overview. Philos Trans R Soc Lond B 357:1153–1170

Thomas MD, Hickling GJ, Coleman JD, Pracy LT (1993) Long-term trends in possum numbers at Pararaki: evidence of an irruptive fluctuation. NZ J Ecol 17:29–34

Tustin KG, Challies CN (1978) The effects of hunting on the numbers and group sizes of Himalayan thar (*Hemitragus jemlahicus*) in Carneys Creek, Rangitata catchment. NZ J Ecol 1:153–157

Wells A, Duncan RP, Stewart GH (2001) Forest dynamics in Westland, New Zealand: the importance of large, infrequent earthquake-induced disturbance. J Ecol 89:1006–1018

13 What Controls Invasion of Indigenous Forests by Alien Plants?

S.K. Wiser and R.B. Allen

13.1 Introduction

Before human arrival (arguably, 1,000 years B.P.), New Zealand was >85 % forested; although this is only 23 % today, indigenous forests still occur over a wide range of conditions between sea level and treeline (Wardle 1991). Diverse, warm-temperate forests dominated by a mixture of evergreen conifer or hardwood tree species occur at low elevations in the northern North Island, and grade into simple forests in the upper-montane zone of South Island and the sub-Antarctic islands. Most high-altitude forests (up to 1,500 m above sea level), and those remaining in drier parts of New Zealand (precipitation down to 800 mm per annum), are dominated by evergreen beech (*Nothofagus*) species. Although invasion of anthropogenically disturbed or human-created vegetation types by alien plants has long been recognized as a serious problem in New Zealand, indigenous forests have historically been considered largely resistant to invasion.

In this chapter, we review the factors controlling alien plant invasion in New Zealand indigenous forests, and compare these patterns with generalities emerging in the ecological literature. Most research has been done in disturbed forest remnants, and forms the primary focus of our review. We also present new data describing alien plant invasion of extensive forest tracts that have not been burned or logged, and so have a tree population structure largely unmodified by human-related impacts. In such forests, however, introduced herbivores may have caused localized forest dieback (see Chap. 21) or modified forest understories (see Chap. 22). First, we describe factors that may predispose forests to invasion, such as disturbance, substrate and soil fertility, and species richness. We then examine the importance of propagule pressure and dispersal modes of aliens, geographic patterns of invasion, and persistence of aliens over time. We treat all alien plant species, including those that may not be currently considered "problem weeds" (sensu Williams and

Ecological Studies, Vol. 186
R.B. Allen and W.G. Lee (Eds.)
Biological Invasions in New Zealand

Timmins 1990). Invasion of early successional stages to forest (e.g., scrub) are reviewed in Chapter 19.

13.2 Role of Disturbance

Internationally, it is widely accepted that disturbance (sensu Sousa 1984) promotes invasion (D'Antonio et al. 1999), but is not a necessary prerequisite (Rejmánek 1989). In New Zealand forests, alien plants preferentially invade, or are more successful in, both human and naturally disturbed areas such as small forest remnants, forest margins, canopy gaps, and landslide scars. This was first recorded by Cockayne (1928), noting invasion of disturbed lowland forests by *Sambucus nigra* and *Hypericum androsaemum*, and Allan (1936) who commented on invasion of lowland forest remnants by *Tradescantia fluminensis*. Wardle (1973) and Williams and Timmins (1990) suggested that extensive areas of closed canopy rendered forest understories resistant to invasion. Some alien plants, however, have invaded extensive tracts of closed forest with little direct anthropogenic disturbance (Wiser et al. 1998). The New Zealand Department of Conservation lists 65 species as invaders of human-disturbed forest, and 10 as invaders of intact forest (Owen 1997).

Invasions of extensive, intact forest tracts may have been underemphasized for two reasons. In forest remnants, alien plants may be prominent in both cover (Williams and Timmins 1990; Wardle 1991) and relative diversity (i.e., comprising a majority of the flora; Healy 1995) whereas in large, relatively intact forest areas, alien plants are often inconspicuous woodland herbs (e.g., Wiser et al. 1998). Moreover, in the 1980s such inconspicuous plants were not considered "weeds", although their presence in forests was known (e.g., Wardle 1991). Second, widespread invasion of intact forests may be a recent phenomenon.

Alien trees generally invade and persist in indigenous forest subject to ongoing canopy disturbance. For example, *Crataegus monogyna*, *Prunus avium* and *Sambucus nigra* readily invade forest margins and damaged forest (Healy 1973). Both *Sambucus* and *Crataegus* seeds are distributed to, and germinate in, intact canopied forest, but seedlings do not survive (Timmins and Williams 1987). Four tree species are possible exceptions to this requirement for ongoing disturbance (Table 13.1). In Peel Forest, Canterbury, *Acer pseudoplatanus* invades forest understories where the canopy is open, and survives there for several years after canopy closure. If the canopy is reopened, seedlings or saplings resume growth (R. Buxton, personal communication). *Pseudotsuga menziesii* can spread from plantations into mature mountain beech forest in Central Otago (Cleary 1982), but elsewhere may spread only where canopies are open or understories sparse (Ledgard 2002). Differences in soil nutrient availability between localities may also explain such spatial

Table 13.1 Alien species recorded as invading the understory of relatively intact indigenous forest

Species	Source[a]	Species	Source[a]
Trees and shrubs		**Graminoids**	
Acer pseudoplatanus	Buxton (personal communication)	*Agrostis capillaris*	NVS databank
		Anthoxanthum odoratum	NVS databank
Archontophoenix cunninghamiana	Cameron (2000)	*Holcus lanatus*	NVS databank
Berberis darwinii	West (personal communication)	**Upright herbs**	
		Ageratina adenophora	Owen (1997)
Euonymus japonicus	Smale and Gardner (1999)	*Ageratina riparia*	Owen (1997)
		Cerastium fontanum	NVS databank
Ilex aquifolium	Webb et al. (1988)	*Cirsium vulgare*	NVS databank
Ligustrum lucidum	Smale and Gardner (1999)	*Conyza albida*	NVS databank
		Digitalis purpurea	NVS databank
Prunus campanulata	Owen (1997)	*Hedychium flavescens*	Owen (1997)
Prunus serrulata	Smale and Gardner (1999)	*Hedychium gardnerianum*	Owen (1997)
		Hieracium lepidulum	Wiser et al. (1998)
Pseudotsuga menziesii	Cleary (1982)	*Hieracium pilosella*	NVS databank
Rhamnus alaternus	Whaley et al. (1997)	*Hieracium praealtum*	NVS databank
Taxus baccata	Webb et al. (1988)	*Hypochoeris radicata*	NVS databank
		Lapsana communis	NVS databank
Vines		*Mycelis muralis*	NVS databank
Asparagus asparagoides	Owen (1997)	*Prunella vulgaris*	NVS databank
Asparagus scandens	Owen (1997)	*Ranunculus repens*	NVS databank
Clematis vitalba	Owen (1997)	*Senecio jacobaea*	NVS databank
Hedera helix	Owen (1997)	*Sonchus oleraceus*	NVS databank
Mat-forming herbs		**Ferns and fern allies**	
Lotus pedunculatus	NVS databank	*Selaginella kraussiana*	Webb et al. (1988)
Tradescantia fluminensis	Owen (1997)		
Trifolium repens	NVS databank		
Vinca major	Owen (1997)		

[a]NVS, National Vegetation Survey Databank, Landcare Research, New Zealand

variability. *Archontophoenix cunninghamiana* establishes and persists in closed forest in the Auckland area up to 300 m from planted adults (Cameron 2000). Finally, Owen (1997) lists *Prunus campanulata* as an invader of intact forest.

Alien woody shrubs readily invade forest remnants; for example, over 20 species occur in the 6.4-ha Riccarton Bush, a conifer forest fragment in the Christchurch city area (Healy 1995). *Rubus fruticosus* and *Leycesteria formosa* occur in natural canopy gaps in South Island West Coast beech forest within 500 m of large clear-cut areas, but are presumed to be shade intolerant and expected to die out as the canopy closes (Wiser 2000). Other shrub species that invade disturbed forests may have the potential to persist. For example, abundant seedlings of *Ligustrum lucidum*, *Euonymus japonicus*, and *Prunus serrulata* occur within a 0.7-ha coastal, broad-leaved forest remnant in the Auckland city area, suggesting a high level of shade tolerance (Smale and Gardner 1999; Table 13.1). In contrast, Webb et al. (1988) describe *Ilex aquifolium* as occurring in forest far from human habitation.

Alien vines, such as *Clematis vitalba*, *Lonicera japonica*, and *Passiflora mollissima*, typically invade disturbed habitats such as early successional forests, forest margins, canopy gaps or clearings (e.g., Williams and Buxton 1995; Baars et al. 1998); their absence from closed forest is often assumed to be due to low light levels. Based on glasshouse experiments, these climbers have relatively low light-compensation points (*Clematis vitalba* and *Lonicera japonica* ~1 % full sunlight, *Passiflora mollissima* ~2 % full sun), but growth and survival increase with light (Baars and Kelly 1996). Light environments in New Zealand lowland forest understories have been measured as ranging from 1 to 5 % full sunlight (McDonald and Norton 1992), so low light can not fully explain why these species are absent from forest understories. Other forms of disturbance can also promote invasion by alien climbers. In 1996, an unusually cold winter killed mature trees and thinned canopies in Southland forest remnants. *Tropaeolum speciosum* rapidly invaded these remnant interiors, whereas previously it was restricted to edges (C. West, personal communication). In contrast, the climber *Asparagus scandens* is shade tolerant and bird-dispersed, and so can invade both intact and disturbed forest (Timmins and Reid 2000). *Actinidia deliciosa* has also recently been recorded growing vigorously in the understory and canopy of mature warm-temperate forest (J. Sullivan and P.A. Williams, unpublished data).

Many herbaceous species invade disturbed forests and margins (e.g., Webb et al. 1988; Williams and Timmins 1990). *Tradescantia fluminensis* and *Hedychium gardnerianum* are two herbaceous invaders with potential to limit tree regeneration (Standish et al. 2001; Williams et al. 2003). *T. fluminensis* is shade tolerant, but grows faster and attains higher biomass with increased light, reaching maximum biomass at 10–15 % full sunlight (Standish et al. 2001). Both species most commonly occur in North Island degraded forest remnants (Williams and Timmins 1990). In contrast, across 250 permanent plots (0.04 ha) in intact Canterbury mountain beech (*Nothofagus solandri* var. *cliffortioides*) forest, the distributions of two alien composites, *Hieracium lepidulum* and *Mycelis muralis*, show little relationship to disturbance as measured by changes in tree biomass (Wiser et al. 1998; Blom 2002).

13.3 Substrate and Soil Fertility

Forests on fertile soils are often more readily invaded than those on low nutrient soils (e.g., Amor and Piggin 1977; Huenneke et al. 1990). Soil fertility was significantly correlated with numbers of alien plant species in forest and scrub reserves in two of eight lowland regions of New Zealand (Timmins and Williams 1991). Certain alien lianes preferentially invade forest fragments having higher soil pH than those where indigenous lianes occur (Baars et al. 1998). Both *Lonicera japonica* and *Clematis vitalba* preferentially establish on recent soils (i.e., alluvial, colluvial or tephras), which typically have high base saturation (Hume et al. 1995; Williams et al. 2001). The preference of *C. vitalba* for fertile soils may explain its absence from beech (*Nothofagus*) forests within its range (Hume et al. 1995). In the Canterbury mountain beech forest study, plots invaded by the herbs *Hieracium lepidulum* or *Mycelis muralis* tended to have higher soil exchangeable Ca, Mg, P, and lower C:N than those not invaded (Wiser et al. 1998; Blom 2002). Experiments also demonstrate the importance of soil fertility. Subtractive nutrient experiments showed suppressed growth of *C. vitalba* when either N, P, Ca, S, or micronutrient suites were lacking (Hume et al. 1995). Pot experiments have demonstrated that the herb *Tradescantia fluminensis* has increased growth at NO_3^- or NH_4^+ levels above 0.1 mol m^{-3} (Maule et al. 1995). Nutrient availability and light often covary in response to disturbance, but few studies partition their independent effects.

13.4 Species Richness

Point-in-time observational studies have often found a positive relationship between alien and indigenous species richness in forest (e.g., Brown and Peet 2003; Stohlgren et al. 2003). The mechanistic basis of this relationship has been debated (Levine and D'Antonio 1999). First, diversity may directly influence invasion, but it also covaries spatially with site conditions and dynamic processes (Wiser et al. 1998; Levine and D'Antonio 1999). Second, the degree to which the invasion itself has influenced species richness can not be determined. One solution is to use time series data to determine how indigenous and alien species richness have changed over time, as well as ancillary site data to segregate those aspects of species richness that do not covary with site conditions. We used the Canterbury mountain beech forest permanent plot data (see above) to relate the numbers of alien, indigenous and total species invading or disappearing from permanent plots to initial alien, indigenous and total species richness (Table 13.2). Although initial indigenous species richness and the number of invading alien species are significantly correlated,

Table 13.2 Spearman's rank correlations between initial alien, indigenous and total species richness and the number of alien, indigenous and total species invading (*new*) or disappearing (*lost*) from 250 permanent (0.04 ha) plots in Canterbury mountain beech forest. Plots were measured in 1985 and 1993 ($P<0.0001$ for all correlations)

1993 status		Initial richness (1985)		
		Alien	Indigenous	Total
Alien	New	0.60	0.53	0.58
	Lost	0.80	0.45	0.55
Indigenous	New	0.56	0.69	0.69
	Lost	0.59	0.87	0.88
Total	New	0.60	0.69	0.71
	Lost	0.73	0.88	0.90

this was considerably weaker than a final (1993) point-in-time correlation (r=0.53 vs. r=0.73). Furthermore, the significant positive correlation between the number of alien species disappearing from plots and initial indigenous species richness (r=0.45) reflects that species-rich plots also have rapid species turnover. Clearly, the relationship between alien and indigenous species richness is dynamic (Davis et al. 2000).

The pattern of increased invasion on relatively species-rich mountain beech plots also holds for the individual herbs *Hieracium lepidulum* and *Mycelis muralis* at two spatial scales – that of 400-m^2 plots and that of 0.75-m^2 subplots (Wiser et al. 1998; Blom 2002). This is consistent with point-in-time findings at large scales (Stohlgren et al. 2003), but contrasts with suggestions that the pattern should be reversed at smaller scales (Levine 2000; Brown and Peet 2003). Logistic regression showed that, for both herbs, species richness explained additional variation in the invasion patterns that could not be explained by a broad range of site variables, such as altitude, slope, soil fertility, or disturbance (Wiser et al. 1998; Blom 2002; Fig. 13.1). This suggests that species richness has a direct positive influence on invasion. Possible explanations include increased temporal and spatial heterogeneity of conditions within species-rich sites (Wiser et al. 1998), or the beneficial effects of arbuscular mycorrhizas. For example, these fungi are more likely to be abundant and diverse in species-rich mountain beech stands than in species-poor ones, where mountain beech, which lacks arbuscular mycorrhizas, may be the only species present (M. Rillig, personal communication).

The positive relationship between indigenous and alien species richness discussed above also occurs across large geographic scales, as shown by analyses of plot data from the National Vegetation Survey (NVS) databank (Wiser et al. 2001). Because plots are distributed unevenly across New Zealand, we selected 777 plots (each 0.04 ha) from ca. 14,000 plots to approx-

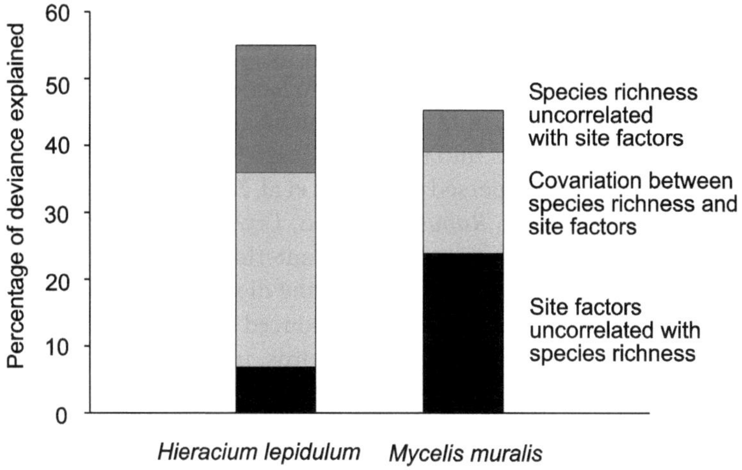

Fig. 13.1 Partitioning the deviance explained by multiple logistic regressions predicting the occurrence of *Hieracium lepidulum* and *Mycelis muralis* in 250 permanent 0.04-ha plots in Canterbury mountain beech forest measured in 1993

imate a representative sample across the country, although some lowland and far northern areas remain undersampled (Hall et al. 2001). On South Island, the presence of alien plants was positively related to indigenous species richness (average richness on invaded plots=23.3±9.8 vs. uninvaded plots=17.6±10.1; $P<0.0001$ using a Wilcoxon rank-sum test). No such relationship emerged for North Island. We conclude our section on invasibility and species richness by suggesting that the commonly evoked, independent *negative* effect of species richness on invasibility is an unimportant mechanism, over a wide range of scales, in New Zealand forests.

13.5 Propagule Pressure and Dispersal

Invasion levels may be low within large tracts of indigenous forest because of their isolation from invader propagules, rather than their intrinsic resistance to invasion (Rejmánek 1989; Timmins and Williams 1991). Proximity to towns is the most important predictor of the number of environmental weeds in New Zealand lowland forest reserves and coastal forest fragments (Timmins and Williams 1991; Sullivan et al. 2005). This is attributed to the high frequency of human visits to reserves, dumping of garden waste, and flocking and foraging of birds that disperse seeds of alien plants. In contrast, most of the herbaceous plants currently found in extensive forest tracts (e.g., *Mycelis muralis*, *Senecio jacobaea*, *Hieracium lepidulum*, *Agrostis capillaris*, *Anthox-*

anthum odoratum; Table 13.1) are readily dispersed composites or graminoids. Dispersal limitation may still be important for these species, however, particularly early in an invasion (Wiser et al. 1998).

Consumption of fleshy fruits by birds can be an important mechanism for spreading alien plant seeds into forests. At least 260 alien species have fleshy fruits that may be bird-dispersed (Williams et al. 2000), including alien woody species (*Prunus cerasifera*, *Rubus fruticosus*, *Leycesteria formosa*, *Ribes sanguineum*, *Sambucus nigra*) recorded in a Canterbury forest reserve (Burrows 1994). Williams and Karl (1996) examined the diets of 500 mist-netted native and introduced birds, and found that introduced and non-endemic birds eat alien fruits more often than do endemic birds, introduced blackbirds being the foremost disperser of alien woody plants' fruits. Introduced blackbirds, thrushes and starlings move between vegetation at different successional stages more frequently than do native birds, which may facilitate invasion of indigenous forests by alien plants. Silvereyes are the most important native disperser of a range of alien plant species, although they are less important at dispersing woody aliens than are introduced birds. Native pigeons may also disperse alien plants (Allen and Lee 1992). Birds will also disperse seeds of species without fleshy fruits, e.g., *Clematis vitalba* fruits are used as nesting materials (C. West, personal communication).

Introduced mammals also disperse alien plants into forests. Consumption of fruits of alien species by Australian brushtail possums (*Trichosurus vulpecula*) was first recorded by Cowan (1990). Seeds of several woody alien plants – *Berberis glaucocarpa*, *Cotoneaster glaucophyllus*, *C. simonsii*, *Pyracantha angustifolia*, *Sorbus hupehensis*, *Crataegus monogyna*, *Leycesteria formosa*, and *Passiflora mollissima* – are germinable after passing through possums (Williams et al. 2000); the latter three species have all been observed in disturbed forests. Dungan et al. (2002) found possums disperse large numbers of *Sambucus nigra* seeds in seral forest in lowland Canterbury. A high proportion of *Leycesteria formosa* and *Solanum aviculare* seeds, which are small, germinate after fruit consumption by ship rats (*Rattus rattus*), whereas rats destroyed most larger-seeded species tested (Williams et al. 2000). Small-seeded species are likely to have several introduced mammalian dispersers. Adherence to mammals or birds may also be important, and is likely the method for dispersal of *Selaginella kraussiana* (C. West, personal communication).

13.6 Geographic Patterns of Invasion

In temperate regions of the world, the number of alien plant species increases geographically with warmer climates, as reflected by decreasing latitude and altitude (e.g., Sax 2001; Stohlgren et al. 2002), but such patterns are largely

Table 13.3 Geographic distribution patterns of alien species in New Zealand indigenous forests[a]

	All New Zealand (n=777)	North Island (n=239)	South Island (n=538)
Percentage of plots containing at least one alien species	17%	19%	16%
Average number of alien species on a plot	0.38±1.12	0.35±0.96	0.39±1.22
Average percentage of species alien on a plot	1.42±4.51	1.03±3.20	1.59±4.98
Rank correlation between number of alien species and latitude	–0.09*	ns	–0.20***
Rank correlation between number of alien species and longitude	0.13**	ns	0.28***
Rank correlation between number of alien species and altitude	ns	–0.13*	ns

[a]Based on data from plots recording total species composition stored in the National Vegetation Survey (NVS) databank; ns, not significant; *, $P<0.01$; **, $P<0.001$; ***, $P<0.0001$

undescribed for New Zealand forests. As an initial step, we used the 777 NVS plots described in Section 13.4 to examine large-scale geographic patterns of alien plant distribution in relatively intact forests. Across New Zealand (North Island and South Island), the number of alien species on plots increased with both decreasing latitude and increasing longitude (i.e., from west to east), but this was largely driven by patterns within South Island (Table 13.3). Within North Island, no geographic patterns were detected, although alien richness and altitude were negatively related. Within South Island, there were stronger relationships between alien richness and latitude or longitude than across the country as a whole, but no relationship to altitude. In South Island, the gradient from west to east corresponds to a decline in rainfall, and an increase in deforestation, agricultural use, and population; in North Island, the deforestation/land use gradient is less related to east–west location.

Regional studies of forest fragments have shown other geographic patterns for certain types of alien plants. Alien lianes are more abundant in forest fragments in the Nelson area than in Banks Peninsula (Baars et al. 1998). Some aliens, such as *Asparagus scandens*, are sensitive to frost, and so typically occur in lowland areas of North Island and northern coastal areas of South Island (Timmins and Reid 2000).

13.7 Temporal Perspectives on Invasion Patterns

Alien plants usually decline in abundance as competitive sorting occurs during succession. The small tree *Crataegus monogyna* initially invaded a Canterbury black beech (*Nothofagus solandri* var. *solandri*) forest remnant in the late 1940s and early 1950s after logging and grazing disturbance. In 1983, age structure analysis suggested the invasion rate had slowed after protection from grazing and with the establishment of native hardwoods. *C. monogyna* was predicted to persist only in stands that continued to be disturbed by logging or grazing (Williams and Buxton 1986). Similarly, after goats were eradicated from Raoul Island, *Alocasia brisbanensis*, the most widespread forest alien, declined substantially as the canopy thickened (West 1996). Other species, such as *Tradescantia fluminensis*, persist after recovery from disturbance, but do not move into undisturbed areas (Maule et al. 1995). *Berberis darwinii*, a shrub of *Nothofagus* forest understories in South America, survives in secondary forests in New Zealand. This suggests that either it persists in forests that have developed since its establishment or it can invade intact forest (McQueen 1993). *Clematis vitalba* establishes in canopy gaps created by windfalls or landslides, and appears to be able to invade adjacent closed forest (Atkinson and Cameron 1993). *C. vitalba* can smother mature trees, thus retaining the open, disturbed conditions that favor its growth (McQueen 1993). Shade-intolerant aliens may persist in the seed bank until conditions are suitable for their germination and establishment. *C. vitalba*, *Passiflora mollissima* and *P. pinnatistipula* produce germinable seed banks (West 1992; Williams and Buxton 1995), and alien species are common in seed banks of Banks Peninsula forest fragments (Partridge 1989).

Aliens that can invade relatively undisturbed forest are more likely to persist than those requiring disturbance to invade. Permanent plot data from three North Island and 10 South Island surveys (stored in the NVS databank) were analyzed to determine changes in the frequency of *Mycelis muralis*, the most widely distributed alien species in New Zealand indigenous forests (Wardle 1991). Plots were remeasured over 5- to 24-year intervals. Frequency of *M. muralis* decreased in two surveys, remained the same in one, and increased in 10 surveys, usually more than doubling in frequency within 10 years (Fig. 13.2). On a more local scale, in the Canterbury mountain beech forest plots, *Hieracium lepidulum* usually persisted once established on a 400-m^2 plot. In 1993, it remained on 85 % of those plots where it occurred in 1970, and on 95 % of those plots where it occurred in 1985 (Wiser et al. 1998). *Mycelis muralis* persisted to a similar level (85 % persistence between 1985 and 1993), whereas turnover among the less common aliens (e.g., *Agrostis capillaris*, *Cerastium fontanum*) is substantially higher (only 31 % of species occurrences in 1985 persisting in 1993).

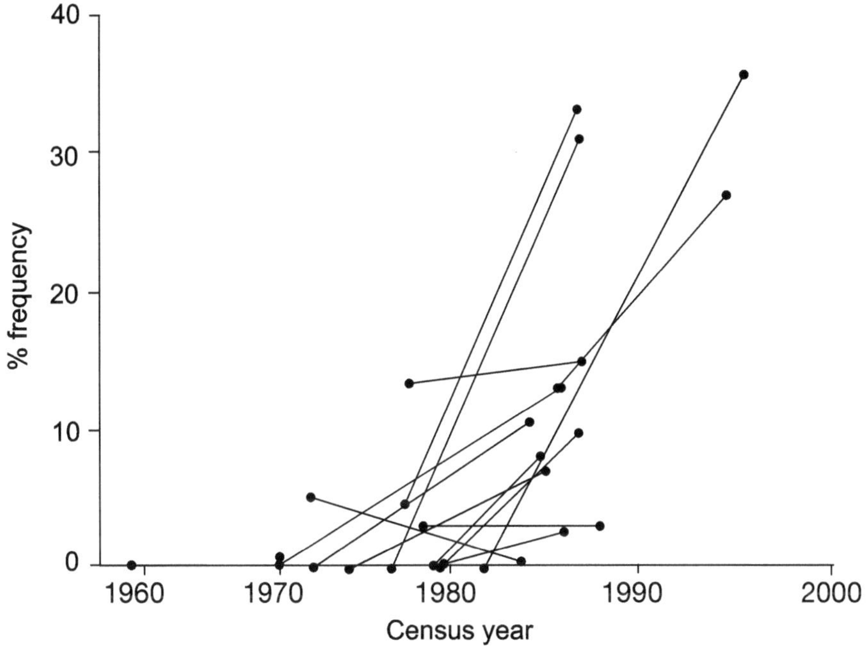

Fig. 13.2 Changes in frequency of *Mycelis muralis* in remeasured plots in 12 vegetation survey areas. Surveys sampled from 30 to 729 (average=211) permanent 0.04-ha plots

As invasions progress, the nature and fit of models explaining an alien's distribution may change, because properties of the invaded communities fluctuate (Davis et al. 2000) and dispersal becomes less limiting. In the *Hieracium lepidulum* invasion of Canterbury mountain beech forest, models explaining which plots were invaded were less accurate at the early stage of invasion (1970), when only a subset of suitable sites were colonized, than at a later stage (1993). At the early stage, distance to the forest margin (and to presumed seed sources) strongly correlated to alien occurrence, suggesting dispersal limitation; later site characteristics were the stronger correlates (Wiser et al. 1998).

13.8 Conclusions and Future Research Directions

Alien species most threatening to indigenous forests are those with the potential to alter forest structure, composition, and functioning; these are generally thought to be woody species, smothering vines, and mat-forming herbs. We need to know whether inconspicuous invaders of intact forest, such as small herbs and grasses, have more novel or subtle negative consequences, for instance, by influencing pollinator assemblages or nutrient uptake through mycorrhizas. How the indigenous and alien flora interact is poorly understood (Williams and Timmins 1990); more long-term studies are needed to elucidate these interactions and assess their consequences.

We require an understanding of the interaction between disturbance regimes and the population biology of aliens, e.g., reproductive age, ability to spread, regeneration requirements, and seed bank persistence. Do alien species invading anthropogenically disturbed forests form self-sustaining populations, or are they replaced by indigenous species as succession proceeds? Although colonization by aliens of naturally disturbed areas (e.g., canopy gaps, landslide scars, periodically flooded creek margins) in otherwise intact indigenous forest has been recorded, the magnitude of this threat is still poorly understood. This may be important because disturbed areas can comprise a significant proportion of the area and biota within otherwise close-canopied forest (e.g., Stewart et al. 1991), and may be of sufficient total size to support metapopulations of alien plants.

We need to clarify the role of introduced animals in dispersing aliens versus their potentially positive role in dispersing indigenous plants, and we need a better understanding of the role of introduced herbivores in controlling forest aliens by grazing. Anecdotal evidence suggests that weed species increase after animal control in ecosystems without trees, but not in forests (Timmins and Geritzlehner 2003). More clarity on the role of humans as dispersers of aliens (e.g., as tourists and gardeners, or via industrial operations such as road construction and forestry activities) is also required.

An understanding of the mechanisms controlling plant invasion in different forests is required if we are to develop a general model of forest invasibility for New Zealand. To date, studies on alien plants in New Zealand forest have largely focused on individual alien species or predictions of numbers of alien species. A more general view of invasibility should integrate propagule pressure, traits of aliens (e.g., growth rates, competitive abilities, dispersal modes), and characteristics of the ecosystems they do, or do not invade. This would allow the question of whether there are suites of characteristics that allow different aliens to invade different types of forests (e.g., dry vs. wet, fertile vs. infertile) to be answered.

References

Allan HH (1936) Indigene versus alien in the New Zealand plant world. Ecology 17:187–193

Allen RB, Lee WG (1992) Fruit selection by birds in relation to fruit abundance and appearance in the naturalised shrub *Berberis darwinii*. NZ J Bot 30:121–124

Amor RL, Piggin CM (1977) Factors influencing the establishment and success of exotic plants in Australia. In: Anderson D (ed) Exotic species in Australia – their establishment and success. Proc Ecol Soc Aust 10:15–26

Atkinson IAE, Cameron EK (1993) Human influence on the terrestrial biota and biotic communities of New Zealand. Trends Ecol Evol 8:447–451

Baars R, Kelly D (1996) Survival and growth responses of native and introduced vines in New Zealand to light availability. NZ J Bot 34:389–400

Baars R, Kelly D, Sparrow AD (1998) Liane distribution within native forest remnants in two regions of the South Island, New Zealand. NZ J Ecol 22:71–85

Blom J (2002) Wall lettuce (*Mycelis muralis*) an exotic invasive plant in New Zealand 1970–1999. BSc Thesis, Umea Universitet, Umea, Sweden

Brown RL, Peet RK (2003) Diversity and invasibility of southern Appalachian plant communities. Ecology 84:32–39

Burrows CJ (1994) Fruit types and seed dispersal modes of woody plants in Ahuriri Summit Bush, Port Hills, western Banks Peninsula, Canterbury, New Zealand. NZ J Bot 32:169–181

Cameron EK (2000) Bangalow pine (*Archontophoenix cunninghamiana*) begins to naturalise. NZ Bot Soc Newslett 60:12–16

Cleary EA (1982) Ecological aspects of Douglas-fir invasion of mountain beech forest on Ben Lomond Reserve, Queenstown, New Zealand. BSc Honours Thesis, University of Otago, Dunedin

Cockayne L (1928) The vegetation of New Zealand, 2nd edn. Die Vegetation der Erde, vol 14. Engelmann, Leipzig

Cowan PE (1990) Fruits, seeds, and flowers in the diet of brushtail possums, *Trichosurus vulpecula*, in lowland podocarp/mixed hardwood forest, Orongorongo Valley, New Zealand. NZ J Zool 17:549–566

D'Antonio CM, Dudley TL, Mack M (1999) Disturbance and biological invasions: direct effects and feedbacks. In: Walker LR (ed) Ecosystems of disturbed ground. Ecosystems of the World, vol 16. Elsevier, Amsterdam, pp 413–452

Davis MA, Grime JP, Thompson K (2000) Fluctuating resources in plant communities: a general theory of invasibility. J Ecol 88:528–534

Dungan RJ, O'Cain MJ, Lopez ML, Norton DA (2002) Contribution by possums to seed rain and subsequent seed germination in successional vegetation, Canterbury, New Zealand. NZ J Ecol 26:121–127

Hall GMJ, Wiser SK, Allen RB, Beets PN, Goulding CJ (2001) Strategies to estimate national forest carbon stocks from inventory data: the 1990 New Zealand baseline. Global Change Biol 7:389–403

Healy AJ (1973) Introduced vegetation. In: Williams GR (ed) The natural history of New Zealand: an ecological survey. Reed, Wellington, pp 170–189

Healy A (1995) Adventive plants. In: Molloy B (ed) Riccarton Bush: Putaringamotu: natural history and management. Riccarton Bush Trust, Christchurch, pp 160–180

Huenneke LF, Hamburg SP, Koide R, Mooney HA, Vitousek PM (1990) Effects of soil resources on plant invasion and community structure in Californian serpentine grassland. Ecology 71:478–491

Hume LJ, West CJ, Watts HM (1995) Nutritional requirements of *Clematis vitalba* L. (Old mans beard). NZ J Bot 33:301–313
Ledgard N (2002) The spread of Douglas-fir into native forests. NZ J For 47(2):36–38
Levine JM (2000) Species diversity and biological invasions: relating local process to community pattern. Science 288:852–854
Levine JM, D'Antonio CM (1999) Elton revisited: a review of evidence linking diversity and invasibility. Oikos 87:15–26
Maule HG, Andrews M, Morton JD, Jones AV, Daly GT (1995) Sun/shade acclimation and nitrogen nutrition of *Tradescantia fluminensis*, a problem weed in New Zealand native forest remnants. NZ J Ecol 19:35–46
McDonald D, Norton DA (1992) Light environments in temperate New Zealand podocarp rainforests. NZ J Ecol 16:15–22
McQueen DR (1993) A review of interaction between naturalised woody plants and indigenous vegetation in New Zealand. Tuatara 32:32–56
Owen SJ (1997) Ecological weeds on conservation land in New Zealand: a database. Working Draft. Department of Conservation, Wellington
Partridge TR (1989) Soil seed banks of secondary vegetation on the Port Hills and Banks Peninsula, Canterbury, New Zealand, and their role in succession. NZ J Bot 3:421–435
Rejmánek M (1989) Invasibility of plant communities. In: Drake JA, Mooney HA, di Castri F, Groves RH, Kruger FJ, Rejmánek M, Williamson M (eds) Biological invasions: a global perspective. Wiley, New York, pp 369–388
Sax DF (2001) Latitudinal gradients and geographic ranges of exotic species: implications for biogeography. J Biogeogr 28:139–150
Smale MC, Gardner RO (1999) Survival of Mount Eden Bush, an urban forest remnant in Auckland, New Zealand. Pac Conserv Biol 5:83–93
Sousa WP (1984) The role of disturbance in natural communities. Annu Rev Ecol Syst 15:353–391
Standish RJ, Robertson AW, Williams PA (2001) The impact of an invasive weed *Tradescantia fluminensis* on native forest regeneration. J Appl Ecol 38:1253–1263
Stewart GH, Rose AB, Veblen TT (1991) Forest development in canopy gaps in old-growth beech (*Nothofagus*) forests, New Zealand. J Veg Sci 2:679–690
Stohlgren TJ, Chong GW, Schell LD, Rimar KA, Otsuki Y, Lee M, Kalkhan MA, Villa CA (2002) Assessing vulnerability to invasion by nonnative plant species at multiple spatial scales. Environ Manage 29:566–577
Stohlgren TJ, Barnett DT, Kartesz JT (2003) The rich get richer: patterns of plant invasions in the United States. Front Ecol Environ 1:11–14
Sullivan J, Timmins SM, Williams PA (2005) Movement of exotic plants into coastal native forests from gardens in northern New Zealand. NZ J Ecol 29(1):1–10
Timmins S, Geritzlehner J (2003) Do weeds respond to pest animal control? DOC Science Internal Series 121. Department of Conservation, Wellington
Timmins SM, Reid V (2000) Climbing asparagus, *Asparagus scandens* Thunb.: a South African in your forest patch. Austral Ecol 25:533–538
Timmins SM, Williams PA (1987) Characteristics of problem weeds in New Zealand's protected natural areas. In: Saunders DA, Arnold GW, Burbidge AA, Hopkins AJM (eds) Nature conservation: the role of remnants of native vegetation. Surrey Beatty, Chipping Norton, NSW, Australia, pp 241–247
Timmins SM, Williams PA (1991) Weed numbers in New Zealand's forest and scrub reserves. NZ J Ecol 15:153–162
Wardle P (1973) Native vegetation. In: Williams GR (ed) The natural history of New Zealand: an ecological survey. Reed, Wellington, pp 155–169
Wardle P (1991) Vegetation of New Zealand. Cambridge Univ Press, Cambridge

Webb CJ, Sykes WR, Garnock-Jones PJ (1988) Flora of New Zealand, vol IV. Naturalised pteridophytes, gymnosperms, dicotyledons. DSIR, Botany Division, Christchurch

West CJ (1992) Ecological studies of *Clematis vitalba* (old man's beard) in New Zealand. DSIR Land Resources Vegetation Rep 736

West CJ (1996) Assessment of the weed control programme on Raoul Island, Kermadec Group. Science and Research Ser 98. Department of Conservation, Wellington

Whaley PT, Clarkson BD, Smale MC (1997) Claudelands Bush: ecology of an urban kahikatea (*Dacrycarpus dacrydioides*) forest remnant in Hamilton, New Zealand. Tane 36:131–155

Williams PA, Buxton RP (1986) Hawthorn (*Crataegus monogyna*) populations in mid-Canterbury. NZ J Ecol 9:11–17

Williams PA, Buxton RP (1995) Aspects of the ecology of two species of Passiflora (*P. mollissima* (Kunth) L. Bailey and *P. pinnatistipula* Cav) as weeds in South Island, New Zealand. NZ J Bot 33:315–323

Williams PA, Karl BJ (1996) Fleshy fruits of indigenous and adventive plants in the diet of birds in forest remnants, Nelson, New Zealand. NZ J Ecol 20:127–145

Williams PA, Timmins SM (1990) Weeds in New Zealand protected natural areas: a review for the Department of Conservation. Science and Research Ser 14. Department of Conservation, Wellington

Williams PA, Karl BJ, Bannister P, Lee WG (2000) Small mammals as potential seed dispersers in New Zealand. Austral Ecol 25:523–532

Williams PA, Timmins SM, Smith JMB, Downey PO (2001) The biology of Australian weeds, 38. *Lonicera japonica* Thumb. Plant Prot Q 16:90–100

Williams PA, Winks CJ, Rijkse WC (2003) Forest processes in the presence of wild ginger (*Hedychium gardnerianum*). NZ J Ecol 27:45–54

Wiser SK (2000) Exotic plants in *Nothofagus* forests managed sustainably for timber production in New Zealand. In: Stewart GH, Benecke U, Hickey J (eds) Sustainable management of indigenous forest: southern hemisphere prospects based on ecological research and silvicultural systems. In: Proc Symp Southern Connection Congr III, Lincoln University, Lincoln, New Zealand, pp 88–94

Wiser SK, Allen RB, Clinton PW, Platt KH (1998) Community structure and forest invasion by an exotic herb over 23 years. Ecology 79:2071–2081

Wiser SK, Bellingham PJ, Burrows LE (2001) Managing biodiversity information: development of New Zealand's National Vegetation Survey databank. NZ J Ecol 25(2):1–17

C Consequences of Alien Invasions

14 Consequences of Brown Trout Invasion for Stream Ecosystems

C.R. Townsend and K.S. Simon

14.1 Introduction

Brown trout (*Salmo trutta*) are particularly successful and well-studied invaders of New Zealand freshwaters. The species was deliberately introduced to the South Island of New Zealand as a sport fish beginning in 1867, and self-sustaining populations are now found in many streams, rivers and lakes in the region (McDowall 1990; Townsend 1996). The invasion of New Zealand is part of a global phenomenon as these fish, prized for the challenge they provide anglers, have been transported from their native range in Europe around the world, and are now among the most widespread of species in cool, fresh waters (Elliott 1994).

Since arriving, brown trout have brought economic and environmental change to New Zealand. Compared to many other invaders, few would contend that brown trout have had a negative economic impact in New Zealand. On the contrary, the recreational salmonid fishery in New Zealand, of which brown trout is a prime part, is conservatively estimated to be worth more than NZ$ 800 million to the nation per year (N. Watson, Otago Fish and Game Council, personal communication). As far as New Zealand's economically important native freshwater fisheries are concerned (principally eels and whitebait, the latter consisting of juveniles of a number of diadromous species in the genus *Galaxias*), we do not know whether trout have had a significant economic impact.

From the point of view of biodiversity, there is no evidence of a global extinction resulting from the introduction of trout to New Zealand. The only freshwater fish to have become extinct since the arrival of trout is the grayling *Prototroctes oxyrhynchus*, but we do not know what role, if any, trout played in its decline. On the other hand, there is considerable evidence to suggest trout have been responsible for local extinctions of native fish (Minns 1990; Crowl et al. 1992) and certain invertebrates (Whitmore et al. 2000). A large body of

Ecological Studies, Vol. 186
R.B. Allen and W.G. Lee (Eds.)
Biological Invasions in New Zealand

research has shown that brown trout also cause changes in streams across multiple levels of ecological organization, including the behavior of individuals, the abundance and distribution of populations of various taxa, the interactions between taxa within communities, and ecosystem processes including energy flux and nutrient cycling. In this chapter, we discuss the impacts brown trout have had at each of these ecological levels, focusing our treatment on the intensive work performed in streams in the Taieri River drainage of South Island.

14.2 Consequences of Brown Trout Invasion for the Behavior of Individuals

Brown trout change the activity and feeding behavior of mayfly nymphs (*Nesameletus* spp. and *Deleatidium* spp.) in streams. *Nesameletus ornatus* collected from a trout stream and placed in small artificial laboratory stream channels without fish were less active during the day than at night, in contrast to those collected from a galaxiid stream where *N. ornatus* were active throughout the day and night (McIntosh and Townsend 1994). In a laboratory study, *Deleatidium* mayflies collected from the Shag River (where both trout and *Galaxis vulgaris* occur) were found to be more nocturnal in behavior when trout were present, rather than *G. vulgaris*. This diurnal shift in feeding behavior was further confirmed by a field-based study of nymphs of *Deleatidium* spp. in five pairs of streams physically similar but contrasting in terms of whether they contained only trout or only galaxiids (*G. eldoni* or *G. depressiceps*; McIntosh and Townsend 1995a). A gut fluorescence technique was used to quantify the amount of algal pigment in the guts of *Deleatidium* collected during the day or night. The mean ratio of algal pigments in night and day mayfly gut samples from galaxiid streams was not different from unity (0.968±0.085), indicating the mayflies foraged equally during the day and night. In contrast, the ratio in trout streams was significantly greater than unity (1.483±0.21), and greater than that observed in galaxiid streams. Thus, *Deleatidium* foraged relatively more at night than during the day in trout streams, the same pattern as that detected in the laboratory. In a further experiment, records were made of *Deleatidium* visible during the late afternoon on the surface of cobbles in artificial channels, containing either no fish, trout, or *G. vulgaris*, placed in a tributary of the Shag River (McIntosh and Townsend 1996). Daytime activity of *Deleatidium* was significantly reduced in the presence of either fish species, but to a greater extent when trout were present. Toward the end of this experiment, six algae-covered cobbles were introduced into each channel to provide patches rich in food. Other cobbles were scrubbed to create food-poor patches. Two days later in the *G. vulgaris* channels, *Deleatidium* were 2.5-fold more abundant on rich than on poor patches whereas in the trout channels abundance was the same on rich and poor patches. Mayfly reluctance to move when trout are present further restricts efficient foraging by *Deleatidium*. *N. ornatus* is quite widely distributed in New Zealand but occurs at relatively low densities in the tributaries of the Taieri River where this work was performed. *Deleatidium*, on the other hand, is a ubiquitous member of New Zealand stream faunas and may comprise up to 90 % of total macroinvertebrate numbers (Scarsbrook and Townsend 1993), and so this effect on grazing behavior of a change in the top predator is profound indeed.

Other work has shown that invertebrates alter their propensity to enter the water column, or drift, in streams. *Deleatidium* had a lower propensity to drift in a trout than in a galaxiid (*G. eldoni*) stream (Williams 2000), and Drinnan (2000) reported that *Deleatidium* reduce their nocturnal drift in response to chemical cues from trout but not galaxiids. Other invertebrates showed greater tendencies to drift at night in a trout than in a *G. eldoni* stream, including a snail (*Potamopyrgus antipodarum*) and larvae of a cased caddisfly (*Olinga feredayi*) and a net-spinning caddisfly (*Aoteapsyche colonica*; Williams 2000).

The mechanism driving these behavioral changes appears to be a difference in the feeding mode of trout and galaxiids. Trout rely principally on vision to capture prey whereas the *Galaxias* species rely on mechanical cues (McIntosh and Townsend 1994, 1995b, 1998). Thus, invertebrates are considerably more at risk of predation from trout during the daylight hours, and this is reflected in the activity patterns that individual invertebrates show if they inhabit a trout stream.

14.3 Consequences of Brown Trout Invasion for Populations

Invasion of brown trout has led to changes in the population structure of fish and invertebrates in some New Zealand streams. In a multi-site study in the Taieri River between December 1989 and March 1990, 198 sites were classified as containing (1) no fish, (2) *Galaxias* only, (3) trout only or (4) both *Galaxias* and trout (Townsend and Crowl 1991). At every site, a variety of physical variables were measured (stream depth, flow velocity, phosphorus concentration in the stream water, % of the streambed composed of gravel, etc.). Trout occurred almost invariably below waterfalls (at least 3 m high) that were large enough to prevent their upstream migration; they tended to occur at low elevations because sites without waterfalls downstream tended to be at lower elevation. Sites containing galaxiids (*G. depressiceps*, *G. eldoni* or *G. anomalus*) and sites with no fish were always upstream of one or several large waterfalls. In allopatry, trout generally attained lower density, but higher biomass than was the case for galaxiids (Townsend 2003). In sympatry, biomass of both fish was reduced, but trout biomass was significantly higher than that of galaxiids (Townsend 2003). The few sites that contained both trout and galaxiids were below waterfalls, at intermediate elevations, and had cobble beds; the unstable nature of these streambeds may have promoted coexistence, but at much reduced densities. A further analysis to determine the variables important in discriminating between land-use classes through which the streams flowed (native forest, pine plantation, native tussock and pastoral) showed that native fish distribution was controlled directly by trout and not, as has sometimes been claimed, by land use (Townsend and Crowl 1991). Allibone (1999) found patterns of presence of trout and absence of non-migratory galaxiids in the upper Waipori River catchment similar to those seen in the Taieri River drainage.

We have experimentally simulated the invasion of brown trout by adding them to a short section of stream (200 m long) that contained *G. depressiceps*, but was bounded upstream by an artificial barrier, together with a large waterfall that prevented trout moving further upstream. The galaxiids disappeared from the reach within 6 months, and have reappeared since trout were removed from the stream upon completion of the experiment (K.S. Simon and C.R. Townsend, unpublished data). In a section of a nearby stream that contained only trout, the trout were removed and replaced with *G. depressiceps*. The galaxiids persisted for the 9-month duration of the experiment, after which they were returned to their original stream.

The impact of brown trout on Taieri galaxiids may be the result of both predation by trout on galaxiids and competition, by interfering with galaxiid foraging or reducing food availability to galaxiids (McDowall 2003). We know of no direct measures of trout predation on galaxiids in streams, but a single small trout in a laboratory aquarium has been recorded consuming 135 *Galaxias* fry in a day. Trout may reduce food availability to galaxiids by reducing invertebrate abundance, and altering invertebrate and galaxiid behavior. They are also aggressive competitors, and an experiment in replicated channels in a real stream showed that *G. vulgaris* was excluded from preferred microhabitats when trout were present (McIntosh et al. 1992), and a laboratory experiment showed that galaxiids (whether *G. depressiceps*, *G. eldoni*, or *G. anomalus*) made fewer successful predation attempts when trout were in the vicinity (Edge et al. 1993). In addition, by inducing invertebrates to become more nocturnal in their activity and less accessible on the streambed surface, trout may further reduce the foraging efficiency of galaxiids (McDowall 2003).

The strength of the effect of trout on galaxiids may depend on fish size and may vary among river systems. In a study of Canterbury streams (further north in South Island), in which brown and rainbow trout (*Onchorhynchus mykiss*) were grouped for analyses and their distributions compared with a different set of galaxiid species (non-migratory *G. vulgaris* and several diadromous species), McIntosh (2000) reported that trout and galaxiids did not coexist where trout were longer than 150 mm (fork length). His results indicate that exclusion of these galaxiids depended on trout size, and reflected the particular dominance of large trout as predators and/or competitors. Reanalysis of the 1990 Taieri River database, however, reveals that the few sites where brown trout and galaxiids were found together contained brown trout whose maximum size was not significantly smaller than at sites where trout occurred alone (Townsend 2003). Moreover, galaxiids co-occurred with trout larger than 150 mm in two streams, an aspect not recorded by McIntosh (2000). The different patterns may reflect the different ecologies of the species involved, or the details of the frequency of waterfalls, or the nature of the bed disturbance regime in the Taieri and rivers further north.

The patterns of trout impact on native fish populations seen in the Taieri River drainage do not hold true for all New Zealand streams or for all species of native fish. The distribution of trout in New Zealand, which ultimately determines where native fish will be exposed to trout, is limited by factors such as temperature and hydrology (e.g., Jowett 1990). Where trout are present, their impacts will depend on the species of native fish originally present and stream type. For example, Jowett et al. (1998) suggested that trout had little effect on the abundance of native fish in streams in Kahurangi National Park in the north of South Island. Those streams were forested and contained primarily diadromous fish, in contrast to the grassland streams containing non-migratory galaxiids in the Taieri drainage. In a survey of streams in the Taieri and Waipori drainages, David et al. (2002) found that giant kokopu (*Galaxias argentus*) frequently occurred in the same streams as did brown trout, but the two species were rarely found in the same stream reach. David et al. (2002) suggested large giant kokopu could competitively exclude trout from some habitats. Eels may also negatively impact on trout populations, via predation, spawning interference or competition for invertebrate prey (Burnet 1968).

An apparent negative relationship between trout and native freshwater crayfish was first suggested many years ago (Thomson 1922), and has been attributed to predation by trout (McDowall 1990). In the Taieri River, the presence of brown trout was negatively correlated with distribution of the crayfish *Paranephrops zealandicus*, both on a catchment-wide basis (Whitmore et al. 2000) and at a local scale in a single tributary (Usio and Townsend 2000). Shave et al. (1994) showed that these crayfish are able to use chemical cues to detect native eels, but not trout.

Among the macroinvertebrates eaten by brown trout, large slow-swimming species such as nymphs of mayflies in the genera *Ameletopsis*, *Oniscigaster* (Hudson 1904), and *Nesameletus* (McIntosh and Townsend 1994) and large carnivorous invertebrates might be expected to be particularly vulnerable and severely affected by the trout invasion. In a comparison of streams in the Upper Waimakariri River system, McIntosh (2002) found that density and biomass of *Nesameletus* were several orders of magnitude lower in streams with trout, compared to streams with *Galaxias vulgaris* or no fish. Moreover, in a channel experiment conducted in a tributary of the Shag River, Flecker and Townsend (1994) found that large carnivorous invertebrates, including *Archichauliodes diversus*, were rarer when trout, rather than *G. vulgaris* were present. Finally, in a survey of two Taieri tributaries, one containing trout and the other *G. eldoni*, Huryn (1998) reported that nine of the 10 largest invertebrate species, including *N. ornatus* and *A. diversus*, were represented by smaller individuals in the trout stream than in the galaxiid stream, presumably a result of strong size-selective predation by trout on these vulnerable prey.

14.4 Consequences of Brown Trout Invasion for Community Interactions

The effects of brown trout at the individual and population levels have implications for interactions among taxa in streams. In the relatively species-poor stream communities in the south of New Zealand, the basal trophic level consists mainly of algae that grow on the streambed. This periphyton is grazed by various insect larvae, which in turn are prey to carnivorous invertebrates, whereas fish are considered the top predators. Experiments involving artificial flow-through channels deployed in a tributary of the Shag River have been used to determine whether trout affect the stream food web differently than is the case for displaced *G. vulgaris*. It is illuminating to consider the results of three different experiments performed with essentially the same experimental design, but at different times: January 1992 and March 1992 (Flecker and Townsend 1994), and February 1993 (McIntosh and Townsend 1996). In each case, three treatments were established in each of several randomized blocks separated by 50–100 m in a 500–800 m stretch of the stream. The treatments were no fish, *G. vulgaris* present, or trout present. Fish lengths (75–120 mm), densities, and biomasses were toward the high end of the natural range recorded in streams in the region, and were selected to provide identical densities and similar biomasses of the two species in the channels. Algae and invertebrates were allowed to colonize the natural substrate within the channels before introducing the fish. After a further 12 or more days, invertebrates and algae were sampled (Fig. 14.1).

The general patterns were similar in all three experiments but there were subtle differences. In comparison to the no-fish control, trout suppressed invertebrate biomass in the first two experiments whereas *Galaxias* did not. In the third experiment, there were no significant differences in invertebrate biomass among the treatments. Algal biomass achieved its highest values in the trout treatments of all the experiments, but this was statistically significant only in the second and third experiments. The most probable reason for

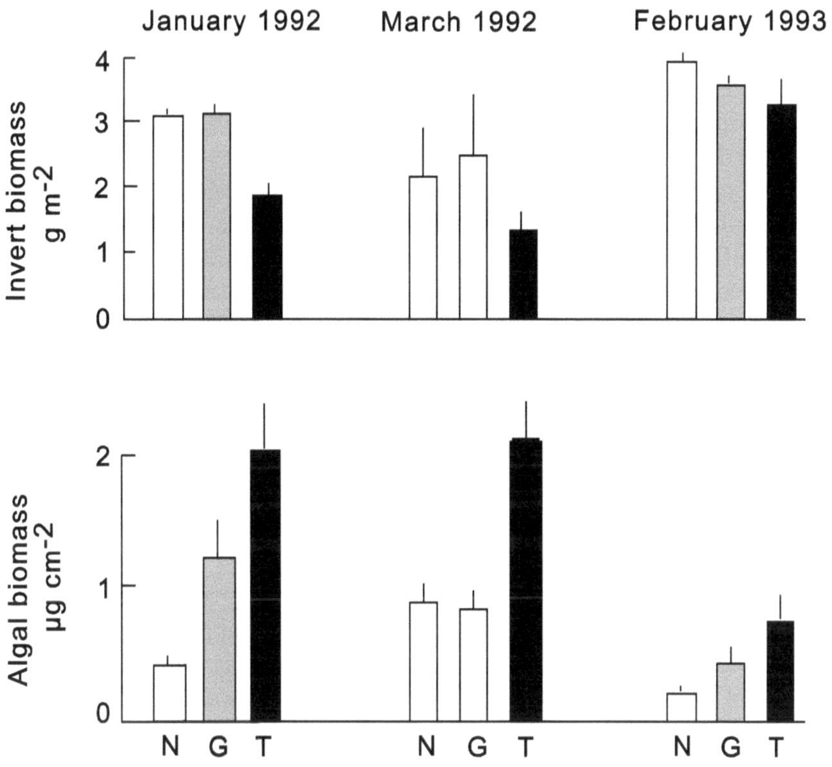

Fig. 14.1 Total invertebrate biomass (g m^{-2}) and algal chlorophyll *a* (μg m^{-2}), ±SE, for no-fish (*N*), *Galaxias* (*G*), and trout (*T*) treatments in three separate experiments in a small New Zealand stream (data for the two 1992 experiments are from Flecker and Townsend 1994; data for the 1993 experiment are from McIntosh and Townsend 1996)

the differences is that both abiotic (e.g., temperature, nitrate concentration) and biotic conditions (algal and invertebrate species available to colonize the channels) will have differed between experimental periods. However, taken together, it is clear that trout do have a more pronounced effect than does *G. vulgaris* on invertebrate grazers and, consequently, on algal biomass.

The indirect effect of trout on algae occurs partly through a reduction in invertebrate biomass (Fig. 14.1) but also because trout restrict the grazing behavior of the invertebrates that are present. Evidence from film records of grazing behavior (McIntosh and Townsend 1994), and of the width of grazing scars (Flecker and Townsend 1994) and the distribution of algae on the sides and tops of stones in experimental channels (McIntosh and Townsend 1996), confirms that grazing invertebrates spend less time out in the open, and feed closer to refuges, when trout are present. The reduced invertebrate biomass in two of the experiments may be a result of higher predation rates by trout than by galaxiids and/or a greater tendency of invertebrates to leave (or not settle in) channels containing trout. Laboratory experiments have shown that the grazing snail *Potamopyrgus antipodarum* exhibits greater drift rates when trout odor is introduced into the channel; on the other hand, *Deleatidium* has a lower propensity to drift in the presence of trout (Williams 2000). The non-significant difference in invertebrate biomass between treatments in the third experiment could be related to the shorter channels used (2 m, rather than 5 m long); local predation losses may be more likely to be obscured by the influx of migrating invertebrates in shorter channels.

These results constitute strong evidence of a trophic cascade, with biomass of plants being increased significantly when a key top predator (trout) is present. In theory, this result is what is expected where there are three (or some other odd number of) trophic levels (Oksanen et al. 1981; Biggs et al. 2000). At first glance, our streams appear to have four trophic levels (fish, carnivorous invertebrates, grazing invertebrates, and algae), with fish as the top predator. However, stable isotope analyses of food webs in streams in the Taieri River drainage indicate that fish occupy the third trophic level by exploiting primarily grazing invertebrates (Simon et al. 2004).

In an attempt to confirm these patterns in non-experimental situations, three pairs of streams, each with one containing trout and the other a galaxiid, were assessed for algal standing crops and algal species composition (Biggs et al. 2000). In two pairs of streams, the biomass of trout was much higher than that of galaxiids, whereas in the third pair the biomass of the two species was similar (uncharacteristically high for the galaxiid, and somewhat low for trout). In all three pairs of streams, the ash-free dry mass of the biofilm of periphyton was, as predicted, greater in the trout than in the galaxiid stream. The density of chlorophyll *a* on the stones was also significantly higher in the trout case in two pairs of streams, but not in the third where trout and galaxiid biomasses were similar. In all stream pairs, the relative abundance of prostrate algae was highest in the galaxiid streams. These species, including *Cocconeis* sp., *Cymbella aspera*, and *Epithemia* spp., are less vulnerable to grazing invertebrates (Steinman 1996) and are expected to be more prominent in streams where grazing is intense (i.e., in the galaxiid streams). The relative abundance of erect algal species (including *Audouinella hermanii*, *Gomphoneis herculeana*, and *Synedra ulna*), considered more vulnerable to grazers, was higher in the trout case in two pairs of streams but was similar in the third pair where trout and galaxiid biomasses were similar. These results provide additional support for the cascading influence of invading trout. However, they also indicate that the consequences of the invasion probably partly depend on the tendency for trout to occur at higher biomass in the streams than that of the galaxiids they replace.

Although we have not directly studied the influence of trout on crayfish, the likely negative impacts on this species deserve some comment. This is because the crayfish *Paranephrops zealandicus* can itself be considered a keystone species, influencing physical processes, such as sedimentation, and biotic processes through consumption and bioturbation (Usio and Townsend 2000; Whitmore et al. 2000). Both descriptive and experimental studies in the Taieri River have shown that crayfish are responsible for a trophic cascade involving negative effects on carnivorous invertebrates (Tanypodinae midge larvae) and consequent positive effects on their prey (Chironominae; Usio and Townsend 2000, 2004). In addition, the omnivorous crayfish have strong direct effects on the abundance of algae and particulate organic matter, both through direct consumption and bioturbation (Usio and Townsend 2001). Crayfish are ecosystem engineers (Jones et al. 1994), and any impacts of trout on crayfish are likely to lead to further knock-on effects in the community.

14.5 Consequences of Brown Trout Invasion for Ecosystem Processes

A strong trophic cascade, such as that produced by the introduced brown trout, would be expected to have consequences at the ecosystem level, but this has not been described before. This is partly because it is difficult to find two communities with contrasting predation regimes but the same physical settings and, in addition, such studies require a particularly large effort and expense (and replication of treatments is usually not feasible). However, the studies above provided the impetus for a detailed energetics investigation of two physically and chemically similar neighboring tributaries of the Taieri River, one containing only trout and the other (because of a waterfall downstream) containing only *G. eldoni* (Huryn 1998). Huryn's (1998) hypothesis was that algal photosynthetic rate would be greater in the trout stream because of a reduced rate of grazing there.

As predicted, net primary production in the trout stream was consistently higher than in the *G. eldoni* stream, and annual net primary production was six times greater in the trout stream. The rate of production of new biomass by grazers in the trout stream was about 1.5 times that in the galaxiid stream, and trout themselves produced new biomass at roughly nine times the rate of *G. eldoni*. Overall, *G. eldoni* consumed only about 18 % of available prey production each year whereas the grazing invertebrates consumed about 75 % of primary production in the galaxiid stream. In stark contrast, trout consumed virtually 100 % of annual invertebrate production in their stream whereas the grazing invertebrates consumed only about 21 % of primary production. In other words, as predicted, there was strong top-down control by trout of invertebrates, and release of the algae to produce and accumulate biomass at a fast rate. This means that there was an annual surplus of algal production (in excess of demand of grazing invertebrates) in the trout stream that was almost 20 times higher than the estimated surplus in the galaxiid stream. Where does this surplus go? Presumably it dies and is sloughed off from the surface of the streambed as fine particulate organic matter to be consumed, at some distance downstream, by microorganisms and detritivorous invertebrates. An ecosystem effect of the trophic cascade may be felt at some distance from the location of the trout.

It is necessary to be cautious when interpreting the results of an unreplicated study (only one trout and one *Galaxias* stream). However, the conclusion that a trophic cascade is responsible for the patterns observed at the ecosystem level can be drawn with some confidence because of the variety of other corroborative studies conducted at the individual, population, and community levels.

The densities of trout and galaxiids in the two streams were very similar but the biomass of brown trout was six times higher than that of galaxiids (as is often the case when trout and galaxiid streams are compared). Regardless of the various underlying mechanisms leading to a trophic cascade, the difference in biomass will contribute further to a large trout effect. As noted above, Huryn's (1998) study revealed a tendency toward smaller size of invertebrates in the stream containing trout as opposed to *G. eldoni*, presumably because of strong size-selective predation by the former. Other things being equal, smaller individuals are more productive (per unit of biomass) than larger ones; in this respect, therefore, trout may be indirectly responsible for a relative increase in productivity of their prey. Perhaps also, by being responsible for an increase in primary productivity, trout gain a food benefit by providing for a greater rate of productivity of their prey.

Finally, we have traced the fate of nitrogen in another pair of streams, one containing trout and the other (because of a downstream waterfall) containing *G. depressiceps*, using a stable isotope tracer (^{15}N). In the trout stream, demand for nitrate was higher, and algal films on the streambed were responsible for a much larger portion of nitrogen uptake and retention, as expected if algal production is enhanced in the presence of trout (Simon et al. 2004). In addition, less nitrogen was retained in grazing invertebrates (*Deleatidium* spp.) in the trout stream. Furthermore, in a comparison of three pairs of *Galaxias* and trout streams, nitrate uptake efficiency was consistently greater in the trout streams (Simon et al. 2004).

14.6 What Mediates the Influence of Brown Trout vs. Native Fish?

The native non-migratory galaxiid species in the Taieri and Shag rivers can cause a weak trophic cascade (perhaps better termed a trophic trickle; Strong 1992), and there was sometimes an increase in algal biomass in the presence of galaxiids, compared with situations without fish. However, brown trout had much more profound and predictable effects. The effect of trout on grazing in our experiments was mediated by its influence on grazer behavior. Differences in the feeding modes of trout (diurnal, visual cues) and galaxiids (diurnal and nocturnal, mechanical cues) are partly responsible for the effects of trout. Some of the impact of trout also appears to be related to a propensity to achieve a higher biomass than do galaxiids. The fact that essentially 100 % of annual invertebrate production was consumed by trout in the energetics study, but only 18 % by galaxiids, indicates that trout may be more efficient at finding and capturing prey. They may also be more efficient than galaxiids at assimilating food and/or converting assimilate into body tissue. Other life-history features may contribute to trout attaining a higher biomass. Trout may be longer lived and grow to a larger size, have more regular recruitment, experience lower mortality, or recolonize sites faster after disturbance (Townsend 2003). Whatever the precise mechanism, trout seem to be consistently able to build populations to the point where food limitation occurs, in contrast to galaxiids (Huryn 1998).

14.7 Lessons for Management

Trout represent an unusual challenge for management. Unlike most invaders in New Zealand, trout provide a large economic benefit. However, we do not know whether the "costs" go beyond changes to ecosystem functioning and local extinctions of non-migratory galaxiids. Note also that most of the research on the impacts of brown trout has been conducted in only a few South Island rivers. Conclusions cannot be extrapolated to species that have not been studied, such as whitebait galaxiids, or to the rest of New Zealand. More research is needed before generalizations can be made about the impacts of trout throughout New Zealand.

Given our current knowledge, however, managers in South Island need to protect native fish refuges above migration barriers. They will need to make landowners and anglers fully aware of the impact of trout on vulnerable native species, and of the need to prevent their further spread. Sites upstream of migration barriers that have been breached in the past by deliberate transport of trout provide opportunities for managers to remove the brown trout and reintroduce the galaxiids (Townsend 2003).

References

Allibone RM (1999) Impoundment and introductions: their impacts on native fish of the upper Waipori River, New Zealand. J R Soc NZ 29:291–299

Biggs BJF, Francoeur SN, Huryn AD, Young R, Arbuckle CJ, Townsend CR (2000) Trophic cascades in streams: effects of nutrient enrichment on autotrophic and consumer benthic communities under two different fish predation regimes. Can J Fish Aquat Sci 57:1380–1394

Burnet AMR (1968) A study of the relationships between brown trout and eels in a New Zealand stream. Fisheries Tech Rep, Fisheries Research Division, MAF, Wellington, New Zealand

Crowl TA, Townsend CR, McIntosh AR (1992) The impact of introduced brown and rainbow trout on native fish: the case of Australasia. Rev Fish Biol Fisher 2:217–241

David BO, Closs GP, Arbuckle CJ (2002) Distribution of fish in tributaries of the lower Taieri/Waipori Rivers, South Island, New Zealand. NZ J Mar Fresh Water 36:797–808

Drinnan KJ (2000) The effects of fish on the behavioural drift of stream invertebrates. MSc Thesis, University of Canterbury, Christchurch, New Zealand

Edge K-A, Townsend CR, Crowl TA (1993) Investigating anti-predator behaviour in three genetically differentiated populations of non-migratory galaxiid fishes in a New Zealand river. NZ J Mar Fresh Water 27:357–363

Elliott JM (1994) Quantitative ecology and the brown trout. Oxford Univ Press, Oxford, UK

Flecker AS, Townsend CR (1994) Community-wide consequences of trout introduction in New Zealand streams. Ecol Appl 4:798–807

Hudson GV (1904) New Zealand neuroptera: a popular introduction to the life-histories and habits of may-flies, dragon-flies, caddis-flies and allied insects inhabiting New Zealand: including notes on their relation to angling. West, Newman, London, UK

Huryn AD (1998) Ecosystem-level evidence for top-down and bottom-up control of production in a grassland stream system. Oecologia 115:173–183

Jones CG, Lawton JH, Shachak M (1994) Organisms as ecosystem engineers. Oikos 69:373–386

Jowett IG (1990) Factors related to the distribution and abundance of brown and rainbow trout in New Zealand clear-water rivers. NZ J Mar Fresh Water 24:429–440

Jowett IG, Hayes JW, Deans N, Eldon GA (1998) Comparison of fish communities and abundance in unmodified streams of Kahurangi National Park with other areas of New Zealand. NZ J Mar Fresh Water 32:307–322

McDowall RM (1990) New Zealand freshwater fishes: a natural history and guide, rev edn. Heinemann Reed, Auckland, New Zealand

McDowall RM (2003) Impacts of introduced salmonids on native galaxiids in New Zealand upland streams: a new look at an old problem. Trans Am Fish Soc 132:229–238

McIntosh AR (2000) Habitat- and size-related variations in exotic trout impacts on native galaxiid fishes in New Zealand streams. Can J Fish Aquat Sci 57:2140–2151

McIntosh AR (2002) Impact of introduced trout on *Nesameletus* mayfly populations measured using electrofishing. Arch Hydrobiol 154:375–392

McIntosh AR, Townsend CR (1994) Interpopulation variation in mayfly antipredator tactics: differential effects of contrasting predatory fish. Ecology 75:2078–2090

McIntosh AR, Townsend CR (1995a) Impacts of an introduced predatory fish on mayfly grazing in New Zealand streams. Limnol Oceanogr 40:1508–1512

McIntosh AR, Townsend CR (1995b) Contrasting predation risks presented by introduced brown trout and native common river galaxias in New Zealand streams. Can J Fish Aquat Sci 52:1821–1833
McIntosh AR, Townsend CR (1996) Interactions between fish, grazing invertebrates and algae in a New Zealand stream: a trophic cascade mediated by fish-induced changes to grazer behaviour? Oecologia 108:174–181
McIntosh AR, Townsend CR (1998) Do different predators affect distance, direction and destination of movements by a stream mayfly? Can J Fish Aquat Sci 55:1954–1960
McIntosh AR, Townsend CR, Crowl TA (1992) Competition for space between introduced brown trout (*Salmo trutta* L.) and a native galaxiid (*Galaxias vulgaris* Stokell) in a New Zealand stream. J Fish Biol 41:63–81
Minns CK (1990) Patterns of distribution and association of freshwater fish in New Zealand. NZ J Mar Fresh Water 24:31–44
Oksanen L, Fretwell SD, Arruda J, Neimelä P (1981) Exploitation ecosystems in gradients of primary productivity. Am Nat 118:240–261
Scarsbrook MR, Townsend CR (1993) Stream community structure in relation to spatial and temporal variation: a habitat template study of two contrasting New Zealand streams. Freshwater Biol 29:347–354
Shave CR, Townsend CR, Crowl TA (1994) Anti-predator behaviours of a freshwater crayfish (*Paranephrops zealandicus*) to a native and an introduced predator. NZ J Ecol 18:1–10
Simon KS, Townsend CR, Biggs BJF, Bowden WB, Frew RD (2004) Habitat-specific nitrogen dynamics in New Zealand streams containing native or invasive fish. Ecosystems 8:777–792
Steinman AD (1996) Effects of grazers on freshwater benthic algae. In: Stevenson RJ, Bothwell ML, Lowe RL (eds) Algal ecology: freshwater benthic ecosystems. Academic Press, San Diego, California, pp 341–373
Strong DR (1992) Are trophic cascades all wet? Differentiation and donor-control in speciose ecosystems. Ecology 73:747–754
Thomson GM (1922) The naturalisation of animals and plants in New Zealand. Cambridge Univ Press, Cambridge, UK
Townsend CR (1996) Invasion biology and ecological impacts of brown trout *Salmo trutta* in New Zealand. Biol Conserv 78:13–22
Townsend CR (2003) Individual, population, community, and ecosystem consequences of a fish invader in New Zealand streams. Conserv Biol 17:38–47
Townsend CR, Crowl TA (1991) Fragmented population structure in a native New Zealand fish: an effect of introduced brown trout? Oikos 61:347–354
Usio N, Townsend CR (2000) Distribution of the New Zealand crayfish *Paranephrops zealandicus* in relation to stream physico-chemistry, predatory fish, and invertebrate prey. NZ J Mar Fresh Water 34:557–567
Usio N, Townsend CR (2001) The significance of the crayfish *Paranephrops* zealandicus as shredders in a New Zealand headwater stream. J Crustacean Biol 21:354–359
Usio N, Townsend CR (2004) Roles of crayfish: consequences of predation and bioturbation for stream invertebrates. Ecology 85:807–822
Whitmore N, Huryn AD, Arbuckle CJ, Jansma F (2000) Ecology and distribution of the freshwater crayfish *Paranephrops zealandicus* in Otago. Science for Conservation 148. Department of Conservation, Wellington, New Zealand
Williams JK (2000) Influence of abiotic and biotic factors on invertebrate drift. MSc Thesis, University of Otago, Dunedin, New Zealand

15 Relative (Un)Importance of Introduced Animals as Pollinators and Dispersers of Native Plants

D. Kelly, A.W. Robertson, J.J. Ladley, S.H. Anderson
and R.J. McKenzie

15.1 Introduction

The island archipelagos of the Pacific are characterized by a recent pulse of species extinctions and declines following the wave of alien introductions brought by people (Steadman 1995). The invasion of mammalian predators in particular has had a devastating effect on keystone vertebrate pollinators and dispersers throughout the region (Cox 1983; Bell 1991; Rainey et al. 1995; Steadman 1997; McConkey and Drake 2002; Meehan et al. 2002). These losses raise the potential for flow-on effects disrupting mutualisms and threatening the plant species that depend on these mutualists (Janzen 1974; Kearns et al. 1998; Robertson et al. 1999; Cox and Elmqvist 2000; Sekercioglu et al. 2004). In some cases, other introduced species may replace the lost endemic species and partially restore the mutualism (Cox 1983; Lord 1991; Paton 1993; Ecroyd 1996). In Hawai'i, the vine ie'ie (*Freycinetia arborea*), which was originally pollinated by four endemic, now-extinct bird species, is currently pollinated by the introduced Japanese silvereye (*Zosterops japonica*; Cox 1983). In Tonga, the Pacific pigeon (*Ducula pacifica*), which colonized Tonga about the same time as people (Steadman 1997), is now the only pigeon present and is the largest avian frugivore, but it fails to match the gape of two extinct *Ducula* spp. As a result, several large-fruited species are now dependent on the sole remaining fruit bat (*Pteropus tonganus*) for dispersal (Meehan et al. 2002).

In New Zealand, introduced rats, stoats and other predators have caused the extinction of a large number of endemic vertebrates and the decline of many others, including birds (Bell 1991; Holdaway 1999), bats (Lloyd 2001), and lizards (Towns et al. 2001). Many of the extinct or declining birds were important frugivores, e.g., moa spp. (Dinornithiformes; all extinct) and

Ecological Studies, Vol. 186
R.B. Allen and W.G. Lee (Eds.)
Biological Invasions in New Zealand

kokako[1] (*Callaeas cinerea*; in severe decline), or pollinators, e.g., stitchbirds (hihi, *Notiomystis cincta*; in severe decline; Atkinson and Millener 1991). Two species of short-tailed bat (family Mystacinidae) were both fruit and nectar feeders (Daniel 1976). The greater short-tailed bat (*Mystacina robusta*) is now extinct, and the lesser short-tailed bat (*M. tuberculata*) is in sharp decline (Lloyd 2001). Lizards have also suffered post-human declines in density (Towns et al. 2001), and it has been suggested that they may also have been important in pollination and dispersal (Whitaker 1987). However, work on offshore islands, where lizard densities remain high, has shown that although lizards do visit some flowers (Eifler 1995), they have limited ability to move between plants, and there is as yet no evidence that they played a significant role in pollination. The evidence is stronger for lizards as effective, albeit short-range, seed dispersers, as in the case of *Coprosma propinqua* (Wotton 2002).

The result of these species losses and declines is a reduced set of vertebrate dispersers and pollinators over much of the New Zealand mainland, but with a more intact fauna on some of the offshore islands that are predator-free. Of the species capable of dispersing the largest fruits (>14 mm diameter), only the New Zealand pigeon (kereru, *Hemiphaga novaeseelandiae*) remains (Clout and Hay 1989). Six plant species have fruits larger than this, and are thus vulnerable to the loss of kereru, which appears to also be in decline (Clout et al. 1995). Several other native birds take smaller fruit, particularly bellbirds (korimako, *Anthornis melanura*), tui (*Prosthemadera novaeseelandiae*), and silvereyes[2] (*Zosterops lateralis*), along with stitchbirds and saddlebacks (tieke, *Philesturnus carunculatus*), which both also take fruit but now persist only on island bird sanctuaries. Four native bird species regularly visit flowers – bellbirds, tui, stitchbirds and silvereyes. Saddlebacks, kaka (*Nestor meridionalis*), and kea (*Nestor notabilis*) are occasional flower-visitors. All except silvereyes have been reduced in density and/or range. As far as is known, no extinct birds were significant flower-visitors (Atkinson and Millener 1991).

The native insect-pollinating fauna is fairly small and simple, with no hawkmoths, few "primitive" bees, and few butterflies (Dugdale 1975; Godley 1979). The most common invertebrate flower-visitors are flies and moths. It is unclear whether these groups have suffered declines; no extinctions of native invertebrate pollinators are known (J. Dugdale, personal communication) but our knowledge of the prehuman and post-human insect pollinators is very sketchy, so losses may have gone unnoticed.

---

[1] Bird names follow Heather and Robertson (1996).

[2] The silvereye has been variously treated as native or introduced, because it first established in New Zealand in 1856. Since it apparently arrived without direct human assistance, we follow Heather and Robertson (1996) in calling it a recently arrived native. It is now widespread (Bull et al. 1985).

There is increasing evidence that the losses of mutualists from much of New Zealand are having an effect on the reproduction of native plants. Mainland–island comparisons suggest that the fruit dispersal of nikau (*Rhopalostylis sapida*), kotukutuku (*Fuchsia excorticata*), and karo (*Pittosporum crassifolium*) is reduced or slowed by a lack of frugivores on the mainland (Anderson 1997; McNutt 1998). The fruit and seed production for six of seven species of bird-pollinated trees and shrubs has been shown to be pollen-limited at some mainland sites (Anderson et al. 2006). For instance, the red mistletoe (*Peraxilla tetrapetala*) is regularly pollen-limited at two South Island sites (Robertson et al. 1999; Kelly et al. 2004), and the climbing fuchsia (*Fuchsia perscandens*) also has very low fruit set at two sites near Christchurch (Montgomery et al. 2001). Both species are pollinated by bellbirds and tui, which at the study sites of both these plants are now uncommon and locally extinct, respectively.

Recent observations of relative abundances might suggest that introduced animals have now become quite important for pollination and dispersal of native plants in New Zealand. It is possible that they might offset the losses of native animals. Although there have been extinctions or severe range and density reductions in important native bird mutualists, a number of introduced plant visitors are now very widespread. The two most widespread of all bird species in New Zealand are the introduced chaffinch (*Fringilla coelebs*) and blackbird (*Turdus merula*; Bull et al. 1985), which visit flowers and fruits (Williams and Karl 1996; Ladley et al. 1997). In fact, 11 of the 15 most widespread birds in New Zealand are introduced (Bull et al. 1985). O'Donnell and Dilks (1994) and Burrows (1994) showed that blackbirds were important for dispersal of various native plants. Burrows said "had it not been for the naturalisation of silvereyes and blackbirds, last century, the link between birds and seed dispersal in Banks Peninsula forests would now be very tenuous" (Burrows 1994, p. 177). Introduced feral bumblebees can be found throughout the country, and honeybees are "probably one of the most common and continuously present insects" (Donovan 1980, p. 109). Butz Huryn (1995) provided a list of 188 native plant species whose flowers are visited by honeybees.

However, a quantitative assessment of the importance of introduced animals has not previously been attempted. There are three possible measures of the importance of introduced species. The least accurate is based on presence/absence – the percentage of introduced species in lists of all visitors to flowers or fruit. The second is quantitative, based on the percentage of visits made by various species. The third and most accurate is effectiveness, which takes into account both frequency and efficacy of visits (e.g., the proportion of flowers successfully pollinated by a particular animal). There are preliminary indications from quantitative information that introduced dispersers may be less important than a "species list" approach suggests: introduced birds around Nelson were more likely than native birds to feed on fruits of introduced plants (Williams and Karl 1996).

The aim of this review is to evaluate the relative importance of introduced animals for pollinating and dispersing native plants, based mainly on the frequency of visits to flowers and fruits. Ideally, this assessment would be based on effectiveness, but there are few data on the effectiveness of individual animals as pollinators or dispersers (e.g., Robertson et al. 2005). Hence, a comprehensive review at that level is not yet possible.

15.2 Approach

We use the term "mutualist" to mean "animal pollinator of flowers or disperser of seeds". We use "flower-visitor" to mean an animal recorded approaching flowers. Such animals may be effective pollinators, may be partial or complete robbers of nectar and/or pollen, or may be searching for other foods among the flowers, with or without incidental pollination. Only when there is clear evidence that the visitor is not achieving any pollination (e.g., kereru eating flowers of a range of different plant species; O'Donnell and Dilks 1994) do we exclude those data. Similarly, for "fruit-visitor" we include all animals visiting fruit, although some may be partial or complete seed predators (e.g., most parrots) or incidental visitors.

To estimate the relative numerical importance of introduced animals, we collated all published studies where the percentages of visits made by all different taxa to native plants were given. Studies that listed the species visiting flowers or fruits without numerical estimates of their importance (e.g., Primack 1983), or that listed native flowers or fruits in the diet of introduced animals (e.g., Butz Huryn 1995; Williams and Karl 1996) were not included. We supplemented the literature with our own unpublished data. Generally, the proportion of "visits" made by each group was measured by the relative length of time spent foraging on fruits and flowers.

In a few cases, several related plants were grouped together (e.g., climbing *Metrosideros* spp.). Bird visitors to plants were identified to species level. In most cases, invertebrate flower-visitors were identified only to broad taxonomic groups, some of which (e.g., Diptera, Coleoptera) may contain a mixture of native and introduced species, although these groups are almost certainly dominated by native species. *Bombus* spp., *Apis mellifera* and *Vespula* spp. were always separated from other flower-visitors. Data in O'Donnell and Dilks (1994) were presented as percentages by columns (birds); we multiplied up to counts, then calculated percentages by rows (plants), excluding plants with fewer than 10 observations. Our own data were obtained either by direct or video observation of groups of plants during the flowering or fruiting season (see Robertson et al. 1999 for details). The data presented here are the proportion of visits made for site–species combinations. Where data from more

Table 15.1 Bird visitors to flowers of native plants, mostly forest species (% of all visits by each bird). Birds are arranged by numerical importance within category (native or introduced). Kereru visit flowers to eat them, so they cannot effect pollination and are not listed[a]

Plant sp.	Site, ref.	Native birds						Introduced birds			
		bel	sil	tui	sti	kak	Other	spa	cha	Other	Total[b]
Alefla	Cra, 1	30	70	0	0	0		0	0		0
Dysspe	Tir, 2	48.5	3.0	45.5	1.2	0	sad 1.8	0	0		0
Dysspe	Wen, 2	0	0	100	0	0		0	0		0
Fucexc	AhS, 1	2.7	97.3	0	0	0		0	0		0
Fucexc	Aka, 1	44.2	54.4	0.5	0	0		0	0.3	ros 0.5	0.8
Fucexc	Koh, 1	1.7	97.9	0	0	0		0.4	0		0.4
Fucexc	Hin, 1	17.5	82.5	0	0	0		0	0		0
Fucexc	MtF, 1	63.5	33.2	3.3	0	0		0	0		0
Fucexc	Pre, 1	15.3	16.6	68.1	0	0		0	0		0
Fucexc	Riti, 1	69.4	0	30.6	0	0		0	0		0
Fucexc	Rku, 1	0	0	100	0	0		0	0		0
Fucexc	Rroa, 1	63.3	0	36.7	0	0		0	0		0
Fucexc	Rua, 1	5.0	93.4	1.7	0	0		0	0		0
Fucexc	Win, 3	37.1	43.9	13.0	0	5.7	kea 0.2, fan 0.1	0	0		0
Gnluc	Tir, 2	51.1	14.9	14.9	0	0	par 19.1	0	0		0
Metexc	Tir, 2	36.5	7.8	50.4	0.3	0	par 1.7, sad 1.8, whi 1.0	0	0	myn 0.3, sta 0.3	1.6
Metexc	Wen, 2	0	16.2	36.5	0	0		37.9	0.6	ros 1.7, myn 2.5, sta 4.7	47.3
Met spp	Win, 3	41.4	28.8	23.0	0	6.9		0	0		0
Metumb	Win, 3	24.5	28.9	15.2	0	17.5	kea 13.9	0	0		0
Per spp	Win, 3	9.1	4.0	3.7	0	80.5	yel 2.7	0	0		0
Percol	Waip, 1	98.7	0	1.3	0	0		0	0		0
Percol	Wak, 1	3	0.2	96.7	0	0		0	0		0
Pertet	Bel, 1	0	0	100	0	0		0	0		0
Pertet	Boy, 1	95.3	0	4.7	0	0		0	0		0
Pertet	Cra, 1	94.4	5.5	0	0	0		0	0.1		0.1
Pertet	Egl, 1	98.5	0	1.5	0	0		0	0		0

Table 15.1 (*Continued*)

Plant sp.	Site, ref.	Native birds						Introduced birds			
		bel	sil	tui	sti	kak	Other	spa	cha	Other	Total[b]
Pertet	Lit, 1	100	0	0	0	0		0	0		0
Pertet	Oha, 1	41.8	25.8	0	0	0		7.4	25.0		32.4
Photen	Cas, 4	52.5	45.8	0	0	0		0	0	sta 1.70	1.7
Photen	Por, 4	20.8	78.7	0	0	0		0	0	sta 0.60	0.6
Pitcra	Tir, 2	81.1	0	18.9	0	0		0	0		0
Pitcra	Wen, 2	0	81.0	14.3	0	0		4.8	0		4.8
Psearb	Mah, 2	0	12.5	87.5	0	0		0	0		0
Psearb	Tir, 2	79.2	1.0	17	0.7	0	whi 2.1	0	0		2.1
Psecra	Win, 3	40.5	39.3	20.2	0	0		0	0		0
Rhasol	Lit, 2	0	0	0	100	0		0	0		0
Rhasol	Tir, 2	52.6	0	0	47.4	0		0	0		0
Rhasol	Wen, 2	0	100	0	0	0		0	0		0
Sopmic	AhV, 1	2.1	96.1	0	0	0		0.2	1.7		1.9
Sopmic	Kow, 1	37.9	0	62.1	0	0		0	0		0
Sopmic	Poh, 1	32.3	62	5.6	0	0		0	0		0
Sopmic	Rroa, 1	67.5	0	32.5	0	0		0	0		0
Sopmic	Tai, 1	0	36.6	0	0	0		27.9	35.4		63.3
Sopmic	Tir, 2	17.0	0	83.0	0	0		0	0		0
Sopmic	Uni, 1	0	18.6	0	0	0		81.4	0		81.4
Sopmic	Waim, 1	0	98.8	0	0	0		0	1.2		1.2
Sopmic	Wen, 2	0	12.7	87.3	0	0		0	0		0
Sopmic	Wen, 1	0	1.7	98.3	0	0		0	0		0
Soppro	Kait, 1	0	100	0	0	0		0	0		0
Vitluc	Tir, 2	79.7	1.9	13	3.9	0		0	0	ros 1.44	1.4
Vitluc	Wen, 2	0	55.5	24.7	0	0		0	0	ros 10.5, myn 9.4	19.9
Weirac	Win, 3	9.8	62.5	13.6	0	14.0		0	0	0	0
Mean		32	31.3	25.5	3.0	2.4		3.1	1.2		5.0

[a] Plant species codes are the first three letters of genus and species names. Full species names are: *Alepis flavida*, *Dysoxylum spectabile*, *Fuchsia excorticata*, *Geniostoma lucida*, *Metrosideros excelsa*, *Metrosideros* spp. (vines), *Metrosideros umbellata*, *Peraxilla* spp. (*colensoi* and *tetrapetala*), *Peraxilla colensoi*, *Peraxilla tetrapetala*, *Phormium tenax*, *Pittosporum crassifolium*, *Pseudopanax arboreus*, *Pseudopanax crassifolius*, *Rhabdothamnus solandri*, *Sophora microphylla*, *Sophora prostrata*, *Vitex lucens*, *Weinmannia racemosa*. Bird species codes are the first three letters of the common name. Full names (with overall mean where not given above) are: bellbird, silvereye, tui, stitchbird, kaka, parakeet spp. 0.4 % (largely yellow-crowned), kea 0.3, saddleback 0.1, whitehead 0.1, yellowhead 0.1, fantail <0.1, house sparrow, chaffinch, eastern rosella 0.3, mynah 0.2, starling 0.1. Full site names as in original reference are: Ahuriri Summit, Ahuriri Valley, Akatarawa, Belgrove, Boyle, Cass, Craigieburn, Eglinton, Hinewai, Kaituna Valley, Kohitere, Kowhai Point, Little Barrier, Mahurangi, Mt. Fyffe, Ohau, Pohangina, Port Hills, Pretty Bridge, Rotoiti, Rotokura, Rotoroa, Ruahine, Taihape, Tiritiri, University of Canterbury, Waimakariri, Waipori, Wakefield, Wenderholm, Windbag. References: 1 Robertson et al. (1999) and unpublished; 2 Anderson (1997); 3 O'Donnell and Dilks (1994); 4 Tisch (1996)

[b] Total visitation by introduced species

than one season were available for a site , we averaged all seasons to provide a single value for that site.

15.3 Bird Flower-Visitors

We found 52 cases covering 18 native plant species from 31 sites throughout New Zealand, visited by 16 bird species of which five (31 %) were introduced (Table 15.1). In 36 cases (69 %), there were no introduced birds recorded visiting flowers. The most often recorded introduced bird, the house sparrow (*Passer domesticus*), made only 3.1 % of all visits, and all introduced species together totaled only 5 %. In contrast, the endemic tui and bellbird and the native silvereye combined made 89 % of all visits. The average for the endemic stitchbird was as high as for sparrows, despite stitchbirds being present at only two island sites, where 10 of the 52 cases were measured. In these 10 cases, stitchbirds made 15.3 % of all flower visits. The key variation among cases was the percentage of flower visits by silvereye versus all other native birds (Fig. 15.1). In only five cases did introduced birds contribute more than

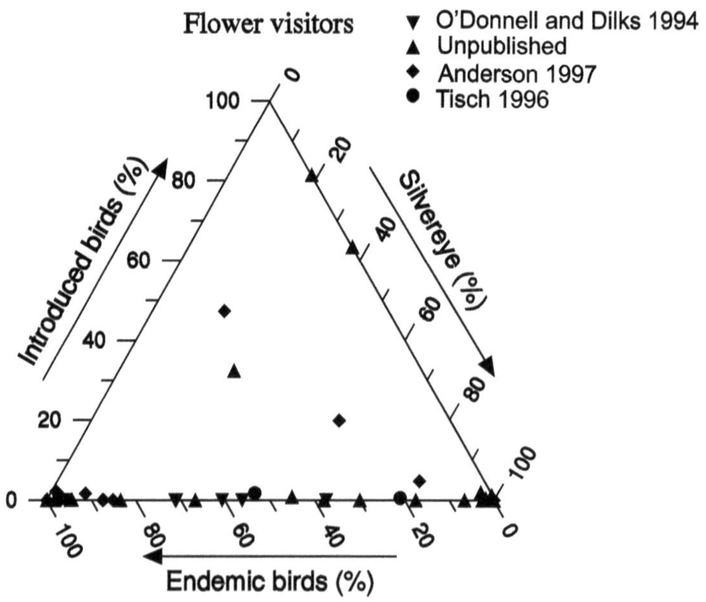

Fig. 15.1 Percentage of avian flower visits made by introduced birds, native silvereyes, and all other native birds (nearly all endemic), in 52 quantitative studies on 18 native plant species. The greater the importance of introduced birds, the higher the data point will be above the baseline. Sites in the *bottom left* are dominated by endemic birds, those in the *bottom right* by silvereyes and in the top by introduced birds. See Table 15.1 for a full listing of bird and plant species

5 % of visits; the most important introduced flower-visitors at these sites were chaffinches and sparrows (Table 15.1). All five cases were from landscapes that have been very heavily modified with land use that is now largely urban or pastoral.

15.4 Bird Fruit-Visitors

The 39 cases covered 32 native plant species from 10 sites from South Westland to Auckland, although two-thirds of the cases came from the Westland site. The plants were visited by 22 bird species, seven of which (32 %) were introduced (Table 15.2). However, 15 of the bird species (including six of the introduced species) were minor fruit-visitors, each responsible for <1 % of total visits. Four native species (silvereye, bellbird, tui, and kereru) accounted for 83.7 % of all visits. There were more cases (than for pollination) with non-trivial contributions by introduced birds (Fig. 15.2), with nine cases (23 %) having more than 5 % of visits by exotics. However, 22 cases (56 %) had no

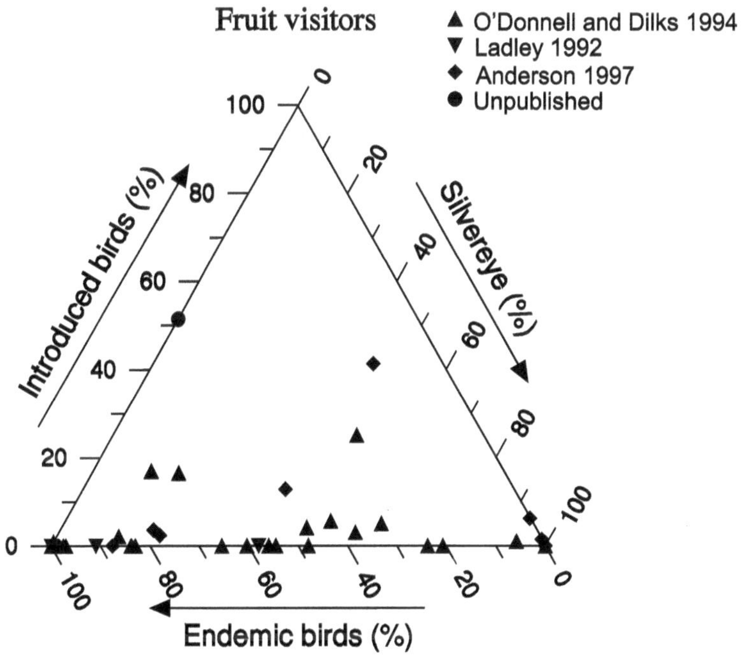

Fig. 15.2 Percentage of visits to fruits on 32 native plant species made by introduced birds, native silvereyes, or all other native birds (nearly all endemic) in 39 studies. The greater the importance of introduced birds, the higher the data point will be above the baseline. See Table 15.2 for a full listing of bird and plant species

Table 15.2 Bird visitors to fruits of native plants, mostly forest species (% of all visits by each bird). Birds are arranged by numerical importance within category (native or introduced)[a]

Plant sp.	Site, ref.	Native birds sil	bel	ker	tui	kak	sad	Other	Introduced birds bla	spa	cha	Other	Total[b]
Alefla	Cra, 1	42	58	0	0	0	0		0	0	0		0.0
Ariser	Win, 2	60	5.9	31	0	0	0		3.1	0	0		3.1
Ascluc	Win, 2	76.2	13.2	0	0	0	0	fan 5.5, tom 5.1	0	0	0		0.0
Carser	Win, 2	17.5	2.8	0	56.5	6.7	0		16.5	0	0		16.5
Copfoe	Win, 3	16.2	50.8	11.8	0	16.7	0	par 2.3, fan 2.3	0	0	0		0.0
Coprob	Tir, 3	20.8	37.8	0	12.4	0	12	par 2.2, whi 10.9, sti 1.6	0	0	0	son 0.8, sta 1.6	2.3
Coprob	Wen, 3	44.5	0	0	14.2	0	0		28.4	6.5	0	son 6.45	41.3
Coprot	Win, 2	44	56	0	0	0	0		0	0	0		0.0
Cop spp	Win, 2	39.7	47.4	9.2	0	0	0	tom 1.9, bro 1.8	0	0	0		0.0
Corarb	Win, 2	64.2	4.9	25.8	0	0	0		5.1	0	0		5.1
Daccup	Win, 2	12.6	7.2	57.8	5.2	10.5	0	par 4.41, tom 0.1, gre 0.1	1.3	0	0.2	red 0.8	2.2
Dacdac	Win, 2	16.3	6.9	72.6	2.8	0	0	bro 1.4	0	0	0		0.0
Dysspe	Tir, 3	1.5	4.4	0	73.5	0	16.2	sti 4.4	0	0	0		0.0
Dysspe	Wen, 3	41	0	0	46.2	0	0		0	10.3	0	myn 2.6	12.8
Fucexc	Win, 2	11.7	4.8	64.1	0	0	0	fan 2.4	17	0	0		17.0
Grilit	Win, 2	93.6	1.0	2.5	0	0	0	gre 0.5, yel 1.5	1.0	0	0		1.0
Hedarb	Win, 2	0	1.7	79.8	0	18.5	0		0	0	0		0.0
Ilemic	Wak, 1	9.1	3.0	0	87.9	0	0		0	0	0		0.0
Myraus	Win, 2	34.6	50.3	9.6	5.5	0	0		0	0	0		0.0
Myrdiv	Win, 2	52	48	0	0	0	0		0	0	0		0.0
Neoped	Win, 2	52	48	0	0	0	0		0	0	0		0.0
Per spp	Win, 2	49.3	7.0	0	0	18.7	0		25.1	0	0		25.1
Percol	Wak, 1	0	49	0	51	0	0		0	0	0		0.0
Pertet	Oha, 4	0	48.6	0	0	0	0		44.6	0	6.8		51.4
Pitcra	Tir, 3	18.9	8.4	0	0	0	58.2	par 1.2, whi 8.4, sti 1.2	0	2.4	0	sta 1.2	3.6

Pitcra	Wen, 3	93.8	0	0	0	0	0		3.1	0	3.1		6.3
Podhal	Win, 2	2.9	3.3	0	0	73.8	0	par 20.0	0	0	0		0.0
Prufer	Win, 2	0	0.6	93.7	0.4	4.5	0		0.9	0	0		0.9
Psearb	Pih, 3	100	0	0	0	0	0		0	0	0		0.0
Psearb	Tir, 3	12.4	21	0	6.8	0	35.3	par 0.9, whi 11.4, sti 12.1	0	0	0		0.0
Psearb	Wtak, 3	98.6	0	0	0	0	0		0	0	0	myn 1.4	1.4
Psecra	Win, 2	53.7	7.3	19.2	0	6.2	0	tom 3.5, bro 1.7, gre 2.8	2.3	0	3.3		5.6
Psewcol	Win, 2	17.1	13.6	69.4	0	0	0		0	0	0		0.0
Rauedg	Win, 2	45.4	7.9	0	10.4	7.8	0	rif 20.6, fan 5.2, tom 2.7	0	0	0		0.0
Rausim	Win, 2	49.6	0.6	36.5	0	0	0	rif 3.1, tom 0.6, bro 2.1, gre 3.3		3.8	0	0.5	4.2
Ripsca	Win, 2	2.5	22.6	66.6	5.5	2.8	0		0	0	0		0.0
Rub spp	Win, 2	100	0	0	0	0	0		0	0	0		0.0
Shedig	Win, 2	79.3	0	19.7	1.0	0	0		0	0	0		0.0
Tupant	Wnui, 1	0	100	0	0	0	0		0	0	0		0.0
Mean		37.8	19.0	17.2	9.7	4.3	3.1		3.9	0.5	0.4		5.1

[a] All codes (except references) are as for Table 15.1, with these additions. Plant species: *Aristotelia serrata*, *Ascarina lucida*, *Carpodetus serratus*, *Coprosma foetidissima*, *C. robusta*, *C. rotundifolia*, *C.* spp. (divaricates), *Coriaria arborea*, *Dacrydium cupressinum*, *Dacrycarpus dacrydioides*, *Griselinia littoralis*, *Heydcarya arborea*, *Ileostylus micranthus*, *Myrsine australis*, *M. divaricata*, *Neomyrtus pedunculata*, *Podocarpus hallii*, *Prumnopitys ferruginea*, *Pseudowintera colorata*, *Raukaua* (was *Pseudopanax*) *edgerlyi*, *Raukaua simplex*, *Ripogonum scandens*, *Rubus* spp., *Schefflera digitata*, *Tupeia antarctica*. Bird species (and means): parakeets 0.8 %, whitehead 0.8, rifleman 0.6, stitchbird 0.5, fantail 0.4, tomtit 0.4, brown creeper 0.2, grey warbler 0.2, yellowhead <0.1, song thrush 0.2, mynah 0.1, starling 0.1, redpoll <0.1. Site names: Piha, Waitakere, Wainui. References: 1 Ladley (1994); 2 O'Donnell and Dilks (1994); 3 Anderson (1997); 4 own unpublished data

[b] Total visitation by introduced species

introduced fruit-visitors, and the average contribution by exotics was only 5 %. By far the most important introduced bird was the blackbird, responsible for 3.9 % of all visits (76 % of introduced bird visits), but the endemic saddleback was almost as high in overall contribution to visits (Table 15.2, Fig. 15.2), despite being extant at only one study site (Tiritiri Matangi Island) where four plant species were studied. Saddlebacks made 30.4 % of all visits to these four species.

15.5 Invertebrate Flower-Visitors

We found 15 cases for insect visitors to 12 plant species (Table 15.3). Of the introduced insects, one group (the Hymenoptera) are important as flower-visitors to native plants. These introduced hymenopterans (the honeybee *Apis mellifera*, four bumblebees *Bombus* spp., and two wasps *Vespula vulgaris* and *V. germanica*) have achieved greater penetration into native systems (Fig. 15.3) than was the case for birds. In only three of the 15 cases were no vis-

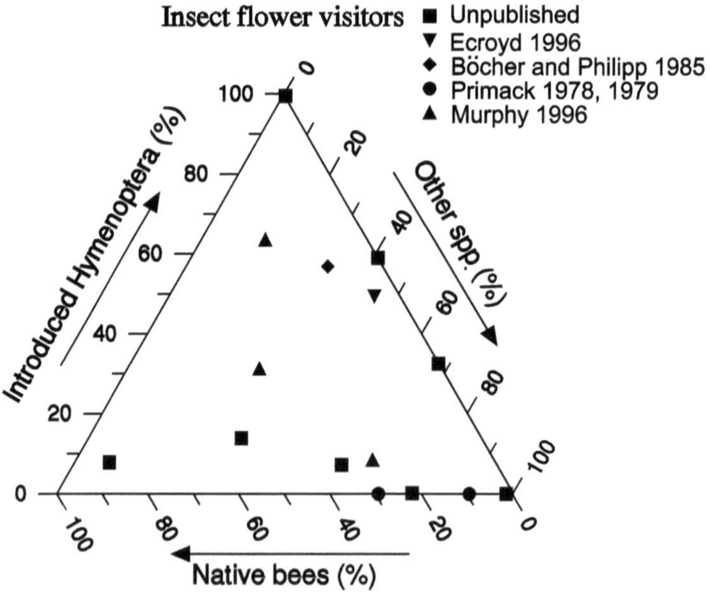

Fig. 15.3 Percentage of visits to flowers of 12 native plant species made by introduced bees and wasps, native bees, or all other invertebrates (mainly Diptera, nearly all native) in 15 studies. The greater the importance of introduced Hymenoptera, the higher the data point will be above the baseline. See Table 15.3 for a full listing of insect and plant species

Table 15.3 Insect visitors to flowers of native plants (% of all visits by insects). Insects are arranged by category (native or introduced)[a]

Plant sp.	Site, ref.	Native insects Nat. bee	Nat. Hym	Coleo	Lepid	Dipt	Unspec.	Introduced insects *Apis*	*Bombus*	*Vespula*	Total[b]
Alefla	Cra, 1	0	0	0	0	67.5	0	0	0	32.5	32.5
Dactay	Mam, 2	5.6	0	25.7	0	19.4	0	0	0	49.3	49.3
Distou	Cas, 3	9.5	0	1.2	1.2	88.1	0	0	0	0	0.0
Hebstr	Ton, 4	22.4	0	0	0.4	13.7	0	51.6	11.8	0.1	63.5
Lepsco	Cas, 5	29.3	0	0	0	69.2	1.3	0	0	0	0
Lepsco	Ton, 4	26.4	0	0	0	63.8	1.3	8.2	0.3	0	8.5
Leugra	MtH, 1	21.9	24.9	2.9	1.0	40.1	8.9	0.2	0	0	0.2
Mimrep	Lak, 6	12.1	0	0	4.8	26.4	0	56.8	0	0	56.8
Percol	Wak, 1	84.5	0	0	0	5.7	2.0	1.9	0	6.0	7.8
Pertet	Cra, 1	33.8	0.2	0	0	35.9	22.8	4.7	2.4	0.1	7.2
Pertet	Oha, 1	52.4	6.9	0	0	16.9	9.8	2.8	11.1	0	13.9
Photen	Ton, 4	39.8	0	0	0	28.9	0	28.6	2.5	0.2	31.3
Raoexi	MtH, 1	1.4	1.3	1.9	1.1	92.3	1.9	0	0	0	0
Soppro	Kait, 1	0	0	0	0	0.6	0	4.9	93.5	0	98.4
Soppro	Tors, 1	0	0	0	0	41.0	0	0	58.9	0	58.9
Mean		22.6	2.2	2.1	0.6	40.6	3.2	10.6	12.0	5.9	28.6

[a] All codes (except references) as for Tables 15.1 and 15.2, with these additions. Plant species: *Dactylanthus taylorii*, *Discaria toumatou*, *Hebe stricta*, *Leptospermum scoparium*, *Leucogenes grandiceps*, *Mimulus repens*, *Raoulia eximia*. Species with entomophilous flowers are Distou, Hebstr, Lepsco, Leugra, Mimrep and Raoexi. Insect visitor taxa as follows: native bees, other native Hymenoptera, Coleoptera*, Lepidoptera*, Diptera*, unspecified* (includes unidentified visitors and taxa not separately listed), *Apis mellifera*, *Bombus* spp., *Vespula* spp. Groups marked * are mainly native but may include some introduced species. Sites: Lake Ellesmere, Mamaku, Mt. Hutt, The Tors, Tongariro. References: 1 own unpublished data; 2 Ecroyd (1996); 3 Primack (1979); 4 Murphy (1996); 5 Primack (1978); 6 Bocher and Philipp (1985)

[b] Total visitation by introduced species

its from introduced bees and wasps recorded, whereas in five cases these insects made more than half of the visits. The rate of visitation by honeybees may depend on the distance to commercial hives (e.g., close at Tongariro, and distant at Mt. Hutt). In some cases, visits from native vertebrates, which are not shown in Fig. 15.3, may be more important than those from insects (e.g., bats to *Dactylanthus*; Ecroyd 1996), but in other cases (e.g., *Hebe stricta*, one site of *Sophora prostrata*), no vertebrates were recorded as visitors. Overall, introduced Hymenoptera were responsible for 29 % of the visits made by insects, which represents a level of relative visitation to flowers that is five times greater than that shown by introduced birds. However, only half of these 12 plant species have "entomophilous" flowers (Table 15.3), and the mean level of visitation by introduced hymenopterans for the entomophilous species was somewhat lower (18 %).

15.6 Implications of Findings

The key result from the data is the relatively low importance of introduced animal mutualists, especially birds. Their quantitative contribution to flower and fruit visits (c 5 %) is much lower than would be expected from a summary of species lists of plant visitors (c 33 %).

Among introduced birds, only blackbirds (for fruit), chaffinches (for flowers), and sparrows (for flowers) make any appreciable percentage of visits, but even the uncommon endemic kaka is more important than any of these widespread and abundant introduced species (Fig. 15.4). Plant–bird mutualisms are now almost entirely dependent on four native birds, with the silvereye and bellbird predominant. Tui make up slightly fewer visits but are significant as the longest-billed species, which are important to long-tubed flowers such as *Sophora* spp. (Anderson 1997). Kereru are important for dispersal of fruits too large (>14 mm) for the other birds to swallow (Clout and Hay 1989), although their overall contribution as mutualists is reduced by their negative role in pollination – they eat the flowers of at least six native plants (O'Donnell and Dilks 1994).

However, there are several limitations to these data. Relative visitation rates do not tell us about the absolute level of visitation to plants, which may be too low for adequate pollination or dispersal. In some cases, the relative contribution of exotic birds rises because the absolute contribution of native birds falls (e.g., *P. tetrapetala* at Ohau; Ladley et al. 1997; Robertson et al. 1999), but this does not mean that the exotic birds are providing adequate service. Also, the effectiveness of different animals as pollinators and dispersers varies widely. Flower-visitors may be partial or complete nectar-robbers, which can reduce the attractiveness of flowers to legitimate pollinators (Irwin and Brody 1999), or cause direct mechanical damage to flowers (Anderson 2003). Fruit-visitors

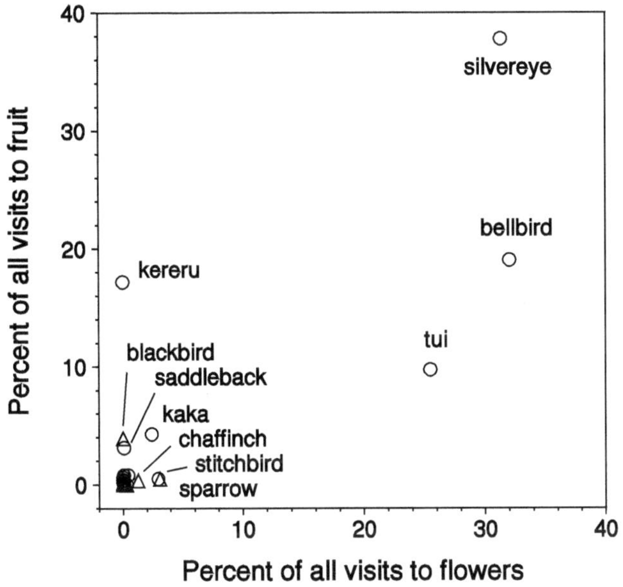

Fig. 15.4 Overall mean contribution by different bird species to flower visits and fruit visits on native plants throughout New Zealand. Introduced birds are marked with *triangles*, native birds with *circles*. Kereru are marked as doing zero pollination, but actually have a negative effect on pollination by eating flowers (see text)

may be partial or complete seed-predators (e.g., native parrots; O'Donnell and Dilks 1994), or may eat the fruit flesh without moving fruits away from the parent (e.g., fruit bats; Rainey et al. 1995).

Direct measures of mutualist effectiveness in New Zealand are scarce but some facts are known at a qualitative level. Short-beaked birds (silvereye, sparrow, chaffinch) are unable to access the nectar in long-tubed native flowers (*Sophora* spp., *Peraxilla* spp., *Fuchsia excorticata*, *Rhabdothamnus solandri*), and act primarily as nectar-robbers (Delph and Lively 1985; Anderson 1997; Ladley et al. 1997). Short-tongued bumblebees (*Bombus terrestris*) rob nectar from long-tubed flowers by biting holes in the corolla (Donovan 1980). Honeybees take nectar from the base of larger flowers (Anderson 1997). Birds with small gapes (especially silvereyes) are unable to swallow larger fruits (Clout and Hay 1989), though by pecking at the flesh they may disperse some seeds of small-seeded large-fruited species such as *Fuchsia excorticata* (Burrows 1994). In general, the effectiveness of most visitors to flowers and fruit is scarcely studied. However, special emphasis should be placed on long-tongued visitors (especially tui, and to a lesser extent bellbirds) to long-tubed flowers, and on large-gaped visitors (especially kereru) to large fruit, as particularly important.

Another gap concerns the role of introduced mammals in pollination and seed dispersal. New Zealand has only two plants apparently adapted to bat pollination, and for these, rats (*Rattus* spp.) or possums (*Trichosurus vulpecula*) may be partial substitutes (Lord 1991; Ecroyd 1996). There are no indications that mammals are important in pollinating other native plants. Contributions by mammals to dispersal are less clear. Rats are largely seed predators but effect some dispersal of small-seeded species (Williams et al. 2000), which are already well serviced by birds including silvereyes. Possums can disperse the seeds of small-fruited plants (Williams et al. 2000; Dungan et al. 2002), but have been tested on only one large-seeded plant, miro (*Prumnopitys ferruginea*), which they did not swallow. Pigs (*Sus scrofa*) eat and destroy large numbers of the seeds of some native plants such as tawa (*Beilschmiedia tawa*), hinau (*Elaeocarpus dentatus*; Thomson and Challies 1988) and matai (*Prumnopitys taxifolia*; Cochrane 1994, p. 28), but there are no data on whether any seeds survive to germinate from the dung. Information is also lacking on other large mammals such as goats (*Capra hircus*), sheep (*Ovis aries*), horses (*Equus caballus*), and cattle (*Bos taurus*), which could potentially provide some dispersal of large-seeded native plants; this warrants further investigation.

The major change in bird–plant mutualisms since human arrival is not a transition from natives to exotic species, but the shift from endemics to the silvereye, which has replaced declining or extinct endemics such as stitchbirds, saddlebacks, and (in Northland) bellbirds. The silvereye is now the most important mutualist, present through 80 % of the country, compared with 45–55 % for tui, bellbirds and kereru (Bull et al. 1985), and responsible for 38 % of all fruit visits and 31 % of all flower visits to native plants. On flower and fruit visits combined, silvereyes are 35 % more important numerically than bellbirds, twice as important as tui, and more than three times as important as kereru. The previous conclusions by Clout and Hay (1989), Lee et al. (1991) and Lord et al. (2002) – that kereru are the most important seed disperser in New Zealand forests – are true only for the small number of large-seeded, fleshy-fruited species in the flora (e.g., miro). The decline of kereru in parts of New Zealand threatens the dispersal of these large fruits (Clout et al. 1995). It is clear from this review that exotic bird species have done very little to help replace declining endemics.

The fewer available data on insect visits to flowers suggest a greater shift toward exotic insects, with slightly more than a quarter of visits made by exotics. Surprisingly, given that *Apis mellifera* is the consummate insect pollinator and is one of New Zealand's commonest insects, other hymenopterans (*Bombus*, *Vespula*) were often more frequent flower-visitors. This trend may increase, as the spread of the varroa mite is expected to reduce feral honeybee numbers (Goodwin 2004).

In conclusion, considering all introduced animals together, their contribution to plant–animal mutualisms with native plants in New Zealand appears to be surprisingly small.

Acknowledgements We thank Victoria Burgess, Alison Beath, and Paul Peterson for help with fieldwork, Claire Murphy for unpublished data, and the Public Good Science Fund for financial support. Ian Atkinson, George Cheely, John Innes and Bill Lee commented on earlier versions of the manuscript and helped to improve it considerably.

References

Anderson SH (1997) Changes in native ecosystem processes: the dynamics of pollination and dispersal in New Zealand forests. MSc Thesis, Auckland University, Auckland, New Zealand

Anderson SH (2003) The relative importance of birds and insects as pollinators of the New Zealand flora. NZ J Ecol 27:83–94

Anderson SH, Kelly D, Robertson AW, Ladley JJ, Innes JG (2006) Birds as pollinators and dispersers: a case study from New Zealand. Acta Zool Sinica (in press)

Atkinson IAE, Millener PR (1991) An ornithological glimpse into New Zealand's prehuman past. Acta XX Congr Int Ornithol I:129–192

Bell BD (1991) Recent avifaunal changes and the history of ornithology in New Zealand. Acta XX Congr Int Ornithol I:195–230

Bocher J, Philipp M (1985) Aspects of the reproductive biology of *Mimulus repens* (Scrophulariaceae) at Lake Ellesmere, Canterbury, New Zealand. NZ J Bot 23:141–149

Bull PC, Gaze PD, Robertson CJR (1985) The atlas of bird distribution in New Zealand. Ornithological Society of New Zealand, Wellington, New Zealand

Burrows CJ (1994) Fruit types and seed dispersal modes of woody plants in Ahuriri Summit Bush, Port Hills, western Banks Peninsula, Canterbury, New Zealand. NZ J Bot 32:169–181

Butz Huryn VM (1995) Use of native New Zealand plants by honeybees (*Apis mellifera* L.): a review. NZ J Bot 33:497–512

Clout MN, Hay JR (1989) The importance of birds as browsers, pollinators and seed dispersers in New Zealand forests. NZ J Ecol suppl 12:27–33

Clout MN, Karl BJ, Pierce RJ, Robertson HA (1995) Breeding and survival of New Zealand Pigeons *Hemiphaga novaeseelandiae*. Ibis 137:264–271

Cochrane CH (1994) Vegetation assessment and its implications for feral goat management in Isolated Hill Scenic Reserve, southern Marlborough. MSc Thesis, University of Canterbury, Christchurch, New Zealand

Cox PA (1983) Extinction of the Hawaiian USA avifauna resulted in a change of pollinators for the Ieie *Freycinetia arborea*. Oikos 41:195–199

Cox PA, Elmqvist T (2000) Pollinator extinction in the Pacific Islands. Conserv Biol 14:1237–1239

Daniel MJ (1976) Feeding by the short-tailed bat (*Mystacina tuberculata*) on fruit and possibly nectar. NZ J Zool 3:391–398

Delph LF, Lively CM (1985) Pollinator visits to floral colour phases of *Fuchsia excorticata*. NZ J Zool 12:599–603

Donovan BJ (1980) Interactions between native and introduced bees in New Zealand. NZ J Ecol 3:104–116

Dugdale JS (1975) The insects in relation to plants. In: Kuschel G (ed) Biogeography and ecology in New Zealand. Junk, The Hague, pp 561–589

Dungan RJ, O'Cain MJ, Lopez ML, Norton DA (2002) Contribution by possums to seed rain and subsequent seed germination in successional vegetation, Canterbury, New Zealand. NZ J Ecol 26:121–128

Ecroyd CE (1996) The ecology of *Dactylanthus taylorii* and threats to its survival. NZ J Ecol 20:81–100

Eifler DA (1995) Patterns of plant visitation by nectar-feeding lizards. Oecologia 101:228–233

Godley EJ (1979) Flower biology in New Zealand. NZ J Bot 17:441–466

Goodwin M (2004) Introduction and spread of varroa in New Zealand. Bee World 85:26–28

Heather BD, Robertson HA (1996) The field guide to the birds of New Zealand. Viking, Auckland, New Zealand

Holdaway RN (1999) Introduced predators and avifaunal extinction in New Zealand. In: MacPhee RDE (ed) Extinctions in near time: causes, contexts, and consequences. Advances in Vertebrate Paleobiology. Kluwer Plenum, New York, pp 189–238

Irwin RE, Brody AK (1999) Nectar-robbing bumble bees reduce the fitness of *Ipomopsis aggregata* (Polemoniaceae). Ecology 80:1703–1712

Janzen DH (1974) The deflowering of Central America. Nat Hist 83:48–53

Kearns CA, Inouye DW, Waser NM (1998) Endangered mutualisms: the conservation of plant-pollinator interactions. Annu Rev Ecol Syst 29:83–112

Kelly D, Ladley JJ, Robertson AW (2004) Is dispersal easier than pollination? Two tests in New Zealand Loranthaceae. NZ J Bot 42:89–103

Ladley JJ (1994) The reproductive ecology of the Loranthaceae mistletoes of New Zealand. MSc Thesis, University of Canterbury, Christchurch, New Zealand

Ladley JJ, Kelly D, Robertson AW (1997) Explosive flowering, nectar production, breeding systems, and pollinators of New Zealand mistletoes (Loranthaceae). NZ J Bot 35:345–360

Lee WG, Clout MN, Robertson HA, Wilson JB (1991) Avian dispersers and fleshy fruits in New Zealand. Acta XX Congr Int Ornithol III:1617–1623

Lloyd BD (2001) Advances in New Zealand mammalogy 1990–2000: short-tailed bats. J R Soc NZ 31:59–81

Lord JM (1991) Pollination and seed dispersal in *Freycinetia baueriana*, a dioecious liane that has lost its bat pollinator. NZ J Bot 29:83–86

Lord JM, Markey AS, Marshall J (2002) Have frugivores influenced the evolution of fruit traits in New Zealand? In: Levey DJ, Silva WR, Galetti M (eds) Seed dispersal and frugivory: ecology, evolution and conservation. CABI, Wallingford, pp 55–68

McConkey KR, Drake DR (2002) Extinct pigeons and declining bat populations: are large seeds still being dispersed in the tropical Pacific? In: Levey DJ, Silva WR, Galetti M (eds) Seed dispersal and frugivory: ecology, evolution and conservation. CABI, Wallingford, pp 381–395

McNutt KL (1998) Impacts of reduced bird densities on pollination and dispersal mutualisms in New Zealand forests. MSc Thesis, Massey University, Palmerston North, New Zealand

Meehan HJ, McConkey KR, Drake DR (2002) Potential disruptions to seed dispersal mutualisms in Tonga, Western Polynesia. J Biogeogr 29:695–712

Montgomery BR, Kelly D, Ladley JJ (2001) Pollinator limitation of seed set in *Fuchsia perscandens* (Onagraceae) on Banks Peninsula, South Island, New Zealand. NZ J Bot 39:559–565

Murphy CL (1996) The impact of honeybees on montane ecosystems within Tongariro National Park. MSc Thesis, Massey University, Palmerston North, New Zealand

O'Donnell CFJ, Dilks PJ (1994) Foods and foraging of forest birds in temperate rainforest, South Westland, New Zealand. NZ J Ecol 18:87–107

Paton DC (1993) Honeybees in the Australian environment: does *Apis mellifera* disrupt or benefit the native biota? Bioscience 43:95–103

Primack RB (1978) Variability in New Zealand montane and alpine pollinator assemblages. NZ J Ecol 1:66–73
Primack RB (1979) Reproductive biology of *Discaria toumatou* (Rhamnaceae). NZ J Bot 17:9–13
Primack RB (1983) Insect pollination in the New Zealand mountain flora. NZ J Bot 21:317–333
Rainey WE, Pierson ED, Elmqvist T, Cox PA (1995) The role of flying foxes (Pteropodidae) in oceanic island ecosystems of the Pacific. In: Racey PA, Swift SM (eds) Ecology, evolution and behaviour of bats. Claredon Press, Oxford, pp 47–62
Robertson AW, Kelly D, Ladley JJ, Sparrow AD (1999) Effects of pollinator loss on endemic New Zealand mistletoes (Loranthaceae). Conserv Biol 13:499–508
Robertson AW, Ladley JJ, Kelly D (2005) The effectiveness of short-tongued bees as pollinators of apparently "ornithophilous" New Zealand mistletoes. Aust Ecol 30:298–309
Sekercioglu CH, Daily GC, Ehrlich PR (2004) Ecosystem consequences of bird declines. Proc Natl Acad Sci USA 101:18042–18047
Steadman DW (1995) Prehistoric extinctions of Pacific Island birds: biodiversity meets zooarchaeology. Science 267:1123–1131
Steadman DW (1997) The historic biogeography and community ecology of Polynesian pigeons and doves. J Biogeogr 24:737–753
Thomson C, Challies CN (1988) Diet of feral pigs in the podocarp-tawa forests of the Urewera Ranges. NZ J Ecol 11:73–78
Tisch PA (1996) Masting: a test of the pollination efficiency and predator satiation hypotheses. MSc Thesis, University of Canterbury, Christchurch, New Zealand
Towns DR, Daugherty CH, Cree A (2001) Raising the prospects for a forgotten fauna: a review of 10 years of conservation effort for New Zealand reptiles. Biol Conserv 99:3–16
Whitaker AH (1987) The roles of lizards in New Zealand plant reproductive strategies. NZ J Bot 25:315–328
Williams PA, Karl BJ (1996) Fleshy fruits of indigenous and adventive plants in the diet of birds in forest remnants, Nelson, New Zealand. NZ J Ecol 20:127–145
Williams PA, Karl BJ, Bannister P, Lee WG (2000) Small mammals as potential seed dispersers in New Zealand. Aust Ecol 25:523–532
Wotton DM (2002) Effectiveness of the common gecko (*Hoplodactylus maculatus*) as a seed disperser on Mana Island, New Zealand. NZ J Bot 40:639–647

16 Trophic Interactions Among Native and Introduced Animal Species

W.A. Ruscoe, G. Norbury and D. Choquenot

16.1 Introduction

To conserve natural ecosystems, we need to understand how biological invasions affect trophic interactions and other key ecosystem processes. Bottom-up hypotheses assume that systems are regulated by nutrient flow from lower trophic levels (White 1984), and that predation by higher trophic levels plays no measurable part in system regulation. Alternative top-down hypotheses assume each trophic level is regulated by the one above, with the top predators being self-regulated. It is accepted that top-down and bottom-up processes can operate together within a single system (Oksanen et al. 1981; Fretwell 1987; Oksanen 1990; Sinclair and Krebs 2001). For example, Krebs et al. (1995, 2001) demonstrated that although vertebrate community composition in the boreal forest was primarily structured by high-level predation, the removal of snowshoe hares (a keystone herbivore) caused the community to collapse. Also, addition of fertilizer resulted in larger increases in plant biomass than that recorded after removal of herbivores (snowshoe hares), demonstrating that bottom-up processes also had important effects in this ecosystem.

New Zealand is one of the few countries that, prior to human occupation, lacked terrestrial mammals, and most exotic mammal introductions have occurred in the 200 years since European colonization (see Chap. 4). These introduced animals compete with native fauna, and impose higher predation rates than those experienced in pre-settlement times. As such, these invasions provide natural experiments to test the relative importance of top-down and bottom-up processes. In this chapter, we describe how these processes structure new species assemblages in two New Zealand ecosystems: beech (*Nothofagus*) forest and drylands.

Ecological Studies, Vol. 186
R.B. Allen and W.G. Lee (Eds.)
Biological Invasions in New Zealand

16.2 Trophic Interactions in Beech Forest Ecosystems

Beech species are distributed throughout New Zealand, either in mixed forest with podocarp and other hardwood species or as the sole dominant canopy taxa. Beech forest canopy dominance is largely unmodified by human-related impacts. Such forest grows from sea level to timberline (approximately 900–1,500 m a.s.l.) but is extensive in montane regions and in the south (Wardle 1984).

16.2.1 A System Driven by Sporadic Resource Input

Beech trees flower and seed heavily at irregular intervals (Wardle 1984). In between these heavy seed years, individual trees may produce some seed, or none at all. At Craigieburn Forest, Canterbury, South Island, annual mountain beech (*N. solandri* var. *cliffortoides*) seed crop size varies in the range 0–12,000 seeds m^{-2} (Allen and Platt 1990). Irregular seeding in many species, including beech, has been related to climate (represented by link 1 in Fig. 16.1), resource

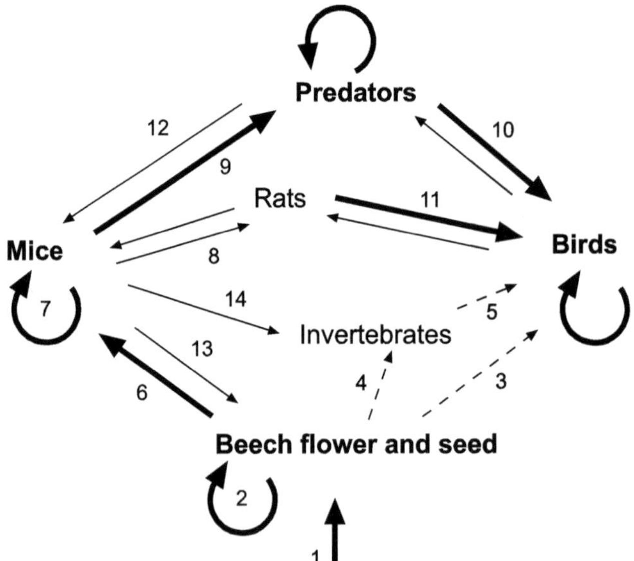

Fig. 16.1 Key trophic relationships in beech forest ecosystems. *Upward arrows* are "bottom-up" processes, and *downward arrows* are "top-down" processes. *Thick-line arrows* indicate strong interactions, and *thin-line arrows* are weak interactions. *Dashed-line arrows* represent native biotic interactions. The *loops* depict intraspecific interactions

availability, reproductive efficiency, predator satiation (link 2), or a combination of these factors (see Kelly and Sork 2002). Whatever the mechanism(s), the large between-year variation in flower and seed production (e.g., Allen and Platt 1990; Richardson et al. 2005) leads to high temporal variation in food availability for a variety of native and exotic animals, and this results in a very dynamic system (cf. Ostfeld and Keesing 2000).

16.2.2 Bottom-Up Processes

The irregular input of a high-quality food resource to the beech forest suggests a system influenced by "bottom-up" processes. The productivity of at least two native birds is linked to beech seeding (link 3 in Fig. 16.1). The native parrots kakariki (*Cyanoramphus auriceps*) and kaka (*Nestor meridionalis*) eat seed while it is green and attached to the tree. In low-seedfall years, kakariki raise only one brood, and kaka do not usually breed at all. However, in years of high seedfall, kakariki raise as many as five broods during a prolonged breeding season, and most kaka raise at least one brood. Breeding continues into autumn and winter, declining only in spring when the remaining viable seed germinates and is no longer available (Elliot et al. 1996; Wilson et al. 1998).

Beech flowering and seeding does not appear to influence the productivity of insectivorous birds, however, even though invertebrates increase in abundance when beech trees flower (links 4 and 5 in Fig. 16.1; Alley et al. 2001). Mohua (*Mohoua ochrocephala*), a small insectivorous passerine in southern beech forests (O'Donnell 1996a, b), attempt to nest at the same rate each year, irrespective of seedfall, unlike the kaka and kakariki. Most pairs monitored in the Eglinton Valley, Fiordland, South Island, raise two broods each year (Elliot 1996), although in other parts of New Zealand only one brood is raised.

Wodzicki (1950) and Riney et al. (1959) first hypothesized that exotic house mice (*Mus musculus*) and kiore (*Rattus exulans*) occasionally reach high numbers in response to unusually heavy beech seeding (link 6 in Fig. 16.1). During low beech-seedfall years, mouse population densities are also generally low. They increase slightly over summer to autumn/winter peaks (King 1982; Murphy 1992), and then usually decrease with cooler temperatures, food shortages, and cessation of breeding. In contrast, during high-seedfall years, mouse populations continue increasing through winter, leading to peak numbers (up to 50 mice ha^{-1}; Ruscoe et al. 2001) in late winter, or in spring or summer (Fitzgerald et al. 1996; Choquenot and Ruscoe 2000; Ruscoe et al. 2003).

The mechanism driving the response of mouse populations to beech seedfall is the sudden increase in food: flowers, seed and invertebrates (e.g., Fitzgerald et al. 1996; Alley et al. 2001; Ruscoe et al. 2005). Choquenot and Ruscoe (2000) used regression modeling to relate seedfall, mouse density, and rat density to instantaneous rates of change in mouse abundance over 25 years (1972–1996) in the Orongorongo Valley, southern North Island. They found that rates of change between autumn (the season in which most beech seed falls) and winter, and between winter and spring were positively related to seedfall (link 6 in Fig. 16.1) and negatively related to mouse density (link 7 in Fig. 16.1). Rates of population change between spring and summer, and between summer and autumn were related to mouse density alone. They concluded that a range of factors associated with seedfall produced a numerical response in mice that was modified by some unidentified density-dependent mechanism that constrained population growth. Plausible mechanisms include social factors, intraspecific competition for available food, predation (links 7 and 12 in Fig. 16.1), and/or the effects of disease or parasites.

Although ship rats (*Rattus rattus*) are found in beech forest, they are not as numerous (up to 4.6 rats ha^{-1}; Ruscoe, personal observations) as house mice, and do not respond to seedfall to the same extent. There is mounting evidence, however, that ship rat numbers may limit mouse abundance in beech forests where they are sympatric (Innes et al. 1995; Choquenot and Ruscoe 2000; Ruscoe et al. 2003; link 8 in Fig. 16.1). However, Innes et al. (1995) showed that when rats were removed, mouse numbers increased faster than would be possible by in situ breeding, suggesting there may be a behavioral as well as a numerical response by mice to the removal of rats.

Stoats in beech forest respond numerically to mouse population increases associated with high-seedfall years (King 1983; O'Donnell et al. 1996; link 9 in Fig. 16.1). Whereas mouse populations respond quickly to seed availability, the timing of stoat reproduction is set by changes in day length, regardless of food supply in autumn and winter (King 1990). Stoats can produce only one litter in spring (around October). However, an increase in food supply leads to a reduction in intra-uterine and nestling mortality, and hence a population increase (King 2002). In years when mice are abundant, juvenile stoats make up 80–90 % of the stoat population in January–February, whereas in low-mouse years, juveniles make up <10 % of the stoat population in summer (Murphy and Dowding 1994; Powell and King 1997; King 2002).

16.2.3 Top-Down Processes

Historically, New Zealand birds, reptiles and invertebrates preyed on seed, insectivorous birds and reptiles preyed on invertebrates, and large hawks and owls preyed on smaller birds, reptiles and invertebrates. Introductions of mustelids, cats and rodents added new predator interactions (and new trophic levels) to the system.

Native ground-dwelling and hole-nesting birds are particularly susceptible to predation by stoats and rats because they have evolved few predator-avoidance behaviors (McLennan et al. 1996; O'Donnell 1996a; link 10 in Fig. 16.1). Mohua were once present in most forest habitats over much of South Island and Stewart Island, but began to decline noticeably around the 1890s and are now present in only 25 % of their former range (see O'Donnell 1996a). The ranges of kakariki and kaka populations have also contracted but in both cases human-induced habitat modification and predation are thought responsible. Even in large, intact forests, populations of all three taxa have declined to such low levels that further local extinctions are possible. Mohua, kaka and kakariki nest in tree hollows up to 20 m from the ground. Both ship rats and stoats are able climbers, and can access a large proportion of nests. Because the nesting holes only have one entrance, incubating females cannot escape, and are taken along with the eggs or chicks (Dilks et al. 2003). During years of high mouse abundance, stoat numbers may increase fivefold, but each stoat still takes the same number of birds as it would in non-mouse (low-stoat) years: there is no prey-switching to mice when mice are plentiful, so native bird populations experience extraordinarily high mortality due to predation. In other words, predation rates on native birds increase because of the numerical response of stoats to mice, not because of the functional response of stoats to birds. This is in contrast to the Northern Hemisphere where good years for rodents often relieve the predation pressure of stoats on birds (King 1980).

Kaka and kakariki breed prolifically in the late summer and autumn in years when beeches flower and then seed, extending their summer breeding into autumn and winter (Fig. 16.2). By the time mouse populations increase during the winter following seedfall, and cause a stoat population increase the following summer, these parrots have finished their extended nesting. Usually, several years pass before the next large *Nothofagus* seed crop is produced, so kaka will not breed again in the following summer, alleviating the risk associated with nesting at a time when stoat numbers are highest. With little seed, kakariki go through a much-reduced "normal" summer breeding season when stoat numbers are high, and only one brood is reared, reducing (but not entirely) adult exposure to nest predation. In contrast, the insectivorous mohua breeds at about the same level every year. There is no increase in breeding in high-seedfall years to offset the high predation rates experienced with the subsequent increase in predator numbers. Additionally, although stoat numbers are low in beech forests in very low-seedfall years, their large home range size makes them a significant predation risk to mohua every year. The long-term survival of mohua populations may be at much higher risk than for some other extant native birds.

Predator impacts compound when there are two high *Nothofagus*-seedfall years in a row; this historically unusual event happened at some locations in 1999 and 2000. Not only were mohua exposed to additional predation pressure, but kaka and kakariki also were at risk of stoat predation in the second seeding year when they attempted to breed during the peak in stoat density. In

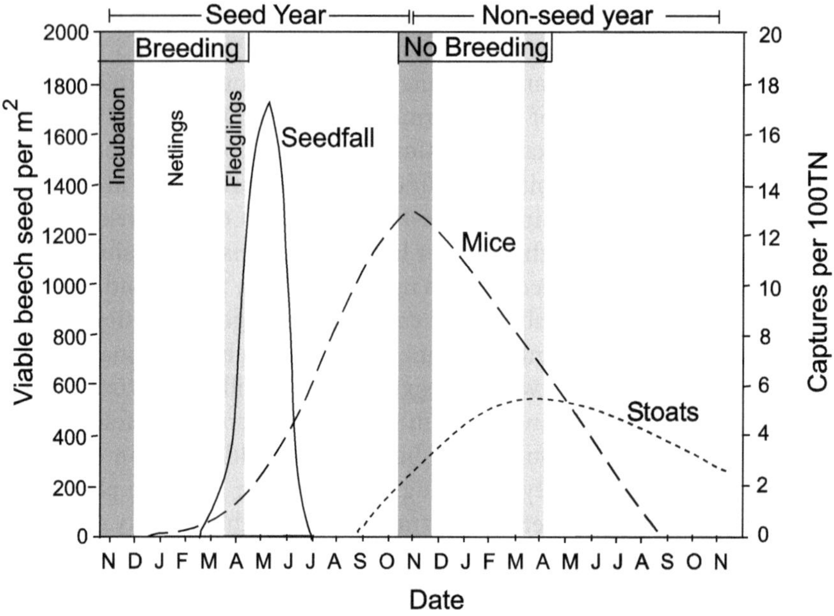

Fig. 16.2 A conceptual model illustrating the timing of kaka breeding, and the response of the mouse and stoat populations to beech seedfall. Reprinted from Wilson et al. (1998), with permission from Elsevier

addition, ship rat populations increased dramatically over these two successive seedfall years, and were responsible for approximately 60% of mohua nest predation events in the Eglinton Valley (Dilks et al. 2003). Under these circumstances, ship rats represent a predation threat equal to, if not greater than, that of the stoat (link 11 in Fig. 16.1). This is alarming, in view of a decadal-level increase in the frequency of moderate to high beech-seedfall years (Richardson et al. 2005).

Although stoats have a major and immediate impact on bird populations, they do not appear to regulate mouse population growth rate in beech forests (Ruscoe et al. 2003; link 12 in Fig. 16.1). This is not surprising, as stoats defend large home ranges (70–250 ha), which prevents the build-up of high stoat numbers relative to mouse numbers (Murphy and Dowding 1994). Also, in single high-seedfall years, by the time stoats are responding numerically to a high mouse population, mouse numbers have peaked and are declining in the summer following seedfall. The mice at this time are undergoing food shortage – therefore, the population will decline, regardless of predation by stoats (Blackwell et al. 2001; Ruscoe et al. 2003).

The extent to which beech masting is sufficient to satiate both native and exotic seed predators will determine whether mice pose a real competitive threat to birds and other native granivores. Recent research has shown that the introduced possum (*Trichosurus vulpecula*) also consumes large amounts of green seed on the tree (Sweetapple 2003). In prehuman New Zealand, there were only birds and invertebrates competing for beech seed, but today rodents and possums place additional predation pressure on seeds (link 13 in Fig. 16.1). This seed removal (Ruscoe et al. 2005), in addition to the effects of other exotic herbivores (deer, pigs, goats), will determine the long-term impacts of introduced herbivores on beech forest dynamics. In reducing plant biomass in forest understories (see Chap. 22), herbivores such as deer may also be exposing birds to higher predation risk through reductions in habitat complexity.

Mice are also known to eat weta (*Hemideina* sp.; Miller and Miller 1995; Fitzgerald et al. 1996). If they are significantly depleting weta populations, then they may pose a direct threat to these native invertebrates (link 14 in Fig. 16.1) as well as indirectly exerting competitive pressure on native insectivores. For example, the native owl (*Ninox novaeseelandiae*) is also known to consume weta (Haw et al. 2001).

16.2.4 Management in Beech Forest Ecosystems

The immediate major conservation threat in the beech forest system is the effect of predation on native biota, particularly forest birds. Although house mice could arguably be described as the keystone invader in this system, current conservation management in southern New Zealand involves intensified predator trapping in years following a high seedfall. For example, stoat trapping has occurred in the Eglinton Valley for the last 10 years. However, in this valley, ship rat numbers have increased in seedfall years (1999, 2000) to beyond levels attained prior to stoat trapping. Concern is growing that this is a result of stoat trapping relieving predation pressure on rats, and, as described above, with negative consequences for conservation. Whether this increase in rat numbers is due to the predator trapping or is the result of an increase in the frequency of beech masting events (Richardson et al. 2005) is unknown, but highlights the need to understand species interactions in a changing world. Although mice are undoubtedly the driver of predator population increases in beech forest, a lack of appropriate technology (e.g., there is public rejection of broad-scale baiting) prevents the large-scale control of rodents on the New Zealand mainland.

16.3 Trophic Interactions in Dryland Ecosystems

New Zealand's dryland or semi-arid ecosystems can be defined as grasslands and shrublands associated primarily with low- to mid-altitude areas (100–1,000 m a.s.l.) in the rain shadow east of the South Island's Southern

Alps. They experience relatively extreme seasonal and diurnal temperatures (daily averages 0–4 °C in the coldest month, and 15–17 °C in the warmest month), low and unpredictable rainfall (annual average 400–600 mm), and frequent soil moisture deficits in summer and autumn (Maunder 1965; Gibbs 1980). There are steep elevational gradients in climate.

16.3.1 A Sequence of Biological Invasions

The first introduced animals arrived ca. 800 years ago, following Polynesian settlement when kiore invaded the drylands (see Chap. 4), although kiore are no longer present in these systems. Burning by Maori removed the woody vegetation (e.g., *Podocarpus*, *Phyllocladus*, *Coprosma*) and converted the ecosystem to mostly grassland (e.g., *Chionochloa*, *Poa*, *Festuca*; McGlone 2001). Burning continued, and was combined with livestock grazing when farming by European people started about 150 years ago. These were later combined with applications of superphosphate fertilizer and oversowing of exotic pasture plants (e.g., *Trifolium*, *Dactylis*, *Lolium*). European rabbits (*Oryctolagus cuniculus*) colonized most of the drylands in about the mid-19th century. The attempted solution to the rabbit problem culminated in other biological invasions – the deliberate release of ferrets (*Mustela furo*), stoats and weasels (*M. nivalis vulgaris*), and the spread of cats (*Felis catus*), in the 1870s. Ferrets and cats now dominate the predator guild in dryland areas. A plethora of other pests – e.g., hedgehogs (*Erinaceus europaeus occidentalis*), possums, magpies (*Gymnorhina tibicen*) – and woody weeds (e.g., *Rosa*, *Cytisus*, *Ulex*) have also invaded. This complex of new and varied trophic and competitive interactions between exotic and indigenous biota has, in combination with habitat depletion, reduced the abundance of much of the indigenous biota in the exotic-dominated dryland ecosystems.

16.3.2 Bottom-Up Processes

Published relationships between climate and the dynamics of ungrazed dryland vegetation are limited to highly disturbed areas (link 1 in Fig. 16.3; Allen et al. 1995; Walker et al. 1999). Rainfall events, during periods of moisture deficit, have their greatest effect on yearly plant growth and species composition at shady sites and on deeper soils where soil moisture-holding capacity is relatively high. Vegetation is more vulnerable to moisture stress if it has been intensively grazed, resulting in greater year-to-year fluctuations compared with ungrazed vegetation. The effects of vegetation on pest herbivore populations (link 2 in Fig. 16.3) are less well understood. As in Australia (Myers et al. 1994), vegetation quality (e.g., nitrogen content) and the duration of the growing season are likely to be the main drivers of rabbit populations. Rabbits do

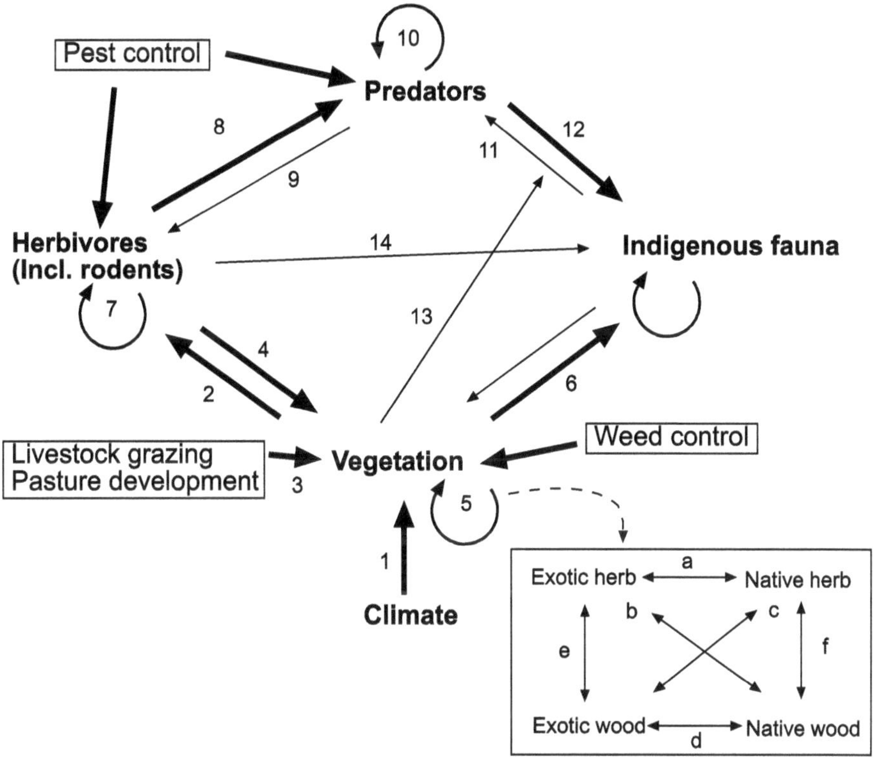

Fig. 16.3 Some of the key trophic interactions in dryland ecosystems. *Upward arrows* are "bottom-up" processes, and *downward arrows* are "top-down" processes. *Thick-line arrows* are what we believe are strong interactions, and *thin-line arrows* are weak interactions. The *loops* depict intraspecific competition and interspecific interactions. Interactions between exotic and native herbaceous and woody vegetation (link 5) are expanded in the *bottom-right box*. The *other boxes* are the human influences of weed and pest control (including rabbit hemorrhagic disease), and pastoral management (e.g., livestock grazing, burning, oversowing of exotic pasture species, fertilizer application). For explanation of *numbered links*, see text

not do well when vegetation is long and rank. This explains why pastoral development, and the subsequent alteration of tall, rank vegetation to low, highly nutritious plants (link 3 in Fig. 16.3), has facilitated rabbit invasions (see Zotov 1938).

There is only one detailed account of bottom-up relationships between vegetation and indigenous fauna (link 6 in Fig. 16.3). White (1991) reported major declines in abundance of native moth species (e.g., Crambidae and Noctuidae) over a 50- to 70-year period as the introduced grass *Agrostis capillaris* dominated the ground flora. In another study, common grassland skinks (*Oligosoma maccanni*, *O. nigriplantare polychroma*) were present above a vegetation threshold of 50 % ground cover (Norbury 2001a). However,

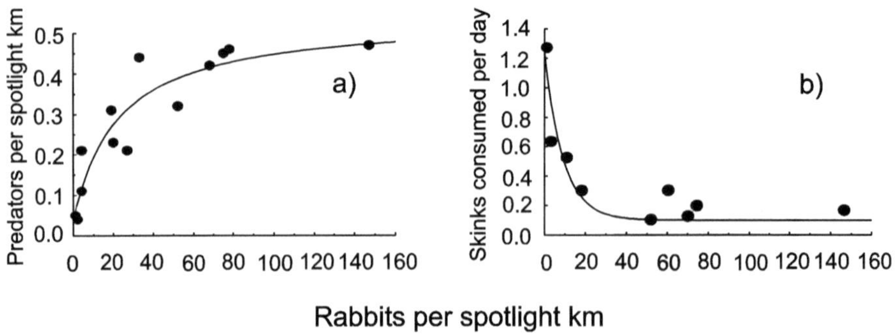

Fig. 16.4a, b Relationships between rabbit numbers in Central Otago and the Mackenzie basin and **a** ferret and cat numbers, and **b** the mean number of skinks consumed per day by ferrets (extracted from Norbury 2001b). Consumption of skinks by cats is similarly related to rabbit abundance, but is higher and more variable than for ferrets. When combined and expressed as a proportion of lizard density, these relationships predict an increasing rate of predation of skinks as skink density declines – this can lead to local extinction

we have only a limited ability to predict assemblages of indigenous fauna resulting from altered vegetation.

Ferrets and cats are generalist predators that depend on introduced mammals as their primary prey, and take native species as their secondary prey (e.g., Smith et al. 1995). Rabbits are the most important primary prey species in this system, and they drive predator populations, bottom-up, both numerically (Fig. 16.4a) and functionally (Fig. 16.4b; link 8 in Fig. 16.3; Norbury 2001a). Predator populations appear to fluctuate independently of indigenous prey, suggesting that link 11 (Fig. 16.3) is weak.

16.3.3 Top-Down Processes

Dryland ecosystem research has focused on the top-down effects of herbivores (rabbits, hares and livestock) on vegetation (links 3 and 4 in Fig. 16.3; e.g., Lord 1990; Allen et al. 1995; Rose et al. 1995; Norbury and Norbury 1996; Meurk et al. 2002). Grazing impacts at high herbivore density can alter the structure and species composition of vegetation from a dominance of taller, perennial species to low, annual and mat-forming browse-resistant plant communities. However, the rate of food consumption by herbivores as vegetation abundance changes (i.e., their functional response) is required for modeling ecosystem function in dryland systems. For degraded vegetation, grazing impacts are often minor, compared with climatic effects (Allen et al. 1995) and invasion of the flatweed *Hieracium pilosella* (links 5a and b in Fig. 16.3; Rose et al. 1995; Meurk et al. 2002).

Because of interactions between exotic and native plants (links 5a–f in Fig. 16.3), there can sometimes be advantages to a subset of native species from some form of grazing because it can suppress palatable exotic weeds (Bellingham and Coomes 2003) that can smother native species (Lord 1990; Walker 2000) and inhibit their associated fauna (White 1991). The problem is that grazing also suppresses recovery of palatable native herbaceous and woody species.

Rabbit hemorrhagic disease virus was illegally released in New Zealand in 1997 to control rabbit populations (see Norbury 2001b). Where the disease has been successful (see Parkes et al. 2002), there have been flow-on effects such as recovery of introduced pasture species, increases in other introduced herbivores (possums and hares; link 7 in Fig. 16.3), and fewer introduced predators (Norbury et al. 2002). The reasons possum and hare numbers have increased are not clear, but could be a function of less need for poisoning of rabbits (and therefore reduced non-target poisoning of possums and hares), less competition for food, less interference competition (e.g., aggressive displacement of possums and hares by rabbits), or release from predation (following the numerical decline in rabbit predators).

There is little evidence from predator-manipulation experiments in dryland ecosystems for top-down effects of predators on rabbit populations (link 9 in Fig. 16.3). Density-dependence has been demonstrated for juvenile ferret survival by Byrom (2002; link 10 in Fig. 16.3). During outbreaks of rabbit hemorrhagic disease, there were no differences in the rate of decline of rabbit populations between sites where predator numbers were reduced and where they were not (Reddiex and Norbury 2002). In theory, predators (particularly cats) in dryland systems can have strong top-down effects on indigenous fauna because predator abundance is driven mainly by rabbits. Predator numbers can therefore be maintained at levels not normally possible without rabbits (sometimes referred to as "hyperpredation"; Courchamp et al. 2000). Norbury (2001a) concluded that the best rabbit management strategy to protect native skinks from predators is to maintain rabbits at low, stable levels. The worst strategy for skinks is likely to be when rabbit numbers fluctuate widely, because this leads to pulses of predators switching to native species following sudden declines in rabbit abundance. There have been no experiments to quantify the top-down effects of predators on indigenous dryland fauna (link 12 in Fig. 16.3) but the indirect evidence (Middlemiss 1995; Norbury 2001a; Sanders and Maloney 2002) suggests they are reasonably strong.

The rabbit–predator interaction appears to flow bottom-up from rabbits to predators, then top-down from predators to indigenous prey. Interaction strengths in the opposite direction appear weak. If they were strong, we would expect some sort of cycling of predators and prey around some equilibrium (May 1972) – we do not see this. Rabbit numbers explode to levels limited mainly by food (not predators), and indigenous prey do the opposite – they continue to spiral to critically low numbers, and often extinction.

The New Zealand indigenous biota is often characterized by low rates of fecundity and growth (e.g., Cree 1994), which predisposes it to overharvest by introduced predators. For some prey populations, their slow rate of recruitment will never be enough to keep pace with predation, and extinction is inevitable without human intervention. For others, this might be true only at low densities – extinction is still one possible outcome, but coexistence of predators and prey is theoretically possible at higher prey densities. A mosaic of local extinction and coexistence is predicted for some common grassland skinks (Norbury 2001a), and is indeed what is observed today for these species.

There is theoretical evidence (e.g., Sih 1987) that increasing habitat complexity can provide refuge for prey from predators (link 13 in Fig. 16.3), and allow these to coexist. Refuge effects have been implicit in predator–prey theory for many years (e.g., Holling 1959) but their importance has not been recognized until recently (Hawkins et al. 1993; Sinclair et al. 1998; Arthur et al. 2003, 2004). Experiments are underway to test refuge effects in dryland systems. They could potentially shift the current foci of predator–prey ecology from the predators and their prey to the biophysical structures in which they interact.

We know very little about introduced rodents (*R. rattus*, *R. norvegicus*, *M. musculus*) in dryland ecosystems, although they do not appear to be abundant. If "superpredators" such as ferrets and cats are regulating rodent populations (through a strong link 9 in Fig. 16.3), then reducing ferret and cat populations may lead to irruptions of rodents. This could result in unpredictable impacts on native species (link 14 in Fig. 16.3). Some dryland systems are dominated by *Chionochloa* spp., and these grasses flower periodically. If rodents respond numerically to these events, then they could lead to higher rates of predation by rodents on native fauna, and these effects might flow into neighboring beech forests (Sect. 16.2).

16.3.4 Management in Dryland Ecosystems

Historically, removal of introduced mammalian predators has been the focus of fauna protection in New Zealand. If the qualitative interaction strengths in Fig. 16.3 are correct, then the removal of predators in dryland ecosystems would affect only one strong interaction – the top-down effects of predation on at-risk fauna (link 12 in Fig. 16.3). Removal of rabbits, on the other hand, potentially affects four strong interactions: the bottom-up effects on predator populations, and therefore the top-down effects these have on indigenous fauna; and the top-down effects of rabbit grazing on vegetation, and therefore the bottom-up effects of vegetation on indigenous fauna (links 4, 6, 8 and 12 in Fig. 16.3). Therefore, rabbit control is arguably a more productive pest control strategy because it restores vegetative habitat while simultaneously reducing predator populations. Rabbits are keystone invasive species in this ecosystem.

Hieracium spp. are highly invasive flatweeds that often out-compete and dominate the herbaceous ground flora (through link 5a in Fig. 16.3; Duncan et al. 1997). This dramatic simplification of habitat structure and composition will have direct negative impacts on indigenous fauna (through link 6 in Fig. 16.3), and indirect negative impacts caused by high predator numbers (sustained by rabbit populations that are favored by open swards of *Hieracium*; links 2 and 8 in Fig. 16.3) and enhanced predator-hunting efficiency (caused by the reduction in refuge for prey; link 13 in Fig. 16.3). As in the case of rabbits, we suggest *Hieracium* is a keystone invasive group in dryland systems. Broadly, succession of grasslands to more woody components is the reciprocal of *Hieracium* invasion, and is likely to significantly benefit indigenous biota overall.

16.4 Generalities and Conclusions

In the absence of exotic mammals, the beech forest system is strongly driven bottom-up. The sporadic heavy seeding of beech trees results in a cascade of population increases in the native fauna, without any known reciprocal effects. In dryland ecosystems, the nutrient pulses occur annually during spring flushes of herbaceous plants. In both the little modified beech forest and the highly human-impacted dryland ecosystems, mammalian introductions have been made at both the herbivore and predator trophic levels. These exotic additions have created strong top-down effects on indigenous fauna because predator abundance (stoat, ferret, and cat) is driven mainly by exotic prey species (mice and rabbits). Predator numbers can reach levels not normally possible without the introduced prey, and this can potentially lead to extinction of the native fauna. The worst scenario for native prey occurs when mice and rabbit numbers fluctuate widely. This leads to acute bouts of predation caused by the increases in predator numbers (in the case of stoats), or as ferrets and cats switch to native species following sudden declines in rabbit abundance. We now know enough about some processes in beech forest and dryland ecosystems to build prototype models that will help to predict the wider effects of controlling introduced species, identify critical knowledge gaps, and ultimately guide management decisions to achieve desired biodiversity outcomes.

Acknowledgements We thank Kim King, John Parkes, John McLennan, Bill Lee and especially Rob Allen for helpful comments on earlier drafts of the manuscript. We also thank Melissa Brignall-Theyer for help with graphics.

References

Allen RB, Platt KH (1990) Annual seedfall variation in *Nothofagus solandri* (Fagaceae), Canterbury, New Zealand. Oikos 57:199–206

Allen RB, Wilson JB, Mason CR (1995) Vegetation change following exclusion of grazing animals in depleted grassland, Central Otago, New Zealand. J Veg Sci 6:615–626

Alley JC, Berben PH, Dugdale JS, Fitzgerald BM, Knightbridge PI, Meads MJ, Webster RA (2001) Responses of litter-dwelling arthropods and house mice to beech seeding in the Orongorongo Valley, New Zealand. J R Soc NZ 31:425–452

Arthur AD, Pech RP, Drew A, Gifford E, Henry S, McKeown A (2003) The effect of increased ground-level habitat complexity on mouse population dynamics. Wildl Res 30:565–572

Arthur AD, Pech RP, Dickman CR (2004) Habitat structure mediates the non-lethal effects of predation on enclosed populations of house mice. J Anim Ecol 73:867–877

Bellingham PJ, Coomes DA (2003) Grazing and community structure as determinants of invasion success by Scotch broom in a New Zealand montane shrubland. Divers Distrib 9:19–28

Blackwell GL, Potter MA, Minot EO (2001) Rodent and predator population dynamics in an eruptive system. Ecol Model 142:227–245

Byrom AE (2002) Dispersal and survival of juvenile ferrets, *Mustela furo*, in New Zealand. J Appl Ecol 39:67–78

Choquenot D, Ruscoe WA (2000) Mouse population eruptions in New Zealand forests: the role of population density and seedfall. J Anim Ecol 69:1058–1070

Courchamp F, Langlais M, Sugihara G (2000) Rabbits killing birds: modelling the hyperpredation process. J Anim Ecol 69:154–164

Cree A (1994) Low annual reproductive output in female reptiles from New Zealand. NZ J Zool 21:351–372

Dilks PJ, Willams M, Pryde M, Fraser I (2003) Large scale stoat control to protect mohua (*Mohoua ochrocephala*) and kaka (*Nestor meridionalis*) in the Eglinton Valley, Fiordland, New Zealand. NZ J Ecol 27:1–9

Duncan RP, Colhoun KM, Foran BD (1997) The distribution and abundance of *Hieracium* species (hawkweeds) in the dry grasslands of Canterbury and Otago. NZ J Ecol 21:51–62

Elliot GP (1996) Productivity and mortality of mohua (*Mohoua ochrocephala*). NZ J Zool 23:229–237

Elliot GP, Dilks PJ, O'Donnell CFJ (1996) The ecology of yellow-crowned parakeets (*Cyanoramphus auriceps*) in *Nothofagus* forest in Fiordland, New Zealand. NZ J Zool 23:249–265

Fitzgerald BM, Daniel MJ, Fitzgerald AE, Karl BJ, Meads MJ, Notman PR (1996) Factors affecting the numbers of house mice (*Mus musculus*) in hard beech (*Nothofagus truncata*) forest. J R Soc NZ 26:237–249

Fretwell SD (1987) Food chain dynamics: the central theory of ecology? Oikos 50:291–301

Gibbs HS (1980) New Zealand soils: an introduction. Oxford Univ Press, Wellington

Haw JM, Clout MN, Powlesland RG (2001) Diet of moreporks (*Ninox novaeseelandiae*) in Pureora Forest determined from prey remains in regurgitated pellets. NZ J Ecol 25:61–67

Hawkins BA, Thomas MB, Hochberg ME (1993) Refuge theory and biological control. Science 262:1429–1432

Holling CS (1959) The components of predation as revealed by a study of small-mammal predation of the European pine sawfly. Cana Entomol 91:293–320

Innes J, Warburton B, Williams D, Speed H, Bradfield P (1995) Large-scale poisoning of ship rats (*Rattus rattus*) in indigenous forests of the North Island, New Zealand. NZ J Ecol 19:5–17

Kelly D, Sork VL (2002) Mast seeding in perennial plants: why, how, where? Annu Rev Ecol Syst 33:427-447

King CM (1980) The weasel *Mustela nivalis* and its prey in an English woodland. J Anim Ecol 49:127–159

King CM (1982) Age structure and reproduction in feral New Zealand populations of the house mouse (*Mus musculus*), in relation to seedfall of southern beech. NZ J Zool 9:467–480

King CM (1983) The relationships between beech (*Nothofagus* sp.) seedfall and populations of mice (*Mus musculus*), and the demographic and dietary responses of stoats (*Mustela erminea*), in three New Zealand forests. J Anim Ecol 52:141–166

King CM (ed) (1990) The handbook of New Zealand mammals. Oxford Univ Press, Auckland

King CM (2002) Cohort variation in the life-history parameters of stoats *Mustela erminea* in relation to fluctuating food resources: a challenge to boreal ecologists. Acta Theriol 47:225–244

Krebs CJ, Boutin S, Boonstra R, Sinclair ARE, Smith JNM, Dale MRT, Martin K, Turkington R (1995) Impact of food and predation on the snowshoe hare cycle. Science 269:1112–1115

Krebs CJ, Boutin S, Boonstra R (eds) (2001) Ecosystem dynamics of the boreal forest: the Kluane project. Oxford Univ Press, New York

Lord JM (1990) The maintenance of *Poa cita* grassland by grazing. NZ J Ecol 13:43–49

Maunder WJ (1965) Climatic character. In: Lister RG, Hargreaves RP (eds) Central Otago. New Zealand Geographical Society, Dunedin, pp 46–48

May RM (1972) Limit cycles in predator-prey communities. Science 177:900–902

McGlone MS (2001) The origin of the indigenous grasslands of southeastern South Island in relation to pre-human woody ecosystems. NZ J Ecol 25(1):1–15

McLennan JA, Potter MA, Robertson HA, Wake GC, Colbourne R, Dew L, Joyce L, McCann AJ, Miles J, Miller PJ, Reid J (1996) Role of predation in the decline of kiwi, *Apteryx* spp., in New Zealand. NZ J Ecol 20:27–35

Meurk CD, Walker S, Gibson RS, Espie P (2002) Changes in vegetation states in grazed and ungrazed Mackenzie Basin grasslands, New Zealand, 1990–2000. NZ J Ecol 26:95–106

Middlemiss A (1995) Predation of lizards by feral house cats (*Felis catus*) and ferrets (*Mustela furo*) in the tussock grasslands of Otago. MSc Thesis. Department of Zoology, University of Otago, Dunedin

Miller CJ, Miller TK (1995) Population dynamics and diet of rodents on Rangitoto Island, New Zealand, including the effect of a 1080 poison operation. NZ J Ecol 19:19–27

Murphy EC (1992) The effects of a natural increase in food supply on a wild population of house mice. NZ J Ecol 16:33–40

Murphy EC, Dowding JE (1994) Range and diet of stoats (*Mustela erminea*) in a New Zealand beech forest. NZ J Ecol 18:11–18

Myers K, Parer I, Wood D, Cooke BD (1994) The rabbit in Australia. In: Thompson HV, King CM (eds) The European rabbit: the history and biology of a successful colonizer. Oxford Univ Press, Oxford, pp 108–157

Norbury G (2001a) Conserving dryland lizards by reducing predator-mediated apparent competition and direct competition with introduced rabbits. J Appl Ecol 38:1350–1361

Norbury G (2001b) Advances in New Zealand mammalogy 1990–2000: lagomorphs. J R Soc NZ 31:83–97

Norbury DC, Norbury GL (1996) Short-term effects of rabbit grazing on a degraded short-tussock grassland in Central Otago. NZ J Ecol 20:285–288

Norbury G, Heyward R, Parkes J (2002) Short-term ecological effects of rabbit haemorrhagic disease in the short-tussock grasslands of the South Island, New Zealand. Wildl Res 29:599–604

O'Donnell CFJ (1996a) Predators and the decline of New Zealand forest birds: an introduction to the hole-nesting bird and predator programme. NZ J Zool 23:213–219

O'Donnell CFJ (1996b) Monitoring mohua (yellowhead) populations in the South Island, New Zealand, 1983–1993. NZ J Zool 23:221–228

O'Donnell CFJ, Dilks PJ, Elliot GP (1996) Control of a stoat (*Mustela* erminea) population irruption to enhance mohua (yellowhead) (*Mohoua ochrocephala*) breeding success in New Zealand. NZ J Zool 23:279–286

Oksanen L (1990) Predation, herbivory, and plant strategies along gradients of primary productivity. In: Grace JB, Tilman D (eds) Perspectives on plant competition. Academic Press, New York, pp 445–474

Oksanen L, Fretwell SD, Arruda J, Niemela P (1981) Exploitation ecosystems in gradients of primary productivity. Am Nat 118:240–261

Ostfeld RS, Keesing F (2000) Pulsed resources and community dynamics of consumers in terrestrial ecosystems. Trends Ecol Evol 15:232–237

Parkes JP, Norbury GL, Heyward RP, Sullivan G (2002) Epidemiology of rabbit haemorrhagic disease (RHD) in the South Island, New Zealand, 1997–2001. Wildl Res 29:543–555

Powell RA, King CM (1997) Variation in body size, sexual dimorphism and age-specific survival in stoats, *Mustela erminea* (Mammalia: Carnivora), with fluctuating food supplies. Biol J Linn Soc 62:165–194

Reddiex B, Norbury G (2002) Interactions of primary and secondary prey in New Zealand dryland ecosystems: evidence for top-down and bottom-up processes. In: Abstr Vol 15th Sci Meet Australasian Wildlife Management Society, 9–11 December 2002, Camden, NSW, p 53

Richardson SJ, Allen RB, Whitehead D, Carswell FE, Ruscoe WA, Platt KH (2005) Temperature and net carbon balance both determine seed production in a temperate tree species. Ecology 86(4):972–981

Riney T, Watson JS, Bassett C, Turbott EG, Howard WE (1959) Lake Monk expedition, an ecological study in Southern Fiordland. DSIR Bull 135:1–75

Rose AB, Platt KH, Frampton CM (1995) Vegetation change over 25 years in a New Zealand short-tussock grassland: effects of sheep grazing and exotic invasions. NZ J Ecol 19:163–174

Ruscoe WA, Goldsmith R, Choquenot D (2001) A comparison of population estimates and abundance indices for house mice inhabiting beech forests in New Zealand. Wildl Res 28:173–178

Ruscoe WA, Choquenot D, Heyward R, Yockney I, Young N, Drew K (2003) Seed production, predators and house mouse population eruptions in New Zealand beech forests. In: Singleton GR, Hinds LA, Krebs CJ, Spratt DM (eds) Rats, mice and people: rodent biology and management. ACIAR, Canberra, Australia, pp 334–337

Ruscoe WA, Elkinton JS, Choquenot D, Allen RB (2005) Predation of beech seed by mice: effects of numerical and functional responses. J Anim Ecol 74(6):1005–1019

Sanders MD, Maloney RF (2002) Causes of mortality at nests of ground-nesting birds in the upper Waitaki Basin, South Island, New Zealand: a 5-year study. Biol Conserv 106:225–236

Sih A (1987) Prey refuges and predator-prey stability. Theor Popul Biol 31:1–12

Sinclair ARE, Krebs CJ (2001) Trophic interactions, community organization, and the Kluane ecosystem. In: Krebs CJ, Boutin S, Boonstra R (eds) Ecosystem dynamics of the Boreal Forest: the Kluane project. Oxford Univ Press, New York, pp 25–48

Sinclair ARE, Pech RP, Dickman CR, Hik D, Mahon P, Newsome AE (1998) Predicting effects of predation on conservation of endangered prey. Conserv Biol 12:564–575

Smith GP, Ragg JR, Moller H, Waldrup KA (1995) Diet of feral ferrets (*Mustela furo*) from pastoral habitats in Otago and Southland, New Zealand. NZ J Zool 22:363–369

Sweetapple PJ (2003) Possum (*Trichosurus vulpecula*) diet in a mast and non-mast seed year in a New Zealand *Nothofagus* forest. NZ J Ecol 27:157–167

Walker S (2000) Post-pastoral changes in composition and guilds in a semi-arid conservation area, Central Otago, New Zealand. NZ J Ecol 24:123–137

Walker S, Wilson JB, Mark AF (1999) Seasonal changes in semi-arid vegetation of Central Otago, New Zealand. NZ J Bot 37:479–493

Wardle JA (1984) The New Zealand beeches: Ecology, utilisation and management. New Zealand Forest Service, Wellington, New Zealand

White TCR (1984) The abundance of invertebrate herbivores in relation to the availability of nitrogen in stressed food plants. Oecologia 63:90–105

White EG (1991) The changing abundance of moths in a tussock grassland, 1962–1989, and 50- to 70-year trends. NZ J Ecol 15:5–22

Wilson PR, Karl BJ, Toft RJ, Beggs JR, Taylor RH (1998) The role of introduced predators and competitors in the decline of kaka (*Nestor meridionalis*) populations in New Zealand. Biol Conserv 83:175–185

Wodzicki KA (1950) Introduced mammals of New Zealand: an ecological and economic survey. DSIR Bull 98

Zotov VD (1938) Survey of the tussock-grasslands of the South Island, New Zealand. NZ J Sci Technol 20A:212–244

17 Keystone Aliens? The Multiple Impacts of Brushtail Possums

M.N. Clout

17.1 Introduction

The Australian brushtail possum (*Trichosurus vulpecula*) was introduced to New Zealand from eastern Australia and Tasmania in the mid-19th century as a furbearer (Clout and Ericksen 2000). It is a solitary, nocturnal, arboreal phalangerid marsupial weighing 2–4 kg. In Australia, these animals have a wide distribution, occupy a range of forested or wooded habitats, and are one of a suite of arboreal marsupials. In New Zealand, they are the sole animal of this type, and are one of the most ecologically disruptive and economically damaging alien mammals.

From the 1860s onward (peaking in the 1890s), there were many importations of brushtail possums to New Zealand to establish a fur industry. These introductions, and subsequent liberations, were carried out by private individuals, acclimatization societies and government agencies. Multiple liberations of New Zealand-bred possums, which continued illegally up to the 1970s, helped to establish the species and accelerated its spread. Possums are now present throughout the main islands of New Zealand, from coast to treeline (Clout and Ericksen 2000). In some lowland forests, they achieve densities of more than 10 ha^{-1}, an order of magnitude greater than typical densities of <1 ha^{-1} in their native Australia. The success of the brushtail possum is attributable to its generalist habits, the lack of competitors, parasites or predators, and the relative abundance of palatable, nutritious vegetation.

Since 1947, brushtail possums have been officially recognized as a pest in New Zealand. It is now widely acknowledged that they not only damage crops and native forests, but also carry bovine tuberculosis and prey on nests of native birds. Considerable and growing resources have been directed toward possum control over the past few decades, but the species remains a serious threat to the livestock industry and to native biodiversity in many parts of the country. This chapter reviews the multiple impacts of brushtail possums in

Ecological Studies, Vol. 186
R.B. Allen and W.G. Lee (Eds.)
Biological Invasions in New Zealand

New Zealand ecosystems, concentrating on their impacts on native biodiversity and concluding with a consideration of their role as "keystone aliens".

17.2 Possums as Herbivores

Brushtail possums were first recognized as pests in New Zealand because of their impacts as terrestrial and arboreal herbivores. In 1919, following reports of browsing damage to crops and forests, the government commissioned biologist Harold Kirk to advise on the impacts of possums on forests and orchards. Kirk (1920) concluded that, whilst damage to orchards and gardens was indisputable, "the damage to New Zealand forests is negligible". He recommended that possums be released in all forest districts away from orchards and gardens. Throughout the 1920s and 1930s, there was continuing controversy over the effects of possums on native forests, but it was not until the 1940s that the first scientific evidence of their impacts was collected (Zotov 1947; Pracy and Kean 1949). This prompted a radical change in official policy. In 1947, the government cancelled all restrictions on the killing of possums, instituted penalties for harboring or liberating them, and legalized the use of poisons for possum control.

Early concerns about possum damage to native forests focused especially on large-scale canopy mortality in southern rata (*Metrosideros umbellata*)/kamahi (*Weinnmania racemosa*) forests. The view that possums, through their selective browsing, were primarily responsible for canopy dieback in these and other forests has since been challenged (Veblen and Stewart 1982; Jane and Green 1983). There is evidence that possum browsing has indeed caused increased levels of canopy dieback, in some rata/kamahi forests (Rose et al. 1992; Payton 2000; but see Chap. 21). However, catastrophic possum-induced canopy dieback does not occur in the majority of forests, most of which are dominated by species that are not particularly palatable to possums. More typically there is a progressive depletion of minor species such as fuchsia (*Fuchsia excorticata*), loranthaceous mistletoes and milktree (*Paratrophis* spp.) that are selectively browsed and killed by possums (Nugent et al. 2001).

Brushtail possums are generalist and opportunistic arboreal folivores, feeding on a wide range of leaves, flowers, fruits and other foods. They have clear preferences for soft-leaved seral species (e.g., fuchsia, *Aristotelia serrata*, *Muehlenbeckia australis*), although the bulk of their diet typically comprises the leaves of abundant canopy trees, such as kamahi, southern rata, pohutukawa (*Metrosideros excelsa*), Hall's totara (*Podocarpus hallii*), kohekohe (*Dysoxylum spectabile*) and tawa (*Beilschmiedia tawa*). Beech (*Nothofagus*) and most podocarps are not favored, nor are many ferns or subcanopy trees such as *Carpodetus serrata* and *Hedycarya arborea* (Nugent et al. 2000). Possum population densities are consequently much lower (ca. 1 ha^{-1}) in beech forests (Clout and Gaze 1984) than in mixed lowland forests (5–10 ha^{-1}), reflecting the variable abundance of preferred food species.

Possums have been described as "reluctant folivores" (Nugent et al. 2000). There is increasing evidence that they achieve highest densities only in habitats where high-energy and nutrient-rich non-foliar foods such as fleshy fruits are seasonally abundant. In such habitats, their impacts on the vegetation are greater because their diet is still mainly foliage, resulting in the defoliation and decline of the most preferred species.

Changes in forest composition and structure due to possums are often confounded with those of other agents, such as introduced deer and natural episodic events, e.g., storm damage and the decline of even-aged stands. Nevertheless, there are several examples of measured changes in forests that have been directly attributed to possums, especially in the wake of population irruptions following colonization of new areas.

Forests in geologically unstable areas are prone to episodic disturbance and tend to contain high proportions of seral shrubs and small trees. Such communities may change relatively rapidly following possum invasion. For example, canopy collapse of kamahi forests in the southern Ruahine Ranges occurred within 15–20 years of possum colonization, including a 68 % reduction in tall forest between 1946 and 1978 (Batcheler 1983). More recently, Rose et al. (1992) reported that an irrupting possum population in the Adams River valley (Westland) in the late 1980s defoliated or killed 50 % of southern rata trees over a period of 2 years.

In forests that have been occupied by possums for longer periods, change is slower, but continues for many decades. Long-term studies in mixed forest of the Orongorongo Valley, Wellington (Campbell 1990; Brockie 1992), have confirmed that selective browsing by possums causes the gradual disappearance of their most preferred food species. For example, fuchsia, a highly preferred possum food, was detected in 32 % of possum stomachs in the Orongorongo Valley in 1946 but was depleted to the extent that it was no longer detectable in the local possum diet 25 years later (Fitzgerald 1976). Between 1969 and 1985, over 70 % of marked kamahi and tawa trees (both of them favored possum foods) died in the Orongorongo Valley, reducing their abundance, whereas less preferred species such as *Hedycarya arborea* and *Alsophila* tree ferns became relatively more abundant (Campbell 1990; Brockie 1992).

In the Orongorongo Valley, possums not only defoliate preferred species, but also suppress the fruiting of nikau palms (*Rhopalostylis sapida*; Cowan 1991) and hinau (*Elaeocarpus dentatus*; Cowan and Waddington 1990), through their destruction of flower-bearing structures, flowers and developing fruits. A graphic illustration of the impacts of possums as herbivores is the rapid change in the health of preferred food species after possums have been reduced by control or eradicated. This has recently been demonstrated for mistletoes (Sessions et al. 2001; Sweetapple et al. 2002), and was evident on Kapiti Island after possum removal (Atkinson 1992; Norton 2000).

17.3 Possums as Predators

There is steadily accumulating evidence that possums can have serious impacts as predators of native animals in New Zealand (Sadleir 2000). They are opportunistic feeders, and readily supplement their primarily herbivorous diet with invertebrates and with birds' eggs and nestlings when available. Insects (including fly larvae, stick insects, cicadas, weta and beetles) have been recorded as regular (albeit usually small) components of possum diet in a wide range of habitats (Clout 1977; Cowan and Moeed 1987; Owen and Norton 1995). Cowan and Moeed (1987) suggest that large nocturnal insects such as weta, stag beetles and large weevils may become depleted where possums are abundant. Dijkgraaf (2002) recorded consistently higher indices of weta abundance where possums had been controlled (see also Veltman 2000), but this effect was confounded by a parallel reduction in rat abundance.

The impacts of possum predation on native land snails are more readily assessed. Meads et al. (1984) recorded high levels of predation on *Powelliphanta* snails in northwest Nelson, which was later attributed to possums. Sadleir (2000) noted high levels of possum damage to *Wainuia urnula* snails, and K. Walker and G. Elliott (New Zealand Department of Conservation) considered that possum predation is having major detrimental effects on populations of at least six species of *Powelliphanta* (see also Veltman 2000).

Possums are increasingly recognized as significant nest predators of a range of native birds, notoriously including the eggs and nestlings of the endangered kokako (*Callaeas cinerea*; Brown et al. 1993). Brown et al. (1993) also noted evidence of possum predation on a range of other birds, including eggs of fantail (*Rhipidura fuliginosa*), chicks of Westland black petrel (*Procellaria westlandica*), and a fledgling harrier (*Circus approximans*). Possums are significant predators of the eggs of kereru (*Hemiphaga novaeseelandiae*; Clout et al. 1995; James and Clout 1996). They have also been identified as predators of eggs of brown kiwi (*Apteryx australis*; McLennan et al. 1996; McLennan 1997), and of kaka (*Nestor meridionalis*) nestlings and an attendant female (Moorhouse et al. 2003). Brushtail possums in their native Australia are significant nest predators of hollow-nesting glossy back-cockatoos (*Calyptorhynchus lathami*; Garnett et al. 1999). In New Zealand, it is highly likely that nest predation by possums has been a significant factor in the decline of birds such as kokako (Innes et al. 1999), and remains a limiting factor for kereru (Clout et al. 1995) and other birds.

Separating the effects of possums on native fauna from those of rats or other introduced mammals is difficult, especially since possum and rat numbers are often reduced by the same management actions. In addition to specific evidence of predation, there are overall negative correlations of possum abundance and abundance of species such as kaka (Veltman 2000). As possums have invaded and become abundant in regions such as South Westland and Northland, vulnerable native fauna such as kaka and kokako have declined. The precise mechanisms involved may not be known (possibly a combination of predation and competition), but the overall relationship is clear.

17.4 Possums as Competitors

Several studies have highlighted the significant dietary overlap between possums and native birds such as kokako (Leathwick et al. 1983; Fitzgerald 1984; Powlesland 1987; Cowan 1990a). Sadleir (2000) considered that these data provided no credible evidence for competition per se, pointing out that it is difficult to separate competitive effects from the more direct and severe effect of predation by possums. Nevertheless, frugivorous birds such as kokako (J. Innes, personal communication) and kereru (Clout et al. 1995; Powlesland et al. 2003) breed more frequently and successfully when large crops of fruit are available, and may not breed at all when fruit supplies are poor. It is also known that possums consume flowers, unripe fruit and ripe fruit, and can severely reduce the fruit crops of nikau (Cowan 1991), hinau (Cowan and Waddington 1990), tawa and other large-fruited species (Dijkgraaf 2002). It is therefore reasonable to conclude that possums may reduce the fruit resources available to frugivorous birds, and possibly affect their breeding success through competition for this key food source.

Effects of possums on nectar resources (through browsing of flowers and reduced vigor of flowering plants) are difficult to demonstrate, but it is certain that *Peraxilla* mistletoes (Wilson 1984), kohekohe and rewarewa (*Knightia excelsa*; Atkinson 1992; Dijkgraaf 2002) all flower more profusely when possum numbers are reduced.

Another potential impact of possums on some native birds is competition for tree hollows or burrows. Possums require these for daytime dens, and some larger native birds (e.g., kaka, kiwi species) use similar-sized hollows or burrows for nesting, and (in the case of kiwi) also for daytime shelter. Individual possums typically use 5–10 dens within their home range, swapping them regularly (Cowan 1989; Day et al. 2000). Use by a possum of the nest site of a bird during incubation or chick rearing is likely to be fatal for that nest, especially since predation of eggs or nestlings may occur as well. Instances of displacement of kiwi from burrows by possums have been recorded (Potter 1989), including one case in which a possum evidently killed an adult kiwi in a fight over burrow occupancy (McLennan 1997).

17.5 Possums as Seed Dispersers

Possums consume a wide range of fleshy fruits (Cowan 1990b), suggesting they may be important seed dispersers of native species (Dungan et al. 2002). Many of the seeds in fruits eaten by possums, especially larger, thin-coated ones such as tawa, are probably destroyed (personal observations), but viable seeds of some species (especially those with woody seeds) can pass intact through the digestive tract. Dungan et al. (2002) claimed that in many areas possums may be the only potential dispersal vector for large-seeded native species. This was disputed by Williams (2003), who argued that a suite of birds (both native and introduced) can also disperse seeds of many fruits exceeding 6 mm in diameter. The significance of possums as dispersers of native species remains uncertain, but they *may* be important to some species with large-seeded fruits (Dungan and Norton 2003; Williams 2003). The negative effects of possums on fruiting of native species such as nikau (Cowan 1991) and hinau (Cowan and Waddington 1990), and in destroying the thin-coated seeds of other species (e.g., tawa) are likely to outweigh any potential benefits from seed dispersal.

Williams et al. (2000) concluded that possums are potentially important seed dispersers of a range of alien species, including some invasive plants such as *Passiflora mollissima* and *Leycesteria formosa*. The relative importance of possums and other species (including introduced rodents) as dispersers of alien plants has yet to be determined.

17.6 Possums as Disease Vectors

Possums are known to carry a range of diseases and endoparasites that can potentially infect humans and other animals (Cowan et al. 2000). Levels of infection of possum populations with these diseases and parasites vary regionally. Bovine tuberculosis (Tb) is fatal to possums, and is prevalent at levels of 1–10 % in possum populations in 28 discrete areas covering ca. 25 % of New Zealand (Coleman and Caley 2000; Coleman and Livingstone 2000). Tb infection of livestock (especially cattle and farmed deer) occurs by environmental contamination and direct contact with dead or dying possums. Infection of wild mammals such as pigs and predators (especially ferrets, *Mustela putorius*) can also occur through their scavenging of possum carcasses. The presence of bovine Tb in possums and other wild mammals is the main obstacle to eradication of this economically important disease from New Zealand (Cowan 2001).

Leptospirosis infection is widespread in possums in central and southern parts of the North Island (affecting up to 80 % of adults), but is apparently absent from the South Island (Cowan et al. 2000). *Cryptosporidium parvum* infection also occurs in wild possums, which are a potential source of human infection. *Giardia intestinalis* infection is widespread among possums, but it is unclear if this poses an infection risk to humans or other animals.

17.7 Risks Posed by Possum Control

Possum control is conducted throughout New Zealand by a variety of methods, including shooting, kill-traps, cyanide paste and various other toxins (especially 1080 (sodium monofluoroacetate) and anticoagulants) that are placed in bait stations or distributed from aircraft. Spurr (2000) notes that most possum control operations involve 1080 poisoning, three-quarters by ground application and one-quarter by aerial distribution. However, most of the possum control conducted by the Animal Health Board is now done by ground-based methods, including increased use of traps and encapsulated cyanide baits (P.E. Cowan, personal communication 2003). The use of anticoagulant baits in bait stations for possum control remains a widespread practice by regional councils and private landowners. Most of these possum control methods pose some threats to other wildlife, including native species.

Most introduced mammals are at some risk from possum control operations (Spurr 2000). Small and medium-sized alien mammals (especially rats) may be caught in leghold traps, and a range of non-target pests (ranging in size from mice to deer) are known to have died from 1080 poison applied for possum control. Red deer (*Cervus elaphus*) populations have been reduced by up to 90 % (Fraser et al. 1995) and ship rat (*Rattus rattus*) populations by 87–100 % (Innes et al. 1995) in aerial 1080 operations. Secondary poisoning of stoats (*Mustela erminea*; Murphy et al. 1999) with 1080, and of feral cats (*Felis catus*) with brodifacoum anticoagulant (Gillies and Pierce 1999) has also been recorded after possum control operations, probably caused by these predators consuming poisoned rodents.

The other animals most at risk from possum control are ground-feeding birds that may become caught in leghold traps or take toxic baits (Spurr 2000), and avian predators that may be secondarily poisoned by feeding on other animals that have consumed toxic baits. Spurr (2000) notes that the most commonly recorded birds caught in leghold traps set for possums are introduced blackbirds (*Turdus merula*) and song thrush (*T. philomelos*), but native birds such as the kiwi (*Apteryx* spp.), weka (*Gallirallus australis*), Australasian harrier and morepork (*Ninox novaeseelandiae*) are also caught. Ground-feeding kiwi and weka are particularly at risk from cyanide paste baits. The impacts of trap- and cyanide-caused deaths on populations of rare species such as kiwi and weka have not been quantified (Spurr 1991), but are of sufficient concern for the New Zealand Department of Conservation to stipulate that traps should be set above ground in the presence of such species.

In all, 19 native and 13 introduced bird species have been found dead after aerial 1080 poisoning operations, mostly after the use of unscreened carrot baits containing small "chaff" fragments (Spurr 2000). Recent aerial operations have shown few, if any, deaths of native birds. For example, all 17 radio-tagged kaka and all 20 radio-tagged kereru survived an aerial 1080 operation against possums in Whirinaki Forest Park in May 2000 (Powlesland et al. 2003). However, after an airdrop of carrot baits at Pureora Forest in 1996, 55 % of 22 color-banded robins (*Petroica australis*) were unrecorded (Powlesland et al. 1999); after a similar operation in 1997, 79 % of 14 tomtits (*P. macrocephala*) disappeared (Powlesland et al. 2000). One year after the 1996 airdrop, robin abundance at the site was higher than it had been before the operation, possibly reflecting the longer-term benefits of pest mammal control.

Native invertebrates are known to feed on possum baits containing 1080 (Sherley et al. 1999) and anticoagulants such as brodifacoum (P. Craddock, personal communication). There is no strong evidence of any direct impacts on invertebrate abundance, but there may be some secondary poisoning risk to ground-feeding insectivorous birds and short-tailed bats (*Mystacina tuberculata*) taking invertebrates that have fed on baits containing 1080 poison (Lloyd and McQueen 2000). The risk of secondary poisoning of insectivorous birds (Booth et al. 2001) has resulted in the cessation of using baits containing brodifacoum at intensive conservation sites (A. Saunders, personal communication).

17.8 Economic Impacts of Possums

The main economic impacts of possums arise from their role as the primary wild vectors of bovine tuberculosis (Tb). Despite recent major increases in possum control expenditure, average levels of Tb incidence in New Zealand cattle and farmed deer remain high, relative to most of the country's trading partners (Coleman and Livingstone 2000). This poses a potential threat to New Zealand's meat and dairy exports, because of the risk of non-tariff barriers being imposed by importing countries. Such trade barriers could potentially cost the country up to NZ$ 1.3 billion per annum in lost earnings (Cowan 2001).

Estimates of the total cost of actual (rather than potential) possum damage are from $ 40–100 million per annum (Cowan 2001). This includes production losses to agriculture (through crop damage and livestock disease) and to forestry (through damage to plantations and erosion control plantings). In addition to production losses, there are major costs of possum control (by the Animal Health Board, Department of Conservation, regional councils and private landowners), which totaled NZ$ 60 million in 2002/2003 (Parkes and Murphy 2003). Research investment on possums is also substantial, totaling NZ$ 9.649 million in 2000/2001 and comprising 71 % of all research expenditure on introduced mammals in New Zealand (Parkes and Murphy 2003). Combined annual costs of possum control and research are therefore of similar magnitude to the costs of damage (Clout 2002). The overall cost of possum damage, control and research is likely to exceed NZ$ 100 million per annum.

17.9 Possums as "Keystone Aliens"

The "keystone species" concept originated from the finding that the removal of top predators on rocky shores led to a loss of local species diversity (Paine 1966). The label of keystone species has since been generally applied to "*a species whose effect is large, and disproportionately large relative to its abundance*" (Power et al. 1996). The strength of effect of a particular species can be measured by the change in community characteristics, such as productivity or species richness, per unit change in the abundance of the species concerned (Mills et al. 1993). The effect may be positive or negative. According to this definition, alien species that have high community impacts (negative or positive) in relation to their abundance could certainly be considered as keystone species. Such "keystone aliens" would include species with a large number of food web connections, whose addition to, or removal from, a community would have major consequences for the persistence of other species and ecosystem processes. In New Zealand, the label of keystone species has been applied to introduced predators, in the context of island restoration (Towns et al. 1997), and to a range of other alien species, including mammalian herbivores, in the general context of conservation management (Payton et al. 2002).

The brushtail possum is certainly a prime candidate for the title of a keystone alien species in many New Zealand forest communities, given the major ecological changes that have been recorded when possums have invaded or been removed from such communities. In lowland forests in particular, possums have multiple food web connections at different trophic levels (Fig. 17.1), and have impacts disproportionate to their abundance. Our understanding of the precise nature and scale of some of these connections and impacts is still incomplete, but the general significance of possums as agents of ecological change is obvious. They act as herbivores (driving changes in forest composition), as predators and competitors (driving changes in the native fauna), and potentially as seed dispersers. In addition to these ecological effects, the presence of brushtail possums, and their role as disease vectors, induces control efforts that can affect the abundance of other alien mammals, and have direct effects on some native fauna, through non-target kills.

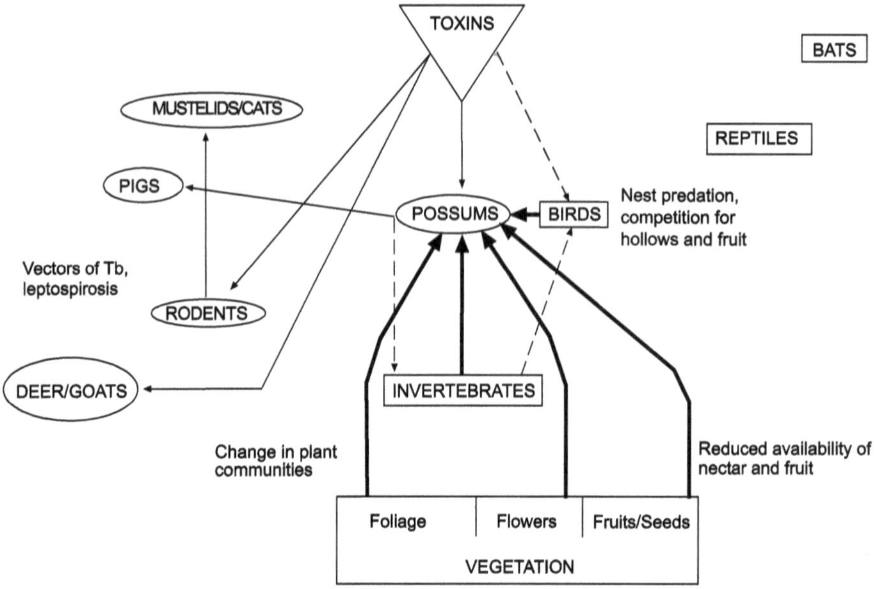

Fig. 17.1 Impacts and food web connections of possums in New Zealand forests, and pathways of toxins used for their control. *Arrows* indicate direction of consumption or energy flow, and *dotted lines* indicate weaker or poorly understood connections. Alien mammals are shown within *ellipses*

17.10 Implications for the Future

Over the past 20 years, possums have almost completed their colonization of the main islands of New Zealand (Clout and Ericksen 2000). In the wake of this geographical expansion, new damage to biodiversity has occurred in the far north and far southwest of New Zealand, and ongoing impacts have continued elsewhere. However, over the same period, possums have been eradicated from several offshore islands, have been excluded from some predator-fenced mainland reserves (e.g., Karori), and have been reduced to almost zero density at several intensively managed "mainland island" sites (e.g., Trounson Kauri Park, Wenderholm Regional Park, Te Urewera, Rotoiti). Recovery of selected native biodiversity components in the absence of possums and other mammals has been demonstrated at such sites (Saunders and Norton 2001). Elsewhere, large-scale possum control programs, using toxins and trapping, have continued for the management of bovine tuberculosis and the protection of conservation values.

In the immediate future (to 2010), it is likely that possum management will continue to be done using current methods of poisoning (including aerial broadcast of toxic baits) and trapping. Further conservation sites, especially peninsulas and forest patches, will be cleared of possums and maintained as possum-free areas behind predator-proof fences. Other conservation areas will continue to be subjected to intensive and sustained possum control, mainly by ground-based poisoning and trapping. Possum management for the control of bovine Tb will also continue at current levels, with the stated aim of "provisional freedom" from bovine Tb (<0.2 % infected cattle and deer herds) by 2013 (NSSC 2002).

In the longer term (after 2010), biological management of possums may become feasible, with the development of disseminating and nondisseminating delivery systems for immunocontraceptive or immunosterilant agents (see Chap. 27). If this eventuates, and is publicly accepted, then possum control could be revolutionized, leading to a capacity to reduce densities over large areas. In conjunction with existing control methods, it may then become possible to progressively eradicate possums from large areas on the mainland. However, for the foreseeable future, possums will persist in most forest ecosystems in New Zealand, with consequent ecological changes. Even at sites where their eradication is achieved, they may leave an "ecological signature" (e.g., through impacts on regeneration of some plants) that will persist for decades. Overall, it is abundantly clear that New Zealand forests and the New Zealand economy would be better off without this invasive marsupial.

Acknowledgements I thank Astrid Dijkgraaf for valuable discussion of brushtail possums as "keystone aliens", Hugh Robertson for information on kiwi, Terry Green for information on kaka, James Russell for assistance with preparation of the figure, and Carola Warner for assistance with manuscript formatting. I am also grateful to Bill Lee and Phil Cowan for constructive comments on an earlier draft.

References

Atkinson IAE (1992) Effects of possums on the vegetation of Kapiti Island and changes following possum eradication. DSIR Land Resources Contract Rep 92/52. Landcare Research, Lincoln, New Zealand

Batcheler CL (1983) The possum and rata-kamahi dieback in New Zealand: a review. Pac Sci 37:415–426

Booth LH, Eason CT, Spurr EB (2001) Literature review of the acute toxicity and persistence of brodifacoum to invertebrates. Science Conserv 177A:1–9

Brockie RE (1992) A living New Zealand forest. David Bateman, Auckland, New Zealand

Brown K, Innes J, Shorten R (1993) Evidence that possums prey on and scavenge birds' eggs, birds and mammals. Notornis 40:169–177

Campbell DJ (1990) Changes in structure and composition of a New Zealand lowland forest inhabited by brushtail possums. Pac Sci 44:277–296

Clout MN (1977) The ecology of the possum *Trichosurus vulpecula* (Kerr) in *Pinus radiata* plantations. PhD Thesis, University of Auckland, Auckland, New Zealand

Clout MN (2002) Ecological and economic costs of alien vertebrates in New Zealand. In: Pimentel D (ed) Biological invasions: economic and environmental costs of alien plant, animal and microbe species. CRC Press, Boca Raton, FL, pp 185–194
Clout MN, Ericksen K (2000) Anatomy of a disastrous success: the brushtail possum as an invasive species. In: Montague TL (ed) The brushtail possum: biology, impact and management of an introduced marsupial. Landcare Research, Lincoln, New Zealand, pp 1–9
Clout MN, Gaze PD (1984) Brushtail possums (*Trichosurus vulpecula* Kerr) in a New Zealand beech (*Nothofagus*) forest. NZ J Ecol 7:147–155
Clout MN, Karl BJ, Pierce RJ, Robertson HA (1995) Breeding and survival of New Zealand Pigeons *Hemiphaga novaeseelandiae*. Ibis 137:264–271
Coleman J, Caley P (2000) Possums as a reservoir of bovine Tb. In: Montague TL (ed) The brushtail possum: biology, impact and management of an introduced marsupial. Landcare Research, Lincoln, New Zealand, pp 92–104
Coleman J, Livingstone P (2000) Fewer possums: less bovine Tb. In: Montague TL (ed) The brushtail possum: biology, impact and management of an introduced marsupial. Landcare Research, Lincoln, New Zealand, pp 220–231
Cowan PE (1989) Denning habits of common brushtail possums, *Trichosurus vulpecular*, in New Zealand lowland forest. Aust Wildl Res 16:63–78
Cowan PE (1990a) Brushtail possum. In: King CM (ed) The handbook of New Zealand mammals. Oxford Univ Press, Auckland, pp 68–98
Cowan PE (1990b) Fruits, seeds, and flowers in the diet of brushtail possums, *Trichosurus vulpecula*, in lowland podocarp/mixed hardwood forest, Orongorongo Valley, New Zealand. NZ J Zool 17:549–566
Cowan PE (1991) Effects of introduced Australian brushtail possums (*Trichosurus vulpecular*) on the fruiting of the endemic New Zealand nikau palm (*Rhopalastylis sapida*). NZ J Bot 29:91–93
Cowan PE (2001) Advances in New Zealand mammalogy 1990–2000: brushtail possum. J R Soc NZ 31:15–29
Cowan PE, Moeed A (1987) Invertebrates in the diet of brushtail possums, *Trichosurus vulpecula*, in lowland podocarp/mixed hardwood forest, Orongorongo Valley, Wellington, New Zealand. NZ J Zool 14:163–177
Cowan PE, Waddington DC (1990) Suppression of fruit production of the endemic forest tree, *Elaeocarpus dentatus*, by introduced marsupial brushtail possums, *Trichosurus vulpecula*. NZ J Bot 28:217–224
Cowan PE, Clark J, Heath D, Stankiewicz M, Meers J (2000) Predators, parasites, and diseases of possums. In: Montague TL (ed) The brushtail possum: biology, impact and management of an introduced marsupial. Landcare Research, Lincoln, New Zealand, pp 82–91
Day TD, O'Connor CE, Matthews LR (2000) Possum social behaviour. In: Montague TL (ed) The brushtail possum: biology, impact and management of an introduced marsupial. Landcare Research, Lincoln, New Zealand, pp 35-46
Dijkgraaf AC (2002) Phenology and frugivory of large-fruited species in northern New Zealand and the impacts of introduced mammals. PhD Thesis, University of Auckland, Auckland, New Zealand
Dungan RJ, Norton DA (2003) Possums as seed dispersers in New Zealand: a response to Williams (2003). NZ J Ecol 27:225–226
Dungan RJ, O'Cain MJ, Lopez ML, Norton DA (2002) Contribution by possums to seed rain and subsequent seed germination in successional vegetation, Canterbury, New Zealand. NZ J Ecol 26:121–128
Fitzgerald AE (1976) Diet of the opossum *Trichosurus vulpecula* (Kerr) in the Orongorongo Valley, Wellington, New Zealand, in relation to food-plant availability. NZ J Zool 3:399–419

Fitzgerald AE (1984) Diet overlap between kokako and the common brushtail possum in central North Island, New Zealand. In: Smith AP, Hume ID (eds) Possums and gliders. Surrey Beatty, Chipping Norton, NSW, Australia, pp 569–573

Fraser KW, Spurr EB, Eason CT (1995) Non-target kills of deer and other animals from aerial 1080 operations. Rod Rifle 16(5):20–22

Garnett ST, Pedler LP, Crowley GM (1999) The breeding biology of the Glossy Black-Cockatoo *Calyptorhynchus lathami* on Kangaroo Island, South Australia. Emu 99:262–279

Gillies CA, Pierce RJ (1999) Secondary poisoning of mammalian predators during possum and rodent control operations at Trounson Kauri Park, Northland, New Zealand. NZ J Ecol 23:183–192

Innes J, Warburton B, Williams D, Speed H, Bradfield P (1995) Large-scale poisoning of ship rats (*Rattus rattus*) in indigenous forests of the North Island, New Zealand. NZ J Ecol 19:5–17

Innes J, Hay R, Flux I, Bradfield P, Speed H, Jansen P (1999) Successful recovery of North Island kokako *Callaeas cinerea wilsoni* populations, by adaptive management. Biol Conserv 87:201–214

James RE, Clout MN (1996) Nesting success of New Zealand pigeons (*Hemiphaga novaeseelandiae*) in response to a rat (*Rattus rattus*) poisoning programme at Wenderholm Regional Park. NZ J Ecol 20:45–51

Jane GT, Green TGA (1983) Episodic forest mortality in the Kaimai Ranges, North Island, New Zealand. NZ J Bot 21:21–31

Kirk HB (1920) Opossums in New Zealand: report on Australian opossums in New Zealand. Appendix to the Journals of the House of Representatives of New Zealand vol H28, pp 1–12

Leathwick JE, Hay JR, Fitzgerald AE (1983) The influence of browsing by introduced mammals on the decline of North Island kokako. NZ J Ecol 6:55–70

Lloyd BD, McQueen SM (2000) An assessment of the probability of secondary poisoning of forest insectivores following an aerial 1080 possum control operation. NZ J Ecol 24:47–56

McLennan J (1997) Ecology of brown kiwi and causes of population decline in Lake Waikaremoana catchment. Conserv Advisory Sci Notes 167. Department of Conservation, New Zealand

McLennan JA, Potter MA, Robertson HA, Wake GC, Colbourne R, Dew L, Joyce L, McCann AJ, Miles J, Miller PJ, Reid J (1996) Role of predation in the decline of kiwi, *Apteryx* spp., in New Zealand. NZ J Ecol 20:27–35

Meads MJ, Walker KJ, Elliott, GP (1984) Status, conservation, and management of the land snails of the genus *Powelliphanta* (Mollusca: Pulmonata). NZ J Zool 11:277–306

Mills LS, Soule ME, Doak DF (1993) The keystone-species concept in ecology and conservation. BioScience 43:219–224

Moorhouse R, Greene T, Dilks P, Powlesland R, Moran L, Taylor G, Jones A, Knegtmans J, Wills D, Pryde M, Fraser I, August A, August C (2003) Control of introduced mammalian predators improves kaka *Nestor meridionalis* breeding success: reversing the decline of a threatened New Zealand parrot. Biol Conserv 110:33–44

Murphy EC, Robbins L, Young JB, Dowding JE (1999) Secondary poisoning of stoats after an aerial 1080 poison operation in Pureora Forest, New Zealand. NZ J Ecol 23:175–182

Norton D (2000) Benefits of possum control for native vegetation. In: Montague TL (ed) The brushtail possum: biology, impact and management of an introduced marsupial. Landcare Research, Lincoln, New Zealand, pp 232–240

NSSC (2002) National Science Strategy Committee for Possum and Bovine Tuberculosis Control: progress and achievements report 2001/2002. Foundation for Research, Science and Technology, New Zealand

Nugent G, Sweetapple PJ, Coleman JD, Suisted PA (2000) Possum feeding patterns: dietary tactics of a reluctant folivore. In: Montague TL (ed) The brushtail possum: biology, impact and management of an introduced marsupial. Landcare Research, Lincoln, New Zealand, pp 10–19

Nugent G, Fraser W, Sweetapple P (2001) Top down or bottom up? Comparing the impacts of introduced arboreal possums and 'terrestrial' ruminants on native forests in New Zealand. Biol Conserv 99:65–79

Owen HJ, Norton DA (1995) The diet of introduced brushtail possums *Trichosurus vulpecula* in a low-diversity New Zealand *Nothofagus* forest and possible implications for conservation management. Biol Conserv 71:339–345

Paine RT (1966) Food web complexity and species diversity. Am Nat 100:65–75

Parkes J, Murphy E (2003) Management of introduced mammals in New Zealand. NZ J Zool 30:335–359

Payton IJ (2000) Damage to native forests. In: Montague TL (ed) The brushtail possum: biology, impact and management of an introduced marsupial. Landcare Research, Lincoln, New Zealand, pp 111–125

Payton IJ, Fenner M, Lee WG (2002) Keystone species: the concept and its relevance for conservation management in New Zealand. Sci Conserv 203:1–29

Potter MA (1989) Ecology and reproductive biology of the North Island brown kiwi (*Apteryx australis mantelli*). PhD Thesis, Massey University, Palmerston North, New Zealand

Power ME, Tilman D, Estes JA, Menge BA, Bond WJ, Mills LS, Daily G, Castilla JC, Lubchenko J, Paine RT (1996) Challenges in the quest for keystones. BioScience 46:609–620

Powlesland RG (1987) The foods, foraging behaviour and habitat use of North Island kokako in Puketi State Forest, Northland. NZ J Ecol 10:117–128

Powlesland RG, Knegtmans JW, Marshall ISJ (1999) Costs and benefits of aerial 1080 possum control operations using carrot baits to North Island robins (*Petroica australis longipes*), Pureora Forest Park. NZ J Ecol 23:149–159

Powlesland RG, Knegtmans JW, Styche A (2000) Mortality of North Island tomtits (*Petroica macrocephala toitoi*) caused by aerial 1080 possum control operations, 1997–1998, Pureora Forest Park. NZ J Ecol 24:161–168

Powlesland RG, Wills DE, August ACL, August CK (2003) Effects of a 1080 operation on kaka and kereru survival and nesting success, Whirinaki Forest Park. NZ J Ecol 27:125–137

Pracy LT, Kean RI (comps) (1949) Control of opossums an urgent problem. NZ J Agric 78:353–358

Rose AB, Pekelharing CJ, Platt KH (1992) Magnitude of canopy dieback and implications for conservation of southern rata-kamahi (*Metrosideros umbellata-Weinmannia racemosa*) forests, central Westland, New Zealand. NZ J Ecol 16:23–32

Sadleir RMFS (2000) Evidence of possums as predators of native animals. In: Montague TL (ed) The brushtail possum: biology, impact and management of an introduced marsupial. Landcare Research, Lincoln, New Zealand, pp 126–131

Saunders A, Norton DA (2001) Ecological restoration at Mainland Islands in New Zealand. Biol Conserv 99:109–119

Sessions LA, Rance C, Grant A, Kelly D (2001) Possum (*Trichosurus vulpecula*) control benefits native beech mistletoes (Loranthaceae). NZ J Ecol 25(2):27–33

Sherley G, Wakelin M, McCartney J (1999) Forest invertebrates found on baits used in pest mammal control and the impact of sodium monofluoroacetate ("1080") on their numbers at Ohakune, North Island, New Zealand. NZ J Zool 26:279–302

Spurr EB (1991) Effects of brushtail possum control operations on non-target bird populations. In: Proc 20th Int Ornithology Congr, 2–9 December 1990, Christchurch, New Zealand, pp 2534–2545

Spurr EB (2000) Impacts of possum control on non-target species. In: Montague TL (ed) The brushtail possum: biology, impact and management of an introduced marsupial. Landcare Research, Lincoln, New Zealand, pp 175–186

Sweetapple PJ, Nugent G, Whitford J, Knightbridge PI (2002) Mistletoe (*Tupeia antarctica*) recovery and decline following possum control in a New Zealand forest. NZ J Ecol 26:61–71

Towns DR, Simberloff D, Atkinson IAE (1997) Restoration of New Zealand islands: redressing the effects of introduced species. Pac Conserv Biol 3:99–124

Veblen TT, Stewart GH (1982) The effects of introduced wild animals on New Zealand forests. Ann Assoc Am Geogr 72:372–397

Veltman C (2000) Do native wildlife benefit from possum control? In: Montague TL (ed) The brushtail possum: biology, impact and management of an introduced marsupial. Landcare Research, Lincoln, New Zealand, pp 241–250

Williams PA (2003) Are possums important dispersers of large-seeded fruit? NZ J Ecol 27:221–223

Williams PA, Karl BJ, Bannister P, Lee WG (2000) Small mammals as potential seed dispersers in New Zealand. Austral Ecol 25:523–532

Wilson PR (1984) The effects of possums on mistletoe on Mt Misery, Nelson Lakes National Park. In: Dingwall PR (comp) Protection and parks: essays in the preservation of natural values in protected areas. Proc Sect A4e, 15th Pacific Science Congr, February 1983, Dunedin. Info Ser 12, Department of Lands and Survey, Wellington, New Zealand, pp 53–60

Zotov VD (1947) Forest deterioration in the Tararuas due to deer and opossum. Trans Proc R Soc NZ 77:162–165

18 Keystone Species: Competition for Honeydew Among Exotic and Indigenous Species

J.R. Beggs and D.A. Wardle

18.1 Introduction

New Zealand is one of the most highly invaded areas on earth, with a large number of aliens. Some invaders were deliberately introduced or arrived as stowaways. The small size, and the dispersal and reproductive modes of many invertebrates make them ideal candidates for stowing away in transported goods. There are an estimated 2,200 exotic invertebrate species established in New Zealand (Ministry for the Environment 1997), and increasing global trade results in new species arriving continually (e.g., Harris and Berry 2001; Hoare 2001; Lester 2005). Despite this, few exotic invertebrates appear to have had a significant impact on New Zealand's indigenous biota (Barlow and Goldson 2002). Exotic social wasps are an exception. Two species of *Vespula* have established in New Zealand, namely German wasps (*Vespula germanica* Fabr., Hymenoptera: Vespidae), and common wasps (*V. vulgaris* L.).

Invasive species can have a range of impacts on the receiving ecosystem, but what happens when the invasive species affects an indigenous keystone species? A keystone species is one whose effect is disproportionately large relative to its biomass (Power et al. 1996). Keystone species are not always of high trophic status, and they can exert effects through a range of mechanisms (Power et al. 1996). Honeydew-producing, endemic scale insects (*Ultracoelostoma assimile* and *U. brittini* Morales, Homoptera; Margarodidae) that infest some of New Zealand's beech (*Nothofagus*) forests (Belton 1978; Morales et al. 1988) may be keystone species because of the strength of their effect on the beech forest community. Hereafter, these forests are referred to as honeydew beech forests.

Although a range of animal species have invaded New Zealand, common wasps are arguably the worst of the invaders in honeydew beech forests in terms of environmental impact. In years of high abundance, wasps consume more than 90 % of the honeydew for about 4 months of the year (Moller et al.

Ecological Studies, Vol. 186
R.B. Allen and W.G. Lee (Eds.)
Biological Invasions in New Zealand

1991a), and compete with indigenous species dependent on this resource. Honeydew provides such a large sugar resource that high densities of wasps are supported, with major flow-on effects throughout the community (Beggs 2001).

In order to evaluate the impact of invasive wasps in honeydew beech forest, it is necessary to understand the primary interactions between native biota in prehuman forests, before the arrival of introduced species. This chapter first reconstructs the prehuman honeydew beech forest food web, and then evaluates the impact of invasive wasps, particularly via competition for honeydew. Finally, it identifies knowledge gaps that must be filled if we are to fully understand the ecosystem consequences of invasive species on a native keystone species.

18.2 Scale Insects as Keystone Species

18.2.1 Honeydew Resource

Southern beeches, of which there are four species in New Zealand, are a major component of about 70 % of New Zealand's indigenous forest; these are especially predominant in the more southern latitudes and in mountainous regions (Wardle 1984). The scale insects *U. assimile* and *U. brittini* infest all New Zealand beech species, although infestation of *N. menziesii* is rare. Not all forests are infested. Where scale insects are present, levels of infestation vary from tree to tree and site to site – estimates of scale insect density range from 100,000 per hectare (Kelly 1990) to 2 million per hectare (Moller et al. 1996b). Even at such densities, scale insects represent only a small fraction of the biomass in a forest. Yet these insects play a pivotal role in the flow of energy because they produce an abundant, year-round renewable resource that is readily available to a range of other species.

Honeydew beech forest is a major ecosystem type in New Zealand, which is currently estimated to cover around 10^6 ha – about 15 % of New Zealand's remaining indigenous forests (Beggs 2001). Prehuman honeydew beech forest is estimated to have covered about 2.7×10^6 ha, and about 40 % of this forest remains today.

Four female and five male stages have been found in the scale insect *U. brittini* (Morales 1991). The first-instar crawler settles in crevices in the bark, inserts its feeding stylet into the phloem cells of the host tree, and builds up a protective capsule. Excess carbohydrates (primarily sucrose, fructose and oligosaccharides; Grant and Beggs 1989) and water are excreted as drops of sugary honeydew at the end of a waxy anal filament that protrudes from the tree. Evaporation leads to increased sugar concentrations, and when the drop is removed, fluid gradually re-accumulates (Moller et al. 1996b). Second instars and third-instar females also produce honeydew. Adult males have wings and are free-living, whereas adult females remain inside capsules where they lay their eggs. The scale insect has overlapping generations, and each generation probably takes more than a year (Morales 1991).

Daily honeydew production ranges from 0.2 to 1.5 mg insect^{-1} 24 h^{-1} (Dungan and Kelly 2003), and estimated annual honeydew production is between 3,500 and 4,500 kg dry weight honeydew ha^{-1} year^{-1} (Beggs et al. 2005). These estimates of production suggest that between ca. 2 % (Dungan and Kelly 2003) and 7 % (Beggs et al. 2005) of net primary productivity (NPP) is released as honeydew annually. However, honeydew production is highly variable between trees (Dungan and Kelly 2003; Beggs et al. 2005), so the proportion of carbon lost as honeydew from an individual tree can be considerably higher, and potentially sufficient to reduce plant growth. Growth rates of trees with high densities of scale insects may be lower than those with fewer scale insects. Alternatively, if honeydew enhances the availability of nutrients in the soil, this may compensate for the loss of carbon and sustain tree growth rates. A further possibility is that scale insects stimulate trees to fix more carbon, thereby promoting tree growth (Kelly et al. 1992).

Honeydew abundance varies seasonally (Moller and Tilley 1989), temporally (Gaze and Clout 1983; Kelly et al. 1992), and with prevailing weather (Moller and Tilley 1989), but it remains the most abundant sugar resource in beech forests because of the paucity of nectar-producing plant species. Parasitic mistletoes (Loranthaceae) may have been an abundant source of nectar in prehuman forests (Ladley et al. 1997), but they flower only for a few months in summer.

Characteristic of honeydew forests is a sooty mould complex, sometimes involving as many as seven species (*Trichopeltheca*, *Capnocybe* and *Capnodium* spp.; Hughes 1976), that grows over any surface where honeydew falls, including on the tree and ground. The sooty mould comprises specialized, dark-colored fungi that proliferate as a result of the honeydew sugars. Although the inter-relationships are largely unstudied, the sooty mould is likely to provide feeding and breeding sites for fungus-feeding arthropods, and refuges for some species (Didham 1993). Furthermore, the sooty mould may have a smothering effect on honeydew beech trees and understory plants in the splash zone (Birch 1937; Gilmour 1965, 1966).

18.2.2 Indigenous Honeydew Consumers

Honeydew is consumed by a wide range of biota, including birds, reptiles, invertebrates, and microbes (Gaze and Clout 1983; Boyd 1987; Beggs and Wilson 1991; Moller et al. 1991a; Didham 1993; Markwell et al. 1993; O'Donnell and Dilks 1994; Wilson et al. 1998; Beggs 2001). Figure 18.1 includes a descrip-

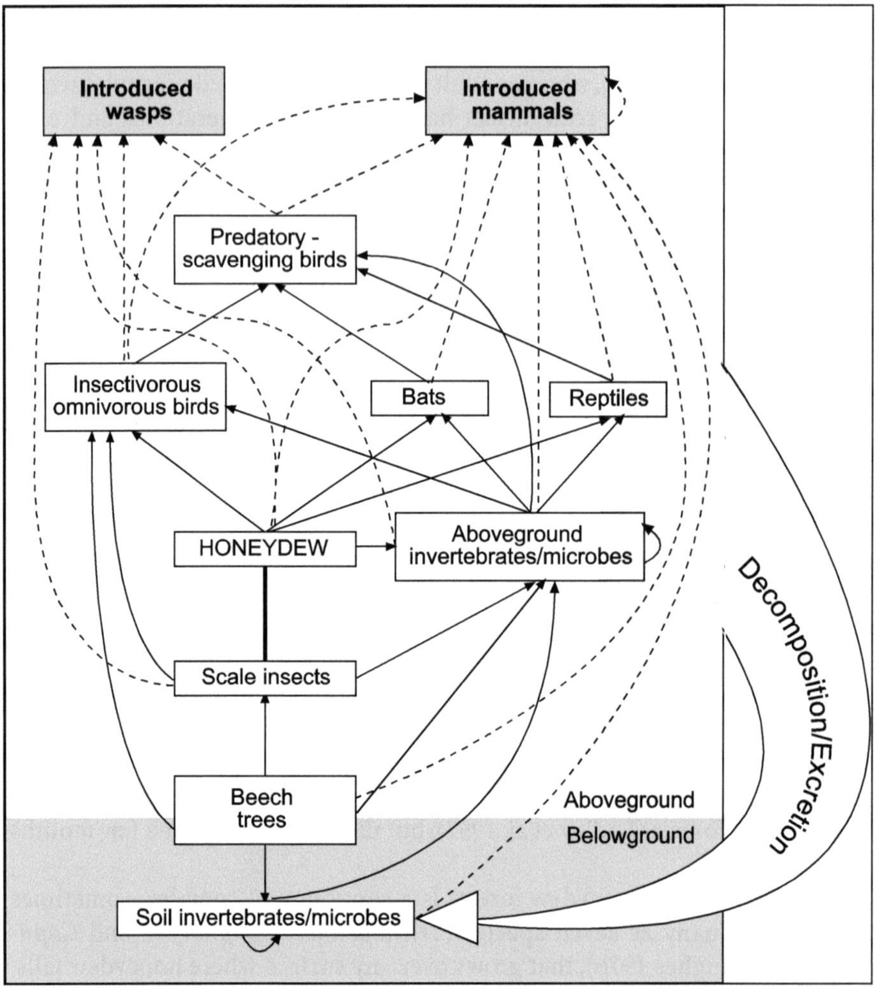

Fig. 18.1 Main trophic interactions, based around the honeydew resource in a prehuman honeydew beech forest, and after the arrival of exotic species. The two *shaded boxes* represent introduced species, whereas the *non-shaded boxes* are primarily native biota. *Dotted lines* represent novel interactions following the arrival of introduced species. Most boxes are species aggregates. *Arrows* point to the consumer groups. The *line* between the scale insect and honeydew denotes that the scale insect produces honeydew

tive food web of the main trophic links for honeydew beech forest prior to the arrival of exotic species. Birds – at least 34 species in this habitat (based on Holdaway et al. 2001, and T. Worthy, personal communication) – dominated the top levels of the food web. There were no social wasps or bees, and the only terrestrial mammals were bats. The link between bats and honeydew (Fig. 18.1) is speculative because bats have not been recorded consuming honeydew. However, short-tailed bats (*Mystacina tuberculata*) are nectarivorous in other habitats (Daniel 1990; Arkins et al. 1999), and it is therefore likely that they do feed on honeydew.

Some indigenous birds in honeydew beech forest are almost solely dependent on honeydew as their energy resource. Kaka (*Nestor meridionalis*), an indigenous forest parrot, spend about 30 % of their foraging time in summer collecting honeydew (Beggs 1988), during which they are able to acquire most of their daily energy requirements (Beggs and Wilson 1991). The rest of their foraging time is spent collecting protein resources (i.e., invertebrates). Similarly, tui (*Prosthemadera novaeseelandiae*) and bellbirds (*Anthornis melanura*) often spend over 80 % of their daily foraging time collecting honeydew (Moller et al. 1996a; Beggs 2001).

Forests with a higher abundance of honeydew have greater densities of tui, bellbirds, and a suite of insectivorous birds: fantails (*Rhipidura fuliginosa*), robins (*Petroica australis*), and unidentified introduced finches (Fig. 18.2; Moller et al. 1988). This suggests that these increased abundances were not causally linked to feeding on honeydew, but perhaps that honeydew promotes the density of invertebrates, or that the ecological conditions that favor the scale insect also favor other invertebrates (Moller et al. 1988).

As well as their consumption aboveground, soil invertebrates and microbes also consume honeydew that falls or is washed to the ground (Fig. 18.1). Scale insects may decrease nutrient availability to beech trees, because honeydew addition to the soil could stimulate microbial activity and hence nutrient immobilization. Alternatively, honeydew addition to the forest floor could increase nutrient availability to plants by adding a relatively labile nutrient pool to the soil. Addition of honeydew from seasonal aphid out-

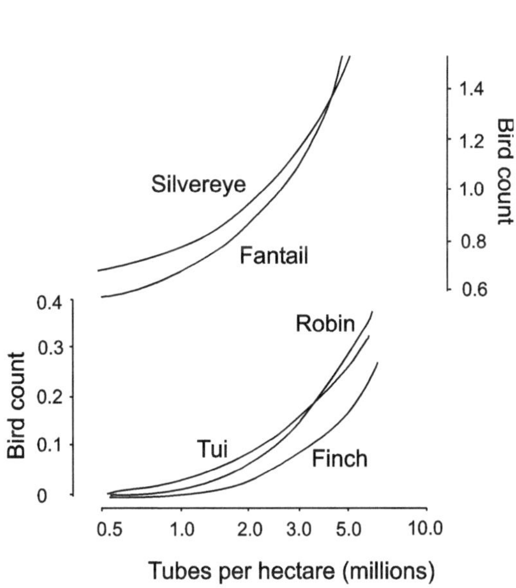

Fig. 18.2 Predicted 5-min bird counts with increasing honeydew abundance. Multiple regression was done for each bird species from measurements made at 23 beech forest sites in northern South Island (from Moller et al. 1988)

breaks in the Northern Hemisphere has been shown to affect nutrient cycling in both these ways (e.g., Pedersen and Bille-Hansen 1995; Stadler et al. 1998; Kindlmann and Stadler 2004; Michalzik and Stadler 2005). It is not known how honeydew influences belowground organisms in New Zealand forests, but given the magnitude of the carbon addition, it is expected that honeydew probably also has important effects on these organisms.

Honeydew beech forests are ecologically unique, because they are the only natural ecosystem worldwide where such large quantities of labile carbon (honeydew) are added year-round. The prehuman food web in these forests reflected both the unusual attributes of New Zealand's biodiversity, such as a lack of mammalian predators, and the pivotal role of honeydew as a food resource.

18.3 Exotic Species

18.3.1 Effect on Food Web Structure

Mammals, such as the Polynesian rat (*Rattus exulans*), were probably the first invaders of honeydew beech forests. There are now a range of mammalian species established, including mustelids (stoats, *Mustela erminea*; ferrets, *M. furo*; weasels, *M. nivalis*), other rats (*R. rattus* and *R. norvegicus*), mice (*Mus musculus*), possums (*Trichosurus vulpecula*), and deer (*Cervus elaphus*). Social wasps were relatively late arrivals, with German wasps arriving in New Zealand in 1945. They were subsequently displaced in honeydew beech forest by the arrival of common wasps in the late 1970s (Harris et al. 1991), reflecting their more efficient harvesting of honeydew (Harris et al. 1994). There are other exotic species too (such as bumblebees (*Bombus* spp.), honeybees (*Apis mellifera*) and six species of birds), but little is known about their densities or impact.

The invasion by mammals and wasps has changed the structure and connectedness of the food web in honeydew beech forest (Fig. 18.1). An additional trophic level has been added, and introduced species are linked to all trophic levels (Fig. 18.1). Predatory birds are no longer the top predator, and many other species, including moa (Dinornithiformes), nesting seabirds (Procellariiformes), tuatara (*Sphenodon punctatus*), and probably short-tailed bats (*Mystacina* spp.), are now absent in these forests. We estimate the number of indigenous bird species to have declined by about 50 % (based on Holdaway et al. 2001, and T. Worthy, personal communication), mostly caused by introduced predators, including humans (Holdaway 1999). It is likely that the diversity of other taxa (e.g., invertebrates) has also been negatively impacted, but fossil remains of many groups are non-existent. The present-day honeydew beech forest community therefore contains an impoverished suite of the former indigenous species, but is augmented by a range of exotic mammals, birds and invertebrates.

18.3.2 Effect on Food Web Interactions

The strength of trophic interactions has also changed with the introduction of exotic species. Wasp populations are larger where there is honeydew, which provides an energy-rich food source that supports the highest recorded densities of common wasps in the world (Thomas et al. 1990; Moller et al. 1991a; Beggs et al. 1998; Beggs 2001). Wasps are now the most abundant consumers of honeydew (Moller and Tilley 1989; Sandlant and Moller 1989; Thomas et al. 1990). Possums consume blackened bark and wood as a consistent part of their diet in honeydew beech forest (Cochrane et al. 2003), perhaps a consequence of feeding on sooty mould (D.A. Norton, personal communication). At the same time, they are probably inadvertently consuming scale insects and other invertebrates, but they appear to have minimal impact on honeydew availability.

Exclusion experiments have shown that wasps reduce the standing crop of honeydew by more than 90 % in late summer and autumn (Beggs and Wilson 1991; Moller et al. 1991a), leaving little for other indigenous and exotic consumers. Wasps cropped honeydew drops so frequently at the peak of the 1987 wasp season (January and March) that insufficient honeydew could be collected for measurements. In the absence of wasps (but with introduced mammals present), the majority of honeydew falls to the ground around the tree (Moller and Tilley 1989). The flow of soluble honeydew carbon to the soil under the trees has been diverted by wasp consumption. However, this type of resource capture would have been much higher by birds and possibly native invertebrates before the arrival of introduced mammals, and wasps may have merely replaced native species as the major consumer of honeydew.

Wasps take honeydew and other food back to their colonies, which are subterranean, located in sunny areas near water (Moller et al. 1991b), but not in the splash zones around honeydew trees (personal observations). Many adult wasps probably die aboveground, and at the end of each wasp season the colony dies, and the nest and any unhatched brood decompose in situ. Furthermore, adult wasps and their larvae respire a substantial (though unknown) proportion of consumed honeydew carbon either aboveground or in their nests. Through these processes, wasps substantially alter the spatial and temporal distribution of energy and nutrients reaching the belowground components of the ecosystem. The consequence of this redistribution of nutrients is unknown but may significantly impair soil processes fuelled by honeydew sugars.

The capture of honeydew by wasps also has consequences for indigenous birds. The behavior of kaka, tui and bellbirds changes when honeydew becomes scarce (Beggs and Wilson 1991; Moller et al. 1996b). They either reduce the time spent feeding on honeydew, or leave the honeydew beech forest to forage in other habitats. These behavioral changes could affect breeding success or survival, but the role of wasps in inducing these effects cannot be easily separated from effects of bird predation by introduced mammals (Beggs 2001).

The availability of honeydew creates a corresponding high demand for nitrogen, which causes wasps to become particularly voracious predators. Wasps consume a large biomass of invertebrates as protein each season, including spiders (30 %), caterpillars (20 %), ants and bees (20 %), flies (15 %), and other invertebrates (15 %; Harris 1991). Predation rates on some invertebrate groups, such as spiders and free-living caterpillars, are extreme, reducing the probability of survival through a wasp season to zero (Toft and Rees 1998; Beggs and Rees 1999). Experiments have demonstrated that wasp abundance needs to be reduced by 80–90 % to conserve vulnerable native invertebrate species (Fig. 18.3). Wasps are also predators of nestling birds (Fig. 18.1; Moller 1990).

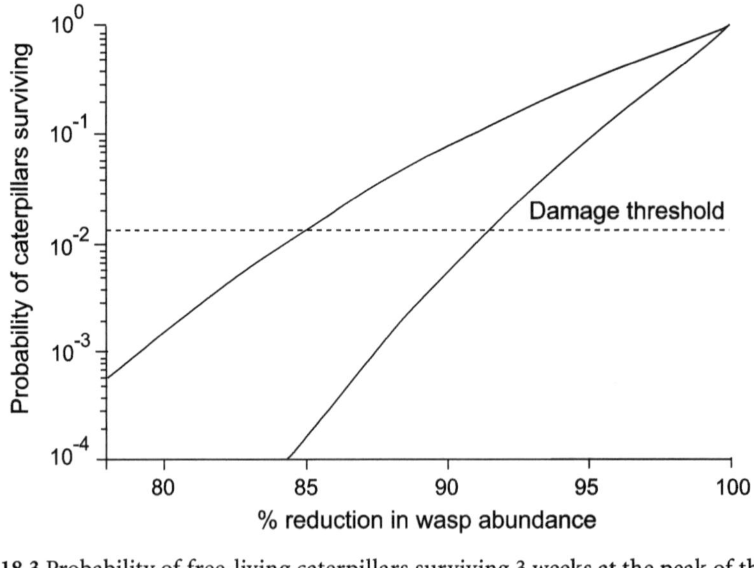

Fig. 18.3 Probability of free-living caterpillars surviving 3 weeks at the peak of the wasp season is predicted given varying percentage reductions in wasp abundance. Caterpillars are assumed to be exposed to wasp predation all day (maximum predation; *lower curve*) or for half a day (minimum predation; *upper curve*). The *horizontal line* gives the ecological damage threshold (where 1.3 % of the caterpillars survive) above which the population of caterpillars is sustainable (from Beggs and Rees 1999)

Wasp density was experimentally reduced at two honeydew beech forest sites (Beggs et al. 1998), and invertebrates collected in Malaise traps (Townes 1972) at these sites and two non-treatment sites, before and after reducing wasp numbers. This experiment demonstrated that wasps reduced the biomass of several invertebrate groups, including native Hymenoptera (Fig. 18.4), leading to an overall reduction in the biomass of indigenous invertebrates (unpublished data). The loss of biodiversity aside, the functional impact on other insectivores and on ecosystem processes is unknown.

Wasps also feed on the juvenile stages of scale insects. In an exclosure experiment, the number of scale insects at the crawler stage was significantly reduced when wasps were abundant (McAllum 1992). Kaka feed on the adult scale insects by stripping back the bark of beech trees (Beggs 1988). However, bark removal assists scale insect colonization by providing new settlement space (McAllum 1992). Kaka were once abundant in beech forest, but the introduction of mammalian predators has greatly reduced their density (Wilson et al. 1998). Thus, the interaction between scale insects, exotic species, and indigenous birds is complex, and it is difficult to predict the net effect of exotic species on the abundance of scale insects.

It is not known if wasps have eradicated some invertebrate prey species, because of a lack of knowledge about the pre-20th century invertebrate fauna in beech forests. The diet of wasps varies seasonally and annually (Harris 1991), suggesting that wasps switch prey species, depending on the abundance of prey. Once a prey species becomes less common, wasps may switch to more abundant prey species; so some prey species may be kept at low density by wasps, but not eradicated.

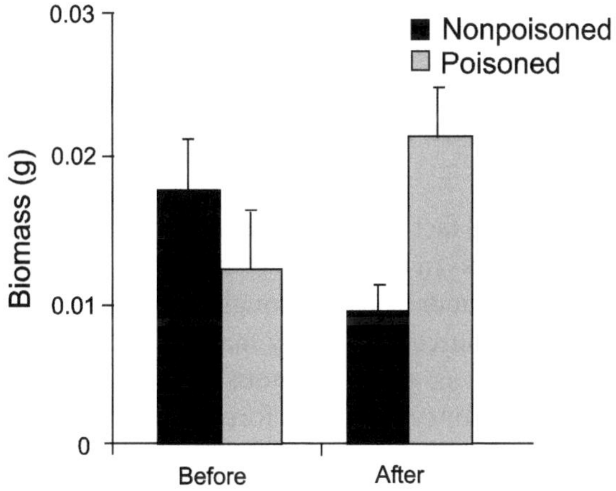

Fig. 18.4 Biomass of indigenous Hymenoptera increased at two sites where introduced *Vespula* wasps were poisoned, compared with biomass at two non-poisoned sites. Insects were caught in Malaise traps before poisoning (1991), and after 4 years of annual poisoning (1995/1996; see Beggs et al. 1998 for details). Standard errors are shown

18.3.3 Potential Effect on Ecosystem Processes

Although previous studies have evaluated aboveground and belowground effects of invasive plant and mammalian species on ecosystem processes (e.g., Ehrenfeld and Scott 2001; Wardle et al. 2001; Ehrenfeld 2003), few studies have applied such an approach to an invasive invertebrate. This is a critical gap because invertebrates drive interactions and feedbacks between producer and decomposer organisms, as well as by their consumers (Scheu 2001; Wardle 2002; Wardle et al. 2004).

Invertebrates can have profound effects on whole-ecosystem function, largely through interfering with keystone processes such as seed dispersal, and litter fragmentation and decomposition. For example, the invasion of South African shrublands by the Argentine ant (*Linepithema humile*) caused a shift in composition of the plant community, with a disproportionate reduction in the densities of large-seeded plants (Christian 2001). Native ant species that provide seed dispersal services were lost from areas invaded by Argentine ants. Likewise, yellow crazy ants (*Anoplolepsis gracilipes*) that have invaded Christmas Island, in the Indian Ocean, have affected ecosystem functioning by attacking and disrupting the migration of a dominant land crab that monopolizes the processing of litter (Green et al. 1999; O'Dowd et al. 2003). There are also studies that illustrate the impact of invasive belowground invertebrates on ecosystem functioning (e.g., flatworms, Boag and Yeates 2001; earthworms, Clements et al. 1991).

Honeydew addition is likely to influence belowground ecosystem processes, particularly the ecological interaction between beech trees and scale insects. The arrival of invasive wasps potentially disrupts these interactions by reducing the flow of carbon to soil microorganisms, possibly leading to changes in soil carbon, densities of soil microorganisms and soil animals, and ultimately the supply of plant-available nutrients. This study system provides an opportunity to considerably add to our knowledge of these types of interactions.

18.4 Conclusions

Despite the fact that honeydew scale insects represent a tiny proportion of total biomass in beech forests, they have a major effect at both the community and ecosystem levels through supplying a year-round, readily available energy resource utilized by many indigenous species. They can therefore be considered as an indigenous keystone species. When introduced wasps arrived in honeydew beech forest, they monopolized the resource to the exclusion of other organisms. This has resulted in a cascade of effects throughout the food web. Wasps reach high densities, and have become a major predator of indigenous invertebrates. They may also disrupt belowground nutrient cycling. The biomass of introduced wasps is greater than that of all the vertebrate predators in these ecosystems combined (Thomas et al. 1990), but wasps are still a relatively small component of the total beech forest biomass. The impact of wasps is disproportionately large relative to their biomass, because they have a major impact on a keystone resource. Wasps have altered the structure, connectedness, and interaction strengths of the food web - they have become a keystone invader.

Our knowledge about prehuman ecological interactions in honeydew beech forest is limited. However, the food web structure was quite different from that of today. Interactions between introduced wasps and mammals are complex, making it difficult to assess the overall changes to the beech forest community. Nevertheless, it is clear that what remains is a highly modified, impoverished community, making it vulnerable to further changes. Much of New Zealand's conservation effort is focused on offshore islands where introduced predators can be eradicated. A crucial point for managing honeydew beech forest is that there are no offshore islands with this type of habitat, so conservation efforts can only occur on the mainland in the presence of the full range of introduced species. Understanding these interactions is therefore critical.

Acknowledgements Thanks to Neal Scott, Louis Schipper and Rachel Standish for discussions on the role of the scale insect in ecosystem processes. The food web diagrams are the result of discussions with many colleagues, including Peter Wilson, Eric Spurr, Richard Toft, Peter Williams, Richard Harris, Colin O'Donnell, Dave Butler and David Norton. Trevor Worthy kindly shared his knowledge on the prehuman fauna of honeydew beech forests. Richard Harris, Bill Lee and Sarah Richardson provided helpful comments on earlier drafts. This work is part of the Invasive Invertebrates in Natural Ecosystems Programme funded by the New Zealand Foundation for Research, Science and Technology.

References

Arkins AM, Winnington AP, Anderson S, Clout MN (1999) Diet and nectarivorous foraging behaviour of the short-tailed bat (*Mystacina tuberculata*). J Zool 247:183–187

Barlow ND, Goldson SL (2002) Alien invertebrates in New Zealand. In: Pimental D (ed) Biological invasions: economic and environmental costs of alien plant, animal, and microbe species. CRC Press, Boca Raton, FL, pp 195–216

Beggs J (1988) Energetics of kaka in a South Island beech forest. MSc Thesis, University of Auckland, Auckland, New Zealand

Beggs J (2001) The ecological consequences of social wasps (*Vespula* spp.) invading an ecosystem that has an abundant carbohydrate resource. Biol Conserv 99:17–28

Beggs JR, Rees JS (1999) Restructuring of Lepidoptera communities by introduced *Vespula* wasps in a New Zealand beech forest. Oecologia 119:565–571

Beggs JR, Wilson PR (1991) The kaka, *Nestor meridionalis*, a New Zealand parrot endangered by introduced wasps and mammals. Biol Conserv 56:23–38

Beggs JR, Toft RJ, Malham JP, Rees JS, Tilley JAV, Moller H, Alspach P (1998) The difficulty of reducing introduced wasp (*Vespula vulgaris*) populations for conservation gains. NZ J Ecol 22:55–63

Beggs JR, Karl BJ, Wardle DA, Bonner KI (2005) Soluble carbon production by honeydew scale insects in a New Zealand beech forest. NZ J Ecol 29(1):105–115

Belton M (1978) The place of the beech scale insect (*Ultracoelostoma assimile*) in the ecology of mountain beech forest. In: Proc Honeydew Sem, 10 August 1978, Advisory Services Division, Ministry of Agriculture and Fisheries, Christchurch, New Zealand, pp 27–37

Birch TTC (1937) A synopsis of forest fungi of significance in New Zealand. NZ J For 4:109–125

Boag B, Yeates GW (2001) The potential impact of the New Zealand flatworm, a predator of earthworms, in western Europe. Ecol Appl 11:1276–1286

Boyd S (1987) Patterns of use of beech honeydew by birds and insects. MSc Thesis, University of Auckland, Auckland, New Zealand

Christian CE (2001) Consequences of a biological invasion reveal the importance of mutualism for plant communities. Nature 413:635–639

Clements RO, Murray PJ, Sturdy RG (1991) The impact of 20 years' absence of earthworms and three levels of N fertilizer on a grassland soil environment. Agric Ecosyst Environ 36:75–85

Cochrane CH, Norton DA, Miller CJ, Allen RB (2003) Brushtail possum (*Trichosurus vulpecula*) diet in a north Westland mixed-beech (*Nothofagus*) forest. NZ J Ecol 27:61–65

Daniel MJ (1990) Order Chiroptera. In: King CM (ed) Handbook of New Zealand mammals. Oxford Univ Press, Oxford, pp 114–137

Didham RK (1993) The influence of honeydew on arthropods associated with beech trees in New Zealand. NZ Nat Sci 20:45–53

Dungan RJ, Kelly D (2003) Effect of host-tree and environmental variables on honeydew production by scale insects (*Ultracoelostoma* sp.) in a high elevation *Nothofagus solandri* forest. NZ J Ecol 27:169–177

Ehrenfeld JG (2003) Effects of exotic plant invasions on soil nutrient cycling processes. Ecosystems 6(6):503–523

Ehrenfeld JG, Scott N (2001) Invasive species and the soil: effects on organisms and ecosystem processes. Ecol Appl 11:1259–1260

Gaze PD, Clout MN (1983) Honeydew and its importance to birds in beech forests of South Island, New Zealand. NZ J Ecol 6:33–37

Gilmour JW (1965) The pathogenic fungi of beech. Beech Forestry in New Zealand, vol 2. New Zealand Forest Service, Forest Research Institute Symp 5, pp 79–84

Gilmour JW (1966) The pathology of forest trees in New Zealand. The fungal, bacterial, and algal pathogens. New Zealand Forest Service, Forest Research Institute Tech Pap 48

Grant WD, Beggs JR (1989) Carbohydrate analysis of beech honeydew. NZ J Zool 16:283–288

Green PT, Lake PS, O'Dowd DJ (1999) Monopolization of litter processing by a dominant land crab on a tropical oceanic island. Oecologia 119:435–444

Harris RJ (1991) Diet of the wasps *Vespula vulgaris* and *V. germanica* in honeydew beech forest of the South Island, New Zealand. NZ J Zool 18:159–169

Harris RJ, Berry JA (2001) Confirmation of the establishment of three adventive ants (Hymenoptera: Formicidae) in New Zealand: *Cardiocondyla minutior* Forel, *Ponera leae* Forel, *Mayriella abstinens* Forel. NZ Entomol 24:53–56

Harris RJ, Thomas CD, Moller H (1991) The influence of habitat use and foraging on the replacement of one introduced wasp species by another in New Zealand. Ecol Entomol 16:441–448

Harris RJ, Moller H, Winterbourn MJ (1994) Competition for honeydew between two social wasps in South Island beech forests, New Zealand. Insectes Sociaux 41:379–394

Hoare RJB (2001) Adventive species of Lepidoptera recorded for the first time in New Zealand since 1988. NZ Entomol 24:23–47

Holdaway RN (1999) Introduced predators and avifaunal extinction in New Zealand. In: McPhee RDE (ed) Extinctions in near time: causes, contexts, and consequences. Advances in Palaeobiology. Kluwer/Plenum Press, New York, pp 189–238

Holdaway RN, Worthy TH, Tennyson AJD (2001) A working list of breeding bird species of the New Zealand region at first human contact. NZ J Zool 28:119–187

Hughes SJ (1976) Sooty moulds. Mycologia 68:693–820

Kelly D (1990) Honeydew density in mixed *Nothofagus* forest, Westland, New Zealand. NZ J Bot 28:53–58

Kelly D, Stirling DJ, Hunt GR, Newell CL, Jarvis CE (1992) Honeydew standing crop and production over 24 hours in *Nothofagus solandri* forest in Canterbury. NZ J Ecol 16:69–75

Kindlmann P, Stadler B (2004) Temporal fluctuations in throughfall carbon concentration in a spruce forest. Ecol Model 176:381–388

Ladley JJ, Kelly D, Robertson AW (1997) Explosive flowering, nectar production, breeding systems, and pollinators of New Zealand mistletoes (Loranthaceae). NZ J Bot 35:345–360

Lester PJ (2005) Determinants for the successful establishment of exotic ants in New Zealand. Diversity Distrib 11(4):279–288

Markwell TJ, Kelly D, Duncan KW (1993) Competition between honey bees (*Apis mellifera*) and wasps (*Vespula* spp.) in honeydew beech (*Nothofagus solandri* var. *solandri*) forest. NZ J Ecol 17:85–93

McAllum P (1992) Interactions of the scale insect and kaka in South Island beech forest. Honours Thesis, University of Canterbury, Christchurch, New Zealand

Michalzik B, Stadler B (2005) Importance of canopy herbivores to dissolved and particulate organic matter fluxes to the forest floor. Geoderma 127:227–236

Ministry for the Environment (1997) The state of the New Zealand environment report. Wellington, New Zealand

Moller H (1990) Wasps kill nestling birds. Notornis 37:76–77

Moller H, Tilley JAV (1989) Beech honeydew: seasonal variation and use by wasps, honeybees, and other insects. NZ J Zool 16:289–302

Moller H, Tilley JAV, Millar IR, Alspach PA, Plunkett GM (1988) Impact of *Vespula* wasps on native insects and birds: first year research report. Contract report to the Department of Conservation, Ecology Division Rep 13. DSIR, Nelson, New Zealand

Moller H, Tilley JAV, Thomas BW, Gaze PD (1991a) Effect of introduced social wasps on the standing crop of honeydew in New Zealand beech forests. NZ J Zool 18:171–179

Moller H, Tilley JAV, Plunkett GM, Clapperton BK (1991b) Nest sites of common and German wasps (Hymenoptera: Vespidae). NZ J Zool 18:121–125

Moller H, Tilley JAV, Bell R, Thomas BW, Toft RJ (1996a) Responses of honeyeater birds to fluctuations in honeydew in a New Zealand beech forest. In: Moller H, Butz Huryn V (eds) Beekeeping and conservation values of protected natural areas. Dunedin, New Zealand, Univ Otago Wildl Manage Rep 51:42–64

Moller H, Toft RJ, Thomas BW, Tilley JAV (1996b) Honeydew renewal rates in beech forests of South Island, New Zealand. In: Moller H, Butz Huryn V (eds) Beekeeping and conservation values of protected natural areas. Univ Otago Wildl Manage Rep 51:65–82

Morales CF (1991) Margarodidae (Insecta: Hemiptera). Fauna of New Zealand, vol 21. DSIR Plant Protection Division, Auckland, New Zealand

Morales CF, Hill MG, Walker AK (1988) Life history of the sooty beech scale (*Ultracoelostoma assimile*)(Maskell), (Hemiptera: Margarodidae) in New Zealand *Nothofagus* forests. NZ Entomol 11:24–37

O'Donnell CFJ, Dilks PJ (1994) Foods and foraging of forest birds in temperate rainforest, South Westland, New Zealand. NZ J Ecol 18:87–107

O'Dowd DJ, Green PT, Lake PS (2003) Invasional 'meltdown' on an oceanic island. Ecol Lett 6(9):812–817

Pedersen LB, Bille-Hansen J (1995) Effects of nitrogen load to the forest floor in sitka spruce stands (*Picea sitchensis*) as affected by difference in deposition and spruce aphid infestations. Water Air Soil Pollut 85:1173–1178

Power ME, Tilman D, Estes JA, Menge BA, Bond WJ, Mills LS, Daily G, Castilla JC, Lubchenco J, Paine RT (1996) Challenges in the quest for keystones. BioScience 46:609–620

Sandlant GR, Moller H (1989) Abundance of common and German wasps (Hymenoptera: Vespidae) in the honeydew beech forests of New Zealand in 1987. NZ J Zool 16:333–343

Scheu S (2001) Plants and generalist predators as links between the below-ground and above-ground system. Basic Appl Ecol 2:3–13

Stadler B, Michalzik B, Müller T (1998) Linking aphid ecology with nutrient fluxes in a coniferous forest. Ecology 79:1514–1525

Thomas CD, Moller H, Plunkett GM, Harris RJ (1990) The prevalence of introduced *Vespula vulgaris* wasps in a New Zealand beech forest community. NZ J Ecol 13:63–72

Toft RJ, Rees JS (1998) Reducing predation of orb-web spiders by controlling common wasps (*Vespula vulgaris*) in a New Zealand beech forest. Ecol Entomol 23:90–95

Townes H (1972) A light-weight malaise trap. Entomol News 83:239–247

Wardle J (1984) The New Zealand beeches: Ecology, utilisation, and management. New Zealand Forest Service, Wellington, New Zealand

Wardle DA (2002) Islands as model systems for understanding how species affect ecosystem properties. J Biogeogr 29:583–591

Wardle DA, Barker GM, Yeates GW, Bonner KI, Ghani A (2001) Introduced browsing mammals in natural New Zealand forests: aboveground and belowground consequences. Ecol Monogr 71:587–614

Wardle DA, Bardgett RD, Klironomos JN, Setälä H, van der Putten WH, Wall DH (2004) Ecological linkages between aboveground and belowground biota. Science 304(5677):1629–1633

Wilson PR, Karl BJ, Toft RJ, Beggs JR, Taylor RH (1998) The role of introduced predators and competitors in the decline of kaka (*Nestor meridionalis*) populations in New Zealand. Biol Conserv 83:175–185

19 Consequences of Alien N_2-Fixers on Vegetation Succession in New Zealand

J.C. McQueen, W.C. Tozer and B.D. Clarkson

19.1 Introduction

Soil nitrogen is one of the most important nutrients determining primary productivity (Vitousek and Howarth 1991; Chapin et al. 1996; Walker and del Moral 2003), and therefore rates of vegetation succession. Nitrogen (N_2)-fixing species are generally thought to be an important part of the process of increasing N levels in an ecosystem during primary succession (Stevens and Walker 1970; Walker and del Moral 2003). In particular, N_2-fixing shrubs are often responsible for rapid accretion of soil N (van Cleve et al. 1971; Vitousek and Walker 1989; Bellingham et al. 2001), but not all seres have a N_2-fixing stage involving vascular plants (Vitousek et al. 1989). For example, N_2-fixing vegetation was dominant in only 28% of 141 primary seres examined by Walker (1993).

Successional trajectories are influenced strongly by the types of plant functional groups represented, e.g., shrubs versus moss/lichen cover (Walker and del Moral 2003). Alien N_2-fixing species, with the ability to change soil fertility, and hence potentially facilitate a different suite of species to a successional sere, might be expected to have far-reaching consequences on natural successional processes, and ultimately ecosystem function. In N-limited systems, an increase in N often results in a decrease in the number of plant species, as fast-growing weed species overgrow and shade slower-growing natives (Aerts and Berendse 1988; Maron and Connors 1996). An increase in fertility can also facilitate the invasion of other weed species, as has been shown after invasion by *Myrica faya* in Hawai'i (Vitousek and Walker 1989). Invasive weeds that alter soil fertility could also be expected to impact greatly on native species that occupy the early stages of primary succession and are tolerant of low nutrients. Such impacts mean plant invasions into natural communities, in particular by woody species, are considered one of the most serious threats to the survival of biological communities (Heywood 1989).

Ecological Studies, Vol. 186
R.B. Allen and W.G. Lee (Eds.)
Biological Invasions in New Zealand

In this chapter, the consequences of alien N_2-fixers on successional sequences under previously forested sites in New Zealand are considered. First, the ecology of the few native N_2-fixers is described. We then discuss the alien woody N_2-fixers now prevalent in New Zealand, and their role and influence on vegetation succession, including future successional trajectories. We have limited the focus of this chapter to the implications of invasive plants on vegetation processes, specifically succession of natural or semi-natural habitats.

19.2 Native N_2-Fixers

The New Zealand flora is depauperate (Danserau 1964; Dawson 1988; Lee 1998), having a poor representation of certain functional types, e.g., cold-tolerant tall trees, nitrogen-fixing woody legumes, and species with "r-selected" characteristics that act as pioneers in successional sequences (Danserau 1964; Lee 1998). These functional gaps in the flora were filled by the evolution of species with wide tolerance (Dawson 1988). N_2-fixing species of *Carmichaelia* and *Discaria* (woody legumes), *Coriaria* (actinorhizal shrubs), and the *Nostoc*-containing *Gunnera*, plus some non-N_2-fixing species now occupy ecological niches that elsewhere are filled by legumes (Silvester 1976). Although the N_2-fixing species have probably played an important part in maintaining the nutrient status of New Zealand soils, they may be less well adapted than many alien species in carrying out this role. They appear unable to displace herbaceous vegetation or rapidly invade open habitats (Danserau 1964). In addition, like much of the New Zealand flora, some of these N_2-fixing species are sensitive to herbivory by introduced animals (Grüner 2003).

Seven N_2-fixing plant genera are indigenous to New Zealand (Table 19.1). These N_2-fixers (predominantly the widespread *Carmichaelia* and *Coriaria* species) can enhance the establishment of other species. *Carmichaelia*, for example, has been shown to increase total N in soil, and facilitate the growth of *Griselinia* and *Weinmannia* (Bellingham et al. 2001). There are 23 endemic species of *Carmichaelia* (Fabaceae) in New Zealand (Heenan and de Lange 1999), and 14 of these are threatened or uncommon (Grüner and Heenan 2001). Most species occur in drought- or frost-prone areas, predominantly alluvial habitats (Wagstaff et al. 1999), and invade disturbed sites in shallow, poorly developed soils (Heenan and de Lange 1999; Wagstaff et al. 1999). Historically, *Carmichaelia* is likely to have fulfilled an important role in natural successional processes at disturbed sites (Wardle 1991; Bellingham et al. 2001). *Carmichaelia* seeds are long-lived and viability is high (100 % viability was found in some species after 24 years of storage; Grüner and Heenan 2001), and stands senesce when around 100 years old, facilitating the establishment of further successional species (Bellingham et al. 2001). However, most *Carmichaelia* species probably have little impact on successional

Table 19.1 Native N_2-fixing genera in New Zealand and their symbionts

Family	Genus (number of species)	Symbiont
Fabaceae	*Carmichaelia* (23)	*Rhizobium*
Coriariaceae	*Coriaria* (9)	*Frankia*
Rhamnaceae	*Discaria* (1)	*Frankia*
Fabaceae	*Chordospartium* (1)	*Rhizobium*
Fabaceae	*Clianthus* (2)	*Rhizobium*
Fabaceae	*Sophora* (8)	*Rhizobium*
Gunneraceae	*Gunnera* (10)	*Nostoc*

Table 19.2 Nitrogen-fixation rates and quantity of N added to soils by native and exotic species in New Zealand. Note that estimations of fixation rates will vary with the methods used, and thus are only indicative

	Vegetation type	N fixation estimation (kg N ha^{-1} $year^{-1}$)	N quantity in soils (kg ha^{-1})	Reference
Native	Lichens and algae	1.5–5		Gadgil (1982), McCurdy (2001)
	Coriaria arborea	150	3,000–4,000	Silvester (1976)
	Carmichaelia angustata	Net N mobilization	–	Wardle and Greenfield (1991)
	Gunnera	72	–	Silvester and Smith (1969)
Exotics	*Trifolium repens* (in pasture)	700		Ball (1982)
	Ulex europaeus	100–200	4,000–5,000[a]	Egunjobi (1969)
	Lupinus arboreus	160	–	Gadgil (1982)
	Alnus species	40–300	–	Silvester (1976)
	Cytisus scoparius	111	–	Watt et al. (2003)

[a] Note that the actual age of the *Ulex europaeus* stands in the study by Egunjobi (1969) was unknown, and therefore the input of N from *U. europaeus* may be overestimated

sequences today because of their limited geographic range and abundance (Thomas and Spurway 2001), susceptibility to herbivory by introduced animals (Grüner 2003), and low measured rates of N_2 fixation (Table 19.2).

There are five common species of *Coriaria* (Coriariaceae) in New Zealand (Good 1930), covering a wide range of habitats (Daly et al. 1972), including scrub, roadsides, screes, dry riverbeds and recent lava flows (Thompson and

Gornall 1995). Mammalian herbivore impacts on *Coriaria* are limited, as it contains the highly toxic substance tutin (Silvester 1976). However, susceptibility to frost (with the exception of the mountainous *C. pteridoides*; Daly et al. 1972), soil moisture deficits (Walker et al. 2003), and a requirement for high light levels (Silvester 1976) may be important limiting factors in its establishment. *Coriaria* has a low seed viability but germination rates are rapid under select conditions (Burrows 1995). *Coriaria arborea* is known to fix large amounts of N (at least 150 kg N ha^{-1} $year^{-1}$), comparable with that of exotic legume species, and can add considerable quantities of N to soils (Silvester 1976; Walker et al. 2003; Table 19.2). Vegetation growing in close association (up to 5 m) with *C. arborea* on recent volcanic basaltic ash and lapilli, Mt. Tarawera, obtains the majority of its N supply from *C. arborea*-fixed N_2 (Tozer 2001). *Coriaria* establishes after the arrival of small herbs and shrubs that ameliorate a harsh physical environment, and facilitates primary succession by inputs of soil nutrients and organic matter (Walker et al. 2003). *Coriaria* regeneration is inhibited at sites where mature *Coriaria* is present, and *Coriaria* is replaced quickly in the successional sequence with broadleaved shrubs, compared with areas where *Coriaria* does not establish (Walker et al. 2003).

The genus *Sophora* (Fabaceae) consists of eight endemic N_2-fixing species, the most widespread being *S. microphylla* and *S. tetraptera* (kowhai; Thomas and Spurway 2001). Distributed throughout New Zealand, these species are limited to riparian or forest edges, and are not vigorous pioneer species (Silvester 1968). *Clianthus puniceus* (kaka beak) is also N_2-fixing, but is rare and threatened (de Lange et al. 1999). The native broom *Chordospartium muritai* (Fabaceae) is confined to a single population of 30 individuals in the northern South Island by rabbit browsing and competition from introduced grasses (Williams et al. 1996). A slow growth rate and short life suggest that this species has never played an important role as a pioneer species in New Zealand.

Discaria toumatou (Rhamnaceae) is important in montane grassland of the South Island high country, where it may facilitate the establishment of shrub species (Bellingham 1998). *D. toumatou* is an actinorhizal, small-leaved, spiny deciduous shrub, having seeds that are dispersed by explosive dehiscence. Because the plants are spiny, they are vulnerable to mammalian grazing only when young.

Gunnera (Gunneraceae) are prostrate, small-leaved herbs (Allan 1961), widely distributed over New Zealand in damp areas, except for *G. hamiltonii* and *G. arenaria* that are found in sand dunes and dune hollows. *Gunnera* is a pioneer species in wet areas, and its significant N_2-fixation ability results from the cyanobacteria symbiont, *Nostoc* (Table 19.2; Silvester and Smith 1969).

19.3 Alien N_2-Fixers

Alien N_2-fixing species generally have high fixation rates, compared with indigenous N_2-fixers (Wardle and Greenfield 1991; Table 19.2). These introduced species have also been observed to out-compete many New Zealand natives (including N_2-fixers) by gathering larger amounts of resources through greater height, more vigorous growth or reproduction, superior dispersal abilities, different environmental tolerances, or resistance to herbivory (Wardle 1991). The consequences of alien N_2-fixers on the trajectory of successional sequences are poorly known. A large number of N_2-fixing taxa have been introduced to New Zealand mainly for agriculture and soil restoration. Over 40 genera of legumes have been introduced, the most speciose genera being *Trifolium* (27), *Racosperma* (15), *Vicia* (12), *Lathyrus* (nine), *Lotus* (eight), and *Medicago* (eight; P.A. Williams, unpublished data). Specific examples of invasive alien N_2-fixers and their effects on vegetation patterns are discussed below.

19.3.1 *Ulex europaeus*

Originally introduced for shelter, *Ulex europaeus* (gorse) was declared a noxious weed in 1900, and is perhaps one of the most visible weeds in the New Zealand landscape, covering 3.5% of the national land area (Rees and Hill 2001). *U. europaeus* produces a continuous canopy, lasting 25–30 years, and soil N accumulation can reach up to 5,000 kg N ha^{-1}, compared with only 3,000 kg N ha^{-1} under the native *Weinmannia racemosa* and *Leptospermum scoparium* (Egunjobi 1969).

Croker (1953) and Druce (1957) recorded vegetation composition in Wellington (cool, wet climate) from differently aged sites, and developed a conceptual view of successional trajectories with, and without gorse following fire. Where *U. europaeus* was the primary colonizer, it was postulated that after about 50 years *U. europaeus* would be replaced by *Pseudopanax arboreus*, followed by a broadleaf forest dominated by *Weinmannia racemosa*. If *Nothofagus truncata* seedlings fail to establish at the pioneer phase, the broadleaved vegetation that develops after *U. europaeus* invasion effectively excludes *Nothofagus* from future successional trajectories (Wilson 1994). Where the site was initially colonized by *L. scoparium*, *Pteridium esculentum* or resprouted *W. racemosa*, or any combination of these, a *W. racemosa*–*Knightia excelsa* forest formed after 55–70 years. If the native N_2-fixer *Coriaria* was present in the pioneer stages, a broadleaved forest appeared in a shorter time (50 years) than under other natives. If other exotics, such as *Erica lusitanica* (Spanish heath) or *Pinus radiata* (pine), also established during the pioneer phase, they could delay the arrival of *W. racemosa* dominance for

30 years, preventing the inclusion of a *Leptospermum scoparium–Pteridium esculentum* stage (Druce 1957).

Some of these postulated successional trajectories have been recently confirmed in the Wellington and Nelson regions (J.J. Sullivan et al., unpublished data). Both *U. europaeus* and *Kunzea ericoides* developed a subcanopy of similar-height broadleaved species; however, the species composition differed, with lower species richness in *U. europaeus* and fewer orchids. The results suggest that the successional trajectories under *U. europaeus* and *K. ericoides* do not converge within a century.

Since Egunjobi's (1967) study, there has been a widely held view that N accumulation associated with *U. europaeus* deflects successional sequences, potentially leading to a different vegetation composition. However, in this study, multiple generations of *U. europaeus* may have existed at the sites, elevating N levels considerably. In addition, the differences in N quantities seen between *U. europaeus* and the native-small-tree *Leptospermum scoparium* stands were only evident in very young stands of *U. europaeus* (<10 years). Older *U. europaeus* stands and 16-year-old *L. scoparium* stands showed no difference in the amount of N in the soil; Williams and Karl (2002) also found no evidence that N accumulation in *U. europaeus* stands exceeded that found under the native *Kunzea ericoides*. Nitrogen availability is likely to be more important to successional developments than are N pools, but no studies have measured this.

In some habitats, *U. europaeus* may dominate long-term in high-disturbance areas where continuous *U. europaeus* germination will prevent establishment of natives (Lee et al. 1986). In addition to its N_2-fixing abilities, *U. europaeus* shows a number of "r-selected" characteristics that facilitate re-establishment following either localized or widespread disturbance, including high seed production, a long-lived (>100 years) seed bank (Hill et al. 2001), and a rapid growth rate (Ogle-Mannering 1995). Establishment of native seedlings occurs as *U. europaeus* stands age and the canopy thins, which requires the absence of disturbance for longer than 30 years. In the northern regions of New Zealand, with faster growth and decomposition, succession is likely to be more rapid. Native woody species in the colder, drier climate of Dunedin were excluded by the high density of *U. europaeus* plants and deep litter, resulting in sparse broadleaved seedlings in *U. europaeus* communities, in spite of close proximity to seed sources (Lee et al. 1986).

As the work of Druce (1957) and others has shown, successional trajectories to indigenous broadleaved forest appear, overall, to be accelerated by the presence of *U. europaeus*, compared with sequences in the absence of N_2-fixers. Broadleaved forest is associated with relatively high soil fertility (McQueen 1993). Thus, one outcome of N_2-fixation by the widespread *U. europaeus* cover in New Zealand may be an increase in species better adapted to nutrient-rich soils. Although *U. europaeus* is generally controlled and eradicated, especially in agricultural settings, it is also noted for its role as a potential “nurse” species in succession to indigenous forest on previously cleared sites (Williams 1979). Current conservation management often leaves *U. europaeus* undisturbed in potentially forested areas, where the aim is a return to forest (Lee et al. 1986).

19.3.2 *Cytisus scoparius*

Cytisus scoparius (broom) is another widespread, invasive weed in New Zealand, capable of producing vast quantities of seed (Williams 1981). In New Zealand, *C. scoparius* has a wide tolerance of soil conditions, grows for most of the year, has two flowering periods at low altitudes, an absence of seed predators, can set seeds over a wide altitudinal range (Williams 1981), and forms monocultural stands for several decades (Williams 1981), although N_2-fixation rates are unknown.

Cytisus scoparius may facilitate the growth of natives in areas of high rainfall where forest remnants occur nearby (McQueen 1993). Succession from a mixed *C. scoparius–U. europaeus* stand to a low forest of *Melicytus ramiflorus* within 50 years has been described by Williams (1983), a transition facilitated by the alien nitrogen-demanding tree, *Sambucus nigra*, which suppressed *C. scoparius*. In much of New Zealand, early-successional species are replaced by bird-dispersed, broadleaved species such as *Coprosma robusta* (karamu) and *Pseudopanax arboreus*. In coastal Canterbury, however, the bird-dispersed exotic *Sambucus canadensis* gradually displaces *C. scoparius*, which dies out after 15 years. The arrival of *S. canadensis* delays the establishment of native forest by 30 years (Williams 1983). The native *Melicytus ramiflorus*, a nutrient-demanding species, can eventually suppress the *S. canadensis* 10–20 years after it arrives in the successional sequence (Williams 1983). The *Melicytus ramiflorus* forest that eventually establishes is associated with high-fertility soils, thus potentially bypassing intermediate successional seres.

Compared with *U. europaeus*, faster establishment of broadleaved shrubs under *C. scoparius* might be expected due to the smaller depth of readily decomposable litter (Williams 1983). However, after 8 years, there were no woody species entering *C. scoparius* stands on the Port Hills (Christchurch; Partridge 1992). Native forest may establish at this site only through amelioration of drought conditions in summer, or after another species (such as the alien elder, *Sambucus canadensis*) provides a microclimate for establishment (Partridge 1992).

In eastern montane grassland, *C. scoparius* competes with the indigenous N_2-fixing shrub *Discaria toumatou* (matagouri; Bellingham 1998), and is maintained by periodic burning. *D. toumatou* facilitates establishment of other exotic shrubs, mainly bird-dispersed species, by providing perch sites. There is no pattern of association between *C. scoparius* and other shrubby species in this area (Williams 1984). Due to its superior growth rates and less-specific site requirements for seed establishment, it is likely that *C. scoparius* will come to dominate after fire (Bellingham 1998).

19.3.3 Other N_2-Fixing Aliens

Other N_2-fixing aliens in New Zealand are also likely to be influencing successional processes. *Lupinus arboreus* (tree lupine) facilitates soil formation through its ability to fix N_2, produce large amounts of leaf litter, and because of its dense root system (Asplin and Fuller 1985). *L. arboreus* improves the stabilization of sand dunes and the formation of organic matter in soils, which allows other, mainly introduced, plant species to establish (Asplin and Fuller 1985), particularly alien pines, which have enhanced growth where they co-exist with L. arboreus (Sprent and Silvester 1973). However, *L. arboreus* suppresses most indigenous foreshore plants, and creates deep shade that restricts succession to native forest (McQueen 1993).

The ridges of the Kaimaumau Gumland, Northland, have been heavily invaded by the Australian N_2-fixer, *Acacia longifolia*. Here, *A. longifolia* is likely to remain the dominant species as native species can not establish through the dense litter layer (McQueen 2000). When the litter layer is physically disturbed, two alien shrubs, *U. europaeus* and *Solanum mauritianum*, establish (McQueen 2000). Although bird-dispersed native species generally replace *U. europaeus*, the paucity of local seed sources near Kaimaumau is likely to exclude natives from the succession (McQueen 2000). Intermittent fire also consolidates *A. longifolia* dominance, preventing further vegetation change.

Gunnera tinctoria, a N_2-fixing megaherb, is a garden escapee (Heenan et al. 2002) that is now well established on slip faces and coastal cliffs of south Taranaki. *G. tinctoria* displaces native plants because shading by its massive leaves (up to several meters wide) prevents establishment of other plants. It is likely that successional trends will be irreversibly altered by *G. tinctoria*'s presence.

19.4 Vegetation Changes Caused by N_2-Fixing Aliens

Growing evidence indicates that vegetation changes caused by alien N_2-fixers are clearly changing successional rates and trajectories in many New Zealand ecosystems. There are two reasons why such impacts may occur widely in New Zealand as a result of the expanding cover of alien N_2-fixing species. First, alien N_2-fixing species appear to dominate across a broader range of environments than do native N_2-fixing species, and as such will influence a greater range of successions. Second, alien N_2-fixing species generally enhance the rates of N input above that of native N_2-fixing species. This may be particularly important in low-N habitats, such as raw mineral surfaces widely formed in New Zealand as a result of landslides, flooding, and volcanic ash deposition. Alien N_2-fixers displace distinctive indigenous species adapted to the early stages of succession and shift compositional trajectories, which may favor the rapid development of either exotic woody species or native broadleaved forest communities, potentially at the expense of other types of native plant community.

The mechanisms whereby N_2-fixing aliens are altering successional rates and trajectories are poorly understood. The presence of a N_2-fixing shrub alters a suite of factors, including soil fertility, light regimes, litter depth, and mycorrhizal assemblages (Walker and del Moral 2003), and many alien species are less susceptible to mammalian herbivory and fire than are native species. In non-forest ecosystems, alien N_2-fixers may have perpetual dominance, especially if disturbances favoring their persistence continue.

19.5 Conclusions

Nitrogen-fixing alien species that alter ecosystem processes change the competitive environment for all organisms in the area (Vitousek 1990). With the presence of alien N_2-fixers over much of the landscape, and the fertility of 48 % of New Zealand's land area enhanced through the widespread application of fertilizers (Hewitt 1999), there is likely to be a marked vegetation homogenization of the landscape. The impacts of alien species are not readily reversible, and alien species are likely to persist when they have long-lived seed banks. In the long term, in those parts of New Zealand where disturbance is frequent, it may well be that native species may never again dominate. These processes are almost certainly leading to a loss of species diversity on both regional and local scales. Reversing these effects will require large-scale human intervention to restore some sort of "natural" community.

Acknowledgements The authors thank the following people for their helpful comments and advice on the manuscript (please note that any misinterpretation of their comments is the authors' responsibility): Bill Lee, Duane Peltzer, Warwick Silvester, Lawrence Walker and Peter Williams. Peter Williams deserves special thanks for making substantial comments on the status of gorse research in New Zealand, and providing us with some of his unpublished research.

References

Aerts R, Berendse F (1988) The effect of increased nutrient availability on vegetation dynamics in wet heathlands. Vegetatio 76:63–69

Allan HH (1961) Flora of New Zealand, vol I. Government Printer, Wellington, New Zealand

Asplin SJ, Fuller J (1985) Study of dune plant communities at the mouth of the Waipoua River. Tane 31:31–39

Ball PR (1982) Nitrogen balances in intensively managed pasture systems. In: Gandar PW, Bertraud DS (eds) Nitrogen balances in New Zealand ecosystems. DSIR, Palmerston North, New Zealand, pp 47–66

Bellingham PJ (1998) Shrub succession and invasibility in a New Zealand montane grassland. Aust J Ecol 23:562–573

Bellingham PJ, Walker LR, Wardle DA (2001) Differential facilitation by a nitrogen-fixing shrub during primary succession influences relative performance of canopy tree species. J Ecol 89:861–875

Burrows CJ (1995) Germination behaviour of the seeds of four New Zealand species of *Coriaria* (Coriariaceae). NZ J Bot 33:265–275

Chapin FS III, Torn MS, Tateno M (1996) Principles of ecosystem sustainability. Am Nat 148:1016–1037

Croker BH (1953) Forest regeneration on the Western Hutt Hills, Wellington. Trans R Soc NZ 81:11–21

Daly GT, Smith BE, Chua S (1972) Carbon and nitrogen fixation in rhizomatous species of *Coriaria*. Proc NZ Ecol Soc 19:65–74

Danserau P (1964) Six problems in New Zealand vegetation. Bull Torrey Bot Club 91:114–140

Dawson J (1988) Forest vines to snow tussocks: the story of New Zealand plants. Victoria Univ Press, Wellington, New Zealand

De Lange PJ, Heenan PB, Given DR, Norton DA, Ogle CC, Johnson PN, Cameron EK (1999) Threatened and uncommon plants of New Zealand. NZ J Bot 37:603–628

Druce AP (1957) Botanical survey of an experimental catchment, Taita, New Zealand. DSIR Bull 124:1–81

Egunjobi JK (1967) A comparison of primary productivity and chemical element cycling in gorse (*Ulex europaeus* L.) and pasture ecosystems. PhD Thesis, Victoria University of Wellington, New Zealand

Egunjobi JK (1969) Dry matter and nitrogen accumulation in secondary successions involving gorse (*Ulex europaeus* L.) and associated shrubs and trees. NZ J Sci 12:175–193

Gadgil RL (1982) Biological nitrogen inputs to exotic forests. In: Gandar PW, Bertraud DS (eds) Nitrogen balances in New Zealand ecosystems. DSIR, Palmerston North, New Zealand, pp 123–130

Good RD'O (1930) The geography of the genus *Coriaria*. New Phytol 29:170–198

Grüner IG (2003) Comparative ecology and conservation of rare native broom, *Carmichaelia* (Fabaceae), South Island, New Zealand. PhD Thesis, University of Canterbury, New Zealand

Grüner I, Heenan PB (2001) Viability and germination of seeds of *Carmichaelia* (Fabaceae) after prolonged storage. NZ J Bot 39:125–131

Heenan PB, de Lange PJ (1999) Reproductive biology, ecology and conservation of *Carmichaelia williamsii* (Fabaceae), a vulnerable legume from New Zealand. Pac Conserv Biol 5:179–190

Heenan PB, de Lange PJ, Cameron EK, Champion PD (2002) Checklist of dicotyledons, gymnosperms, and pteridophytes naturalised or casual in New Zealand: additional records 1999–2000. NZ J Bot 40:155–174
Hewitt AE (1999) Our valuable soils. NZ Soil News 47:98–102
Heywood VH (1989) Patterns, extents and modes of invasions by terrestrial plants. In: Drake JA, Mooney HA, di Castri F, Groves RH, Kruger FJ, Rejmánek M, Williamson M (eds) Biological invasions: a global perspective. Wiley, Chichester, pp 31–60
Hill RL, Gourlay AH, Barker RJ (2001) Survival of *Ulex europaeus* seeds in the soil at three sites in New Zealand. NZ J Bot 39:235–244
Lee WG (1998) The vegetation of New Zealand – functional, spatial, and temporal gaps. In: Lynch R (ed) Ecosystems, entomology and plants. R Soc NZ Misc Ser 48:91–101
Lee WG, Allen RB, Johnson PN (1986) Succession and dynamics of gorse (*Ulex europaeus* L.) communities in the Dunedin Ecological District South Island, New Zealand. NZ J Bot 24:279–292
Maron JL, Connors PG (1996) A native nitrogen-fixing shrub facilitates weed invasion. Oecologia 105:302–312
McCurdy SM (2001) Lichens and their potential as a nitrogen source in a regenerating forest in Tongariro National Park, New Zealand. MSc Thesis, University of Waikato, Hamilton, New Zealand
McQueen DR (1993) A review of interaction between naturalised woody plants and indigenous vegetation in New Zealand. Tuatara 32:32–49
McQueen JC (2000) *Acacia longifolia* invasion in the Kaimaumau Gumland, Northland, New Zealand. MSc Thesis, University of Waikato, Hamilton, New Zealand
Ogle-Mannering M (1995) Establishment of woody species in the gorse-forest succession. PhD Thesis, University of Otago, Dunedin, New Zealand
Partridge TR (1992) Successional interactions between bracken and broom on the Port Hills, Canterbury, New Zealand. J Appl Ecol 29:85–91
Rees M, Hill RL (2001) Large-scale disturbances, biological control and the dynamics of gorse populations. J Appl Ecol 38:364–377
Silvester WB (1968) Nitrogen fixation by root nodules of *Coriaria*. PhD Thesis, Lincoln University, New Zealand
Silvester WB (1976) Ecological and economic significance of the non-legume symbiosis. In: Newton WE, Nyman CJ (eds) Proc 1st Int Symp Nitrogen Fixation. Washington State University Press, Pullman, WA, pp 489–507
Silvester WB, Smith DR (1969) Nitrogen fixation by *Gunnera-Nostoc* symbiosis. Nature 224:1231
Sprent JI, Silvester WB (1973) Nitrogen fixation by *Lupinus arboreus* grown in the open and under different aged stands of *Pinus radiata*. New Phytol 72:991–1003
Stevens PR, Walker TW (1970) The chronosequence concept and soil formation. Q Rev Biol 45:333–350
Thomas MB, Spurway MI (2001) Kowhai (*Sophora* species) and other nitrogen-fixing plants of New Zealand. Combined Proc Int Plant Propagators Soc 51:94–97
Thompson PN, Gornall RJ (1995) Breeding systems in *Coriaria* (Coriariaceae). Bot J Linnean Soc 117:293–304
Tozer W (2001) Nitrogen isotope ratios in nitrogen limited communities: determining origins of nitrogen. MSc Thesis, University of Waikato, Hamilton, New Zealand
Van Cleve K, Viereck LA, Schlentner RL (1971) Accumulation of nitrogen in alder (*Alnus*) ecosystems near Fairbanks, Alaska. Artic Alpine Res 3:101–114
Vitousek PM (1990) Biological invasions and ecosystem processes: towards an integration of population biology and ecosystem studies. Oikos 57:7–13
Vitousek PM, Howarth RW (1991) Nitrogen limitation on land and in the sea: how can it occur? Biogeochemistry 13:87–115

Vitousek PM, Walker LR (1989) Biological invasion by *Myrica faya* in Hawai'i: plant demography, nitrogen fixation, ecosystem effects. Ecol Monogr 59:247–265

Vitousek PM, Matson PA, van Cleve K (1989) Nitrogen availability and nitrification during succession: primary, secondary, and old-field seres. Plant Soil 115:229–239

Wagstaff SJ, Heenan PB, Sanderson MJ (1999) Classification, origins, and patterns of diversification in New Zealand Carmichaelinae (Fabaceae). Am J Bot 86:1346–1356

Walker LR (1993) Nitrogen fixers and species replacements in primary succession. In: Miles J, Walton DWH (eds) Primary succession on land. Blackwell, Oxford, pp 249–272

Walker LR, del Moral R (2003) Primary succession and ecosystem rehabilitation. Cambridge Univ Press, Cambridge

Walker LR, Clarkson BD, Silvester WB, Clarkson BR (2003) Colonization dynamics and facilitative impacts of a nitrogen-fixing shrub in primary succession. J Veg Sci 14:277–290

Wardle P (1991) Vegetation of New Zealand. Cambridge Univ Press, Cambridge

Wardle DA, Greenfield LG (1991) Release of mineral nitrogen from plant root nodules. Soil Biol Biochem 23:827–832

Watt MS, Clinton PW, Whitehead D, Richardson B, Mason EG, Leckie AC (2003) Aboveground biomass accumulation and nitrogen fixation of broom (*Cytisus scoparius* L.) growing with juvenile *Pinus radiata* on a dryland site. For Ecol Manage 184:93–104

Williams PA (1979) The role of gorse in the regeneration of native bush on the Port Hills. Botany Division DSIR Rep 307. Landcare Research Library, Lincoln, New Zealand

Williams PA (1981) Aspects of the ecology of broom (*Cytisus scoparius*) in Canterbury, New Zealand. NZ J Bot 19:31–43

Williams PA (1983) Secondary vegetation succession on the Port Hills Banks Peninsula, Canterbury, New Zealand. NZ J Bot 21:237–247

Williams PA (1984) Flowering currant (*Ribes sanguineum*) shrublands in the lower Waitaki Valley, South Canterbury. NZ J Agric Res 27:473–478

Williams PA, Karl BJ (2002) Birds and small mammals in kanuka (*Kunzea ericoides*) and gorse (*Ulex europaeus*) scrub and the resulting seed rain and seedling dynamics. NZ J Ecol 26:31–41

Williams PA, Norton DA, Nicholas JM (1996) Germination and seedling growth of an endangered native broom, *Chordospartium muritai* A.W. Purdie (Fabaceae), found in Marlborough, South Island, New Zealand. NZ J Bot 34:199–204

Wilson HD (1994) Regeneration of native forest on Hinewai Reserve, Banks Peninsula. NZ J Bot 32:373–383

20 Hidden Effects: the Belowground Consequences of Introduced Browsing Mammals in New Zealand Forests

D.A. Wardle

20.1 Introduction

Over the past decade, there has been an increasing interest in understanding the linkages between aboveground and belowground components of ecosystems. This includes consideration of how the producer and decomposer subsystems influence one another, and the consequences of this for ecosystem properties (van der Putten et al. 2001; Wardle 2002). The organisms of both the aboveground and belowground components are organized into food webs, and there has been growing recent recognition of both the aboveground consequences of belowground food web relationships (e.g., Laakso and Setälä 1999; Scheu et al. 1999), and the belowground consequences of aboveground food web associations (reviewed in Wardle 2002; Bardgett and Wardle 2003). Foliar herbivory represents the primary trophic relationship in the aboveground food web, and has been shown in several ecosystems to operate as an important ecosystem driver (e.g., McNaughton et al. 1988; Pastor et al. 1988). Appreciation of how foliar herbivory affects the decomposer subsystem may increase our understanding of the controls of decomposition and nutrient supply in ecosystems.

Several species of browsing mammals were liberated throughout New Zealand between the 1770s and 1920s, including feral goats (*Capra hircus*) and deer (particularly European red deer, *Cervus elaphus scoticus*; see Chap. 4). Prior to this, no browsing mammals were present in New Zealand. New Zealand's native megaherbivores, moas (Aves, Dinornithiformes), became extinct following Polynesian colonization of New Zealand 800–1,000 years ago. New Zealand presents an almost unique opportunity to investigate the ecosystem consequences of the introduction of an entire functional group of alien organisms (forest-dwelling browsing mammals) to an ecosystem from which that group was previously absent.

Ecological Studies, Vol. 186
R.B. Allen and W.G. Lee (Eds.)
Biological Invasions in New Zealand

In this article, the mechanistic basis of how foliar herbivory affects the decomposer subsystem is first considered. The possible belowground effects of introduced browsing mammals (notably deer and goats) in New Zealand "natural" forests are then discussed within this context. Finally, the conservation implications of the belowground consequences of herbivory by browsing mammals in New Zealand forests are evaluated.

20.2 How Herbivores Affect Decomposers

20.2.1 Mechanisms

Foliar herbivory, and defoliation, influence organisms in the decomposer subsystem mainly through determining the quantity and quality of resources that are returned to the soil. This can operate through a number of mechanisms, and at a range of spatial and temporal scales, and these can have both positive and negative effects on decomposer organisms and processes (Table 20.1).

In the short term, defoliation often induces plants to allocate more resources belowground. Several experimental studies have found that defoliation of herbaceous plants results in a greater rate of exudation of carbon into the rhizosphere and associated rooting zone (e.g. Bokhari and Singh 1974; Holland et al. 1996). This may explain why defoliation or herbivory often stimulates soil organisms even when total root productivity is impaired (e.g., Mawdsley and Bardgett 1997; Mikola et al. 2001). Increased root carbon exudation following herbivory has been proposed as promoting "fast" cycling of nutrients and selection for bacteria in the rooting zone, as opposed to "slow" cycles caused by more recalcitrant substances and domination by fungi (Bardgett et al. 1996). Whether this is a widespread phenomenon for woody plant species has yet to be tested.

In the longer term, herbivores can affect the resources entering the decomposer subsystem (and, therefore, decomposer organisms) through altering net primary productivity (NPP). In a quantitative literature synthesis, Milchunas and Lauenroth (1993) found that herbivore effects on aboveground NPP in grasslands were usually negative, but that many exceptions existed. Furthermore, this synthesis found that comparable numbers of available data sets showed reductions and enhancements of root mass in response to herbivore exclusion. In grassland systems at least, intermediate levels of herbivory often optimize NPP, although the importance and evolutionary significance of this has been debated (e.g., McNaughton 1985; Belsky 1986). The mechanistic basis through which herbivores can optimize NPP are poorly understood, although there is growing evidence that NPP can be maximized by herbivores when their effects on the circulation of nutrients

Table 20.1 Summary of mechanisms by which foliar herbivores may positively or negatively impact upon the quantity and quality of resources entering the decomposer subsystem

Resources	Effect category	Positive effects of herbivores	Negative effects of herbivores
Quantity of resources	Effects on rhizosphere carbon exudation	Enhancement of rhizosphere exudation	(None)
	Effects on net primary productivity (NPP)	Grazing optimization of NPP	Reduction of NPP through loss of tissues, or through increased exposure of residual foliage
Quality of resources	Return of feces and urine	Resources returned to soil in a form more labile than from plant litter	Reduced carbon-to-nitrogen ratio of fecal pellets
	Effects on litter quality through physiological response by plant	Increased concentrations of nutrients and reduced secondary metabolites in shoot and root litter	Induced defenses resulting in greater concentrations of secondary metabolites in litter; reduced litter nutrient concentrations
	Effects on litter quality through influencing vegetation succession	Grazing optimization of flora preventing establishment by later successional plant species with poorer litter quality	Acceleration of succession through browsed plant species being disadvantaged, and replacement by species with poorer litter quality

within the system is considered (Dyer et al. 1993; de Mazancourt and Loreau 2000). It is expected that less productive systems, in which plants are unable to adequately compensate for loss of foliage, should only show negative responses of NPP to herbivory. The fact that foliar herbivory can either increase or reduce NPP, depending upon both context and intensity, has important consequences for the quantity of resources entering the soil as plant litter, and suggests that both positive and negative responses of decomposers to aboveground herbivores are possible.

In addition to mechanisms involving resource quantity, herbivores also affect the decomposer subsystem through altering resource quality. The most important mechanism of this type involves plant-derived carbon and nutrients being returned to the soil as feces and urine, rather than as plant litter. This is likely to be most important in highly productive systems (in which >50% of aboveground NPP can be consumed by herbivores), and unimportant in very unproductive systems (where <1% is consumed; McNaughton et al. 1989). Because carbon and nutrients in fecal material are generally far more accessible to decomposers than those in plant litter, this mechanism has been proposed as being of importance in both grasslands (e.g., Tracy and Frank 1998) and productive forests (e.g., Molvar et al. 1993). There are also instances in which fecal pellets may result in nutrient immobilization and retardation of soil processes, although this has been demonstrated only for invertebrate herbivore systems (e.g., Lovett and Ruesink 1995).

Herbivores can also alter the quality of resources entering the soil through inducing physiological changes in the plant. Herbivory has been shown to increase the concentration of nutrients in both leaf tissues (e.g., Ruess and McNaughton 1987) and root tissues (e.g., Seastedt 1985), either through reallocation strategies in response to defoliation, or through improved nutrient availability resulting from enhanced activity of soil organisms in the rooting zone (see Bardgett et al. 1998). This can be associated with reduced concentrations of secondary metabolites in leaves following browsing. Such mechanisms can have important consequences for the quality of litter produced by plants; for example, Kielland et al. (1997) and Kielland and Bryant (1998) found that mammalian browsing of trees in Alaska enhanced the quality and decomposition rates of litter produced by the trees. However, severe defoliation events can also result in increased concentrations of secondary metabolites and reduced nutrient concentrations in subsequently produced foliage (Tuomi et al. 1990). This could conceivably reduce plant litter quality, although this mechanism has seldom been investigated with regard to the decomposer subsystem (but see Findlay et al. 1996 with regard to invertebrate herbivores). The fact that foliar herbivory can both increase and decrease the quality of litter produced by plants can be explained in terms of either site fertility (where herbivory may promote litter quality in nutrient-replete conditions but reduce it in infertile conditions) or intensity of herbivory (with quality promoted under intermediate herbivory but reduced under more severe herbivory).

Herbivores can also induce longer-term effects on the quality of litter entering the decomposer subsystem through influencing plant succession. This is because, across plant species, palatability of foliage to herbivores and decomposability of leaf litter are driven by similar suites of traits, such as foliar concentrations of nutrients and secondary compounds, leaf structural and physical attributes, and plant carbon allocation strategies (Grime et al. 1996; Wardle et al. 1998). Therefore, plant species that are more palatable generally produce litter of higher quality, and which is more favorable to decomposers. Again, both positive and negative effects of herbivores on decomposers are possible through this mechanism. Herbivory can stall succession by maintaining vegetation that is optimized by browsing. Here, the existing vegetation derives a competitive advantage from herbivory, which enables it to prevent establishment by later-successional species that are less palatable and produce poorer-quality litter (see Augustine and McNaughton 1998). Conversely, when herbivores adversely affect palatable vegetation, this can accelerate succession, encouraging domination by unpalatable plant species that produce poorer litter quality. This has been shown for the "moose, microbes and boreal forest" study of Pastor et al. (1988), in which herbivory by moose (*Alces alces*) was found to encourage replacement of deciduous tree species by spruce (*Picea abies*), which has much poorer litter quality. This, in turn, resulted in retardation of soil microbial activity and rates of carbon and nitrogen mineralization.

20.2.2 Evidence from Exclosure Studies

It is apparent from the above discussion that a range of mechanisms can be involved in determining the net response of the decomposer subsystem to foliar herbivory. If this is indeed the case, then we should expect to find a spectrum of responses of decomposers and soil processes to herbivory in nature, the direction of which will be determined in each case by the net balance of several mechanisms. Long-term exclosure studies represent a unique tool for investigating the belowground effects of mammalian herbivore exclusion in the field, with reasonable control for all other confounding extrinsic factors. Synthesis of the available literature on long-term exclosure studies provides evidence confirming that large mammalian herbivores can cause a range of effects on the belowground subsystem (Wardle 2002). For example, exclusion of browsing mammals has been shown to have effects on decomposer organisms that are both positive (e.g., Bardgett et al. 1997) and negative (e.g., Pastor et al. 1988), effects on diversity of specific decomposer groups that are both positive (e.g., Stark et al. 2000) and negative (e.g., Suominen 1999), effects on carbon and nitrogen mineralization that are both positive (e.g., Kielland et al. 1997) and negative (e.g., van Wijnen et al. 1999), and effects on ecosystem carbon and nitrogen sequestration that are both positive (e.g., Pastor et al. 1988) and negative (e.g., Derner et al. 1997). Much remains to be understood about what types of site conditions predispose the decomposer subsystem to show a positive or negative response to foliar herbivory. However, likely site factors include soil fertility, the functional types of vegetation present, whether herbivores are naturally present at current densities, and the species of mammals involved. Furthermore, exclosure studies are likely to yield different results depending on the length of time that the exclosures have been set up for, especially given that herbivore effects on nutrient cycling and plant productivity may often represent transient phenomena (de Mazancourt and Loreau 2000).

20.3 Browsing Mammals and Decomposers in the New Zealand Context

20.3.1 Plant Functional Responses to Browsing Mammals

The mechanistic basis of how introduced browsing mammals in New Zealand forests may influence the belowground subsystem has been little investigated, although some insights can be gained from previous literature suggesting that palatable and unpalatable forest plant species differ in key functional traits. In fenced exclosure studies, browsing mammals in New Zealand forests are found to consistently reduce palatable species (meaning that optimization of palatable species does not appear to occur), and frequently promote unpalatable species through competitive release from the palatable species. In particular, several studies have found browsing mammals to generally encourage domination in the vegetation browse layer by unpalatable monocotyledonous plants, ferns, and small-leaved species at the expense of faster-growing, palatable, broad-leaved species, though exceptions exist (e.g., Conway 1949; Wallis and James 1972; Smale et al. 1995). The preference of browsing mammals for palatable species with large leaves and high specific leaf areas (SLAs) is consistent with literature suggesting that plants with these attributes are typically poorly defended and produce nutrient-rich foliage, in contrast to species that produce smaller, longer-lived leaves (Grime 1979; Coley et al. 1985). Because palatability and decomposability are governed by a similar suite of plant traits (e.g., SLA, nutrient content; Cornelissen et al. 1999), those plant species reduced by browsers in exclosure studies should produce litter that is more readily decomposed than do those that are promoted.

Further evidence emerges from studies involving chemical analysis of foliage and plant litter samples. For example, a synthesis of studies involving analysis of rumen contents and fecal pellets from forest browsing mammals (Forsyth et al. 2002) found that some plant species were consistently preferred whereas others were consistently avoided. Leaves from preferred plant species had statistically significantly lower concentrations of lignin than those that were avoided; leaf nitrogen concentrations were greater for preferred species, but this was not statistically significant. Both lignin and nitrogen are well known as controls of herbivory and decomposition. Furthermore, Wardle et al. (2002) found that plant species reduced by browsing mammals in exclosure studies differed substantially from those promoted by browsers in terms of concentrations of phenolics, condensed tannins, and lignin (though not nitrogen or phosphorus). Litter from those plant species reduced by browsers was also decomposed more slowly, although release rates from litter of nitrogen or phosphorus did not differ between favored and unfavored species. The evidence therefore points clearly to browsing mammals reducing plant species that produce litter of high quality, and which is likely to favor decomposer activity, and encouraging domination by plant species with recalcitrant litter. However, it is important to note, in the light of Table 20.1, that this mechanism represents only one route by which decomposers may be influenced by browsing mammals, and other mechanisms that operate in opposite directions may also be involved.

20.3.2 Belowground Evidence from Exclosure Plots

Over the period 1997–1999, 30 fenced exclosure plots, established mainly by the former New Zealand Forest Service between 1961 and 1985, were sampled to evaluate the effects of certain browsing mammals (several deer species, goats) on community and ecosystem properties, both aboveground and belowground. These span 12 degrees of latitude (from Waipoua Forest in Northland to Stewart Island), range from near sea level to just below timberline, and represent most of New Zealand's major natural forest types. The main findings of this work are now briefly described; further detail is presented by Wardle et al. (2001).

Browsing mammals were found to significantly alter plant community composition at most locations, reducing densities of palatable, large-leaved, dicotyledonous plant species and frequently promoting unpalatable species. There was also a consistent pattern of reduction of vegetation density and plant species diversity in the browse layer. However, several belowground components did not show a corresponding unidirectional response across locations to browsing mammals (Fig. 20.1). Averaged across all 30 locations, some measures of resource base quality were marginally significantly greater inside than outside exclosures (i.e., concentrations of nitrogen in the litter layer, and soluble carbon in the humus), whereas others were unresponsive to browsing mammals (nutrient concentrations in the humus, and pH of substrate; Wardle et al. 2001). Furthermore, the soil microbial biomass and all the groups of microfauna (main trophic groups: nematodes, rotifers, copepods, tardigrades) showed highly idiosyncratic responses to browsers, being strongly stimulated at some sites and strongly reduced at others. This indicates that a range of mechanisms, which operate in opposite directions, are probably involved in determining soil biotic responses, with different mechanisms dominating at different sites (Wardle 2002; Table 20.1). For many groups of organisms, the magnitude of response to browsing across the 30 locations was significantly correlated with soil nutrient status and macroclimatic variables, pointing to large-scale differences in extrinsic driving factors across sites in determining which mechanisms dominated. Although such explanations can justify the nature of idiosyncratic results identified, other variables could potentially also play a role in generating intersite differences. Firstly, different exclosure sites supported different browsing mammal species, and the possibility exists for differences in browse preference among

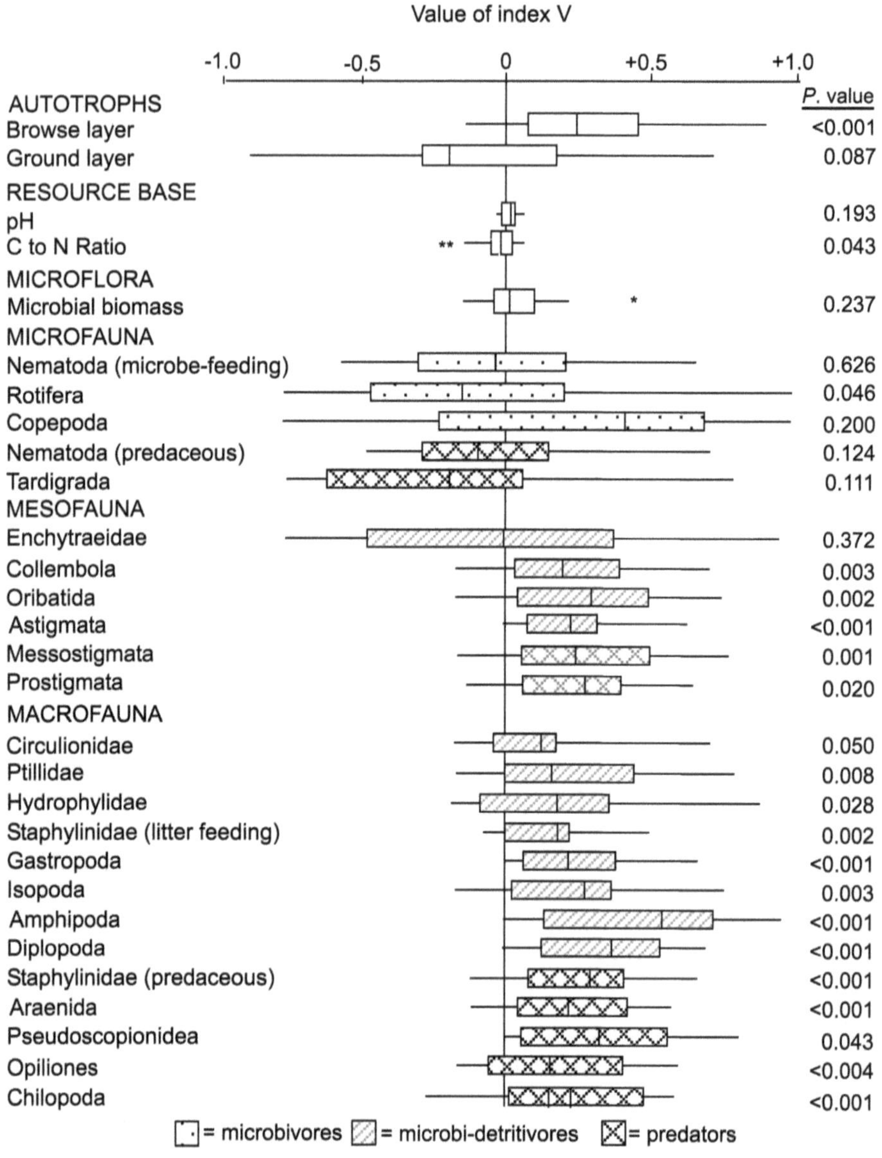

Fig. 20.1 Box-and-whisker plots summarizing data for the response to browsing mammals of resources, autotrophs, and components of the decomposer food web in the litter layer, as assessed by measurements performed inside and outside fenced exclosure plots for each of 30 locations throughout New Zealand (see Wardle et al. 2001). The index *V* (Wardle et al. 2001), determined for each response variable at each location, becomes increasingly negative if the value is increasingly greater outside the exclosure relative to inside it, and increasingly positive if it becomes increasingly greater inside the exclosure than outside it; the index ranges from –1 to +1, with 0 indicating no difference. For each variable, the *box* encompasses the middle half of the data (values of *V*) between the first and third quartiles (i.e., 15 of the 30 locations); the *bisecting line* is at the value of the median, and the *horizontal line outside the box* represents the typical range of data val-

species (Fraser 1996). Secondly, most exclosures have been implemented only after mammals were already having a major impact in the forest, and the exclosure effect may be influenced to some extent by the magnitude of change that had already occurred in the forest (Allen et al. 1984).

A very different pattern was found, however, for larger (mesofaunal and macrofaunal) soil invertebrates occupying the litter layer (Fig. 20.1). Here, the effect was highly unidirectional, with most locations showing adverse effects of browsing mammals on invertebrate densities. This suggests that the mechanistic basis of browsing mammal effects on these soil animals differs from the mechanisms through which browsers affect microflora and microfauna. The adverse effect of browsers on larger soil animals is unlikely to be a direct consequence of the effects of browsers on vegetation properties. This is because, across the 30 locations, no statistically significant relationships were found between the magnitude of soil faunal response to browsing and the magnitude of vegetation response to browsing (rather, faunal response to browsing across sites was related to a range of other variables, including microclimatic, soil nutrient, and tree stand properties). One possible mechanism through which browsers affect soil animals may involve physical effects of their hoof pressure, and associated treading and scuffing activity. This could explain why microfaunal and microbial organisms, which occupy the aqueous portion of the litter layer and are intimately involved with the leaf substrate, were not consistently adversely affected by browsing mammals, whereas larger soil animals, which are free living and more exposed to extrinsic soil factors, were consistently adversely affected.

Community composition of each of five soil faunal groups (humus-dwelling nematodes, litter-dwelling nematodes, diplopods, gastropods, and beetles) was significantly altered by browsing mammals for the majority of locations. Furthermore, taxonomic diversity within each of these groups was often altered by browsers, although both positive and negative effects were noted at different locations. Averaged among all 30 locations, taxonomic richness of two of these groups (gastropods and diplopods) was significantly reduced by browsers, whereas the Shannon-Weiner diversity index for beetles was promoted by browsers. Across these locations, the magnitude of effects of browsers on the diversity of only one faunal group (humus-dwelling nematodes) was significantly correlated with the magnitude of browser effects on plant diversity. The generally poor relationship between responses of plant diversity and responses of faunal diversity is consistent with literature suggesting that different plant species have vastly different effects on the diversity of soil biota, and that this may override any effect that changes in plant diver-

---

Fig. 20.1 (*Continued*)
ues. *Asterisks* indicate outlier values. *P* values are for paired *t*-tests comparing the significance of difference of the value of the variable inside vs. outside the exclosure across the 30 locations. Reproduced with permission from the Ecological Society of America

sity may directly exert on the diversity of soil organisms (Wardle et al. 1999). However, the magnitude of browser effects on the diversity of two further groups (gastropods and diplopods) was correlated with browser effects on the microhabitat diversity (i.e., the diversity of twig and leaf material) in the litter layer.

The effects of browsers on key belowground ecosystem properties were either minor or idiosyncratic. Carbon mineralization rates in the soil were weakly, if at all, influenced by browsers. Browsing mammals significantly affected soil storage of carbon and nitrogen on an areal basis for several locations, although these effects were idiosyncratic and again reflect the different mechanisms through which browsers can affect the decomposer subsystem. This means that effects of browsers on carbon and nitrogen storage are likely to be context dependent. In total, this exclosure survey suggests that the introduction of browsing mammals to New Zealand has wide-ranging effects at both community and ecosystem level. For soil processes, soil microflora, and soil microfauna, these effects are likely to be idiosyncratic, with both positive and negative effects being possible. However, consistently negative effects of browsers are apparent for indigenous plant communities (e.g., understory vegetation density and diversity) and most groups of litter-dwelling mesofauna and macrofauna.

20.4 Conservation Implications

There is a growing awareness of the need to manage natural habitats for conservation purposes with an ecosystem-level, rather than a single-species focus. Given that ecosystems are driven by both producer and decomposer subsystems, as well as by the relationships between the two, any ecosystem-level approach to conservation must therefore consider decomposer organisms and the processes they carry out. The exclosure survey outlined above provides clear evidence that introduced browsing mammals in New Zealand forests have important, and previously unappreciated, effects on the belowground subsystem. These effects cannot be predicted by observations of vegetation responses to browsing mammals; for some of the 30 locations, browsing mammals had strong belowground effects even when there was a neutral effect of browsers on the vegetation itself. The effects of browsers on mesofaunal and macrofaunal invertebrate groups were consistently adverse, suggesting that larger soil organisms play a much less important role in systems with browsing mammals than in those without. Although the ecological role of these soil animals in New Zealand forests remains largely unexplored, studies in other temperate forests frequently point to their significance in influencing carbon and nutrient cycling, regulation of nutrient supply from the soil, and ultimately plant growth and nutrient acquisition (reviewed by Lavelle 1997;

Wardle 2002). The possibility therefore remains that the impacts of browsing mammals on the decomposer subsystem may have important effects on ecosystem processes and aboveground properties that become apparent only in the longer term.

Gradually, browsing mammals could conceivably alter ecosystem processes by changing the functional composition of understory vegetation. Browsers tend to avoid consuming plant species that produce poor-quality litter (Wardle et al. 2002), and in New Zealand this includes seedlings of most canopy tree species (Wardle et al. 2002; Coomes et al. 2003), although browsing mammal species differ in the extent to which they browse tree seedlings. Thus, in the short term at least, the effects of browsers are likely to be more important in determining the types of shrubs that dominate in the understory than in the regeneration of forest canopy species and tree species composition. However, much remains unknown about the roles that palatable shrub species, which produce high-quality litter, play in forest ecosystem processes. The rapid rates of nutrient release from the litter of such species may be important in maintaining nutrient availability, and in accelerating the decomposition of recalcitrant forest canopy litter through litter-mixing effects (cf. Seastedt 1984). Browsing mammals would reverse these types of effects by removing palatable species from the community. Although this mechanism is clearly unimportant at some sites studied by Wardle et al. (2001), i.e., those in which browsing mammals enhanced the rates of soil processes despite removing species that produce high-quality litter, the potential remains for it to be important in others. In any case, adverse effects of browsing mammals on nutrient supply rates in forests as a result of this mechanism, and consequent effects on forest composition detectable only in the long term, cannot be ruled out.

The development of an understanding of how introduced browsing mammals affect community and ecosystem properties and processes, both aboveground and belowground, is important in evaluating how much of a problem these animals pose in forests, and therefore what level of management should be implemented to reduce their populations. The data presented by Wardle et al. (2001) suggest that their effects are far more wide ranging than simply altering vegetation composition. However, much remains unknown about how the current effects of mammals in New Zealand forests represent a deviation from those of the moa species that occupied the herbivore browser guild prior to human settlement about 800–1,000 years ago. The precise timing of the extinction of the moas has been debated (cf. Anderson 2000; Holdaway and Jacomb 2000), although they were certainly extinct centuries before the introduction of browsing mammals to New Zealand. Our knowledge on the importance of moas in New Zealand forests is far from clear (cf. Atkinson and Greenwood 1989; McGlone and Clarkson 1993), and although they probably exerted sufficient feeding pressure to induce morphological adaptations in some components of the New Zealand flora (Bond et al. 2004), the effects of their browsing were probably less than those exerted by browsing mammals (McGlone and Clarkson 1993). Furthermore, if, as proposed by Wardle et al. (2001), adverse effects of browsing mammals on soil invertebrates result from trampling, then it is relevant that disturbance in the litter layer caused by these mammals probably is much more severe than that caused by moas. Calculations performed by Duncan and Holdaway (1989) demonstrated that foot pressure exerted by browsing mammals was between 2.5 and 3 times higher than that exerted by the largest moa species. They also pointed out that the ungulate hoof acts as a chisel, with the hoof edge scraping the substratum, whereas feet of ratites are probably more flexible and cause little damage to the soil. Therefore, it would appear that, regardless of their relative effects on vegetation, browsing mammals are likely to exert much stronger effects on soil mesofauna and macrofauna than did the moa. It is even conceivable that some litter-dwelling invertebrates in New Zealand are much less adapted to foot pressure from ungulates than are comparable litter-dwelling fauna from regions that support an indigenous forest ungulate fauna.

Goals to "restore" New Zealand's natural forests to their prehuman condition are unattainable, given that functionally important components of the indigenous fauna are extinct and contemporary substitutes do not exist. However, for management purposes (including prioritization of management), it is important that we better understand the impacts of alien organisms in the current environment on communities of indigenous organisms, as well as the processes that they carry out. Little is known about the levels of introduced deer and goats that could be tolerated within such systems without significant reductions in indigenous organisms resulting, although it appears that even low densities of browsing mammals can significantly reduce populations of soil mesofauna and macrofauna. The results of Wardle et al. (2001) suggest that the minimum densities of browsing mammals above which adverse effects on the belowground subsystem are observed are probably much lower than the minimum densities required to show demonstrable effects on vegetation.

20.5 Conclusions

It is apparent that browsing mammals do not simply exert influences upon vegetation, but also indirectly affect the decomposer subsystem through a variety of mechanisms that can operate both positively and negatively. Several areas remain to be investigated with regard to the ecosystem impacts of introduced browsing mammals in New Zealand, including effects that are manifested belowground. For example, little is known about the relative resistance to herbivory of productive and unproductive ecosystems; plants in productive systems are potentially more capable of compensating for biomass loss due to herbivory, but conversely, intensity of herbivory is usually much greater in productive systems. Furthermore, little is known about the extent to which ecosystem effects of different sympatric species of browsing mammals substitute for one another (and therefore whether reductions of one species would simply be matched by increased impacts of another); such information is critical when considering single-species versus multiple-species pest management strategies (cf. Forsyth et al. 2000). Although the study of Wardle et al. (2001) provides clear evidence of strong effects of browsing mammals on soil invertebrates, much remains unknown about how these changes in soil invertebrate populations and food web structures in turn affect soil processes and nutrient supply for the aboveground biota. We are only beginning to understand the ways in which introduced browsing mammals can drive belowground organisms and processes, but such information is essential if we are to use ecosystem-level approaches to underpin conservation strategies and priorities in New Zealand's forests.

Acknowledgements The work described in this article is supported by the Foundation for Research, Science and Technology, and the New Zealand Marsden Fund. Thanks to my collaborators on the exclosure study, Gary Barker, Karen Bonner and Gregor Yeates, who have contributed to the development of ideas presented in this article. Rob Allen and two anonymous reviewers made helpful comments on the text.

References

Allen RB, Payton IJ, Knowlton JE (1984) Effects of ungulates on structure and species composition in the Urewera forests as shown by exclosures. NZ J Ecol 7:119–130

Anderson A (2000) Less is moa. Science 289:1472–1473

Atkinson IAE, Greenwood RM (1989) Relationships between moas and plants. NZ J Ecol suppl 12:67–96

Augustine DJ, McNaughton SJ (1998) Ungulate effects on the functional species composition of plant communities: herbivore selectivity and plant tolerance. J Wildl Manage 62:1165–1183

Bardgett RD, Wardle DA (2003) Herbivore-mediated linkages between aboveground and belowground communities. Ecology 84:2258–2268

Bardgett RD, Hobbs PJ, Frostegård Å (1996) Changes in soil fungal:bacterial biomass ratios following reductions in the intensity of management of an upland grassland. Biol Fertil Soils 22:261–264

Bardgett RD, Leemans DK, Cook R, Hobbs PJ (1997) Seasonality of the soil biota of grazed and ungrazed hill grasslands. Soil Biol Biochem 29:1285–1294

Bardgett RD, Wardle DA, Yeates GW (1998) Linking above-ground and below-ground interactions: how plant responses to foliar herbivory influence soil organisms. Soil Biol Biochem 30:1867–1878

Belsky AJ (1986) Does herbivory benefit plants? A review of the evidence. Am Nat 127:870–892

Bokhari UG, Singh JS (1974) Effects of temperature and clipping on growth, carbohydrate reserves, and root exudation of western wheatgrass in hydroponic culture. Crop Sci 14:790–794

Bond WJ, Lee WG, Craine JM (2004) Plant structural defenses against browsing birds: a legacy of New Zealand's extinct moas. Oikos 104:500-508

Coley PD, Bryant JP, Chapin FS III (1985) Resource availability and plant antiherbivore defense. Science 230:895–899

Conway MJ (1949) Deer damage in a Nelson beech forest. NZ J For 6:66–67

Coomes DA, Allen RB, Forsyth DM, Lee WG (2003) Factors preventing the recovery of New Zealand forests following control of invasive deer. Conserv Biol 17:450–459

Cornelissen JHC, Pérez-Harguindeguy N, Díaz S, Grime JP, Marzano B, Cabido M, Vendramini F, Cerabolini B (1999) Leaf structure and defence control decomposition rate across species and life form in regional floras on two continents. New Phytol 143:191–200

De Mazancourt C, Loreau M (2000) Effect of herbivory, and plant species replacement on primary production. Am Nat 155:735–754

Derner JD, Briske DD, Boutton TW (1997) Does grazing mediate soil carbon and nitrogen accumulation beneath C4, perennial grasses along an environmental gradient? Plant Soil 191:147–156

Duncan K, Holdaway R (1989) Footprint pressures and locomotion of moas and ungulates and their effects on the New Zealand indigenous biota through trampling. NZ J Ecol suppl 12:97–101

Dyer MI, Turner CL, Seastedt TR (1993) Herbivory and its consequences. Ecol Appl 3:10–16

Findlay S, Carreiro M, Krischik V, Jones CG (1996) Effects of damage to living plants on leaf litter quality. Ecol Appl 6:269–275

Forsyth DM, Parkes JP, Hickling GJ (2000) A case for multi-species management of sympatric herbivore pest impacts in the central Southern Alps, New Zealand. NZ J Ecol 24:97–103

Forsyth DM, Coomes DA, Nugent G, Hall GMJ (2002) Diet and diet preferences of introduced ungulates (Order: Artiodactyla) in New Zealand. NZ J Zool 29:323–343

Fraser KW (1996) Comparative rumen morphology of sympatric sika deer (*Cervus nippon*) and red deer (*C. elaphus scoticus*) in the Ahimanawa and Kaweka Ranges, central North Island, New Zealand. Oecologia 105:160–166

Grime JP (1979) Plant strategies and vegetation processes. Wiley, Chichester

Grime JP, Cornelissen JHC, Thompson K, Hodgson JG (1996) Evidence of a causal connection between anti-herbivore defence and the decomposition rate of leaves. Oikos 77:489–494

Holdaway RN, Jacomb C (2000) Rapid extinction of the moas (Aves: Dinornithiformes): model, test, and implications. Science 287:2250–2254

Holland JN, Cheng W, Crossley DA Jr (1996) Herbivore-induced changes in plant carbon allocation: assessment of below-ground C fluxes using carbon-14. Oecologia 107:87–94

Kielland K, Bryant JP (1998) Moose herbivory in taiga: effects on biogeochemistry and vegetation dynamics in primary succession. Oikos 82:377–383

Kielland K, Bryant JP, Ruess RW (1997) Moose herbivory and carbon turnover of early successional stands in interior Alaska. Oikos 80:25–30

Laakso J, Setälä H (1999) Sensitivity of primary production to changes in the architecture of belowground food webs. Oikos 87:57–64

Lavelle P (1997) Faunal activities and soil processes: adaptive strategies that determine ecosystem function. Adv Ecol Res 27:93–132

Lovett GM, Ruesink AE (1995) Carbon and nitrogen mineralisation from decomposing gypsy moth frass. Oecologia 104:133–138

Mawdsley JL, Bardgett RD (1997) Continuous defoliation of perennial ryegrass (*Lolium perenne*) and white clover (*Trifolium repens*) and associated changes in the composition and activity of the microbial population of an upland grassland soil. Biol Fertil Soils 24:52–58

McGlone MS, Clarkson BD (1993) Ghost stories: moa, plant defences and evolution in New Zealand. Tuatara 32:1–21

McNaughton SJ (1985) Ecology of a grazing ecosystem: the Serengeti. Ecol Monogr 55:259–294

McNaughton SJ, Ruess RW, Seagle SW (1988) Large mammals and process dynamics in African ecosystems. BioScience 38:794–800

McNaughton SJ, Oesterheld M, Frank DA, Williams KJ (1989) Ecosystem-level patterns of primary productivity and herbivory in terrestrial habitats. Nature 341:142–144

Mikola J, Yeates GW, Barker GM, Wardle DA, Bonner KI (2001) Effects of defoliation intensity on soil food-web properties in an experimental grassland community. Oikos 92:333–343

Milchunas DG, Lauenroth WK (1993) Quantitative effects of grazing on vegetation and soils over a global range of environments. Ecol Monogr 63:327–366

Molvar EM, Bowyer RT, Van Ballenberghe V (1993) Moose herbivory, browse quality, and nutrient cycling in an Alaskan treeline community. Oecologia 94:472–479

Pastor J, Naiman RJ, Dewey B, McInnes P (1988) Moose, microbes and the boreal forest. BioScience 38:770–777

Ruess RW, McNaughton SJ (1987) Grazing and the dynamics of nutrients and energy regulated microbial processes in the Serengeti grasslands. Oikos 49:101–110

Scheu S, Theenhaus A, Jones TH (1999) Links between the detritivore and the herbivore system: effects of earthworms and Collembola on plant growth and aphid development. Oecologia 119:541–551

Seastedt TR (1984) The role of microarthropods in decomposition and mineralization processes. Annu Rev Entomol 29:25–46

Seastedt TR (1985) Maximization of primary and secondary productivity by grazers. Am Nat 126:559–564

Smale MC, Hall GMJ, Gardner RO (1995) Dynamics of kanuka (*Kunzea ericoides*) forest on south Kaipara Spit, New Zealand, and the impact of fallow deer (*Dama dama*). NZ J Ecol 19:131–141

Stark S, Wardle DA, Ohtonen R, Helle T, Yeates GW (2000) The effect of reindeer grazing on decomposition, mineralization and soil biota in a dry oligotrophic Scots pine forest. Oikos 90:301–310

Suominen O (1999) Impact of cervid browsing and grazing on the terrestrial gastropod fauna in the boreal forests of Fennoscandia. Ecography 22:651–658

Tracy BF, Frank DA (1998) Herbivore influence on soil microbial biomass and nitrogen mineralization in a northern grassland ecosystem: Yellowstone National Park. Oecologia 114:556–562

Tuomi J, Niemelä P, Sirén S (1990) The Panglossian paradigm and delayed inducible accumulation of foliar phenolics in mountain birch. Oikos 59:399–410

Van der Putten W, Vet LEM, Harvey JA, Wäckers FL (2001) Linking above- and belowground multitrophic interactions of plants, herbivores, pathogens, and their antagonists. Trends Ecol Evol 16:547–554

Van Wijnen HJ, van der Wal R, Bakker JP (1999) The impact of herbivores on nitrogen mineralization rate: consequences for salt-marsh succession. Oecologia 118:225–231

Wallis FP, James IL (1972) Introduced animal effects and erosion phenomena in the northern Urewera forests. NZ J For 17:21–36

Wardle DA (2002) Communities and ecosystems: linking the aboveground and belowground components. Princeton Univ Press, Princeton

Wardle DA, Barker GM, Bonner KI, Nicholson KS (1998) Can comparative approaches based on plant ecophysiological traits predict the nature of biotic interactions and plant species effects in ecosystems? J Ecol 86:405–420

Wardle DA, Bonner KI, Barker GM, Yeates GW, Nicholson KS, Bardgett RD, Watson RN, Ghani A (1999) Plant removals in perennial grassland: vegetation dynamics, decomposers, soil biodiversity, and ecosystem properties. Ecol Monogr 69:535–568

Wardle DA, Barker GM, Yeates GW, Bonner KI, Ghani A (2001) Introduced browsing mammals in New Zealand natural forests: aboveground and belowground consequences. Ecol Monogr 71:587–614

Wardle DA, Bonner KI, Barker GM (2002) Linkages between plant litter decomposition, litter quality, and vegetation responses to herbivores. Funct Ecol 16:585–595

21 Distinguishing Natural Processes from Impacts of Invasive Mammalian Herbivores

P.J. BELLINGHAM and W.G. LEE

21.1 Introduction

Mammalian herbivores have diverse direct effects on New Zealand's vegetation, ranging from seed predation and depletion of palatable species locally (Campbell and Atkinson 2002), through to defoliation of forest canopies (Meads 1976) and the restricted demise of forests (McKelvey 1995). Compaction, nutrient transfer, and compositional changes in the forest understory may also have long-term consequences on ecosystem structure and function (Wardle et al. 2001). Immediate effects, soon after invasion, are often striking (e.g., dramatic reduction of biomass and plant species diversity in forest understories; Mark and Baylis 1975), and there has been a strongly held perception that introduced herbivores are the principal, even the sole, cause of phenomena such as hill country erosion and lowland flooding, and the sparse nature of subcanopy layers of many forests. Much of the history of introduced wild animal control in New Zealand has been variously based on damage to soil and vegetation apparently caused by introduced mammals.

New Zealand's history from the Pliocene to the present day has involved prolonged climatic fluctuations, marine inundation and tectonic uplift, which have caused massive extinction events and major changes in the biota (see Chap. 2). Even during the Holocene, regional volcanic eruptions, earthquakes, flooding and cyclonic storms have initiated large-scale changes in vegetation composition and structure. The chronic consequences of recent biological invasions need to be understood and interpreted against such a background of constant natural change and disturbance (Dayton et al. 1998). Knowledge of underlying ecosystem dynamics and processes is required so that whole-ecosystem consequences of these invasions can be recognized and, if possible, remedied. Ecologists have struggled to understand ecosystem-level impacts of invasions (Zavaleta et al. 2001), and to readily distinguish natural changes

Ecological Studies, Vol. 186
R.B. Allen and W.G. Lee (Eds.)
Biological Invasions in New Zealand

in ecosystem properties from those caused specifically by invading biota (e.g., Veblen and Stewart 1982).

In this chapter, we examine three ecological issues that have been debated in New Zealand regarding the relative roles of introduced herbivorous mammals versus natural processes – upland soil erosion, forest regeneration, and mass dieback in forest canopies. They represent ecosystem- to community-level impacts assumed to be caused by herbivorous mammals released in the 19th century when Europeans first settled. All issues concentrate on the management of introduced mammals in indigenous forests, which have been the foci for wild animal control. Importantly, each issue has an extended history of ecological research and controversy. Here, we assess what progress has been made in the last 20 years in evaluating the extent to which invasive mammals influence or cause these phenomena.

21.2 Soil Erosion – Introduced Herbivores or Natural Processes?

New Zealand is characteristically a steep, mountainous country, and soil and substrate instability are evident as landslides and gullying. The considerable sediment loads carried by major rivers reflect the impacts of this landscape volatility. In the early 20th century, after 50 years of European colonization and extensive clearance of natural vegetation for pastoralism and settlements, concern was expressed about the extent of bare soil in eastern lowland and upland areas of the South Island, and about landslides and gully erosion in North Island hill country forest. Perceptions about the importance of forests for controlling water runoff, and pervasive fears about the loss of potentially productive soil as well as the downstream consequences of sedimentation for agriculture and urban settlement were derived largely from emerging evidence in the Northern Hemisphere identifying the catastrophic consequences of deforestation (McKelvey 1995). Land management to sustain soil and water values was embedded in legislation, and several government departments embarked on ambitious soil protection programs based on the experimental planting of a great range of soil-stabilizing early-successional trees imported from the Northern Hemisphere.

Overgrazing by feral and domestic mammals was suspected to be a critical factor in causing accelerated rates of natural erosion, although the evidence was largely anecdotal. In the eastern South Island, exposed areas of bare soil, active landslides and gully formation, and dust clouds associated with strong föhn winds were all interpreted as evidence of human-induced erosion (Hewitt 1996). European pastoralists certainly exploited the extensive, tall tussock grasslands created mostly by Polynesian burning of *Nothofagus* and mixed conifer–broadleaved forest, by overstocking the grasslands with sheep and cattle, and practising near-annual burning to improve access and forage quality (O'Connor 1982). Episodic eruptions of feral rabbit populations, facilitated by the grazing-induced transition from tall tussock grasslands to an open short-tussock community, contributed significantly to creating a sparse vegetation cover.

In the North Island hill country, the depletion of the forest understory by introduced mammals (goats, deer, and possum) was seen as the main cause of landslides and streambank collapse in catchments. The reduction of plant biomass by browsers was considered to reduce rainfall interception and increase surface water runoff. Subsequent root failure after plant death was also thought to exacerbate slope instability (Grant 1985). Such impacts were initially one of the major reasons for these animals being considered pests.

New Zealand is one of the more dynamic regions on the globe, due to its position across an active plate boundary (Australian–Pacific) that diagonally bisects the country (Whitehouse and Pearce 1992). Frequent large earthquakes, and surface faulting and folding, are the consequence of deformational strain caused by subducting plates and transcurrent plate movements, and compressional uplift rates of up to 10 mm $year^{-1}$ are occurring in the Southern Alps. The orientation of these mountains across the prevailing westerly oceanic airstream results in very high precipitation (>6,000 mm $year^{-1}$) along the exposed western ranges. This interaction between tectonism and climate is increasingly seen to explain much of the landscape instability in New Zealand.

Active screes and rockfalls in eastern South Island mountains were often thought to be erosional features created following the removal of forest on steep hill slopes, actively expanding due to herbivore grazing of adjoining vegetation. However, Whitehouse et al. (1980), using dating based on rock weathering rinds, demonstrated that some screes had been active since well before European pastoralism. Moreover, some appeared to pre-date human colonization of New Zealand, a view supported by the presence of endemic scree biota. Although fire and animal grazing may have removed the associated vegetation and encouraged marginal spread (Wardle 1991), the impact on the primary processes causing scree formation (steep slopes, rock type, glaciation, climate) was unlikely to be related directly or indirectly to human activities and introduced animal grazing.

The role of extreme rainfall events in the stability of alluvial systems at the catchment scale has also been highlighted. According to Grant (1985), erosional periods appear to be related to long-term, meridional, atmospheric circulation patterns that move warm air over New Zealand, resulting in increases in rainfall and the frequency of major floods. During the period of European occupation, the amount of sediment deposited in alluvial systems in forested catchments of the southern North Island declined, suggesting that feral animal populations had minimal negative impact on sediment supply resulting from this type of land instability. Grant (1991) recognized that forest disturbance is associated with an increase in slips and gully erosion, but suggested that these processes are not caused by animal browsers. Rather, they are correlated with other causes of forest dieback, namely wind, drought, heavy snowfalls, and rainstorms. In the South Island, rainfall and storm events are also important drivers of erosion rates. Basin sediment yield in catchments receiving more than 2,500 mm $year^{-1}$ precipitation exceeded 1,800 t km^{-2} $year^{-1}$, more than twice the levels found in drier areas (McSaveney and Whitehouse 1989).

The driest areas (<1,000 mm annual rainfall) of Central Otago and Canterbury have long been used as examples of vegetation depletion after 150 years of overgrazing by domestic and feral mammals, coupled with excessive burning. In montane and lowland areas, sparse, stunted, open vegetation, soil pedestals, truncated soil profiles, and gravel pavements have been used to identify areas of severe sheet and wind erosion. McSaveney and Whitehouse (1989) cautioned against assuming any direct relationship between animal grazing and the percentage of bare ground, and again emphasized the antiquity of many of these landscape features and the idiosyncratic recovery of natural vegetation following cessation of grazing.

Radionuclide ^{137}Cs markers in soil horizons stemming from the era of global atmospheric atomic weapons testing allowed Hewitt (1996) to provide quantitative measures of soil erosion over the last ca. 40 years in these dry environments. Reduction in topsoil depth was patchy, and was greatest (8.5 mm per decade) on ridge crests and north-facing upper slopes. However, south-facing aspects in general, and lower parts of north-facing slopes showed a net gain (0.08 and 0.16 mm per decade, respectively). These results suggest that wind erosion is redistributing soil locally, and that soil loss is restricted to upper parts of ridge crests, comprising approximately 24 % of the landscape. Wind sheet erosion of topsoil at this scale is undoubtedly attributable to animal grazing and vegetation depletion, as these sites did support woody vegetation in the past.

Past deforestation in New Zealand by humans has affected the stability of fluvial systems, especially flood frequency, and today greatly influences the rates of soil and gully erosion in catchments (Trustrum et al. 1999). As the impact of Cyclone Bola in 1988 demonstrated, it is only when indigenous forest is completely transformed to grassland pasture that detectable differences occur in the susceptibility of landscapes to shallow landslides during storm events (Marden and Rowan 1993). The obvious patchy depletion of canopy and understory biomass caused by introduced herbivorous mammals must modify interception, evapotranspiration rates, and ultimately water runoff, but the magnitude of this in relation to natural erosion processes is certainly small, and at this stage largely undetectable.

21.3 Do Grazing Ungulate Mammals Cause Sparse Regeneration of some Canopy Trees?

Introduced grazing ungulate mammals have indisputably impoverished many of New Zealand's forest understories, and even cause the local extinction of some plants. Fenced exclosure plots throughout New Zealand forests often have greater understory biomass than do adjacent controls where grazing mammals are present (Wardle et al. 2001). In several regions, grazing by mammals leads to selective removal of palatable species, for both understory plants and seedlings of canopy trees (e.g., Wardle 1984; Smale et al. 1995). Because many existing canopy trees established well before the introduction of grazing mammals (i.e., more than ca. 230 years ago), the long-term consequences of herbivory for the structure and persistence of forests remain uncertain (Lee 1998). Nonetheless, concern is often expressed that the maintenance of future forest canopies is jeopardized because these mammals graze seedlings and saplings of some canopy tree species.

Conceptual models based on idealized seedling abundance in mammal-free areas have been proposed where the relative abundance of seedlings of both palatable and less palatable tree species varies solely according to the abundance of deer (Nugent et al. 2001a). Such models do not incorporate other important determinants of seedling abundance and growth, e.g., light availability, soil fertility, proximity of seed parents (Grime 1979), all of which vary in space and time (Coomes et al. 2003). Although a few New Zealand canopy trees appear to exhibit continuous recruitment beneath their own canopies, most seldom regenerate under parents or closed canopies, and regeneration occurs elsewhere in time and/or space. New Zealand conifers rarely regenerate below parent trees (Wardle 1978), and typically recruit after disturbance that eliminates former canopies (Ogden and Stewart 1995). This disturbance-dependent recruitment pattern also occurs in several common angiosperm canopy trees (e.g., Wardle and Guest 1977; Stewart and Veblen 1982). Mammals graze seedlings and saplings of certain angiosperm tree species preferentially (e.g., *Griselinia littoralis*, *Pseudopanax lessonii* and *Raukaua simplex*; e.g., Nugent et al. 2001a). An essential challenge for ecologists is to determine whether poor regeneration by some tree species, especially under forest canopies, is attributable to a species' regeneration niche, dispersal limitation, or to its selective elimination by recently introduced grazing mammals, or an interaction between these factors.

Vertebrate herbivory is not an evolutionary novelty in New Zealand forests (Holdaway and Jacomb 2000; Anderson 2002). Extinction of moa (seven species of large ratite birds, Dinornithiformes) would have altered the relative abundance of canopy and understory species (see Chap. 11). If moa were a selective force, then the current relative abundance of forest canopy species in general, and particularly those that lack a divaricate life form, may be an artifact of the post-moa pre-mammal period (ca. 1500 A.D.–1800 A.D.) in New Zealand (Caughley 1989). Similarly, where ungulates now graze forest understories, some divaricating plants (those with dense, caged canopies) are often disproportionately common (e.g., Smale et al. 1995).

It is particularly difficult to distinguish the extent to which grazing mammals alone influence seedling abundance of canopy trees. Seedlings of canopy tree species usually represent only a small proportion of the diet of grazing mammals such as ungulates, especially in recently established populations (Parkes 1993; Yockney and Hickling 2000; Nugent et al. 2001b). Changes in the abundance of seedlings of most common Stewart Island canopy trees over 18 years were more strongly related to local changes in abundance of parent trees and canopy cover than to changes in deer density (Bellingham and Allan 2003). However, dietary preferences differ among the various exotic grazing mammals. For example, seedlings of *Nothofagus* species seldom feature in the diet of red deer (*Cervus elaphus scoticus*) whereas they are a major component of diets of sympatric sika deer (*Cervus nippon*; Fraser 1996). Sika deer can reduce the abundance of seedlings of *Nothofagus* trees in open-canopied stands to the extent that a regeneration "gap" is apparent (e.g., Allen and Allan 1997).

In forests, shrubs and herbs generally feature more prominently in ungulate diets than do canopy tree seedlings, but again considerable selectivity is evident (e.g., Wardle 1984). Understory species highly palatable to grazing mammals are most abundant at fertile sites (e.g., alluvial floodplains), forest margins, and in early-successional vegetation. These sites may in turn support greater densities of herbivores. However, nutrient-rich sites may also be more resilient to grazing, because of greater productivity (Coomes et al. 2003). For example, a cohort of the palatable shrub *Brachyglottis rotundifolia* germinated and grew rapidly in mild-climate coastal forests of Stewart Island after poisoning of white-tailed deer, but recruitment ceased as deer numbers rebuilt (Bellingham and Allan 2003). However, reduced light under the developing canopy may have also limited seedling establishment.

Grazing mammals can exert strong influences early in natural successions. Under canopies of *Kunzea ericoides*, deer can reduce or even locally eliminate seedlings and saplings of many angiosperm tree species that would otherwise replace the *K. ericoides* canopy (Payton et al. 1984; Smale et al. 1995). However, grazing mammals alone may not be driving regeneration failure of some canopy trees. For example, regeneration of many trees under a *K. ericoides* canopy was expected after the extermination of goats (*Capra hircus*) on Great Island, Three Kings Islands, northern New Zealand, in 1946. However, many former canopy tree species had still not regenerated 36 years later, notably those dispersed by the pigeon *Hemiphaga novaeseelandiae*, which had become locally extinct (Cameron et al. 1987). During successions, the selective removal by grazing mammals of understory species, especially those with higher-quality litter, may also retard belowground processes – a further influence in successions (Wardle et al. 2002; see Chap. 20). For example, high-quality litter of the early-successional nitrogen-fixing shrub *Carmichaelia odorata* can enhance the growth of seedlings of some later-successional tree species (Bellingham et al. 2001). Reduction in abundance or biomass of this shrub, which is grazed selectively by chamois (*Rupicapra rupicapra*; Yockney and Hickling 2000), may in turn influence future forest composition.

Other successional consequences of grazing by mammals may be manifest in the promotion of regeneration of unpalatable species (Wardle 1984; Stewart and Burrows 1989; Nugent et al. 2001a; Coomes et al. 2003), often with poorer litter quality (Wardle et al. 2002). Grazing mammals can promote the abundance of understory herbs resistant to grazing (e.g., the sward grass *Microlaena avenacea*; Moore and Cranwell 1934), or that are unpalatable (e.g., ferns such as *Blechnum discolor*; Stewart 1986). Resulting dense understories can modify the regeneration niche of canopy tree species. For example, regeneration of *Nothofagus* spp. can be restricted to raised surfaces such as logs and root mounds (June and Ogden 1975; Wardle 1984).

How might we better assess the impacts of grazing mammals? In a few instances, we can compare forest structure and dynamics in invaded and uninvaded forests (e.g., invaded coastal forests of Stewart Island and uninvaded islands off its coast; Stewart and Burrows 1989). More comparative studies are required, although opportunities for paired comparisons are limited. Current exclosure plots suffer through inadequate replication and non-random placement (Coomes et al. 2003); both of these issues should be addressed in establishing exclosures in future. Many assessments rely solely on counts of seedlings (e.g., Stewart et al. 1987; Nugent et al. 2001a), and very few have yet followed recruitment, growth, and mortality of seedlings within and beyond the browsing height of grazing mammals – this detail is essential to determine whether regeneration is adequate to replace current canopies. Finally, more rigorous experimental studies are required to determine the consequences of reduction in grazing animal numbers. Such experiments should ensure that treatment areas are comparable in advance (e.g., in rainfall, soils, vegetation). To date, studies of tree seedling demography in experiments to reduce numbers of grazing mammals have rarely lasted 10 years. Even 10 years is inadequate to assess whether the growth and survival of seedlings is sufficient to replace canopies. Realistic assessment of changes induced by grazing mammals will be achieved only with a long-term commitment to maintain treatments and measure response variables.

21.4 Do Brushtail Possums Cause Dieback of Canopy Trees?

Cool temperate evergreen rainforests dominate the high-rainfall areas of the central western South Island. Several angiosperm and conifer tree species form the canopy, and *Nothofagus* spp. are absent (Wardle 1991). In montane regions, the dominant canopy tree species, notably *Metrosideros umbellata* and *Weinmannia racemosa*, are prone to die over areas often exceeding 10 ha. Because these, and other canopy trees such as *Podocarpus hallii*, prone to dieback feature in the diet of brushtail possums (*Trichosurus vulpecula*; Fitzgerald and Wardle 1979), some authors (e.g., Batcheler 1983) have stated unequivocally "that the possum is responsible for dieback in (these) forests". Dieback episodes in certain valleys have sometimes coincided with invasion by possums (e.g., Rose et al. 1992). As a consequence, substantial sums have been spent over the past 40 years on aerial poisoning of forested catchments to kill possums. Early investigations pointed out the potential role of insects and pathogens as agents of tree death (Chavasse 1955). More recently, other biotic agents and wind creating canopy openings have been shown to cause the death of canopy trees (Payton 1988, 1989).

Veblen and Stewart (1982) proposed alternative explanations for the mechanisms of dieback in these forests. Building on earlier studies of forest structure (e.g., Wardle 1980), they noted that most canopy trees show evidence of episodic recruitment, many establishing on primary seres and seldom under their own canopy. Veblen and Stewart (1982) pointed out that young stands of tree species that are browsed by possums (e.g., *Metrosideros umbellata*) are not prone to dieback, and argued that widespread dieback of mature trees was a successional process, i.e., senescence of cohorts that had arisen during earlier disturbances. If this were so, then it would require evidence of past disturbances that could have presented widespread opportunities for recruitment of cohorts of trees that are presently dying. Recently, four different cohorts of canopy trees in these forests were shown to have arisen widely in the last 550 years on surfaces created by earthquakes (Wells et al. 2001). In addition, more localized disturbances give rise to other opportunities for succession, resulting in a mosaic of vegetation in different successional stages in these valleys (Wardle 1980; Reif and Allen 1988). Stands with dieback of canopy trees can have a wide range of diameters (Allen and Rose 1983), and are therefore likely to contain trees of various ages. This suggests that the cause of dieback is not a simple relationship between development and later death of single cohorts of trees.

Underlying geology fundamentally influences soil development and vegetation growth in these rainforests. Whereas canopy dieback is prominent in forests on more fertile and unstable schist and gneiss (Allen and Rose 1983), it is less pronounced in forests developed on greywacke, and is scarcely evident on infertile and comparatively stable granite (found west of the plate boundary; Reif and Allen 1988). Additionally, stronger faulting in some schist and gneiss valleys results in a greater frequency of early-successional communities arising on landslides (e.g., Allen et al. 1991).

Forest dieback may be conspicuous because dead spars of many species, including *M. umbellata*, *P. hallii* and *Libocedrus bidwillii*, decompose slowly and can remain visible in the canopy for decades after tree death, giving "an erroneous impression of mortality rates and causes" (Batcheler 1983). Data from permanent forest plots established representatively in montane Westland in the 1970s now form an appropriate basis for quantitative assessments of canopy tree population dynamics. Recruitment rates of one species prone to dieback, *W. racemosa*, match or slightly exceed its mortality rates in four Westland montane forests (Fig. 21.1). Mortality rates of *M. umbellata* exceed its recruitment rates in some catchments (Fig. 21.1), but this is not universal, and this species generally shows very slow rates of population turnover (0.4 % $year^{-1}$ across these four catchments). In contrast, mortality rates of *P. hallii* greatly exceed its recruitment rates (nil in two forests). Mortality rates of *P. hallii* in Westland montane forests are generally much higher than its mortality rates in other New Zealand forests in which possums are also present (Bellingham et al. 1999).

As time elapses, the case for causally linking dieback directly to the time of invasion of a site by possums weakens. For example, in the Kokatahi Valley, dieback was first observed in the 1940s–1950s, and was considered to be due to peak possum populations at that time. However, canopy dieback continues 50 years later in other parts of that valley, making the link increasingly tenuous between a peak of possum numbers during invasion and the incidence of dieback. The four catchments in Fig. 21.1 have different histories of possum invasion – from long-established at Kokatahi (peak population 1950s) to more recent invasion at Copland (peak population 1970s) and control (from intensive to nil) – yet population trends in some tree species that they browse, such as *W. racemosa*, appear to be independent of both invasion and control histories (Fig. 21.1). Indeed, there is so far no strong evidence to demonstrate poi-

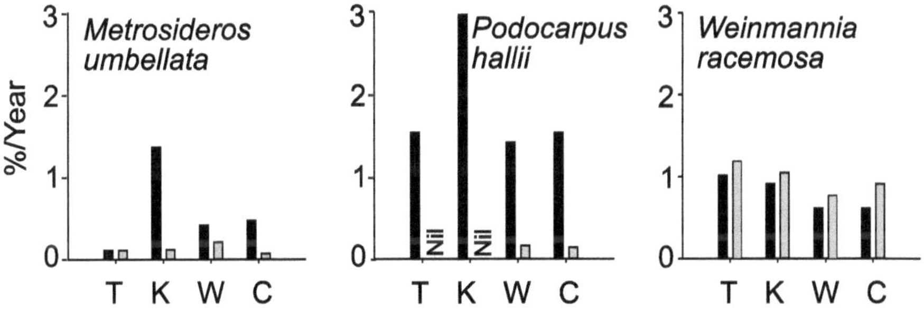

Fig. 21.1 Mortality (*closed bars*) and recruitment (*shaded bars*) rates (% per year) for stems≥10 cm dbh of three canopy trees (*Metrosideros umbellata*, *Podocarpus hallii* and *Weinmannia racemosa*) in four montane rainforests in central Westland, New Zealand (increasing latitude, *left to right*). Forests were sampled using 20x20 m plots: Taramakau (*T*; 10 plots, 1979–1992, 13.1 years), Kokatahi (*K*; 22 plots, 1972–2002, 29.9 years), Whitcombe (*W*; 23 plots, 1972–1999, 27.1 years), Copland (*C*; 45 plots, 1979–1992, 13.1 years); Kokatahi and Whitcombe forests were sampled randomly, Taramakau and Copland forests with subjective plot placement in old-growth stands. Rates were calculated using logarithmic models (Sheil et al. 1995), summing all stems across all plots

soning of possums has altered the mortality rates of canopy tree species in New Zealand.

Further evidence against possum browsing as the primary cause of dieback in these species comes from forests where possums are absent. For example, dieback of *M. umbellata* occurs in coastal forests of Stewart Island, which have indeed been invaded by possums, but it also occurs in uninvaded forests on nearby Bench Island (Veblen and Stewart 1980). In sub-Antarctic Auckland Island forests, not invaded by possums, pollen records show local evidence of two episodes of collapse of a *M. umbellata* canopy during the past 4,000 years (McGlone et al. 2000), and its replacement on both occasions by many of the same understory shrubs that replace *M. umbellata* canopies in Westland montane forests. Dieback also occurs in closely related species in other countries, notably in *Metrosideros polymorpha* forests in Hawai'i where there are no arboreal herbivores to invoke as causal agents. Akashi and Mueller-Dombois (1995) support a patch dynamic interpretation of these forests related to forest development at sites of different fertility and drainage.

Although the evidence for direct effects of possums on the death of canopy trees in these forests is weak, possums may have an ecologically significant role in dieback by weakening trees through repeated browsing (Payton 2000). In addition to accelerating natural senescence, possums may also reduce reproductive output prematurely, either by reducing tree vigor or by browsing flowers (e.g., Cowan and Waddington 1990). During the last decade, possums have been shown to be significant carnivores, preying on eggs, nestlings and adults of native birds (Cowan 2001). The population declines of frugivorous birds (see Chap. 17) may then be limiting dispersal of the many fleshy-fruited species in New Zealand forests.

Evidence that suggests possums can browse subcanopy and early-successional trees to death, or even to the point of local extinction, is much stronger. *Fuchsia excorticata* is a rapidly growing, deciduous, early-successional tree, widespread in New Zealand and prominent in possum diets. In Westland montane forests, there is strong coincidence of mortality of *F. excorticata* with possum invasions (Pekelharing et al. 1998), as has been recorded elsewhere (e.g., Brockie 1992), and now adult trees are either very rare or locally extinct (e.g., all adult trees found in permanent plots in the Whitcombe Valley in 1972 had died by 1999). The loss of this deciduous species, and possible reductions by possums of populations of another semi-deciduous early-successional tree, *Aristotelia serrata*, and of a nitrogen-fixing early-successional shrub, *Coriaria arborea*, at some sites (Brockie 1992) may alter soil development in primary successions toward communities dominated by early-successional plants with lower foliar nutrient levels and slower decomposition rates. Therefore, the long-term effects of possums may emerge primarily in forests that develop on new surfaces.

More persuasive evidence to link widespread death of canopy trees to browsing by possums is required from replicated long-term experiments that demonstrate the effects of possums on a range of plant life-history traits, from reproductive output (e.g., Cowan and Waddington 1990) to canopy tree growth and mortality for a range species. Especially for species such as *Podocarpus hallii*, for which there are clear trends of high mortality and poor recruitment over decades, treatments could include long-term elimination of possums at stand scales. Such experiments need to be designed also to take account of influences on stand growth, including site fertility, stability, and successional status.

21.5 Conclusions

Despite 50 years of studies suggesting multiple and complex causes for changes in New Zealand vegetation, including the contribution of introduced herbivores, some current documents ignore these interpretations. For example, possums were claimed recently to "have caused the complete collapse of the canopy within 15–20 years of their arrival" in western South Island rainforests (Green 2003). Therefore, even now, there is an unquestioning acceptance that invasive mammals are the primary agent of many changes in these ecosystems. Management is usually directed at reducing their numbers, and often does not measure whether the perceived change is altered. On the other hand, ascribing changes solely to natural processes, and discounting the role of invasive biota as a significant influence can be equally naïve and negligent for the conservation of indigenous biodiversity. Recently, Worthy and Holdaway (2002) argued strongly for evidential ecological realism, as opposed to adopting anecdotally based ecological doctrines for prioritizing the impact of invasive species. From the 1940s to the 1970s, the decline and extinction of New Zealand's vertebrates was attributed largely to natural processes, and introduced mammalian predators were considered to have had an insignificant role. Worthy and Holdaway (2002) present a compelling case, from both fossil and contemporary evidence, that the current period of extinctions in the vertebrate fauna is due almost entirely to the introduction of predatory mammals. Although our knowledge about the ecological effects of invasive herbivorous mammals remains limited, this is not a reason to discontinue efforts to reduce their numbers, or even aim for their local elimination. However, to set priorities for where best to gain benefits from controlling invasive mammals, we contend that well-designed experiments are required, combined with improving our knowledge of the whole-ecosystem effects of invaders.

Acknowledgements We appreciate critical comments from Rob Allen, Allan Hewitt, Susan Walker and Janet Wilmshurst.

References

Akashi Y, Mueller-Dombois D (1995) A landscape perspective of the Hawaiian rain forest dieback. J Veg Sci 6:449–464

Allen RB, Allan C (1997) Mountain beech forest dynamics in the Kaweka Range and the influence of browsing mammals. Sci Conserv 44:1–23

Allen RB, Rose AB (1983) Regeneration of southern rata (*Metrosideros umbellata*) and kamahi (*Weinmannia racemosa*) in areas of dieback. Pac Sci 37:433–442

Allen RB, Reif A, Hall GMJ (1991) Elevational distribution of conifer-broadleaved hardwood forests on South Island, New Zealand. J Veg Sci 2:323–330

Anderson A (2002) A fragile plenty: pre-European Maori and the New Zealand environment. In: Pawson E, Brooking T (eds) Environmental histories of New Zealand. Oxford Univ Press, Auckland, New Zealand, pp 19–34

Batcheler CL (1983) The possum and rata-kamahi dieback in New Zealand: a review. Pac Sci 37:415–426

Bellingham PJ, Allan CN (2003) Forest regeneration and the influences of white-tailed deer (*Odocoileus virginianus*) in cool temperate New Zealand rain forests. For Ecol Manage 175:71–86

Bellingham PJ, Stewart GH, Allen RB (1999) Tree species richness and turnover throughout New Zealand forests. J Veg Sci 10:825–832

Bellingham PJ, Walker LR, Wardle DA (2001) Differential facilitation by a nitrogen-fixing shrub during primary succession influences relative performance of canopy tree species. J Ecol 89:861–875

Brockie RE (1992) A living New Zealand forest. David Bateman, Auckland, New Zealand

Cameron EK, Baylis GTS, Wright AE (1987) Vegetation quadrats 1982–1983 and broad regeneration patterns on Great Island, Three Kings Islands, northern New Zealand. Rec Auckland Inst Mus 24:163–185

Campbell DJ, Atkinson IAE (2002) Depression of tree recruitment by Pacific rat (*Rattus exulans* Peale) on New Zealand's northern offshore islands. Biol Conserv 107:19–35

Caughley G (1989) New Zealand plant-herbivore systems: past and present. NZ J Ecol suppl 12:3–10

Chavasse CGR (ed) (1955) Mortality in rata/kamahi protection forests – Westland. New Zealand Forest Service Rep, Wellington

Coomes DA, Allen RB, Forsyth DM, Lee WG (2003) Factors preventing the recovery of New Zealand forests following control of invasive deer. Conserv Biol 17:450–459

Cowan PE (2001) Advances in New Zealand mammalogy 1990–2000: brushtail possum. J R Soc NZ 31:15–29

Cowan PE, Waddington DC (1990) Suppression of fruit production of the endemic forest tree, *Elaeocarpus dentatus*, by introduced marsupial brushtail possums, *Trichosurus vulpecula*. NZ J Bot 28:217–224

Dayton PK, Tegner MJ, Edwards PB, Riser KL (1998) Sliding baselines, ghosts, and reduced expectations in kelp forest communities. Ecol Appl 8:309–322

Fitzgerald AE, Wardle P (1979) Food of the opossum *Trichosurus vulpecula* (Kerr) in the Waiho Valley, south Westland. NZ J Zool 6:339–345

Fraser KW (1996) Comparative rumen morphology of sympatric sika deer (*Cervus nippon*) and red deer (*C. elaphus scoticus*) in the Ahimanawa and Kaweka Ranges, central North Island, New Zealand. Oecologia 105:160–166

Grant PJ (1985) Major periods of erosion and alluvial sedimentation in New Zealand during the Late Holocene. J R Soc NZ 15:67–121

Grant PJ (1991) Disturbance in the forests of the Ruahine Range since 1770. J R Soc NZ 21:385–404

Green WQ (2003) The use of 1080 to control possums and other pests: a resource document. Animal Health Board and Department of Conservation, Wellington, New Zealand
Grime JP (1979) Plant strategies and vegetation processes. Wiley, Chichester
Hewitt AE (1996) Estimating surface erosion using ^{137}Cs at a semi-arid site in Central Otago, New Zealand. J R Soc NZ 26:107–118
Holdaway RN, Jacomb C (2000) Rapid extinction of the moas (Aves: Dinornithiformes): model, test, and implications. Science 287:2250–2254
June SR, Ogden J (1975) Studies on the vegetation of Mount Colenso, New Zealand. 3. The population dynamics of red beech seedlings. Proc NZ Ecol Soc 22:61–66
Lee WG (1998) Vegetation of New Zealand – functional, spatial, and temporal gaps. R Soc NZ Misc Ser 48:91–101
Marden M, Rowan D (1993) Protective value of vegetation on Tertiary terrain before and during Cyclone Bola, East Coast, North Island, New Zealand. NZ J For Sci 23:255–263
Mark AF, Baylis GTS (1975) Impact of deer on Secretary Island, Fiordland, New Zealand. Proc NZ Ecol Soc 22:19–24
McGlone MS, Wilmshurst JM, Wiser SK (2000) Lateglacial and Holocene vegetation and climate change on Auckland Island, Subantarctic New Zealand. Holocene 10:719–728
McKelvey P (1995) Steepland forests: a historical perspective of protection forestry in New Zealand. Canterbury Univ Press, Christchurch, New Zealand
McSaveney MJ, Whitehouse IE (1989) Anthropic erosion of mountain land in Canterbury. NZ J Ecol suppl 12:151–163
Meads MJ (1976) Effects of opossum browsing on northern rata trees in the Orongorongo Valley, Wellington, New Zealand. NZ J Zool 3:127–139
Moore LB, Cranwell LM (1934) Induced dominance of *Microlaena avenacea* (Raoul) Hook. f., in a New Zealand rain-forest area. Rec Auckland Inst Mus 1:219–238
Nugent G, Fraser W, Sweetapple P (2001a) Top down or bottom up? Comparing the impacts of introduced arboreal possums and 'terrestrial' ruminants on native forests in New Zealand. Biol Conserv 99:65–79
Nugent G, Fraser KW, Asher GW, Tustin KG (2001b) Advances in New Zealand mammalogy 1990–2000: deer. J R Soc NZ 31:263–298
O'Connor KF (1982) The implications of past exploitation and current developments to the conservation of South Island tussock grasslands. NZ J Ecol 5:97–107
Ogden J, Stewart GH (1995) Community dynamics of the New Zealand conifers. In: Enright NJ, Hill RS (eds) Ecology of the southern conifers. Melbourne Univ Press, Melbourne, pp 81–119
Parkes JP (1993) Feral goats: designing solutions for a designer pest. NZ J Ecol 17:71–83
Payton IJ (1988) Canopy closure, a factor in rata (*Metrosideros*) – kamahi (*Weinmannia*) forest dieback in Westland, New Zealand. NZ J Ecol 11:39–50
Payton IJ (1989) Fungal (*Sporothrix*) induced mortality of kamahi (*Weinmannia racemosa*) after attack by pinhole borer (*Platypus* spp.). NZ J Bot 27:359–368
Payton IJ (2000) Damage to native forests. In: Montague TL (ed) The brushtail possum. Manaaki Whenua Press, Lincoln, New Zealand, pp 111–125
Payton IJ, Allen RB, Knowlton JE (1984) A post-fire succession in the northern Urewera forests, North Island, New Zealand. NZ J Bot 22:207–222
Pekelharing CJ, Parkes JP, Barker RJ (1998) Possum (*Trichosurus vulpecula*) densities and impacts on fuchsia (*Fuchsia excorticata*) in South Westland, New Zealand. NZ J Ecol 22:197–203
Reif A, Allen RB (1988) Plant communities of the steepland conifer-broadleaved hardwood forests of central Westland, South Island, New Zealand. Phytocoenologia 16:145–224

Rose AB, Pekelharing CJ, Platt KH (1992) Magnitude of canopy dieback and implications for conservation of southern rata-kamahi (*Metrosideros umbellata* – *Weinmannia racemosa*) forests, central Westland, New Zealand. NZ J Ecol 16:23–32

Sheil D, Burslem DFRP, Alder D (1995) The interpretation and misinterpretation of mortality rate measures. J Ecol 83:331–333

Smale MC, Hall GMJ, Gardner RO (1995) Dynamics of kanuka (*Kunzea ericoides*) forest on South Kaipara Spit, New Zealand, and the impact of fallow deer (*Dama dama*). NZ J Ecol 19:131–141

Stewart GH (1986) Forest dynamics and disturbance in a beech/hardwood forest, Fiordland, New Zealand. Vegetatio 68:115–126

Stewart GH, Burrows LE (1989) The impact of white-tailed deer *Odocoileus virginianus* on regeneration in the coastal forests of Stewart Island, New Zealand. Biol Conserv 49:275–293

Stewart GH, Veblen TT (1982) Regeneration patterns in southern rata (*Metrosideros umbellata*) – kamahi (*Weinmannia racemosa*) forest in central Westland, New Zealand. NZ J Bot 20:55–72

Stewart GH, Wardle JA, Burrows LE (1987) Forest understorey changes after reduction in deer numbers, northern Fiordland, New Zealand. NZ J Ecol 10:35-42

Trustrum NA, Gomez B, Page MJ, Reid LM, Hicks DM (1999) Sediment production, storage and output: the relative role of large magnitude events in steepland catchments. Z Geomorphol NF suppl 115:71–86

Veblen TT, Stewart GH (1980) Comparison of forest structure and regeneration on Bench and Stewart Islands, New Zealand. NZ J Ecol 3:50–68

Veblen TT, Stewart GH (1982) The effects of introduced wild animals on New Zealand forests. Ann Assoc Am Geogr 72:372–397

Wardle P (1978) Regeneration status of some New Zealand conifers, with particular reference to *Libocedrus bidwillii* in Westland National Park. NZ J Bot 16:471–477

Wardle P (1980) Primary succession in Westland National Park and its vicinity, New Zealand. NZ J Bot 18:221–232

Wardle J (1984) The New Zealand beeches: ecology, utilisation, and management. New Zealand Forest Service, Wellington, New Zealand

Wardle P (1991) Vegetation of New Zealand. Cambridge Univ Press, Cambridge

Wardle J, Guest R (1977) Forests of the Waitaki and Lake Hawea catchments. NZ J For Sci 7:44–67

Wardle DA, Barker GM, Yeates GW, Bonner KI, Ghani A (2001) Introduced browsing mammals in New Zealand natural forests: aboveground and belowground consequences. Ecol Monogr 71:587–614

Wardle DA, Bonner KI, Barker GM (2002) Linkages between plant litter decomposition, litter quality, and vegetation responses to herbivores. Funct Ecol 16:585–595

Wells A, Duncan RP, Stewart GH (2001) Forest dynamics in Westland, New Zealand: the importance of large, infrequent earthquake-induced disturbance. J Ecol 89:1006–1018

Whitehouse IE, Pearce AJ (1992) Shaping the mountains of New Zealand. In: Soons JM, Selby MJ (eds) Landforms of New Zealand, 2nd edn. Longman Paul, Auckland, New Zealand, pp 144–160

Whitehouse IE, McSaveney MJ, Chinn TJ (1980) Dating your scree. J Tussock Grasslands Mountain Lands Inst Rev 39:15–24

Worthy TH, Holdaway RN (2002) The lost world of the moa: prehistoric life of New Zealand. Canterbury Univ Press, Christchurch, New Zealand

Yockney IJ, Hickling GJ (2000) Distribution and diet of chamois (*Rupicapra rupicapra*) in Westland forests, South Island, New Zealand. NZ J Ecol 24:31–38

Zavaleta ES, Hobbs RJ, Mooney HA (2001) Viewing invasive species removal in a whole-ecosystem context. Trends Ecol Evol 16:454–459

D Management of Biological Invasions

22 Animal Control and Ecosystem Recovery

D.A. Coomes, A.F. Mark and J. Bee

22.1 Introduction

The New Zealand Government is committed to halting the decline in indigenous biodiversity and restoring the country's degraded ecosystems (Anonymous 2000; Logan 2001; Saunders and Norton 2001). In recognition that biological invaders are continuing to degrade New Zealand's ecosystems, the Government spends at least NZ$ 40 million annually on the control of mammalian pests. A key question is whether these resources are being used effectively. Parkes (1990) suggests this question cannot be fully resolved with the information currently available, because a policy flaw has been to confuse goals (protection of the native biota) with means (the removal of invasive animals and plants). For example, the National Parks Act requires managers to "eradicate, as far as possible", all invasive organisms from parks. Such a requirement endorses managers to kill as many pests as their resources allow, but does not oblige them to decide whether their true goal – halting the decline of indigenous biodiversity – has been achieved. Reporting on numbers of animals removed might be justified if a close correlation could be demonstrated between control effort and ecosystem recovery, but there are many reasons to question whether the impacts of biological invasion are reversible. Therefore, it is important to identify the circumstances under which successful recovery is likely.

We start this chapter by briefly defining the terms "recovery" and "restoration", and then consider some of the reasons why recovery of native species should not be taken for granted following animal control operations. Then, we examine the extent to which vegetation has recovered following herbivore control in various New Zealand ecosystems, focusing on the management of feral deer, goats, rabbits, and domestic cattle populations.

These studies highlight substantial differences in recovery rates among the various datasets. Finally, we attempt to draw some conclusions from our literature review on whether resources invested in animal control are being used effectively.

Ecological Studies, Vol. 186
R.B. Allen and W.G. Lee (Eds.)
Biological Invasions in New Zealand

22.2 Defining Ecosystem Restoration and Recovery

In its most ideal sense, ecosystem restoration can be defined as "the process of intentionally altering a site to establish a defined, indigenous, historic ecosystem. The goal of this process is to emulate the structure, function, diversity, and dynamics of the specified ecosystem" (see Mansfield 1996). Many people have pointed out that such aspirations may be impossible to attain, because (1) we often know little about the functioning of historical ecosystems, (2) extinction of ecosystem components may preclude the possibility of a return to a former state, and (3) restoration of this sort may be prohibitively expensive (Simberloff 1990; Hobbs and Norton 1996; Towns 2002). Despite these difficulties, it is increasingly recognized that an ecosystem-based approach to conservation is essential, as species-focused approaches often ignore food-web dynamics, an understanding of which is critical for effective management (Hobbs and Norton 1996; Saunders and Norton 2001). Ecosystem recovery is a type of ecological restoration in which the expectation is that the system will regain desirable attributes through "natural" processes (Meffe and Carroll 1997) – in our examples, once the "pressure" imposed by exotic herbivores on the native organism is reduced by pest control. Conservation management of habitats in New Zealand is usually founded on the principle of natural recovery following pest control, whereas programmes aimed at restoring critically endangered species involve other interventions, such as captive breeding, and relocation onto predator-free reserves. Natural recovery is the focus of this chapter.

22.3 What Are the Impediments to Ecosystem Recovery?

22.3.1 Multi-Trophic Interactions

Most impacted sites in New Zealand are invaded by a multitude of exotic species, which have altered the structure of food webs. Changes in the abundance of one species indirectly affects the abundances of other species separated by more than one link in the food web (e.g. "trophic cascades"; Polis et al. 2000). Thus, the removal of a pest species does not necessarily cause other populations to increase or decrease in a way which might be expected intuitively when direct links are considered (Zavaleta et al. 2001).

The greatest impediments to recovery would seem to result from the fact that many native species are the secondary prey (or by-catch) of exotic predators which feed primarily on exotic prey. The rate of consumption of secondary prey is not strongly regulated by their own density, but rather by interactions between primary prey and predators (Holt and Lawton 1994; Norbury

2001). As a consequence, the secondary prey lack a density-dependent mechanism by which to recover from rarity – as is the case in simple prey–predator models (Holt and Lawton 1994; Choquenot and Parkes 2001). Indeed, shared natural enemies can result in inverse density dependence, in which an individual's chance of predation increases as its population size declines, carrying a high risk of extinction (Holt and Lawton 1994). One example of such predator–prey interactions in New Zealand is that of stoats (*Mustela erminea*), the main predator in *Nothofagus* forests, which feed primarily on mice (*Mus musculus*) and also on native birds; another example being cats and ferrets (*Mustela furo*), which are the main predators in dry grassland areas, feeding primarily on rabbits (*Oryctolagus cuniculus*) and rodents (*Rattus* spp.), and secondarily on native birds and skinks (*Oligosoma* spp.; see Chap. 16).

Secondary predation may impede recovery if the removal of a generalist predator from a food web leads to a rapid increase in the population of its exotic prey, without benefiting the native organism of conservation interest. For example, extermination of Pacific rats (*Rattus exulans*) on Motuopao Island led to an increase in the abundance of an exotic snail, which had detrimental consequences for the native species of snail (Towns et al. 1997). Feral cats (*Felis catus*) prey primarily on rats where these co-occur (up to 95 % of food intake on islands where rats are present but rabbits are absent), so their eradication could lead to increased rat populations, and to greater rat predation of other flightless bird species (Karl and Best 1982; Zavaleta et al. 2001). There are several examples of exotic plants responding to release from herbivores. Removal of rabbits from short-tussock grasslands led to rapid increases in the biomass of exotic grasses, with less response from native tussock species (Norbury et al. 2002). Similarly, Moore (1976) reported an expansion of sweet briar (*Rosa rubiginosa*) following rabbit control on Molesworth, and Benecke (1967) documented the spread of wilding pines (*Pinus* spp.) into tussock grasslands where sheep grazing was reduced. Radcliffe (1986) found that goats are effective at preventing the spread of gorse (*Ulex europaeus*), especially after fire.

Secondary predation also impedes recovery in systems in which the populations of prey and predators are influenced by pulsing of resources (Holt and Lawton 1994; Ostfeld and Keesing 2000; Norbury 2001), leading to increased predation when predator numbers are temporarily inflated by a superabundance of prey, or when primary prey populations crash, leaving predators in search of alternative foods. For example, sudden declines in rabbit abundance (brought about by poisoning and disease) resulted in a marked increase in skink predation by cats and ferrets in lowland tussock grasslands (Norbury 2001). Irruptions in predator–prey populations are a feature of many New Zealand ecosystems because a high proportion of plants are mast seeding, meaning that heavy-seed crops occur at infrequent intervals, with little production in the intervening years (Schauber et al. 2002). Consequently, the energy supply into the granivore-based food chain is highly variable, an aspect particularly well documented for *Nothofagus* forests (see Chap. 16). A wider issue is that some threatened native prey populations are more reliant on heavy seeding years to reproduce than others – for example, the kakapo (*Strigops habroptilus*) seem to breed only when resources are plentiful, whereas the mohua (*Mohoua ochrocephala*) breed every year. Native species dependent on heavy seeding events are particularly vulnerable to predators whose populations are also driven by those events (Barlow and Choquenot 2002).

Recovery is also impeded when more than one exotic organism preys upon a native organism, as single-species control will not necessarily relieve the pressure on that species (Forsyth et al. 2000). For example, stoats and rats both prey on birds (Innes and Hay 1991). A population of 80 mohua in Fiordland National Park lost 53 individuals during the very cold winter of 1996, and the population did not recover in the following 2 years (Dilks 1999). Two large beech-mast-seeding events in 1999 and 2000 led to huge increases in ship rat (*Rattus rattus*) and stoat numbers, resulting in the local extinction of mohua. In this case, mohua nests were being monitored by video cameras, and all recorded predation events were by rats, rather than by stoats (P.J. Dilks, personal communication).

Finally, selective herbivory by an invasive mammal can shift the abundance of native species in the food web, creating alternative states which do not revert to the original. The removal of deer-preferred species from forests is thought to enable the spread of less-preferred forest species as a result of reduced competition (another example of "apparent competition"; Connell 1990; Holt and Lawton 1994). Once these less-preferred species occupy space, they may effectively prevent the re-establishment of the preferred species following deer control (Wardle 1984), and may alter the quality of litter entering the belowground food chain (Wardle et al. 2001).

22.3.2 Residual Pest Densities Remain too High to Allow Recovery

Native species will not recover unless exotic organisms are kept at low enough densities for sufficiently long for the recruitment rates of native species to exceed their mortality rates (Parkes 1996). Needless to say, the best opportunities for recovery arise from total eradication (see Chap. 26). However, where one-off operations cannot achieve eradication, the only positive alternative is to manage pests on a sustained-control basis, and this is by far the most common strategy employed (Sweetapple and Burns 2002; Parkes and Murphy 2003). By definition, control means that some invasive organisms persist, albeit at reduced density.

Managers need to know when, where, and how often to control exotics in order to maximize conservation benefit (Hone 1994). For example, Innes et al. (1999) showed that recovery of North Island kokako (*Callaeas cinerea*), an endangered forest bird, depends upon keeping possum (*Trichosurus vulpecula*) and ship rats at low levels to increase nest success, but such intensive management is costly. Recent modelling work suggests that pulsing pest control may be a viable alternative to continual control (Basse et al. 2003). The rationale is that kokako adults are long-lived (>20 years) and much less heavily preyed upon than juveniles, so recruitment into the adult population does not have to occur every year for the population to persist. Control in 3 out of every 10 years seems sufficient.

22.3.3 Global or Local Extinction of Ecosystem Components

In the strictest sense, restoration is impossible where original ecosystem components are now globally extinct. Most prominently, New Zealand ecosystems suffered the irrevocable loss of an entire herbivore–predator system with the extinction of moa species, and consequent extinction of the Haast eagle which preyed upon them (Bond et al. 2004). Where extinction of ecosystem components is local, rather than global, natural recovery may be greatly impeded. Browsing by introduced deer, for example, may eventually result in the elimination of plant species from patches of forest and, without local seed sources, such species may be unable to re-establish. Seed limitation may be exacerbated in New Zealand by the loss of native pollinators and seed dispersers as a consequence of predation (Clout and Hay 1989), and by a lack of long-lived seeds in the soil (Enright and Cameron 1988).

22.4 Vegetation Recovery Following Herbivore Control in New Zealand

22.4.1 Recovery on Lands Managed for Conservation

Deer

As the abundance of deer increased rapidly following colonization, there was highly selective foraging on understory plants in forests (e.g. Riney et al. 1959; Mark and Baylis 1975), and on a variety of grasses and herbs in grasslands. A subset of the New Zealand flora largely disappeared from the browse layer, primarily those species with relatively large leaves containing little lignin and non-digestible fibre, although the association between any one trait and palatability is weak (Wardle et al. 2001; Forsyth et al. 2002). Divaricating branches and spines on herbs may have provided some physical defences against extinct avian herbivores, but afforded less protection against ungulates (Bond et al. 2004). Not all vegetation types were equally affected; ecosystems on infertile soils or xeric sites would not have contained many nutritious species for deer to eat (Coomes et al. 2003a). The question of whether deer are responsible for ongoing changes is more debatable. Bellingham and Allan (2003) showed that some changes in Stewart Island forests over the last 30 years could be associated with ongoing white-tail deer herbivory, whereas other changes appeared unrelated to deer effects. Husheer et al. (2003) report that sika deer have reduced regeneration in a beech forest, and may cause the development of browse-tolerant herbaceous turfs, but the compositional changes were not strong. Coomes et al. (2003b) showed that a national decline of 18 % in the number of small trees (i.e. those of 2–5 cm diameter) was generally associated with the loss of deer-palatable species, but it should be noted that a non-preferred species (*Nothofagus solandri*) showed the greatest decline in numbers, so factors other than herbivores must contribute to the observed effects.

Whether ecosystems will recover from such changes following deer control is context dependent. Alpine grasslands, which occupy a 500-m-wide low-alpine zone above the climatic treeline, were highly degraded by red deer (*Cervus elaphus scoticus*) and other ungulate grazers by the 1960s, especially in remote areas where hunting was ineffective in limiting exotic animal numbers. Government "cullers" employed as early as the 1940s to deal with some of the worst-affected areas were superseded in the late 1960s by very effective helicopter hunting which virtually eliminated the alpine deer populations. Recovery of alpine grasslands has been monitored in Mt. Aspiring National Park using a series of 88 permanent photographic points in several grassland types. These plots were established in 1970 and 1973, and then re-monitored in 1977, 1986, 1991 and 1999, in conjunction with brief semi-quantitative community descriptions (Mark 1989). This information provides some basis for assessing the degree of ecosystem recovery, against baseline information on grassland composition at the time of European settlement from qualitative descriptions by early, mostly amateur, botanists (Poppelwell 1916), together with observations of the existing plant cover on bluffs inaccessible to the feral mammals.

The alpine grasslands have generally shown distinct recovery since ungulates (deer, and chamois, *Rupicapra rupicapra*) have been reduced, particularly of the dominant tussock, forb and shrub species. The most striking recovery was the midribbed snow tussock *Chionochloa pallens* type, but all grassland types showed increases in deer-palatable species. More generally, it can be cautiously concluded that the subalpine scrub and low-alpine snow tussock grasslands may now be close to their "pre-deer" state in terms of both structure and composition (see also Rose and Platt 1987). The retention of the tussock species as subdominants in the degraded grasslands is likely to have been a major factor in facilitating recovery. Two factors could undermine the obvious recent recovery trend: firstly, higher deer numbers, because of a decline in hunting with a reduced global market for feral animals, and secondly, the current spread, at least locally, of the invasive herb *Hieracium*. A valley-bottom site (705 m a.s.l.) which was added to the national park alpine grassland monitoring programme in 1991 shows this herb is increasing rapidly in cover at the expense of a range of indigenous species.

Effective recovery of forest ecosystems depends upon deer numbers being brought down for sufficiently long periods to allow the regeneration of palatable species, and there are many reasons why recovery may not occur even then (Coomes et al. 2003a). A good place to test whether long-term control leads to forest recovery is the Murchison Mountains, in Fiordland National Park, where deer densities have been kept low for the last 25 years (Nugent and Sweetapple 1989; Coomes et al., unpublished data). In the valley bottoms, deer removed many preferred species from the browse layer, but these species persisted as established trees: by 1975, palatable species comprised only 10 % of all seedlings but around 20 % of tree basal area, and virtually none of those seedlings survived beyond a few centimetres height. Over the last 25 years, there have been some signs of recovery, with a doubling in number of preferred seedlings, but very few seedlings have grown beyond 135 cm in height (Fig. 22.1a). The lack of larger seedlings could simply reflect slow rates of recruitment and growth in forest understory conditions, or may indicate that the residual population of deer continues to exert a strong browsing pressure (Nugent et al. 2001). Evidence from four valley-bottom exclosure plots, established around 25 years ago, points to the latter explanation. Within these

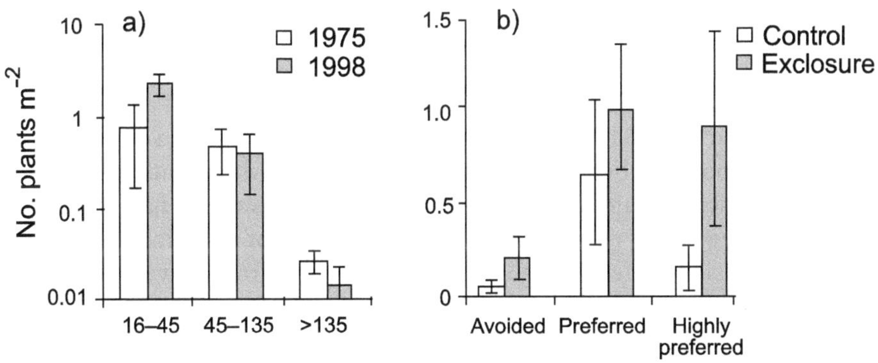

Fig. 22.1a Recovery of forests with sustained deer control in Fiordland National Park. There were small but statistically significant increases in the number of seedlings of deer-preferred species on the forest floor in 1998 compared with 1975, but no changes were observed for taller seedlings. **b** Many more tall seedlings (>135 cm height, <2.5 cm stem diameter) of highly preferred trees were found inside exclosure plots than in neighbouring control plots, indicating that deer still hamper recovery, despite their lower numbers

exclosures, large seedlings of preferred species are so profuse that they form a dense layer, under which smaller seedlings and herbs are suppressed. In stark contrast, the forest immediately outside the exclosures has virtually no highly preferred saplings (Fig. 22.1b). It is likely that the large deer herds which roamed the area until the 1960s had removed the majority of preferred understory plants by the time their population was reduced, and it now takes only a few deer to prevent recovery.

Recovery following deer control may well depend on the extent of degradation. For example, deer control on Secretary Island (Fiordland National Park) was started within a few years of the animal's arrival, and the process was partially arrested (Mark and Baylis 1975). These results highlight the importance of eradicating any new populations of deer which establish in previously deer-free areas, such as in Taranaki and Northland (Fraser et al. 2003). Recovery may also depend on vegetation type. For example, high-altitude forests have few palatable canopy trees (<1 % of basal area) and, in the Murchison Mountains, have not changed in either structure or composition over the last 25 years, despite the near-eradication of deer.

Goats

Goats are generalist feeders which are held responsible for pushing several insular plant species to the verge of extinction (Parkes 1990), increasing the abundance of non-preferred species such as the grass *Microlaena avenacea* on forest floors (Parkes 1990), turning closed forest into open woodland (Parkes 1993), and altering habitat to the detriment of native animals, such as the frog *Leiopelma hochstetteri*. First liberated in New Zealand by Captain Cook in 1773, goats now occupy about 15 % of the land (Parkes 1990). Goats were eradicated from Raoul Island (in the Kermadec group) in 1984 after a protracted culling campaign, largely to protect the endemic biota, including the dominant tree (*Metrosideros kermadecensis*) which formed a third of the goats' diet (Parkes 1984). Parkes (1984) records that terrestrial seedlings and epicormic shoots had become common again by 1982, when goats were on the verge of eradication, and noted other preferred species which had started to regenerate. Vegetation composition also showed signs of reverting to its probable pre-invasion state after goats were eradicated from Cuvier Island in the1960s (Atkinson 1988), and coastal forest regenerated on the Three Kings Islands following goat eradication in 1946 (Wright and Cameron 1990).

On the mainland, goat control can halt the losses of preferred plants if it is initiated soon after colonization. For example, goats colonized Mt. Egmont National Park in about 1910, and have been culled annually since 1924, maintaining the resident population at about 3,000 since 1981 (Parkes 1990; Forsyth et al. 2003). Gut-content surveys show that the culled animals consume a high proportion of their preferred foodstuffs, such as hen and chicken fern (*Asplenium bulbiferum*), which are much less common in heavily browsed areas (Mitchell et al. 1987). In contrast, infrequent control may not provide sufficient opportunities for vegetation to recover, especially under closed forests. Parkes (1990) reported ongoing regeneration failure of preferred species in forests near Gisborne, and this despite goat culling over a 20-year period.

Seedling recruitment and growth are much higher in canopy gaps created by the death of mature trees (e.g. Sweetapple and Burns 2002). Even if low densities of goats and deer prevent regeneration under closed forests, the opening up of those forests by a large-scale disturbance event (e.g. a windstorm or earthquake) might produce such a sudden increase in seedling numbers that a proportion will grow beyond the browse layer and establish as a new cohort of trees. Manipulation of pest densities led to increases in the regeneration success of preferred species in canopy gaps scattered through North Island forests. Consequently, pulsed control of terrestrial herbivores soon after major disturbance events would allow an even greater proportion of preferred species to escape the browse layer. Since some palatable tree species survive for many centuries, even infrequent opportunities for recruitment are sufficient for their long-term persistence.

22.4.2 Recovery of Farmed Montane Grasslands

Much of New Zealand's land surface has been transformed by agriculture, and now hosts ecosystems comprised primarily of exotic species. For example, montane sites in the rain shadow of the Southern Alps were *Chionochloa* tussock grasslands when Europeans arrived, but were previously open forests until burnt by Maori settlers (McGlone 2001). *Chionochloa* species have a general tolerance of fire but a clear intolerance to heavy grazing, particularly during the early period of recovery from fire (Payton and Mark 1979), when the re-growth foliage is relatively nutritious (Payton et al. 1986; Mark 1994). Consequently, a combination of frequent burning and heavy grazing by sheep (*Ovis aries*) and rabbits rapidly degraded the montane tussock grasslands. These grasslands were further degraded in the 1950s, during which the government subsidized aerial over-sowing of exotic grasses, and topdressing (O'Connor 1982). Nevertheless, native species still persist in such farmland, and rocky outcrops within the agricultural matrix support native species. A key question is whether these partially transformed landscapes will ever recover if grazing pressure and burning are reduced.

Grazing pressure on montane tussock grasslands has declined in recent years as a consequence of diseases introduced to control feral rabbits. Populations of rabbit have fallen by about 50 % on average, and by more than 90 % at a few sites, as a result of rabbit haemorrhagic disease, which reached New Zealand in 1997 (Norbury et al. 2002). As a consequence, there was a 40–500 % increase in the biomass accumulation of exotic grasses (particularly *Agrostis capillaris, Anthoxanthum odoratum* and *Holcus lanatus*) and exotic herbs (particularly *Hieracium lepidulum*), but slower-growing native tussocks (*Festuca novaezelandiae* and *Poa colensoi*) have not responded at anywhere near the same extent (Norbury et al. 2002). Other studies show that reduced grazing pressure can be detrimental to native plant species in the long term, because they tend to be smothered by the faster-growing exotics (Walker 2000).

Small reserves established in the late 1960s provide important insights on the "recovery" potential of montane grasslands (Mark 1985). These areas were fenced to prevent further grazing by domestic stock, and burning was suppressed. There has been a threefold increase in the biomass indices of narrow-leaved snow tussock (*Chionochloa rigida*) over the past 30 years in one such reserve (Black Rock Reserve), primarily because established tussocks have recovered. Seedling establishment remains slow, particularly within dense swards of exotic grass. Of the two persistent exotic sward grass species, *Anthoxanthum odoratum* is clearly still declining as the snow tussock canopy recovers, whereas *Agrostis capillaris* is much more resistant and will probably persist indefinitely. Some low shrub species also increased in biomass index (Dickinson et al. 1992). Plant diversity declined from a mean of 53 taxa in 1971 (range 49–60), to 41 (range 34–44) after 16 years, and to 40 (range 28–46) over the following 14 years. Although most of the exotic species originally present still persist, they are generally declining, particularly the rosette species (Mark and Dickinson 2003). The removal of grazing mammals from Campbell Island provides another example of the strong recovery of native tussock species (Molloy and Dingwall 1990).

Yet, natural recovery of tussock grasslands is far from certain. A rapid decline in species richness across many tussock grasslands is being reported, most notably a decline in small herbaceous plants other than *Hieracium*. These changes seem unrelated to local variation in management, and are occurring without any obvious explanation (Duncan et al. 2001). The widespread rise in abundance of *Hieracium* spp. is another factor which might prevent effective recovery (e.g. Duncan et al. 2001), although some sites are more susceptible than others (Duncan et al. 1997; Meurk et al. 2002).

22.5 Conclusions

Parkes (1990) reiterated that the most common flaw in invasive organism control is to confuse goals with means. The complexities of food-web interactions provide reasons to believe that removal of invasive organisms will not necessarily lead to ecosystem recovery, so the confusion of means with goals has important implications. Quantification of the extent to which reducing pest "pressure" leads to conservation benefits needs to be explicitly incorporated into the evolving set of tools used in New Zealand for conservation planning (e.g. Stephens et al. 2002), and could alter the prioritisation of management activities.

Tussock grasslands and island vegetation sometimes appear to recover rapidly following the control or eradication of herbivores, although there is a lack of baseline data. In contrast, recovery under closed forests following long-term deer and goat control is slow. Seedling growth under closed forest is slow, and few seedlings survive, irrespective of browsing pressure. It appears that even low densities of herbivores can prevent regeneration under these conditions, because herbivores eat almost all seedlings of preferred species which grow into the browse layer (e.g. Wardle 1984; Parkes 1990; Nugent et al. 2001; Coomes et al. 2003a). Pulsed control of herbivores should be considered as an option for managing forests, because only infrequent recruitment is required for the persistence of these long-lived species. It would make sense for peak hunting effort to coincide with periods after large-scale disturbances, after which seedling recruitment and growth rates are much higher (e.g. Sweetapple and Burns 2002), providing greater opportunities for individuals to escape the browse layer.

Restoration is a practical subject guided by general scientific theories. Success is contingent upon many poorly understood factors, including the complex interactions within food webs, the idiosyncrasies of individual species, and the influx of new invasive species. Effective management therefore requires knowledge of both the invasive and native species. We have described the effects of herbivore control on vegetation, because this subject is relatively well documented, but much less is known about the recovery of other ecosystem components (see Chap. 20). Simulation modelling might bring about a greater understanding of ecosystem responses to various intensities of pest control (e.g. Choquenot and Parkes 2001; Coomes et al. 2003a, b). Monitoring before and after control will also assist with the provision of information for future management options (Dickinson et al. 1992), and sharpen debate over the options and objectives of management.

Acknowledgements This chapter benefited immeasurably from detailed reviews by John Parkes and Peter Bellingham.

References

Anonymous (2000) The New Zealand biodiversity strategy: February 2000. Department of Conservation and Ministry for the Environment, Wellington, New Zealand

Atkinson IAE (1988) Opportunities for ecological restoration. NZ J Ecol 11:1–12

Barlow ND, Choquenot D (2002) Predicting the impact and control of stoats: a review of modelling approaches. Science for Conservation 191. Department of Conservation, Wellington, New Zealand

Basse B, Flux I, Innes J (2003) Recovery and maintenance of North Island kokako (*Callaeas cinerea wilsoni*) populations through pulsed pest control. Biol Conserv 109:259–270

Bellingham PJ, Allan CN (2003) Forest regeneration and the influences of white-tailed deer (*Odocoileus virginianus*) in cool temperate New Zealand rain forests. For Ecol Manage 175:71–86

Benecke U (1967) The weed potential of lodgepole pine. Tussock Grasslands Mountain Lands Inst Rev 13:36–42

Bond WJ, Lee WG, Craine JM (2004) Plant structural defences against browsing birds: a legacy of New Zealand's extinct moas. Oikos 104:500–508

Choquenot D, Parkes J (2001) Setting thresholds for pest control: how does pest density affect resource viability? Biol Conserv 99:29–46

Clout MN, Hay JR (1989) The importance of birds as browsers, pollinators and seed dispersers in New Zealand forests. NZ J Ecol suppl 12:27–33

Connell JH (1990) Apparent versus 'real' competition in plants. In: Grace JB, Tilman D (eds) Perspectives on plant competition. Academic Press, New York, pp 9-25

Coomes DA, Allen RB, Forsyth DM, Lee WG (2003a) Factors preventing the recovery of New Zealand forests following control of invasive deer. Conserv Biol 17:450–459

Coomes DA, Duncan RP, Allen RB, Truscott J (2003b) Disturbances prevent stem size-density distributions in natural forests from following scaling relationships. Ecol Lett 6:980–989

Dickinson KJM, Mark AF, Lee WG (1992) Long-term monitoring of non-forest communities for biological conservation. NZ J Bot 30:163–179

Dilks P (1999) Recovery of a mohua (*Mohoua ochrocephala*) population following predator control in the Eglinton Valley, Fiordland, New Zealand. Notornis 46:323–332

Duncan RP, Colhoun KM, Foran BD (1997) The distribution and abundance of *Hieracium* species (hawkweeds) in the dry grasslands of Canterbury and Otago. NZ J Ecol 21:51–62

Duncan RP, Webster RJ, Jensen CA (2001) Declining plant species richness in the tussock grasslands of Canterbury and Otago, South Island, New Zealand. NZ J Ecol 25(2):35–47

Enright NJ, Cameron EK (1988) The soil seed bank of a kauri (*Agathis australis*) forest remnant near Auckland, New Zealand. NZ J Bot 26:223–236

Forsyth DM, Parkes JP, Hickling GJ (2000) A case for multi-species management of sympatric herbivore pest impacts in the central Southern Alps, New Zealand. NZ J Ecol 24:97–103

Forsyth DM, Coomes DA, Nugent G, Hall GMJ (2002) Diet and diet preferences of introduced ungulates (Order: Artiodactyla) in New Zealand. NZ J Zool 29:323–343

Forsyth DM, Hone J, Parkes JP, Reid GH, Stronge D (2003) Feral goat control in Egmont National Park, New Zealand, and the implications for eradication. Wildl Res 30:437–450

Fraser KW, Parkes JP , Thomson C (2003) Management of new deer populations in Northland and Taranaki. Science for Conservation 212. Department of Conservation, Wellington, New Zealand

Hobbs RJ, Norton DA (1996) Towards a conceptual framework for restoration ecology. Restor Ecol 4:93–110
Holt RD, Lawton JH (1994) The ecological consequences of shared natural enemies. Annu Rev Ecol Syst 25:495–520
Hone J (1994) Analysis of vertebrate pest control. Cambridge Univ Press, Cambridge
Husheer SW, Coomes DA, Robertson AW (2003) Long-term influences of introduced deer on the composition and structure of New Zealand *Nothofagus* forests. For Ecol Manage 181:99–117
Innes JG, Hay JR (1991) The interactions of New Zealand forest birds with introduced fauna. Acta Congr Int Ornithol 20:2523–2533
Innes J, Hay R, Flux I, Bradfield P, Speed H, Jansen P (1999) Successful recovery of North Island kokako *Callaeas cinerea wilsoni* populations, by adaptive management. Biol Conserv 87:201–214
Karl BJ, Best HA (1982) Feral cats on Stewart Island; their foods, and their effects on kakapo. NZ J Zool 9:287–293
Logan H (2001) Gondwana invaded: an address on distinctive features of managing indigenous biodiversity in protected areas in New Zealand. J R Soc NZ 31:813–818
Mansfield B (1996) Ecosystem restoration on mainland New Zealand. IUCN World Conservation Congress, Montreal, http://www.doc.govt.nz/Conservation
Mark AF (1985) The botanical component of conservation in New Zealand. NZ J Bot 23:789–810
Mark AF (1989) Responses of indigenous vegetation to contrasting trends in utilizations by red deer in two southwestern New Zealand national parks. NZ J Ecol suppl 12:103–114
Mark AF (1994) Effects of burning and grazing on sustainable utilization of upland snow tussock (*Chionochloa* spp.) rangelands for pastoralism in South Island, New Zealand. Aust J Bot 42:149–161
Mark AF, Baylis GTS (1975) Impact of deer on Secretary Island, Fiordland, New Zealand. Proc NZ Ecol Soc 22:19–24
Mark AF, Dickinson KJM (2003) Temporal responses over 30 years to removal of grazing from a mid-altitude snow tussock grassland reserve, Lammerlaw Ecological Region, New Zealand. NZ J Bot 41:655–688
McGlone MS (2001) The origin of the indigenous grasslands of southeastern South Island in relation to pre-human woody ecosystems. NZ J Ecol 25:1–15
Meffe GK, Carroll CR (1997) Principles of conservation biology, 2nd edn. Sinauer, Sunderland, MA
Meurk CD, Walker S, Gibson RS, Espie P (2002) Changes in vegetation states in grazed and ungrazed Mackenzie Basin grasslands, New Zealand, 1990–2000. NZ J Ecol 26:95–106
Mitchell RJ, Fordham RA, John A (1987) The annual diet of feral goats (*Capra hircus* L.) in lowland rimu-rata-kamahi forest on eastern Mount Taranaki (Mt Egmont). NZ J Zool 14:179–192
Molloy LF, Dingwall PR (1990) World heritage values of New Zealand islands. In: Towns DR, Daugherty CH, Atkinson IAE (eds) Ecological restoration of New Zealand islands. Conservation Sciences Publ 2. Department of Conservation, Wellington, New Zealand, pp 194–206
Moore LB (1976) The changing vegetation of Molesworth Station, New Zealand, 1944 to 1971. DSIR Bull 217:1–118
Norbury G (2001) Conserving dryland lizards by reducing predator-mediated apparent competition and direct competition with introduced rabbits. J Appl Ecol 38:1350–1361

Norbury G, Heyward R, Parkes J (2002) Short-term ecological effects of rabbit haemorrhagic disease in the short-tussock grasslands of the South Island, New Zealand. Wildl Res 29:599–604

Nugent G, Sweetapple P (1989) The impact of three deer hunting regimes in northeastern Fiorldland. NZ J Ecol 12:33–46

Nugent G, Fraser W, Sweetapple P (2001) Top down or bottom up? Comparing the impacts of introduced arboreal possums and 'terrestrial' ruminants on native forests in New Zealand. Biol Conserv 99:65–79

O'Connor KF (1982) The implications of past exploitation and current developments to the conservation of South Island tussock grasslands. NZ J Ecol 5:97–107

Ostfeld RS, Keesing F (2000) Pulsed resources and community dynamics of consumers in terrestrial ecosystems. Trends Ecol Evol 15:232–237

Parkes JP (1984) Feral goats on Raoul Island. II. Diet and notes on the flora. NZ J Ecol 7:95–101

Parkes JP (1990) Feral goat control in New Zealand. Biol Conserv 54:335–348

Parkes JP (1993) Feral goats: designing solutions for a designer pest. NZ J Ecol 17:71–83

Parkes JP (1996) Integrating the management of introduced mammal pests of conservation values in New Zealand. Wildl Biol 2:179–184

Parkes J, Murphy E (2003) Management of introduced mammals in New Zealand. NZ J Zool 30:335–359

Payton IJ, Mark AF (1979) Long-term effects of burning on growth, flowering, and carbohydrate reserve in narrow-leaved snow tussock (*Chionochloa rigida*). NZ J Bot 17:43–54

Payton IJ, Lee WG, Dolby R, Mark AF (1986) Nutrient concentrations in narrow-leaved snow tussock (*Chionochloa rigida*) after spring burning. NZ J Bot 24:529–537

Polis GA, Sears ALW, Huxel GR, Strong DR, Maron J (2000) When is a trophic cascade a trophic cascade? Trends Ecol Evol 15:473–475

Poppelwell DL (1916) Botanical results of an excursion to the upper Makarora Valley and the Haast Pass, supported by a list of the species observed. Trans Proc NZ Inst 49:161–166

Radcliffe JE (1986) Gorse – a resource for goats? NZ J Exp Agric 14:399–410

Riney T, Watson JS, Bassett C, Turbott EG, Howard WE (1959) Lake Monk expedition, an ecological study in southern Fiordland. DSIR Bull 135:1–75

Rose AB, Platt KH (1987) Recovery of northern Fiordland alpine grasslands after reduction in the deer population. NZ J Ecol 10:23–33

Saunders A, Norton DA (2001) Ecological restoration at Mainland Islands in New Zealand. Biol Conserv 99:109–119

Schauber EM, Kelly D, Turchin P, Simon C, Lee WG, Allen RB, Payton IJ, Wilson PR, Cowan PE, Brockie RE (2002) Masting by eighteen New Zealand plant species: the role of temperature as a synchronizing cue. Ecology 83:1214–1225

Simberloff D (1990) Reconstructing the ambiguous: can island ecosystems be restored? In: Towns DR, Daugherty CH, Atkinson IAE (eds) Ecological restoration of New Zealand islands. Conservation Sciences Publ 2. Department of Conservation, Wellington, New Zealand, pp 37–51

Stephens T, Brown D, Thornley N (2002) Measuring conservation achievement: concepts and their application over the Twizel area. Sci Conserv 200:1–114

Sweetapple PJ, Burns BR (2002) Assessing the response of forest understoreys to feral goat control with and without possum control. Sci Conserv 201:1–33

Towns DR (2002) Korapuki Island as a case study for restoration of insular ecosystems in New Zealand. J Biogeogr 29:593–607

Towns DR, Simberloff D, Atkinson IAE (1997) Restoration of New Zealand islands: redressing the effects of introduced species. Pac Conserv Biol 3:99–124

Walker S (2000) Post-pastoral changes in composition and guilds in a semi-arid conservation area, Central Otago, New Zealand. NZ J Ecol 24:123–137

Wardle JA (1984) The New Zealand beeches. Ecology, utilisation and management. New Zealand Forest Service, Wellington, New Zealand

Wardle DA, Barker GM, Yeates GW, Bonner KI, Ghani A (2001) Introduced browsing mammals in New Zealand natural forests: aboveground and belowground consequences. Ecol Monogr 71:587–614

Wright AE, Cameron EK (1990) Vegetation management on northern offshore islands. In: Towns DR, Daugherty CH, Atkinson IAE (eds) Ecological restoration of New Zealand islands. Conservation Sciences Publ 2. Department of Conservation, Wellington, New Zealand, pp 221–239

Zavaleta ES, Hobbs RJ, Mooney HA (2001) Viewing invasive species removal in a whole-ecosystem context. Trends Ecol Evol 16:454–459

23 Biological Control: Reducing the Impact of Invasive Weeds and Pests, or just Another Source of Alien Invaders?

S.V. Fowler and T.M. Withers

23.1 Introduction

Biological control to target an alien invasive pest is inevitably contentious because the biocontrol agent itself increases the number of alien species (Keey 2003). Historically, careless introductions for biological control have exacerbated the ecological problems associated with introduced species. Introductions of generalist vertebrate predators in New Zealand were early errors: releasing stoats to control rabbits in the 19th century had catastrophic effects on the distinctive biota (Thomson 1922). However, biocontrol introductions for the past 130 years in New Zealand have almost exclusively targeted arthropod pests and weeds. In this chapter, we limit our discussion to deliberately introduced invertebrates and pathogens; agents that can show a high degree of specificity to their target pests. We begin by reviewing the global debate that has challenged the environmental safety of biocontrol introductions. Then, for biological control programs against weeds and insect pests in turn, we review safety and success, and discuss whether successful programs have resulted in environmental benefits. Next, we attempt to answer a specific criticism – that biocontrol introductions are swamping the indigenous fauna of New Zealand. Finally, we discuss whether biocontrol is at all worth the risk.

23.2 Challenge to the Environmental Safety of Biological Control

The environmental safety of introduced biocontrol agents has been increasingly challenged following the influential review by Howarth (1983) titled

Ecological Studies, Vol. 186
R.B. Allen and W.G. Lee (Eds.)
Biological Invasions in New Zealand

"Classical biological control: panacea or Pandora's box?". The ensuing debate has been lively, with rebuttals from the biological control community (e.g., Funasaki et al. 1998). However, more detailed analyses and ecological studies have now provided evidence of some non-target effects from insects introduced for the biocontrol of various pests and weeds (Howarth 1991; Simberloff and Stiling 1996; Louda et al. 1998; Thomas and Willis 1998; Henneman and Memmott 2001). Prior to the 1980s, many insect parasitoids and predators introduced for control of pest arthropods worldwide were intentionally not host specific (so that reservoir populations of the agent would be maintained on non-targets even when the target pest was rare). Thus, in one food web study (Henneman and Memmott 2001), substantial infiltration of the native leafmining moth community in montane forest in Hawai'i involved only relatively generalist parasitoids that were released in biological control programs before 1945. Modern biological control programs against arthropod targets use more-specialist predators (e.g., Booth et al. 1995) or parasitoids (Cameron et al. 1993; Sands 1997; Charles and Allan 2002). With introductions for the biological control of weeds, there have always been concerns, initially economic, about the possible negative consequences from agents attacking non-target plants. Consequently, host range testing has been a major part of programs from an early date. Despite this testing, there are examples of non-target attack by insect herbivores introduced for biological control of weeds (McFadyen 1998). Worldwide, these represent perhaps 3 % of the nearly 400 releases of agent species in the past 100 years (Julien and Griffiths 1999; Fowler et al. 2000). Most are either minor, transitory or were predicted from the host range testing (Fowler et al. 2000). In the USA, two weed biocontrol agents do attack non-target native plants. Firstly, the weevil *Rhinocyllus conicus*, released in the USA in the 1960s for biological control of alien thistle weeds, is attacking native *Cirsium* spp. thistles, with potentially harmful effects on their populations (Louda et al. 1997). This was predictable from host range testing, but was not considered important in the 1960s, highlighting the irreversibility of biological control, and that environmental concerns have altered over time. Secondly, the pyralid moth *Cactoblastis cactorum* is attacking endangered native *Opuntia* spp. in Florida (Bennett and Habeck 1992). The arrival of *C. cactorum*, following its deliberate release in several nearby Caribbean islands, raises the issue of cross-border effects of biological releases (Pemberton 2000).

Could evolutionary changes in host range occur after an approved, apparently low-risk biocontrol agent is released? Evolutionary theory tells us that expansions in host range can happen (Futuyma 2000). Indeed, there are examples of rapid expansions in the host range of herbivorous insects, but this does not appear to be a common phenomenon in narrowly host-specific herbivores (van Klinken and Edwards 2002). There are exceptions, however, and the phylogenies of some genera of herbivorous insects also suggest that monophagous species can evolve to attack new plant species (Futuyma 2000).

For weed and arthropod biocontrol agents, there are no proven examples yet where post-release expansions in host range have occurred. Most, if not all examples appear to be temporary "spill-over" effects when the attack levels on the target weed are extremely high, rather than an evolutionary increase in host range (Marohasy 1996). Currently, it appears that the risk of host-specific insects evolving to attack new target species is very low (van Klinken and Edwards 2002).

23.3 Classical Biological Control of Weeds

23.3.1 History and Safety Record of Biological Control of Weeds

Since 1929, 39 species of biocontrol agents have been released into New Zealand with the aim of suppressing 14 weed species (including one native plant species, on which the one agent released failed to establish; Fowler et al. 2000; Landcare Research, unpublished data). In all, 27 agent species established, seven failed, and five are too recent to assess, giving an establishment rate of 79 %. To discuss the safety of these agents, we divide the potential harmful effects into the direct impacts of herbivorous biocontrol agents attacking non-target plants, and the subtler, "knock-on" indirect effects in food webs.

Direct Non-Target Effects

A retrospective analysis of all insect herbivore introductions for biological control of weeds in New Zealand (Fowler et al. 2004) showed that all insect herbivore species underwent host range testing using appropriate methods. These tests concentrated on economically significant alien plants, and did not always include indigenous plant species. Research to look for non-target effects of weed biocontrol agents in New Zealand is ongoing. To date, there are only two examples where New Zealand native species have been attacked: (1) cinnabar moth (*Tyria jacobaeae*) larvae can be found feeding on some native fireweeds, e.g., *Senecio minimus*, after defoliating their normal alien host, ragwort *S. jacobaea* (Fowler et al. 2004); and (2) the old man's beard leafmining fly (*Phytomyza vitalbae*) has been reared in low numbers from the native *Clematis foetida* growing at several sites in New Zealand (Paynter et al. 2004). Host range testing of cinnabar moth showed that larvae can feed on groundsel *Senecio vulgaris* and *Cineraria maritima* (Miller 1970). Hence, some feeding on native fireweeds that are closely related to ragwort, but were not included in the host range tests, is not unexpected. For the old man's beard leafminer, the sporadic and low level of attack on *C. foetida* appears consistent with the prediction from the host range tests that it might occasionally mine other *Clematis* species, but can only persist on the target weed, *C. vitalba* (Hill et al. 2001). There are several other examples of minor damage to a few non-native plant species (Fowler et al. 2000). Most of these were predictable from the original host range testing, and occur only when large populations of the biocontrol agents fully, or nearly fully, exploit their target weed in a given habitat. However, there are two cases where host range testing did not predict that agents would attack other alien plants closely related to the target weed. In both these cases, persistent levels of attack represent significant new trophic interactions. The first of these is the broom seed beetle, *Bruchidius villosus*, which has been reared from field-collected pods of the alien plant tagasaste, *Chamaecytisus palmensis* (the latter closely related to the target weed, broom *Cytisus scoparius*; Syrett 1999). The second case is the gorse pod moth *Cydia succedana*, which has been reared from field-collected pods of several alien Fabaceae in the same family as that of the target host, gorse *Ulex europaeus* (Fowler et al. 2004; Paynter et al. 2004). Research is ongoing to establish why host range testing failed to predict these host shifts, but in both cases there does not appear to have been an expansion of host range (Haines et al. 2004; Landcare Research, unpublished data).

Indirect Non-Target Effects

Currently, we know of no indirect effects from weed biocontrol agents that are mediated via, for example, increased populations of natural enemies that exploit the introduced agent. Food webs suggest that such knock-on effects should indeed occur (Landcare Research, unpublished data), but there is no suggestion that they are likely to be major effects unless a "keystone species" became affected in some way. Such food webs also indicate that knock-on effects could be potentially complex, difficult to measure, and not necessarily always negative. For example, there are scenarios where some indigenous species could benefit indirectly from the introduction of a biocontrol agent. The potential for indirect effects in food webs from biocontrol agents is the subject of increasing debate (Pearson and Callaway 2003; Thomas et al. 2004).

Successful reduction of a target weed might have some harmful knock-on effects on certain components of the native biota. For example, the recent application to release biocontrol agents against the alien weed mist flower, *Ageratina riparia*, evaluated a report that the endemic and threatened kiwi, *Apteryx* spp., sometimes used dense stands of mist flower to escape predation by dogs. Possible minor negative effects of successful mist flower biological control were duly acknowledged, but it was also noted that there are many native plant species that provide effective cover for kiwi (Fowler et al. 2000). Of course, the suppression of an alien environmental weed is expected to have far greater beneficial knock-on effects through potentially allowing the return of ecosystems dominated by native biota.

23.3.2 Success Rate of Classical Biological Control of Weeds

Global estimates of the success of biological control against weed species vary from 17 to 39 %, and the success rates of individual species of agents range from 8 to 29 % (Fowler 2000). However, many programs summarized in these global databases were under-resourced, under-monitored, or given insufficient time, which diminishes the overall success rate (McFadyen 1998; Fowler et al. 2000). In contrast, a recent analysis of weed biological control in South Africa rated over 80 % of programs fully or partially successful (Hoffmann 1995). Full success was defined as biological control suppressing the weed to a level such that other control methods were unnecessary. Partial success occurred when biological control made a significant contribution to weed suppression, but other control methods were still necessary at times. Applying these criteria to weed biological control in New Zealand is difficult because monitoring of success has been poor (Fowler et al. 2000). However, we probably have one complete success (St. John's wort, *Hypericum perforatum*) and four partial successes (nodding thistle *Carduus nutans*, ragwort *S. jacobaeae*, Mexican devil weed *Ageratina adenophora*, and alligator weed *Alternanthera philoxeroides*; Cameron and Wigley 1989; Syrett et al. 1991; Hayes 1999; Fowler et al. 2000; McGregor 2001). The program against Californian thistle, *Cirsium arvense*, has been classified as a failure (Fowler et al. 2000). For all the other biological control programs against weeds in New Zealand, it is too soon to classify them as successes or failures. Although the data are sparse, New Zealand may have a full or partial success rate in alien weed biological control programs of around 80 % (five full or partial successes; one failure; seven too soon to assess). Within the six programs where enough time has elapsed to assess success, it appears that five individual agent species have had substantial impacts, whereas a further eight have failed to impact on the weed to any significant degree, giving an individual agent success rate in New Zealand of 38 % (Landcare Research, unpublished data).

23.3.3 Environmental Benefits from the Suppression of Invasive Alien Weeds by Classical Biological Control

Benefits to the indigenous biota have probably occurred from biological control of environmental weeds such as alligator weed and Mexican devil weed in New Zealand, but no data are available from these programs. For programs against heather (*Calluna vulgaris*) and hieracium (*Hieracium* spp.), invading native habitats, biocontrol agents have been established (Landcare Research, unpublished data) and are being monitored, but it is too soon to detect benefits. In the forested Waitakere Hills near Auckland, where mist flower is being suppressed by the introduced white smut fungus *Entyloma ageratinae*, encouraging preliminary results show that native plants are colonizing plots containing damaged mist flower (Barton et al. 2004).

23.4 Classical Biological Control of Insects

23.4.1 History and Safety Record of Insect Biological Control

Without native natural enemies, many accidentally introduced pests became significant problems to early farmers in New Zealand. With few pesticides available, biological control was often the only solution to damaging pests. The first purposeful biocontrol introduction into New Zealand was *Coccinella undecimpunctata*, a coccinellid predator released in 1874 to attack aphids (Cameron et al. 1993).

Cameron et al. (1993) reported that 187 species of predators and parasitoids were purposely released into New Zealand against both native and alien pests up to 1991, although 124 of these apparently failed to establish. Since then, another 12 species have been released, with the targets being solely alien pests, and eight of these have established (Forest Research, unpublished data). So, in total, 71 alien insect biocontrol agents have established, although this value may be inaccurate, because the identity of past-imported insects can be uncertain (e.g., Berry 1998), and parts of the New Zealand fauna, such as the Hymenoptera, are poorly known (Berry 2005). Consequently, we cannot ascertain the exact number of released species of predators and parasitoids, and compare these to self-introduced incursions. Overall in New Zealand, the rate at which insect predator and parasitoid species have established in the country to date is approximately 36 % (Forest Research, unpublished data). This is similar to the reported world average of 34 % (Cameron et al. 1993).

Direct Effects on Native Targets

Until the 1980s, native species regarded as pests were often targets for biocontrol introductions in New Zealand. Indeed, the endemic scarabid grass grub, *Costelytra zealandica*, a pest of pastures, has had more alien natural enemies released against it in New Zealand than any other biological control target: from 1921 to 1981, 24 species, known to be natural enemies of other scarabids worldwide, were released (Cameron and Wigley 1989). None established, so there have been no negative effects. The same is true for releases made against endemic porina moths (*Wiseana* spp.; Cameron 1989), and the endemic nuisance pests black flies (called sandflies in New Zealand), *Austrosimulium australense* and *A. ungulatum* (Crosby 1989). Several alien biocontrol agents, which were released to control native noctuid moths, have established. The target moths include tomato fruitworm *Helicoverpa armigera* ssp. *conferta*, green looper *Chrysodeixis eriosoma*, greasy cutworm *Agrotis ipsilon*, and tropical armyworm *Spodoptera litura*, which are also native to other parts of Australasia, the Pacific, or even further afield. Given the continued impact on agriculture and horticulture by larval feeding of these moths, it is difficult to imagine that biological control is a threat to their populations.

Direct Effects on Non-Targets

Prior to 1940, biological control programs often imported natural enemies with relatively wide host ranges, such as predatory insects, and considered that some non-target attack was advantageous because it provided reservoir populations of the biocontrol agent even when the target pest was rare. For instance, *Coccinella septempunctata* was introduced as a predator of aphids, of which New Zealand has endemic species. It may be fortunate this species failed to establish (Cameron 1989). Parasitoids are usually much more host specific than predators (Sands 1997), but early introductions of these were made in New Zealand even when native species were expected to be attacked (Barron et al. 2003). A well-studied example is the tachinid *Trigonospila brevifacies*, introduced in 1967 for biological control of the Australian leafroller *Epiphyas postvittana*. Observations suggested it would attack native leafrollers, but some of these were considered major pests (Thomas 1989), so the introduction went ahead. Recent research shows that *T. brevifacies* attacks many non-target Lepidoptera in the families Oecophoridae, Tortricidae, Pterophoridae, Geometridae and Stathmopodidae in modified and indigenous ecosystems, with parasitism rates from 16 to 80 %. It is quite likely that *T. brevifacies* may be reducing population sizes of uncommon native Lepidoptera (Munro and Henderson 2002).

The contrast between earlier introductions of less-host-specific natural enemies and recent use of more-species-specific parasitoids is demonstrated in the biological control program against the sitona weevil *Sitona discoideus*. In 1982, the braconid parasitoid *Microctonus aethiopoides* was introduced into New Zealand (Stufkens and Farrell 1989) without laboratory host-specificity testing. It is now attacking at least 17 different species of non-target weevils in eight different genera in New Zealand (Barratt et al. 1997, 2000), including some native weevils with limited distributions in alpine habitats, which the parasitoid was not expected to inhabit. However, although negative effects seem likely, the actual impacts of *M. aethiopoides* on the populations of rare native weevils is extremely challenging to quantify, and a predictive model is being developed (B. Barratt, personal communication). Another biocontrol agent, *Microctonus hyperodae*, released in 1992 against Argentine stem weevil *Listronotus bonariensis*, was subjected to more-stringent host-specificity testing (Goldson et al. 1992). These tests predicted some attack on a native weevil, *Irenimus aequalis*, but permission to release was nevertheless given because the species had become a minor pest on alien host plants (Kuschel 1969). More recent research (Barratt et al. 1997) confirmed the earlier test results (Goldson et al. 1992), but also found very low levels of parasitism on a further four native species. Low levels of attack on native weevils could be viewed as an unavoidable cost that can be weighed up against any economic benefits gained from control of the pest weevil.

Indirect Effects

Where direct non-target attack by biocontrol agents occurs, there may also be ensuing indirect effects. For instance, *M. aethiopoides* is parasitizing the receptacle weevil *Rhinocyllus conicus* (Barratt et al. 1997; Murray et al. 2002), itself released for biological control of the alien pasture weed nodding thistle, *Carduus nutans*. Levels of parasitism of 0–17 % have been recorded, which may be too low to disrupt the contribution receptacle weevil makes to the biological control of nodding thistle (Murray et al. 2002).

Some "knock-on" effects in native food webs are also likely: for example, the introduced tachinid fly *T. brevifacies* is now the most abundant leafroller parasitoid in New Zealand, with a host range that overlaps with those of 12 native parasitoid species. This introduced biocontrol agent conceivably could be competitively excluding native parasitoid species in native forest habitats in New Zealand (Munro and Henderson 2002).

23.4.2 Success Rate of Classical Biological Control of Insects

Cameron et al. (1993) report that partial to complete success has been achieved in biological control programs targeting 24 insect and mite pests of field, fruit or greenhouse crops, pasture, and forestry. The following summary is taken from their comprehensive review and analysis. Success was assessed as complete, substantial or partial, following the criteria of DeBach (1964): complete success means that pest outbreaks are rare and other control methods are necessary; substantial success results in less pronounced economic or environmental benefits and/or other control methods are occasionally required; partial success is claimed when other control methods are commonly needed but that biological control augments these other methods, resulting in satisfactory control.

Complete biological control is claimed for five alien insects pests: *Icerya purchasi* (cottony cushion scale), *Eriococcus coriaceus* (gum tree scale), *Gonipterus scutellatus* (gum tree weevil), *Phyllonorycter messaniella* (oak leafminer), and *Phylacteophaga froggatti* (leaf blister sawfly; Cameron et al. 1993; Forest Research, unpublished data). No economic data are available on the benefits to New Zealand of these programs, in part because the control is so complete that pest outbreaks never occur. Indeed, Cameron et al. (1993) emphasize that the benefit to New Zealand of these programs is underrated because of the lack of outbreaks to remind people how serious and costly many of these pests were. Substantial success is claimed for 12 other pests, and the economic gains from some of these have been quantified. For example, the benefits from the control of *Metopolophium dirhodum* (rose-grain aphid) are estimated at NZ$ 0.3–5 million per year from yield increases and reduced chemical control costs (Grundy 1989; Farrell and Stufkens 1990). For the noctuid *Mythimna separata*, the estimated savings from substantial biological control are NZ$ 4.5–10 million per year (Hill and Allan 1989). A partial success level has been assessed for a further seven target insect pests. With the 12 releases of biocontrol agents against insect pests since 1989, it is too soon to assess their level of success. Given that 57 pests in productive sectors were targeted by biocontrol releases from 1874 to 1989, the 24 complete, substantial, or partial successes give an overall success rate of 42 % (Cameron et al. 1993; Forest Research, unpublished data).

23.4.3 Environmental Benefits from the Suppression of Alien Invasive Insects Using Classical Biological Control

To date, there is only one example of a classical biological program in New Zealand where the pest species were targeted in part because they are threats to the indigenous biota: three parasitic wasps, *Sphecophaga vesparum burra*, *S. v. vesparum* and *S. orientalis*, were released against the European wasps *Vespula vulgaris* and *V. germanica* in the 1990s (Donovan et al. 2002). Since the first two parasitoids have established in New Zealand, wasp populations have fluctuated, but whether biological control plays a role remains uncertain (e.g., Donovan et al. 2002).

Other benefits to New Zealand's indigenous biota may have accrued where a generalist pest, which could have been a threat to native plant species, has come under biological control, e.g., greenhouse thrips, *Heliothrips haemorrhoidalis*, or cottony cushion scale, *Icerya purchasi*. Classical biological control of the Australian lymantriid *Teia anartoides* (painted apple moth) will be one of the few tools available to protect susceptible native host plants, should the current eradication attempt fail.

23.5 Are Biological Control Introductions Flooding the Indigenous Biota?

For a start, we look at biological control introductions in terms of the number of species introduced deliberately, compared to accidental arrivals. In fruit crops, 121 species of parasitoids and predators have arrived accidentally, compared with only 24 deliberate biological control establishments (Charles 1998). Five encyrtid parasitoids of alien *Eucalyptus* psyllids have arrived accidentally whereas none have been purposely released (Withers 2001). Only one encyrtid parasitoid of mealybugs (Pseudococcidae) has been released, compared to the seven accidentally established (Charles and Allan 2002). There are 112 alien Lepidoptera that have arrived accidentally in New Zealand (Dugdale 1988; Hoare 2001), compared with six established biocontrol agents. In summary, Emberson (2000) estimated that deliberately released biocontrol agents comprised less than 3 % of the total number of alien insect species in New Zealand – over 97 % were accidental arrivals.

In terms of species numbers, deliberately released biocontrol agents are of minor concern, compared with accidental arrivals. Furthermore, the accidental arrivals do not go through any safety screening, and so include species with wide host ranges. Going beyond species counts, it is true that some biocontrol agents have become extremely common insects. For example, Miller (1925) commented that the gorse seed weevil was one of the most abundant insect species in New Zealand. However, we have no indications that this weevil is having detrimental effects on the native biota by competition for resources. Overall, evidence indicates that deliberately introduced biocontrol agents of insect pests or weeds are not swamping the indigenous biota of New Zealand, and that accidental introductions pose the more serious threats.

23.6 Discussion: Is Biological Control a Cure for or Source of Invasions?

A biological control introduction is a deliberately created alien invasion. The question is whether such actions are justified by the benefits, both economic and environmental, and whether the risk of negative effects is sufficiently low. We know there are a few cases where deliberately released weed biocontrol agents are attacking native plants (Simberloff and Stiling 1996; Louda et al. 1998). Unfortunately, insect predators and parasitoids introduced to suppress arthropod pests have a poorer record than that of introduced weed biocontrol agents. Indeed, prior to the 1980s, a wider host range was actually perceived as an advantage for biocontrol agents for insects, because reservoirs of the agent would exist even if the target pest was locally rare. Not surprisingly, there is a much longer list of native arthropods in New Zealand that are attacked by deliberately released biocontrol agents (Barratt et al. 1997; Munro and Henderson 2002). However, whether the current levels of attack are detrimental to the populations of these indigenous species is not known. Modern biological control programs are required to use more tightly host-specific natural enemies, so future impacts on non-targets should be minimal. Whether evolutionary changes to host range could occur is a subject of much debate and research. However, evidence to date suggests that host-specific insect herbivores or natural enemies have a very low risk of expanding their fundamental host ranges (van Klinken and Edwards 2002; Louda et al. 2003). Even without direct non-target attack, ecologists know that more-subtle effects occur in complex food webs, although whether these are significant is open to debate (Henneman and Memmott 2001; Louda et al. 2003). Unraveling the scale and significance of any knock-on effects, particularly in advance of a biological control release, is likely to present a significant challenge to experimental ecology (Fowler et al. 2000).

Given that we appreciate that biological control introductions entail some risk, we need to ask if the benefits outweigh the risks. There have been many biological control successes against important insect pests of New Zealand's productive systems, and a few cases where assessments have been made of the economic gain from increased yields and reduced use of pesticides (Cameron et al. 1993). However, the most spectacular biological control successes remain underrated because the targeted pests, such as the cottony cushion scale or apple woolly aphid, are now under such permanent and effective control that they no longer cause any problems (Cameron et al. 1993). There have probably been direct benefits to the indigenous flora of New Zealand from the biological control of polyphagous pests such as cottony cushion scale, which would certainly include many native plants in its host range, but these remain unstudied. For weeds, we are confronted with a relative paucity of hard data. For the five claimed biological control successes against alien weeds, there are limited high-quality data to prove that the biological control was successful, and that other factors were not responsible (Fowler 2000). There are few economic data to show the benefit to New Zealand of any weed biological control programs. Therefore, it is hardly surprising that there are also few data on the potential environmental benefits from these programs, although monitoring in the biological control program against mist flower is producing promising evidence that native plant species are replacing the weed after suppression by the deliberately released white smut fungus (Barton et al. 2004).

Many of the alien incursions include very serious pests with the potential to cause ecological devastation. Lack of action to control significant environmental pests and weeds in future will have a huge detrimental effect on New Zealand's biota, which arguably will dwarf the potential negative effects of carefully selected and tested biological control introductions. For many of these widespread environmental pests, considered and regulated biological control introductions offer one of the few cost-effective and environmentally acceptable tools – biological control may be "risky but necessary" (Thomas and Willis 1998).

Acknowledgements We extend our thanks to many contributing colleagues, particularly the opinions of John Bain, Barbara Barratt, Jo Berry, John Charles, Karyn Froud, and Vanessa Munro. Richard Hill gave helpful comments to an earlier version of the manuscript. The research was funded by the New Zealand Foundation for Research, Science and Technology.

References

Barratt BIP, Evans AA, Ferguson CM, Barker GM, McNeill MR, Phillips CB (1997) Laboratory nontarget host range of the introduced parasitoids *Microctonus aethiopoides* and *M. hyperodae* (Hymenoptera: Braconidae) compared with field parasitism in New Zealand. Environ Entomol 26:694–702

Barratt BIP, Goldson SL, Ferguson CM, Phillips CB, Hannah DJ (2000) Predicting the risk from biological control agent introductions: a New Zealand approach. In: Follett PA, Duan JJ (eds) Nontarget effects of biological control. Kluwer, Norwell, MA, pp 59–75

Barron MC, Barlow ND, Wratten SD (2003) Non-target parasitism of the endemic New Zealand red admiral butterfly (*Bassaris gonerilla*) by the introduced biological control agent *Pteromalus puparum*. (Biol Control 27:329–335

Barton J, Webster R, Boow J, Ragiel K, Edenborough K, Whaley K (2004) Evaluating the flow-on effects of the biological control agents for mist flower (*Ageratina riparia*) on plant succession. In: Cullen JM (ed) Proc 11th Int Symp Biological Control of Weeds, 27 April–2 May 2003, Canberra, Australia. CSIRO Entomology, Canberra, pp 487–492

Bennett FD, Habeck DH (1992) *Cactoblastis cactorum*: a successful weed control agent in the Caribbean, now a pest in Florida? In: Delfosse ES, Scott RR (eds) Proc 8th Int Symp Biological Control of Weeds, 2–7 February 1992, Canterbury, New Zealand. DSIR/CSIRO, Melbourne, pp 21–26

Berry JA (1998) The bethyline species (Hymenoptera: Bethylidae: Bethylinae) imported into New Zealand for biological control of pest leafrollers. NZ J Zool 25:329–333

Berry JA (2005) Checklist of New Zealand Hymenoptera. www.landcareresearch.co.nz/research/biodiversity/invertebratesprog/hymenoptera/checklist_index.asp (March 2005)

Booth RG, Cross AE, Fowler SV, Shaw RH (1995) The biology and taxonomy of *Hyperaspis pantherina* (Coleoptera: Coccinellidae) and the classical biological control of its prey, *Orthezia insignis* (Homoptera: Ortheziidae). Bull Entomol Res 85:307–314

Cameron PJ (1989) *Wiseana* spp., porina (Lepidoptera: Hepialidae). In: Cameron PJ, Hill RL, Bain J, Thomas WP (eds) A review of biological control of invertebrate pests and weeds in New Zealand 1874 to 1987. CAB and DSIR, Oxon, UK, pp 39–42

Cameron PJ, Wigley PJ (1989) *Costelytra zealandica* (White), grass grub (Coleoptera: Scarabaeidae). In: Cameron PJ, Hill RL, Bain J, Thomas WP (eds) A review of biological control of invertebrate pests and weeds in New Zealand 1874 to1987. CAB and DSIR, Oxon, UK, pp 9–16

Cameron PJ, Hill RL, Bain J, Thomas WP (1993) Analysis of importations for biological control of insect pests and weeds in New Zealand. Biocontrol Sci Technol 3:387–404

Charles JG (1998) The settlement of fruit crop arthropod pests and their natural enemies in New Zealand: an historical guide to the future. Biocontrol News Info 19:47N–58N
Charles JG, Allan DJ (2002) An ecological perspective to host-specificity testing of biocontrol agents. NZ Plant Protect 55:37–41
Crosby TK (1989) *Austrosimulium australense* (Schiner), New Zealand black fly and *Austrosimulium ungulatum* Tonnoir, West Coast black fly (Diptera: Simuliidae). In: Cameron PJ, Hill RL, Bain J, Thomas WP (eds) A review of biological control of invertebrate pests and weeds in New Zealand 1874 to 1987. CAB and DSIR, Oxon, UK, pp 375–379
DeBach P (1964) Biological control of insect pests and weeds. Chapman and Hall, London
Donovan BJ, Havron A, Leathwick DM, Ishay JS (2002) Release of *Sphecophaga orientalis* Donovan (Hymenoptera: Ichneumonidae: Cryptinae) in New Zealand as a possible 'new association' biocontrol agent for the adventive social wasps *Vespula germanica* (F.) and *Vespula vulgaris* (L.) (Hymenoptera: Vespidae: Vespinae). NZ Entomol 25:17–25
Dugdale JS (1988) Lepidoptera – annotated catalogue, and keys to family-group taxa. Fauna of New Zealand 14, DSIR, Wellington, New Zealand
Emberson RM (2000) Endemic biodiversity, natural enemies, and the future of biological control. In: Spencer N, Nowierski R (eds) Proc 10th Int Symp Biological Control of Weeds, 4–14 July 1999, Montana State University, Bozeman, pp 875–880
Farrell JA, Stufkens MW (1990) The impact of *Aphidius rhopalosiphi* (Hymenoptera: Aphidiidae) on populations of the rose grain aphid (*Metopolophium dirhodum*) (Hemiptera: Aphididae) on cereals in Canterbury, New Zealand. Bull Entomol Res 80:377–383
Fowler SV (2000) Trivial and political reasons for the failure of classical biological control of weeds: a personal view. In: Spencer N, Nowierski R (eds) Proc 10th Int Symp Biological Control of Weeds, 4–14 July 1999, Montana State University, Bozeman, pp 169–172
Fowler SV, Syrett P, Hill RL (2000) Success and safety in the biological control of environmental weeds in New Zealand. Austral Ecol 25:553–562
Fowler SV, Gourlay AH, Hill RL, Withers TM (2004) Safety in New Zealand weed biocontrol: a retrospective analysis of host specificity testing and the predictability of impacts on non-target plants. In: Cullen JM (ed) Proc 11th Int Symp Biological Control of Weeds, 27 April–2 May 2003, Canberra, Australia. CSIRO Entomology, Canberra, pp 265–270
Funasaki GY, Lai P-Y, Nakahara LM, Beardsley JW, Ota AK (1998) A review of biological control introductions in Hawaii: 1890 to 1985. Proc Hawaiian Entomol Soc 28:105–160
Futuyma DJ (2000) Potential evolution of host range in herbivorous insects. In: Van Driesche, RG, Heard T, McClay AS, Reardon R (eds) Host-specificity testing of exotic arthropod biological control agents: the biological basis for improvement in safety. USDA Forest Service Bull, Morgantown, West Virginia, pp 42–53
Goldson SL, McNeill MR, Phillips CB, Proffitt JR (1992) Host specificity testing and suitability of the parasitoid *Microctonus hyperodae* (Hym.: Braconidae, Euphorinae) as a biological control agent of *Listronotus bonariensis* (Col.: Curculionidae) in New Zealand. Entomophaga 37:483–498
Grundy TP (1989) An economic analysis of biological control of rose-grain aphid in New Zealand. Agribusiness and Economics Unit, Lincoln College, Canterbury, New Zealand
Haines ML, Syrett P, Emberson RM, Withers TM, Fowler SV, Worner SP (2004) Ruling out a host range expansion as the cause of the unpredicted non-target attack of tagasaste

(*Chamaecytisus palmensis*) by *Bruchidius villosus*. In: Cullen JM (ed) Proc 11th Int Symp Biological Control of Weeds, 27 April–2 May 2003, Canberra, Australia. CSIRO Entomology, Canberra, pp 271–276

Hayes LM (1999) A review of the biological control of weeds programme in the Canterbury Region, 1987–1999. Landcare Research Contract Rep LC9900/011, Landcare Research, Lincoln, New Zealand

Henneman ML, Memmott J (2001) Infiltration of a Hawaiian community by introduced biological control agents. Science 293:1314–1316

Hill MG, Allan DJ (1989) *Mythimna separata* (Walker), cosmopolitan armyworm (Lepidoptera: Noctuidae). In: Cameron PJ, Hill RL, Bain J, Thomas WP (eds) A review of biological control of invertebrate pests and weeds in New Zealand 1874 to 1987. CAB and DSIR, Oxon, UK, pp 107–109

Hill RL, Wittenberg R, Gourlay AH (2001) Biology and host range of *Phytomyza vitalbae* and its establishment for the biological control of *Clematis vitalba* in New Zealand. Biocontrol Sci Technol 11:459–473

Hoare RJB (2001) Adventive species of Lepidoptera recorded for the first time in New Zealand since 1988. NZ Entomol 24:23–47

Hoffmann JH (1995) Biological control of weeds: the way forward, a South African perspective. In: Stirton CH (ed) Weeds in a changing world. BCPC Symp Proc no 64. British Crop Protection Council, Brighton, pp 77–98

Howarth FG (1983) Classical biocontrol: panacea or Pandora's box? Proc Hawaiian Entomol Soc 24:239–244

Howarth FG (1991) Environmental impacts of classical biological control. Annu Rev Entomol 36:485–509

Julien MH, Griffiths MW (comps) (1999) Biological control of weeds: a world catalogue of agents and their target weeds, 4th edn. CABI, New York

Keey G (2003) The changing world underfoot. Forest Bird 307:32–35

Kuschel G (1969) The genus *Catoptes* Schonherr and two species oblitae of Fabricius from New Zealand. NZ J Sci 12:789–810

Louda SM, Kendall D, Connor J, Simberloff D (1997) Ecological effects of an insect introduced for the biological control of weeds. Science 277:1088–1090

Louda SM, Simberloff D, Boettner G, Connor J, Kendall D (1998) Insights from data on the nontarget effects of the flowerhead weevil. Biocontrol News Info 19:70N–72N

Louda SM, Pemberton RW, Johnson MT, Follett PA (2003) Nontarget effects – the Achilles' heel of biological control? Retrospective analyses to reduce risk associated with biocontrol introductions. Annu Rev Entomol 48:365–396

Marohasy J (1996) Host shifts in biological weed control: real problems, semantic difficulties or poor science? Int J Pest Manage 42:71–75

McFadyen REC (1998) Biological control of weeds. Annu Rev Entomol 43:369–393

McGregor PG (2001) Biological control of ragwort: assessing the impact of ragwort flea beetle (*Longitarsus jacobaeae* (Coleoptera: Chrysomelidae)). Landcare Research Contract Rep LC0102/019, Landcare Research, Lincoln, New Zealand

Miller D (1925) Forest and timber insects of New Zealand. NZ State Forest Service Bull 2. Government Printer, Wellington, New Zealand

Miller D (1970) Biological control of weeds in New Zealand 1927-48. New Zealand Department of Scientific and Industrial Research Info Ser 74. Government Printer, Wellington, New Zealand

Munro VMW, Henderson IM (2002) Nontarget effect of entomophagous biocontrol: shared parasitism between native Lepidopteran parasitoids and the biocontrol agent *Trigonospila brevifacies* (Diptera: Tachinidae) in forest habitats. Environ Entomol 31:388–396

Murray TJ, Barratt BIP, Ferguson CM (2002) Field parasitism of *Rhinocyllus conicus* Froelich (Coleoptera: Curculionidae) by *Microctonus aethiopoides* Loan (Hymenoptera: Braconidae) in Otago and South Canterbury. NZ Plant Protect 55:263–266

Paynter QE, Fowler SV, Gourlay AH, Haines ML, Harman HM, Hona SR, Peterson PG, Smith LA, Wilson-Davey JRA, Winks CJ, Withers TM (2004) Safety in New Zealand weed biocontrol: a nationwide survey for impacts on non-target plants. NZ Plant Protect 57:102–107

Pearson DE, Callaway RM (2003) Indirect effects of host-specific biological control agents. Trends Ecol Evol 18:456–461

Pemberton RW (2000) Predictable risk to native plants in weed biological control. Oecologia 125:489–494

Sands DPA (1997) The 'safety' of biological control agents: assessing their impact on beneficial and other non-target hosts. Mem Museum Victoria 56:611–615

Simberloff D, Stiling P (1996) How risky is biological control? Ecology 77:1965–1974

Stufkens MW, Farrell JA (1989) *Sitona discoideus* Gyllenhal, sitona weevil (Coleoptera: Curculionidae). In: Cameron PJ, Hill RL, Bain J, Thomas WP (eds) A review of biological control of invertebrate pests and weeds in New Zealand 1874 to 1987. CAB and DSIR, Oxon, UK, pp 27–28

Syrett P (1999) Broom beetle gets taste for Tagasaste. Biocontrol News Info 20:51N

Syrett P, Grindell JM, Hayes LM, Winks CJ (1991) Distribution and establishment of two biological control agents for ragwort in New Zealand. In: Proc 44th New Zealand Weed and Pest Control Conf, New Zealand Plant Protection Society, 13–15 August 1991, Tauranga, New Zealand, pp 292–295

Thomas WP (1989) *Epiphyas postvittana* (Walker), lightbrown apple moth (Lepidoptera: Tortricidae). In: Cameron PJ, Hill RL, Bain J, Thomas WP (eds) A review of biological control of invertebrate pests and weeds in New Zealand 1874–1987. CAB and DSIR, Oxon, UK, pp 187–195

Thomas MB, Willis AJ (1998) Biocontrol – risky but necessary? Trends Ecol Evol 13:325–329

Thomas MB, Casula P, Wilby A (2004) Biological control and indirect effects. Trends Ecol Evol 19:61

Thomson GM (1922) The naturalisation of animals and plants in New Zealand. Cambridge Univ Press, Cambridge

Van Klinken RD, Edwards OR (2002) Is host-specificity of weed biological control agents likely to evolve rapidly following establishment? Ecol Lett 5:590–596

Withers TM (2001) The colonisation of eucalypts in New Zealand by Australian insects. Austral Ecol 26:467–476

24 Strategies to Reduce Predation on Bird Populations

J.A. McLennan

24.1 Introduction

The defence of isolation for remote islands has no fallback position. It is all or nothing, akin to virginity, with no intermediate state. The slather following the invasion of New Zealand by people and other mammals is well known internationally, an exemplar of the tragic consequences of ill-considered introductions and the impacts of novel predators on native fauna. In the last 2,000 years, approximately 74 species of birds, mainly endemics, have been lost from the greater New Zealand region (Holdaway et al. 2001). Extinction rates have been particularly high among groups with high levels of taxonomic distinctiveness, among species which were either entirely flightless or only capable of weak flight, and among species whose eggs were small enough to be eaten by rats (Holdaway 1999; Worthy and Holdaway 2002). Geographically, rates of loss have been highest in North Island (51 % of breeding bird species).

In many places in New Zealand, few traces of the prehuman avifauna now remain. The once abundant petrel (Procellariidae) colonies on mainland New Zealand have been reduced to a few scattered remnants (Holdaway 1999; Holdaway et al. 2001). The story is similar for the extant, endemic land and freshwater birds, 81 % of species now being rare, uncommon, or only locally abundant (Heather and Robertson 1996). No corner of mainland New Zealand, no matter how remote, has escaped the impacts of introduced mammals. Nowhere is it possible to revisit the past. In the remote uplands of Northwest Nelson, excavations at the site of a laughing owl (*Sceloglaux albifacies*) nest show that 55 % of its avian prey species are, like the owl itself, no longer there (Worthy 2001). And yet, for the remainder of the biodiversity, that landscape remains ostensibly just as it was when laughing owls hunted over it, perhaps as recently as 100 years ago.

Today, the avian community of the New Zealand countryside is dominated by introduced species from Europe and native species from Australia, with

Ecological Studies, Vol. 186
R.B. Allen and W.G. Lee (Eds.)
Biological Invasions in New Zealand

few endemics among them. The most vulnerable species have gone: flighted species have taken the place of flightless ones, real mammals have taken the place of mammal-like birds, and introduced birds have taken the place of extinct natives. However, the loss of biological distinctiveness is not over. The endemic component of the avifauna will continue to be diluted by self-introduced colonists from Australia (and perhaps elsewhere), and further extinctions are still probable, despite determined conservation efforts to prevent them. Some 23 species/subspecies are classified as endangered or rare (Crawley 1982), and a much larger number are still suffering predation-driven declines (Clout et al. 1995; McLennan et al. 1996; O'Donnell 1996; Grant et al. 1997; Wilson et al. 1998; Innes et al. 1999).

Legal protection of species and preservation of remaining habitats are not sufficient on their own to prevent further losses (Clout and Saunders 1995). Rather, it is the management of predator impacts which will largely determine the extent to which avian biological distinctiveness is maintained over the coming decades. This chapter reviews recent developments in predator control, the relationships between impacts and predator abundance, and the strategies which are being adopted to conserve and restore some of the remaining parts of the prehuman avifauna of New Zealand.

24.2 Maintaining the Defence of Isolation

24.2.1 Offshore Islands

A number of species in New Zealand owe their existence to quirks of geography. At least six species of land birds lost on North Island, South Island or Stewart Island (referred to here as the mainland) retained a hold on offshore islands, their defence of isolation at least partially intact, albeit on a microscale. In a few instances, this last stand also failed. The lighthouse keeper's cat exterminated Stephens Island wren (*Traversia lyalli*) in 1894, and Stead's bush wren (*Xenicus longipes*) was lost when rats (*Rattus rattus*) invaded Big South Cape Island, its last refuge, in 1964. The translocation of six individuals to Kaimohu Island failed to save the species, though some individuals persisted until 1972 (Heather and Robertson 1996). Stead's bush wren is New Zealand's most recent avian extinction.

Translocation – the intentional release of a species in a new, safe location – is arguably the country's central conservation strategy. Since the pioneering efforts of Richard Henry (Hill and Hill 1987), some 42 taxa have been moved from mainland New Zealand to offshore islands, or from one island to another, to reduce exposure to predators and establish additional populations (Saunders 1994). In the period from 1992 to 1995 alone, there were at least 407 translocation events (Girardet 2000), chiefly involving weka (*Gallirallus australis*), saddleback (*Philesturnus carunculatus*), kiwi (*Apteryx* spp.), stitchbird (*Notiomystis cincta*) and brown teal (*Anas aucklandica*). Many translocations have failed (e.g. great spotted kiwi, *Apteryx haastii*, from Nelson to Little Barrier Island in 1915) but others have been spectacularly successful and have saved many species from probable extinction (e.g. the undocumented but probable transfer of little spotted kiwi, *A. owenii*, from South Island to Kapiti Island in the 1920s).

Parallel developments in techniques for eradicating rats (*Rattus* spp.), cats (*Felis catus*), feral pigs (*Sus scrofa*), possums (*Trichosurus vulpecula*) and stoats (*Mustela erminea*) have enabled previously infested islands to be reclaimed (Veitch 1995), increasing their value as refuges, and expanding the range of islands available for occupation by endangered wildlife. Rat eradication programmes are now being attempted on large islands (DOC 2002) such as Kapiti (20 km^2; 1996) and Campbell (113 km^2; 2001), a staggering advance on the 9–20 ha islands which were being attempted in the late 1980s. If technical advances continue at their present rate, islands which are formidable challenges now, such as Great Barrier (300 km^2) and Rakiura (Stewart Island; 1,680 km^2), could be cleared of rats within the next 20–50 years.

24.2.2 Predator-Proof Fencing

Since the 1950s, fences of varying designs have been used throughout the world to exclude predators from localized areas with high conservation values (Short et al. 1992; Kuchling 1997). Their use in New Zealand has been limited but is now increasing, initiated primarily by a growing interest in species restoration on the mainland. The 250-ha reserve at Karori on the outskirts of Wellington is currently the largest of its type, its 8.6-km predator-proof fence protecting a variety of species formerly confined to offshore islands. For the first time in more than a century, people in Wellington can hear kiwi calling at night, and saddleback by day.

The Karori example has inspired several similar initiatives, mainly small (<2 ha) exclosures built specifically to protect only one species (mainly juvenile kiwi) for a few months of each year. A private initiative to ring-fence Maungatautari Mountain in the Waikato is now well advanced. If it proceeds beyond the planning stage, a predator-free habitat equivalent in size to Little Barrier Island (30 km^2) will be created in the middle of what is now a highly modified and heavily populated agricultural landscape. This venture signals a marked shift in thinking among the general public, and the beginnings of a new conservation development. Wildlife management is no longer regarded as the sole responsibility of government agencies. Local communities are instead initiating their own projects and restoring parts of the mainland to an ecological status equivalent to that of predator-free offshore islands. This development would have been dismissed as fanciful as little as a decade ago.

There is no doubt that the current interest in predator-proof enclosures will grow over the coming decades. Indeed, it is surprising that they are not used more often now to protect localized concentrations of threatened species, such as nesting waders on riverbeds, and seabirds in mainland habitats. They are used extensively in marsupial recovery programmes in Australia, where they are generally erected across the neck of peninsulas to maximize benefit/cost ratios (Short et al. 1992; Short and Smith 1994). New Zealand's ragged coastline affords many equivalent opportunities here, and several projects are now underway in the Marlborough Sounds.

24.2.3 Captive Management – Protection at Critical Stages

The strategy of using captive-reared (or bred) birds to establish self-sustaining populations in the wild has had limited success internationally (Beck et al. 1994) but it is an integral part of endangered species management in New Zealand. The takahe (*Porphyrio hochstetteri*) captive-rearing programme, started in 1957, has contributed significantly to the survival of populations in the wild (Eason and Willans 2001), and is one of the longest-running and most successful captive-rearing and release programmes in the world. It has become the exemplar for numerous other endemic species, including the brown teal, blue duck (*Hymenolaimus malacorhynchos*), Campbell Island teal (*Anas aucklandica*), kokako (*Callaeas cinerea*), shore plover (*Thinornis novaeseelandiae*), black stilt (*Himantopus novaezelandiae*), kakapo (*Strigops habroptilus*), kaka (*Nestor meridionalis*), kiwi, North Island weka (*Gallirallus australis greyi*) and stitchbird.

New species are being added to the list each year, including several species of seabirds (Eason and Willans 2001). This trend will continue, fuelled by a growing demand for founders to restock areas cleared of predators, and the need to protect some species from predators during brief but critical periods of life.

24.3 Living with Predators

It is difficult to imagine that any of the 12 species of predatory mammal now well established and widely distributed in mainland habitats (Table 24.1) will ever be eradicated totally, despite the promise of new control techniques based on genetically modified organisms (Robinson et al. 1999). In places unprotected by fences (the greater part of the mainland), vulnerable bird species will continue to co-exist with a diverse mix of predators, just as they

Table 24.1 Dates of introduction (years A.D.) and typical mass of mammalian predators in New Zealand. Data from Daniel and Baker (1986) and Holdaway (1999)

Species	Year	Mass (g)
Rattus exulans	ca. 150	80–90
R. norvegicus	1792	250–300
R. rattus	1860	120–150
Mus musculus	ca. 1830	12–15
Felis catus	1830s	1,000–5,000
Canis familiaris	1300	20,000?
Mustela furo	1882	450–1,000
M. erminea	1886	160–475
M. nivalis	1885	80–150
Trichosurus vulpecula	1858	1,500–4,500
Sus scrofa	1792	45,000–150,000
Erinaceus europaeus	1907	800–1,100

have over the past 100–150 years. Predator-driven declines will also continue as before. Options and techniques for reducing impacts to sustainable levels are considered in the following sections.

24.3.1 Predation Systems in Mainland Habitats

Australian and New Zealand predation systems have much in common. In both countries, endemic prey suffer hyperpredation (Holt 1977, 1984) from introduced carnivores, sustained at high levels of abundance by one or more species of common prey. The red fox, feral cat/rabbit/marsupial system in Australia (Pech et al. 1995) has its equivalents here in the (1) stoat/rodent/endemic birds system in indigenous forests (King 1983; Murphy and Dowding 1995), (2) ferret/feral cat/rabbit/nesting wader system on braided riverbeds (Rebergen et al. 1998), and (3) ferret/feral cat/rabbit/skink system in grasslands (Norbury 2001). In all cases, an introduced prey indirectly facilitates the extinction of a native prey by enabling a shared predator to increase in population size. Taylor (1979) was the first to describe the process of hyperpredation in New Zealand. More recently, Imber et al. (2000) described a classic example on Whale Island, where the take of eggs and chicks of grey-faced petrels (*Pterodroma macroptera gouldi*) by rats increased to unsustainable levels following the introduction of rabbits.

In many hyperpredation systems, native prey are merely picked off as bycatch (secondary prey) when predators chance upon them. Often, the predators vastly outnumber their native prey, as exemplified by stoats and juvenile kiwi in mainland forests (McLennan 1997). In the latter stages of decline, the native prey may appear so infrequently in the predator's diet that the predator itself is sometimes dismissed as the agent of decline. Diet can give a false impression of impact.

Predation impacts on a particular prey species are the product of a predator's functional response (the number of prey items consumed per unit of time) and numerical response (density or rate of increase) to the abundance of that prey (Pech et al. 1995). To reduce impacts, one or both of the predator's responses have to be manipulated, either directly or indirectly. In theory at least, a small change in a predator's functional response can sometimes lead to a marked change in prey abundance – though there are few concrete examples of this prediction materializing in the field. In New Zealand, managers typically reduce predator abundance to reduce predation impacts, and it often works. The functional responses of predators are generally ignored, although they too are often manipulated, usually unwittingly, in many pest control programmes. The potential they offer for reducing impacts is largely untapped, possibly because they are hard to manage, require detailed knowledge of particular systems, are species- and possibly site-specific, and (in some cases at least) exist only as predictions of theoretical models. They are nevertheless reviewed below, as part of the mix of strategies potentially available for reducing predation impacts, outside of complete isolation.

24.3.2 Manipulating Functional Responses to Reduce Predation Impacts

The functional responses of predators can be manipulated in a variety of ways to reduce predation losses to sustainable levels. The manipulations are generally done over short periods of time, within the context of relatively stable predator numbers. Some techniques, however, such as the provision of prey refuges, may produce much longer-lasting benefits.

Supplementary or Diversionary Feeding

Supplementary feeding aims to temporarily deflect predation away from threatened species to enhance survival rates during critical periods (generally the breeding season). Overseas studies show it is most successful in ecological communities with a relatively small guild of predators (Greenwood et al. 1998). It has never been formally tested in New Zealand, although diversionary feeding has been done unwittingly on numerous occasions, and it also occurs naturally from time to time. Large-scale possum and rabbit control operations, conducted over several weeks, temporarily provide vast quantities of alternative food for scavenging predators, sufficient to strongly alter their normal patterns of activity and consumption of native prey. The periodic seeding of beech (*Nothofagus*) has much the same effect, by increasing the abundance of rodents, and thus the food supply of stoats in beech forest habitats (King 1983). The irruption of rodents in the first few months following seedfall is analogous to the windfall induced by pest control, with the same potential to generate temporary diet switches in stoats, away from native prey.

Diversionary feeding via pest control offers the greatest potential for protecting endangered wildlife, simply because it is short-lived (and thus may not generate a numerical response from predators) and can be timed precisely to produce maximum benefit. As yet, however, the strategy is largely untested and is not part of any conservation programme.

Habitat Manipulations

Habitat manipulations which create refuges for native prey either (1) reposition the predator's functional response curve, so that some of the prey are always exempt from predation or (2) change the shape of the curve, by making the prey more difficult or costly to catch. Predator-proof nest boxes are an example of the former outcome, with the potential to benefit threatened hole-nesting birds such as the kaka, parakeets (*Cyanoramphus* spp.) and yellowhead (*Mohoua ochrocephala*). Ungulate control, promoting recovery of the forest understory, could well achieve the latter outcome for species which spend considerable time feeding on the ground. Cats appear to be able to drive red-crowned parakeets extinct only on offshore islands in which ungulates (usually goats, *Caprus hircus*) have opened up the vegetation (Veitch and Bell 1990). Extensive ungulate control in mainland habitats could well provide similar benefits for a range of other species, by making the habitats less suitable for ground-hunting mammals.

Increasing the Abundance of Threatened Prey to Reduce Predator Impacts

One model of Pech et al. (1995; reproduced here in Fig. 24.1a) describes a hyperpredation system in which a predator depends primarily on alternative prey but takes a species of native prey as a small, consistent by-catch. The proportion of the native prey population eaten by predators increases progressively as it declines, so the response curve of the predators is inversely density dependent at all densities of the prey. When the prey declines (for whatever reason) below a certain critical density (NB in Fig. 24.1a), mortality due to predation exceeds recruitment and the prey decline to extinction. At densities above NB, recruitment gains exceed predation losses, and the prey increases towards carrying capacity (K in Fig. 24.1a).

This model may be relevant to at least one of New Zealand's endemic birds, the tui (*Prosthemadera novaeseelandiae*), where declines appear to be occur-

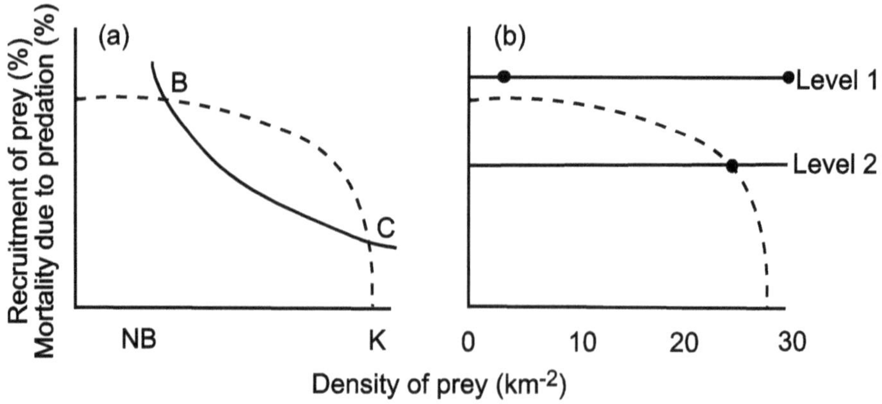

Fig. 24.1a, b Schematic representation of two predator–prey systems described by Pech et al. (1995). In **a**, predators take a constant number of prey, so mortality due to predation (*solid line*) increases as prey declines. If densities fall below *NB*, predation losses exceed prey recruitment rates (*dotted line*) and prey declines to extinction. In **b**, predators take a constant proportion of prey at all prey densities. This model is appropriate for kiwi in North Island forests. *Level 1* is indicative of losses in areas with no predator control. The *points* shown are for Waikaremoana (3.3 adults km^{-2}; 52 % predation loss of juveniles; unpublished data) and Purua, Northland (ca. 30 adults km^{-2}; 49 % predation loss of juveniles; H. Robertson, personal communication). *Level 2* indicates reduced losses in the presence of a controlled predator population, and the recovery of the kiwi population back to *C*

ring sporadically, and picking up speed as numbers dwindle. In such species (if more exist at all), a single burst of management could lift densities above NB, and initiate recovery. Any approach which either increased prey densities in the field (e.g. captive rearing and release), improved prey recruitment rates (e.g. supplementary feeding of prey), or lowered predator densities would achieve the desired outcome. (The second and third approaches shift NB to the left by changing the intersection points of the recruitment and predation curves).

It is clear, however, that this model is not applicable to North Island brown kiwi. In this species, the percentage take of juveniles by stoats is consistently high and constant across the full range of kiwi densities now extant in mainland forests (Fig. 24.1b). Here, the most feasible option for retaining kiwi in the wild is to reduce predator densities to levels which allow recruitment to exceed mortality (level 2 in Fig. 24.1b). Methods for reducing predators are discussed in the section below.

24.3.3 Manipulating Predator Densities to Reduce Predator Impacts

Indirect Manipulations

Hyperpredation systems work both ways – predator populations decline when the populations of their primary prey decline (King 1983; Newsome et al. 1989; Norbury and McGlinchy 1996). Nowhere in New Zealand has a population of primary prey been deliberately and systematically controlled to reduce predation impacts on threatened fauna. Indeed, they have been controlled for other reasons, either because they are an economic threat (e.g. rabbits, *Oryctolagus cuniculus*), predators in their own right (rats), or non-target casualties in control campaigns directed against other animals (e.g. rats in possum control operations).

In the short term, predation impacts on native prey usually increase following sudden and substantial declines in primary prey, because of prey switching (Norbury et al. 1998). In the longer term, impacts generally decrease, because the overall decline in predator numbers is usually more than sufficient to offset any increase in native prey consumption by the ones which remain.

Life-history traits mean that it is generally more difficult and expensive to suppress primary prey than it is to suppress the predators themselves. Primary prey have high rates of increase and usually recover quickly from control operations – sometimes in as little as 4–6 months (Innes et al. 1995). Prolonged (but not permanent) control of rabbits was achieved following the introductions of myxomatosis into Australia, and rabbit haemorrhagic disease into Australia and New Zealand. The development of similar biocontrol agents for rodents would profoundly benefit threatened birds in New Zealand forests. Until such time (if this is ever possible), direct control of predators is likely to be the best management option.

Direct Manipulations

In most situations, managers directly manipulate predator abundance to reduce predation impacts on endangered wildlife. The technique works if predators are reduced to appropriate levels for sufficient lengths of time, and several species of prey may benefit in the treatment area. An arsenal of methods exists for killing predators, and new techniques, refinements, toxins and delivery systems are emerging every year (see Chap. 27).

Five key questions are common to all control campaigns:

1. which predators should be targeted?
2. how much should densities be reduced to decrease impacts to sustainable levels?

3. how long should control be maintained?
4. how often should it be repeated? and
5. on what scale should control be undertaken to maintain viable populations of threatened prey?

These questions are considered below, chiefly in relation to kiwi (*Apteryx mantelli*) and kokako, two threatened occupants of North Island forests.

Targeting the Right Predators

It is self-evident that control programmes must target the predators responsible for a species decline, but identifying them in the first place is not always easy. In New Zealand's mainland habitats, as many as nine different species are potential candidates, working either alone or in combination. Teasing out their relative impacts requires detailed study, especially when different predators are involved at different life stages. In recent years, infrared filming has been used extensively to identity nest raiders, revealing the previously unsuspected and substantial impacts of possums in forests (Brown et al. 1993), and hedgehogs in coastal and riparian habitats (Sanders and Maloney 2002).

Some generalities have emerged after a decade of detailed studies. Groups of birds with common attributes usually share the same predators – open-nesting forest birds (ship rats, possums and stoats); hole-nesting forest birds and flightless species (stoats and ship rats); large flightless birds in their adult stage of life (ferrets and domestic dogs); and nesting waders on beaches and riverbeds (ferrets, feral cats and hedgehogs). It is noteworthy that most birds are taken by at least two different predator species, so control efforts aimed at the dominant member of the cluster may do nothing more than allow another member to take over. Substitution appears to have happened in Fiordland, where the role of ship rats as nest predators of yellowhead has increased markedly following intensive control of stoats (Macalister 2002). The result may in part be attributable to an increase in rat abundance – but either way, it illustrates both the complexity of some systems, and the dangers of targeting only one of several sympatric predators.

Timing of Control Operations and Threshold Predator Densities for Species Survival

Most of the endemic birds still extant in mainland forests suffer intense predation at only one stage of life, or during one part of the year. Predation bottlenecks are punctuated by periods of reprieve, some lasting for months and sometimes for years. In most birds, the nesting period is typically the time of greatest risk. Recruitment failure is therefore the most common cause of decline, exacerbated in some species by the additional loss of incubating adults (usually females) and the development of gender-biased adult populations (Greene and Fraser 1998; Innes et al. 1999). In North Island brown kiwi, juveniles suffer intense predation in their first 20 weeks of life, but thereafter become too large for stoats to tackle (McLennan 1997).

When rats and/or stoats are the primary predators, ongoing control operations are generally required throughout the entire period of vulnerability to provide adequate protection. Both species recover rapidly from a knockdown, either through natural increase or immigration (King and McMillan 1982; Innes et al. 1995). Possums have a lower rate of increase and recover more slowly, so a single operation may provide protection for several years (Innes et al. 1999).

The extent to which predators must be reduced to arrest population declines is known for few species. In kiwi, kokako and kereru (*Hemiphaga novaeseelandiae*), and probably all threatened forest birds, the relationship between survival (of a critical life stage such as eggs, chicks, or independent juveniles) and predator abundance is described by a steeply descending decay curve (Fig. 24.2). Impacts are not directly proportional to predator abundance, relatively few predators can inflict great harm, and predators have to be reduced substantially from average densities in order to benefit threatened species. In Te Urewera National Park, intensive trapping failed to protect young kiwi in years when stoats were numerous, but did so in years when stoats were scarce (unpublished data). Many stoats were caught in plague years but immigrants quickly replaced them, and average densities remained consistently above the threshold required for kiwi survival. Overall, juvenile survival was inversely related to predator catch rate. Managers often report high predator catches with some pride, yet for birds such as the kiwi they are clearly cause for concern, rather than celebration.

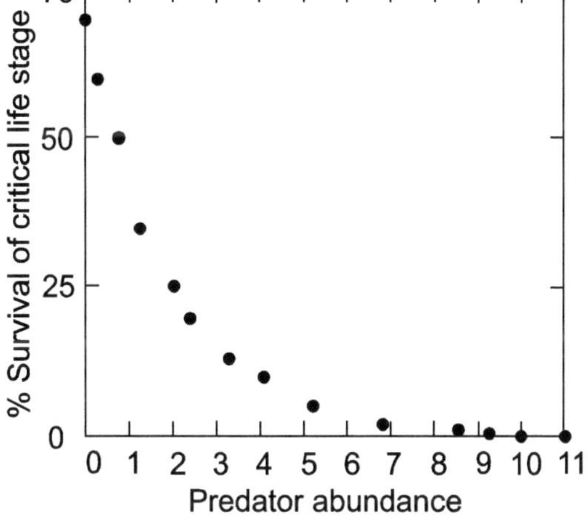

Fig. 24.2 Relationship between survival and predator abundance (arbitrary scale) in kiwi, kokako and kereru. In kiwi, survival rates of independent juveniles are insufficient to maintain populations when stoats exceed a threshold density of two per square kilometre (Basse et al. 1999). In kokako and kereru, fewer than 30 % of nesting attempts fledge young when the catch rate of possums exceeds five per 100 trap nights (Innes et al. 1999; J. Innes, unpublished data)

Methods which merely crop predator populations will produce few or no benefits (Fig. 24.2). As the kiwi example illustrates, timing is often also critical, with success more probable when control operations coincide with natural troughs in predator abundance. Even good techniques can fail in some years. One of the unfortunate realities of New Zealand predation systems is that predators often peak at the time of year when native prey species are most vulnerable. For this reason alone, reducing predators below acceptable damage thresholds is often difficult to accomplish.

Frequency of Control

There is no rule of thumb for determining how often predator control should be undertaken to maintain a population of a threatened species. The answer varies from one species to the other, depending on their rates of mortality and recruitment in the absence of predators, their mortality rates in the presence of predators, and the extent to which those losses can be reduced by predator control.

Annual control is not required in long-lived species with low rates of adult mortality. In kokako, 3 "control-on" years in every 10 appear to be sufficient to maintain a population (Basse et al. 2003), irrespective of how it is applied (3 on, 7 off, or vice versa). The same is true of North Island brown kiwi, where recruitment rates in "on" years are generally more than double the average rate (19%) required for population maintenance (McLennan et al. 1996). In both of these examples, the adults are largely immune from predation, so persist during periods of low or no recruitment. Adult kiwi in deep forest habitats may live for 20–40 years (McLennan et al. 1996). Species which have no immunity from predators in particular seasons, stages of life, or habitats have now probably disappeared altogether from mainland habitats. Brown teal (*Anas aucklandica chlorotis*) might be the last exception – and its decline seems to be unstoppable, even in areas subjected to continuous predator control.

Scale of Control

Even the most ardent optimist would agree there is no foreseeable end to the reign of mammalian predators on mainland New Zealand. The costs of control and eradication (within fenced areas) are so large that only a small portion of the total landscape will ever be restored to a (near) predator-free state. The general decline of threatened endemic birds in mainland habitats will therefore continue, until eventually they persist only in protected areas, surrounded by large tracts of available but unoccupied habitat. Clearly, protected areas have to be large enough to support viable populations of the species they aim to protect. The area requirements of individual species therefore ultimately determine both the size of reserves, and the scale of the predator-control operations needed to create and maintain them.

Isolated populations face two potential threats: the progressive loss of genetic variance, and dispersal-induced failure. The latter threat is unique to populations on the mainland. It is caused by emigration out of a source population, but no back-migration into it (Dias 1996). Theoretical analyses (Franklin 1980; Lande and Barrowclough 1987) and experimental tests (Reed and Bryant 2000) indicate genetic variance is maintained permanently in isolated populations with more than 500–5,000 breeding individuals. The threat of dispersal-induced failure also diminishes with increasing population size, abating altogether at about 3,000 individuals in the North Island brown kiwi (Basse and McLennan 2003). In most species, a population capable of surviving genetic threats is probably also large enough to survive dispersal losses, though this has not been tested.

For kiwi and other large forest birds, protected areas probably need to be in the order of 3,750–37,500 ha for long-term population persistence. This calculation is simply the product of mean territory size of individual pairs (ca. 15 ha) and the number of pairs required for viability. Species with smaller territories and relatively short dispersal distances (e.g. robins, *Petroica australis*) can undoubtedly get by with much less. Biogeographic studies of species persistence on offshore islands support these theoretical estimates. Kiwi, for example, persisted on Little Barrier (30 km^2) and D'Urville (162 km^2) islands following their isolation from greater New Zealand some 10,000 years ago (East and Williams 1984), but failed on smaller islands.

In general, the species with the largest area requirements are those most in need of help (East and Williams 1984). Ongoing, successful predator control over tens of thousands of hectares is at the limits of current technology. Nevertheless, the boundaries are being pushed back each year, in part because of the New Zealand Department of Conservation's "mainland island" programme (e.g. DOC 2000, 2002), and the opportunities this presents for testing and developing new technologies. It is simply a matter of time before predator control is possible on a scale which benefits all of the remaining extant endemic birds in New Zealand.

24.4 Conclusions

The extraordinary avifaunal community which greeted the first human colonists belongs to the past. It cannot be restored because only part of it remains. Some 40 % of its constituent species are now extinct. Even today, further losses are still probable. The rat invasion of Big South Cape Island in 1964 is a salutary reminder of how quickly the fortunes of some threatened species can change, and how tenuous the defence of isolation really is. Breaches of this type are infrequent – but they will happen again, both offshore, and on the mainland in virtual islands protected by fences.

For many mainland species, the worst is not yet over. They will continue to decline in abundance and range, despite being widespread now and seemingly permanent features of the landscape. The tens of thousands of kiwi, kereru and kaka surviving today belie the substantial declines they have already suffered, and the ones which are yet to come. Still, amidst the declines, there are also growing signs of recovery. Evidently, the process of change has two faces – some species have been rescued from almost certain extinction; predators have been eliminated from numerous offshore islands and parts of the mainland; better techniques and strategies are being developed to reduce predator impacts in mainland forests; and local communities are increasingly initiating their own restoration programmes. Biological techniques based on genetically modified organisms hint at the possibility of potent new methods of control, and the opportunity to manage predators on a scale undreamed of today. Although the worst is not yet over, the current conservation status of New Zealand's endemic birds could well be bettered in years to come.

Acknowledgements I thank Mick Clout, Grant Norbury and Bill Lee for improving the first draft of this manuscript, and John Innes and Hugh Robertson for providing unpublished data.

References

Basse B, McLennan JA (2003) Protected areas for kiwi in mainland forests of New Zealand: how large should they be? NZ J Ecol 27:95–105

Basse B, McLennan JA, Wake GC (1999) Analysis of the impact of stoats, *Mustela erminea*, on northern brown kiwi, *Apteryx mantelli*, in New Zealand. Wildl Res 26(2):227–237

Basse B, Flux I, Innes J (2003) Recovery and maintenance of North Island kokako (*Callaeas cinerea wilsoni*) populations through pulsed pest control. Biol Conserv 109:259–270

Beck BB, Rapaport LG, Stanley Price MR, Wilson AC (1994) Reintroduction of captive-born animals. In: Olney PJS, Mace GM, Feistner ATC (eds) Creative conservation: interactive management of wild and captive animals. Chapman and Hall, London, pp 265–286

Brown K, Innes J, Shorten R (1993) Evidence that possums prey on and scavenge birds' eggs, birds and mammals. Notornis 40:169–177

Clout MN, Saunders AJ (1995) Conservation and ecological restoration in New Zealand. Pac Conserv Biol 2:91–98

Clout MN, Denyer K, James RE, McFadden IG (1995) Breeding success of New Zealand pigeons (*Hemiphaga novaeseelandiae*) in relation to control of introduced mammals. NZ J Ecol 19:209–212

Crawley MC (1982) Presidential address: wildlife conservation in New Zealand. NZ J Ecol 5:1–5

Daniel M, Baker A (1986) Collins guide to the mammals of New Zealand. Collins, Auckland, New Zealand

Dias PC (1996) Sources and sinks in population biology. Trends Ecol Evol 11:326–330

DOC (2000) Boundary Stream Mainland Island project report 1996–1998. September 2000. Department of Conservation, East Coast Hawke's Bay Conservancy, Gisborne, New Zealand

DOC (2002) Statement of Intent 2002–2005. Department of Conservation, Wellington, New Zealand

Eason DK, Willans M (2001) Captive rearing: a management tool for the recovery of the endangered takahe. In: Lee WG, Jamieson IG (eds) The takahe: fifty years of conservation management and research. University of Otago Press, Dunedin, New Zealand, pp 80–95

East R, Williams GR (1984) Island biogeography and the conservation of New Zealand's indigenous forest-dwelling avifauna. NZ J Ecol 7:27–35

Franklin IR (1980) Evolutionary change in small populations. In: Soule ME, Wilcox BA (eds) Conservation biology: an evolutionary-ecological perspective. Sinauer, Sunderland, MA, pp 135–149

Girardet S (2000) Tools for protecting endangered species: eradication, translocation, triangulation. PhD Thesis, University of Auckland, New Zealand

Grant AD, Powlesland RG, Dilks PJ, Flux IA, Tisdall CJ (1997) Mortality, distribution, numbers and conservation of the Chatham Island Pigeon (*Hemiphaga novaeseelandiae chathamensis*). Notornis 44:65–77

Greene TC, Fraser JR (1998) Sex ratio of North Island kaka (*Nestor meridionalis septentrionalis*), Waihaha Ecological Area, Pureora Forest Park. NZ J Ecol 22:11–16

Greenwood RJ, Pietruszewski DG, Crawford RD (1998) Effects of food supplementation on predation of duck nests in upland habitat. Wildl Soc Bull 26:219–226

Heather BD, Robertson HA (1996) The field guide to the birds of New Zealand. Viking, Auckland, New Zealand

Hill S, Hill J (1987) Richard Henry of Resolution Island – a biography. John McIndoe and New Zealand Wildlife Service, Dunedin, New Zealand

Holdaway RN (1999) Introduced predators and avifaunal extinction in New Zealand. In: MacPhee RDE (ed) Extinctions in near times: causes, contexts, and consequences. Kluwer/Plenum Advances in Palaeobiology, New York, pp 189–238

Holdaway RN, Worthy TH, Tennyson AJD (2001) A working list of breeding bird species of the New Zealand region at first human contact. NZ J Zool 28:119–187

Holt RD (1977) Predation, apparent competition, and the structure of prey communities. Theor Population Biol 12:197–229

Holt RD (1984) Spatial heterogeneity, indirect interactions, and the coexistence of prey species. Am Nat 124:377–406

Imber M, Harrison M, Harrison J (2000) Interactions between petrels, rats and rabbits on Whale Island, and effects of rat and rabbit eradication. NZ J Ecol 24:153–160

Innes J, Warburton B, Williams D, Speed H, Bradfield P (1995) Large-scale poisoning of ship rats (*Rattus rattus*) in indigenous forests of the North Island, New Zealand. NZ J Ecol 19:5–17

Innes J, Hay R, Flux I, Bradfield P, Speed H, Jansen P (1999) Successful recovery of North Island kokako *Callaeas cinerea wilsoni* populations, by adaptive management. Biol Conserv 87:201–214

King CM (1983) The relationships between beech (*Nothofagus* sp.) seedfall and populations of mice (*Mus musculus*), and the demographic and dietary responses of stoats (*Mustela erminea*), in three New Zealand forests. J Anim Ecol 52:141–166

King CM, McMillan CD (1982) Population structure and dispersal of peak-year cohorts of stoats (*Mustela erminea*) in two New Zealand forests, with especial reference to control. NZ J Ecol 5:59–66

Kuchling G (1997) Managing the last survivors: integration of *in situ* and *ex situ* conservation of *Pseudemydura umbrina.* In: Van Abbema J (ed) Proc Conf Conservation, Restoration and Management of Tortoises and Turtles, 11–16 July 1993. New York Turtle and Tortoise Society, New York, pp 339–344

Lande R, Barrowclough GF (1987) Effective population size, genetic variation, and their use in population management. In: Soule ME (ed) Viable populations for conservation. Cambridge Univ Press, Cambridge, pp 87–123

Macalister A (2002) Mohua in trouble. NZ Geogr 57:8–10

McLennan J (1997) Survival at Waikaremoana. Forest Bird 283:16–21

McLennan JA, Potter MA, Robertson HA, Wake GC, Colbourne R, Dew L, Joyce L, McCann AJ, Miles J, Miller PJ, Reid J (1996) Role of predation in the decline of kiwi, *Apteryx* spp., in New Zealand. NZ J Ecol 20:27–35

Murphy EC, Dowding JE (1995) Ecology of the stoat in *Nothofagus* forest: home range, habitat use and diet at different stages of the beech mast cycle. NZ J Ecol 19:97–109

Newsome AE, Parer I, Catling PC (1989) Prolonged prey suppression by carnivores – predator-removal experiments. Oecologia 78:458–467

Norbury G (2001) Conserving dryland lizards by reducing predator-mediated apparent competition and direct competition with introduced rabbits. J Appl Ecol 38:1350–1361

Norbury G, McGlinchy A (1996) The impact of rabbit control on predator sightings in the semi-arid high country of the South Island, New Zealand. Wildl Res 23:93–97

Norbury GL, Norbury DC, Heyward RP (1998) Behavioral responses of two predator species to sudden declines in primary prey. J Wildl Manage 62:45–58

O'Donnell CFJ (1996) Predators and the decline of New Zealand forest birds: an introduction to the hole-nesting bird and predator programme. NZ J Zool 23:213–219

Pech RP, Sinclair ARE, Newsome AE (1995) Predation models for primary and secondary prey species. Wildl Res 22:55–64

Rebergen A, Keedwell R, Moller H, Maloney R (1998) Breeding success and predation at nests of banded dotterel (*Charadrius bicinctus*) on braided riverbeds in the central South Island, New Zealand. NZ J Ecol 22:33–41

Reed D, Bryant EH (2000) Experimental tests of minimum population size. Anim Conserv 3:7–14

Robinson AJ, Jackson RJ, Merchant JC, Kerr PJ, Holland MK, Lawson MA, Shellam GR (1999) Biological control of pest mammals in Australia – progress towards virally vectored immunocontraception for the mouse and rabbit. Recent Adv Microbiol 7:63–98

Sanders MD, Maloney RF (2002) Causes of mortality at nests of ground-nesting birds in the Upper Waitaki Basin, South Island, New Zealand: a 5-year video study. Biol Conserv 106:225–236

Saunders A (1994) Translocations in New Zealand: an overview. In: Serena M (ed) Reintroduction biology of Australian and New Zealand fauna. Surrey Beatty, Chipping Norton, NSW, Australia, pp 43–46

Short J, Smith A (1994) Mammal decline and recovery in Australia. J Mammal 75:288–297

Short J, Bradshaw SD, Giles J, Prince RIT, Wilson GR (1992) Reintroduction of macropods (Marsupialia: Macropodioalea) in Australia – a review. Biol Conserv 62:189–204

Taylor RH (1979) How the Macquarie Island parakeet became extinct. NZ J Ecol 2:42–45

Veitch CR (1995) Habitat repair: a necessary prerequisite to translocation of threatened birds. In: Serena M (ed) Reintroduction biology of Australian and New Zealand fauna. Surrey Beatty, Chipping Norton, NSW, Australia, pp 97–104

Veitch CR, Bell BD (1990) Eradication of introduced animals from the islands of New Zealand. In: Towns DR, Daugherty CH, Atkinson IAE (eds) Ecological restoration of

New Zealand islands. Conservation Sciences Publ 2. Department of Conservation, Wellington, New Zealand, pp 137–146

Wilson PR, Karl BJ, Toft RJ, Beggs JR, Taylor RH (1998) The role of introduced predators and competitors in the decline of kaka (*Nestor meridionalis*) populations in New Zealand. Biol Conserv 83:175–185

Worthy TH (2001) A fossil vertebrate fauna accumulated by laughing owls (*Sceloglaux albifacies*) on the Gouland Downs, northwest Nelson, South Island. Notornis 48:225–233

Worthy TH, Holdaway RN (2002) The lost world of the moa: prehistoric life of New Zealand. Canterbury Univ Press, Christchurch, New Zealand, and Indiana Univ Press, Bloomington, IN

25 Setting Priorities for the Management of Marine Pests Using a Risk-Based Decision Support Framework

B.M. Forrest, M.D. Taylor and J. Sinner

25.1 Introduction

At least 148 marine species have been accidentally introduced into New Zealand, with a further 4 deliberate introductions (Cranfield et al. 1998). A number of these threaten New Zealand's environmental, economic, social and cultural resources, with changes in patterns of trade meaning that further incursions of unwanted organisms are inevitable (Taylor et al. 1999). In recognition of such threats, the Biosecurity Strategy for New Zealand expands the traditional focus from terrestrial and freshwater issues to also emphasize management of risks from marine pest species. This chapter outlines the key elements of a decision support framework that will contribute to this goal by providing a systematic and transparent mechanism for identifying and analyzing risks, and prioritizing management objectives in the marine environment.

Our framework is based on the risk management process described by Sinner and Gibbs (1998), which involves four stages: risk identification, risk assessment, analysis of risk treatment options, and risk evaluation. In this chapter, we provide a brief overview of key steps and considerations for the risk identification stage, and focus more on methodological approaches for the latter three stages. We build on lessons learned in developing a management strategy for the Asian kelp *Undaria pinnatifida* in New Zealand (Sinner et al. 2000), and reveal some of the peculiarities of bioinvasion and pest management in marine environments that contrast terrestrial and freshwater systems. Our underlying premise is that a logical starting point in setting management priorities for marine pests is to consider the values we wish to protect from adverse impacts. This approach is particularly relevant to those having an interest in the protection of areas that are geographically defined at local and regional scales, such as aquaculture sites and Marine Protected

Ecological Studies, Vol. 186
R.B. Allen and W.G. Lee (Eds.)
Biological Invasions in New Zealand

Areas, but the same logic can also be applied at greater spatial scales. Hence, in the sections below we describe a framework that allows: marine biosecurity risks to be identified in an explicit fashion; the probabilities that lead to a pest infestation estimated; the consequences of infestation at pest density assessed; and priorities to be established through comparison of the feasibility, benefits and costs of risk management.

25.2 Risk Identification

The risk identification process is an information gathering phase that we have subdivided into four key steps (Fig. 25.1). Our framework first requires that values are identified, and high-value areas (HVAs) prioritized in a defensible way. For decisions post-border, we suggest that priorities should be determined based on comparisons that are segregated according to the types of values being considered, such as distinguishing commercial aquaculture from marine conservation values. It would be more appropriate to assess the latter, for example, within the context of other conservation initiatives (including non-biosecurity initiatives), so that the maximum benefit for conservation is achieved within the available budget of the relevant organization. However, we acknowledge that management interventions (especially at a national scale) may have benefits across the different environmental, economic, social

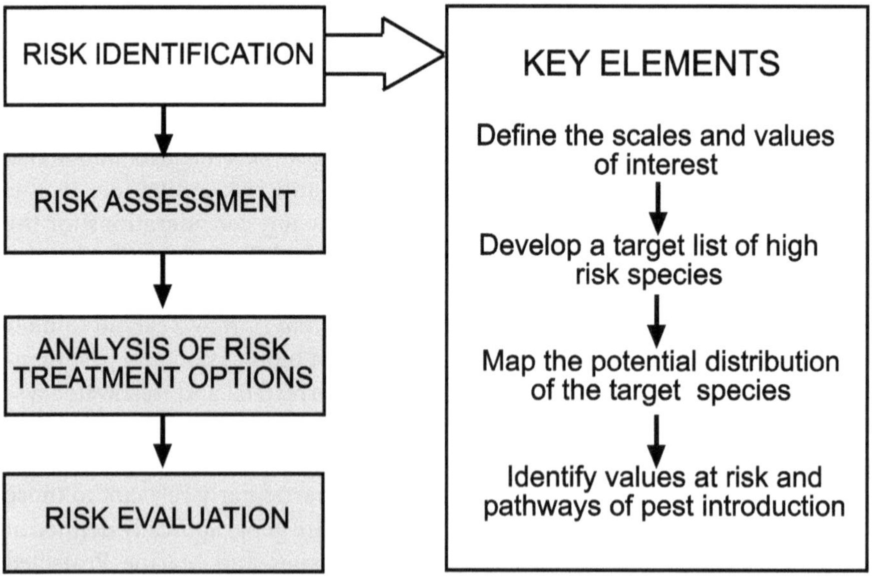

Fig. 25.1 Key elements of the risk management framework and risk identification stage

and cultural value sets, in which case a process would be required to determine the measures having the greatest benefits overall.

The second step, developing a target list of high-risk pests, is a precursor to making predictions about their potential distribution, and thus the values that they threaten. A target list should be based on explicit selection criteria (e.g., Hewitt and Hayes 2002; Hayes et al. 2002; Hayes and Sliwa 2003). Screening for pests based on their common biological characteristics, which is an approach used in terrestrial and freshwater weed management (Groves et al. 2001), may have merit for some groups of marine species (e.g., Nyberg and Wallentinus 2005). However, this approach may not be generally feasible for marine environments where the more idiosyncratic features of particular species often facilitate their success as invaders (Forrest et al. 1997; Ruiz and Hewitt 2002). Until better screening tools are developed, identifying potentially high-risk pests based on their invasiveness or impacts elsewhere is a useful starting point, and one that can motivate stakeholder interest, even though this approach may not encompass the full suite of high-risk species (McEnnulty et al. 2001; Simberloff 2003).

Given a target list, the potential distribution of each pest species in a recipient area assists in the identification of values at risk and the pathways to HVAs. A simple approach to estimate this distribution is to evaluate the 'match' between an organism's natural tolerances (e.g., temperature) and the environmental conditions in the recipient area (Smith et al. 1999). However, this type of assessment should be seen as conservative because experience in both terrestrial and marine systems has shown that it may underestimate actual pest distribution (e.g., Mack 1996; Floc'h et al. 1996).

The final step in the process, shown in Fig. 25.1, is to identify the pathways by which target species might be introduced into areas considered high priority for protection. Natural dispersal via water currents is likely to be particularly important in the local or regional spread of pests, especially those with planktonic larval stages. At these scales and greater, the importance of numerous human-mediated invasion pathways is also well recognized, with particular risks for the New Zealand marine environment posed by ballast water discharge (Hay et al. 1997; Inglis 2001), fouled hulls (e.g., Coutts and Taylor 2004), vessel sea chests (Coutts et al. 2003), and transfer of contaminated aquaculture equipment or shellfish seed-stock (Forrest and Blakemore 2002). It is important to note that nominally minor or unrecognized pathways can also pose significant risks in some circumstances (e.g., Hay and Dodgshun 1997; Coutts 2002), and need to be accounted for in the risk management process.

25.3 Risk Assessment

Following risk identification, the risk assessment stage estimates the likelihood of each target species being introduced to each HVA and becoming established at pest density, and the associated consequences. For this purpose, we propose a standard 'chain of events' approach that combines these elements to determine the level of threat posed by pest species, with a process to rank risks according to the level of importance attached to each HVA. This essentially provides an index for each HVA whose score reflects the 'unmanaged risk', which can be represented in simplistic terms as:

$$R_{Uij}=P_{Iij}xP_{PDij}xV_{j}xI_{ij} \qquad (25.1)$$

where R_{Uij} is the unmanaged risk from species *i* in area *j*, which is the expected value of damage from the pest in the absence of measures to reduce the likelihood of introduction or to respond to an incursion, P_{Iij} the probability of introduction of species *i* to area *j*, P_{PDij} the probability that, once introduced, species *i* will reach pest density in area *j*, V_j the total value at risk in area *j*, and I_{ij} the consequences of establishment at pest density of species *i* in area *j*, in terms of the proportion of the values at risk that are lost due to the pest.

The highest values of R_{Uij} represent the greatest risks. The approach is hierarchical, in that each main component can be broken into increasingly detailed parts, given the availability of sufficient information. In the analysis of risk treatment options, these implied priorities are re-ranked taking account of the feasibility, efficacy, and costs of management. We recognize the importance of incorporating measures of uncertainty throughout this process but restrict our discussion here to the logic of our approach.

25.3.1 Likelihood of Introduction: P_{Iij}

P_{Iij} represents the likelihood that a target species will be transported to an HVA during a given time frame, either by natural or by human-mediated pathways. With respect to human-mediated pathways, this assessment can be a significant undertaking, as exemplified by risk assessment approaches for ballast water alone (e.g., Hayes and Hewitt 1998; Hayes 2002). For a broad decision-making tool, more simplistic approaches may be needed (e.g., Aurand et al. 2000; Hayes et al. 2002). We suggest that effort is made to at least separate the likelihood of target pest introduction (e.g., based on qualitative scores) into the key pathways, because management interventions would typically address specific pathways in order to reduce the probability. For example, in a situation where key pathways are identified as hull fouling (HF), ballast water (BW), aquaculture (AQ), and natural spread (NS), the probability of introduction can be expressed as:

$$P_{Iij}=f\,(P_{I/HFij}, P_{I/BWij}, P_{I/AQij}, P_{I/NSij}) \tag{25.2}$$

The nature of the function f for calculating the overall probability P_{Iij} of at least one introduction during a selected time period depends on the relationship between the individual probabilities. A probability $P_{I/UEij}$ can also be used to represent the possibility of introduction via some unexpected or unanticipated pathway. This identifies residual risk that is not being managed, even though it will not affect the relative management priorities that emerge from the analysis. In most cases, it will be reasonable to assume that the probabilities are independent of each other but not mutually exclusive. In this case, the probability of at least one event is 1 minus the probability that none of them will occur (Snedecor and Cochran 1980). Thus, the expression is:

$$P_{Iij}=1-[(1-P_{I/HFij})x(1-P_{I/BWij})x(1-P_{I/AQij})x(1-P_{I/NSij})x(1-P_{I/UEij})] \tag{25.3}$$

25.3.2 Likelihood of Establishment at Pest Density: P_{PDij}

Environmental matching analyses made during the risk identification stage will provide rudimentary guidance on the likelihood of pest establishment in a recipient area, but prediction of infestation levels and hence potential impacts will be more difficult (Williamson 2001). An adequate knowledge of underlying invasion processes, likely infestation levels, and density-dependent effects is seldom available, with a general consensus that even with detailed study the prospect of making reliable predictions of invasion success is remote (e.g., Lawton and Brown 1986; Vermeij 1996; Kareiva et al. 1996; Forrest and Taylor 2002). Furthermore, knowledge of the general attributes of species and recipient communities that may influence the likelihood of success may not assist with prediction of whether a particular species will invade a particular locality, and to what extent (Lawton and Brown 1986; Simberloff 1989; Lodge 1993). Determination of the likelihood that an invader will reach pest density will therefore continue to rely on expert judgment. This can be formalized by providing categories for considering likely success based on factors such as (1) invader attributes (e.g., extent of prior invasion success, reproductive potential, and dispersal mode), (2) physical attributes of the recipient environment that may affect invasion success, such as regimes of temperature, salinity, wave exposure, space availability, and substratum suitability, and (3) biotic attributes of the recipient environment that may affect invasion success, such as the presence of grazers, predators, or competitors. Alternatively, in the absence of information, one can assign the same value to P_{PDij} for all species and sites, so that the evaluation of relative priorities is not influenced by this parameter, but this default approach may ignore potentially useful information.

25.3.3 Consequences of Establishment: $V_j \times I_{ij}$

The third main component needed to determine unmanaged risk, R_{Uij}, is estimation of the severity of consequences of an introduced organism reaching pest density in a given HVA. This reflects not only the level of infestation, but also the type of values affected and the level of importance attached to an HVA. In the case of the former, for example, infestation by a conspicuous invader at a density resulting in only minor ecological effects could have impacts that are more than minor if the location were highly valued for its natural character (i.e., the pest density threshold depends on the type of values being considered). For current purposes, we assume that different types of values will be addressed separately. The model does, however, account for the fact that the consequences of a given pest density will be greater for HVAs of greater value.

The term V_j, the total value at risk in a given area, essentially applies a weighting factor to the unmanaged risk score, giving greater weight to HVAs of relatively high value. The term I_{ij}, the consequences of establishment at pest density, provides further weighting according to the proportion of the values at risk that could be lost due to the pest. V_j could be expressed on any relevant scale – for example, a dollar scale for commercial values, or a 1–5 scale representing values of local through to international conservation significance, with qualitative scores assigned to I_{ij} to represent a scale from negligible to catastrophic consequences (e.g., Wotton and Hewitt 2004). Once V_j and I_{ij} have been determined, Eq. (25.1) can be calculated to estimate values for R_{Uij} to represent *relative* risk across species–site combinations.

25.4 Analysis of Risk Treatment Options

Treatment options for invasive marine species are primarily (1) management of spread to minimize the introduction of target species to HVAs, and (2) surveillance and response to new infestations. Clearly, 'no intervention' may also be valid in some circumstances – for example, where the costs of intervention outweigh the benefits, where the risks are negligible, or where they are essentially unmanageable. A further option may be mitigation of adverse impacts, an example being closure of coastal shellfish resources for harvesting because of blooms of toxin-producing microalgae (Rhodes et al. 2001).

Development of effective measures to manage marine pests is at an early stage. Even for measures that are technically feasible, high costs and other constraints often preclude their implementation. Within our framework, consideration of the likely effectiveness of management leads to an assessment of the residual threat posed by managed risk, R_{Mij}. Costs of management, C_{Mij}, must also be estimated to enable evaluation of which measures provide the most value for money. The level of managed risk and associated costs can be expressed as follows:

$$R_{Mij}=P'_{Iij}xP_{PDij}xV_jxI_{ij}x(1-P_{SCij}) \quad (25.4)$$

$$C_{Mij}=C_{SMij}+(P'_{Iij}xP_{PDij}xC_{SCij}) \quad (25.5)$$

where R_{Mij} is the managed risk from species i in area j, which is the expected value of damage from the pest despite measures to reduce the likelihood of introduction and respond to any incursion (i.e., residual risk), and P'_{Iij} the reduced probability of the introduction of species i in area j, after feasible measures to manage spread have been implemented; P_{PDij}, V_j and I_{ij} are defined as per Eq. (25.1), P_{SCij} is the probability of successful control of an incursion of species i in area j, C_{Mij} the expected cost of management measures to reduce the risk from species i in area j, C_{SMij} the cost of measures to manage spread that could be implemented to reduce the likelihood of introduction of species i to area j, and C_{SCij} the expected cost of incursion response to an introduction of species i to area j, i.e., the cost of incursion response discounted by the probability of an incursion.

Equation (25.4) is similar to Eq. (25.1), but requires consideration of the reduced risk of pest introduction to an HVA through management of spread (P'_{Iij}), and determination of the feasibility of management measures and the likelihood of successful control (P_{SCij}), so that the probability that control measures will fail ($1-P_{SCij}$) can be incorporated into the expression of residual risk. The terms V_j and I_{ij} are independent of the other terms, and remain the same as in Eq. (25.1). For simplicity, we assume that the probability of establishment at pest density (P_{PDij}) also remains the same, even though the likelihood of pest introduction may have decreased (i.e., $P'_{Iij}=P_{Iij}$). This reflects the level of uncertainty (even for many well-studied pests) regarding the relationship between inoculum pressure and subsequent establishment.

The analyst has to exercise judgment about which management measures to include in the model. Situations will invariably arise where only one of P'_{Iij} or P_{SCij} will be relevant or meaningful. For example, for New Zealand's sub-Antarctic islands (highly valued for conservation reasons), Sinner et al. (2000) demonstrated that managing pathways for *Undaria* would greatly reduce the risk of the seaweed's incursion, but that surveillance and incursion response were not feasible because of the isolated and rugged nature of the islands, i.e., P_{SCij} was treated as zero, and no incursion response was contemplated. Similarly, there may be reasons to consider control of well-established pests to densities that avoid adverse effects, where management of spread is clearly pointless (i.e., $P'_{Iij}=P_{Iij}$). In effect, the analyst must determine which package of measures to evaluate for each species–site combination, based on what appears to be the most feasible. Alternatively, more than one package of measures may be compared for a given species–site (e.g., Sinner et al. 2000). These points are further considered below.

25.4.1 Reducing the Risk of Introduction Through Management of Spread: $P'I_{ij}$

An assessment of the relative importance of natural vs. human-mediated spread of a pest is central to decisions regarding the need for management of anthropogenic pathways, and Fig. 25.2 outlines a screening tool that could be used for this purpose. Although the timescales in Fig. 25.2 are arbitrary, they are included to highlight the principle that the more vulnerable a locality is to natural spread, the less likely that management of human-mediated pathways will be worthwhile. Clearly, however, the extent to which management measures are considered necessary or desirable, especially in the 'medium priority' categories shown in Fig. 25.2, will depend on the values at stake. For example, a commercial aquaculture locality vulnerable to natural spread within a matter of a few years may be of such high value that it is worth evaluating the feasibility of managing human-mediated pathways to reduce the risk of pest introduction, perhaps to provide sufficient time for the industry to adapt or to enable development of effective incursion response measures.

Where analysis following Fig. 25.2 suggests further evaluation of anthropogenic pathways is important, one then considers whether management is likely to be feasible, because effective management strategies may have major costs that limit their usefulness. In southern New Zealand, for example, marine farmers adopted a voluntary ban on movements of aquaculture equipment and shellfish seed-stock, the aim being to prevent the transfer of *Undaria* to a region where a management program for the seaweed was in place, but they incurred costs from lost production when seed-stock from an alternative source was unavailable.

Often, management will need to focus on measures to reduce, rather than eliminate the spread of target species – for example, by limiting contamination of transfer mechanisms (e.g., through control of pest populations in source regions), pathway monitoring for target pests, and generic pathway management measures that may have added benefits beyond the target pest species (e.g., hull cleaning and anti-fouling, ballast water exchange). The ability to implement any or all of these will depend on factors such as the frequency and complexity of human-mediated pathways, the characteristics of the pest species or life-stage transported, the availability and cost of effective management measures, and the willingness of owners and operators of transfer mechanisms to partake in a management program.

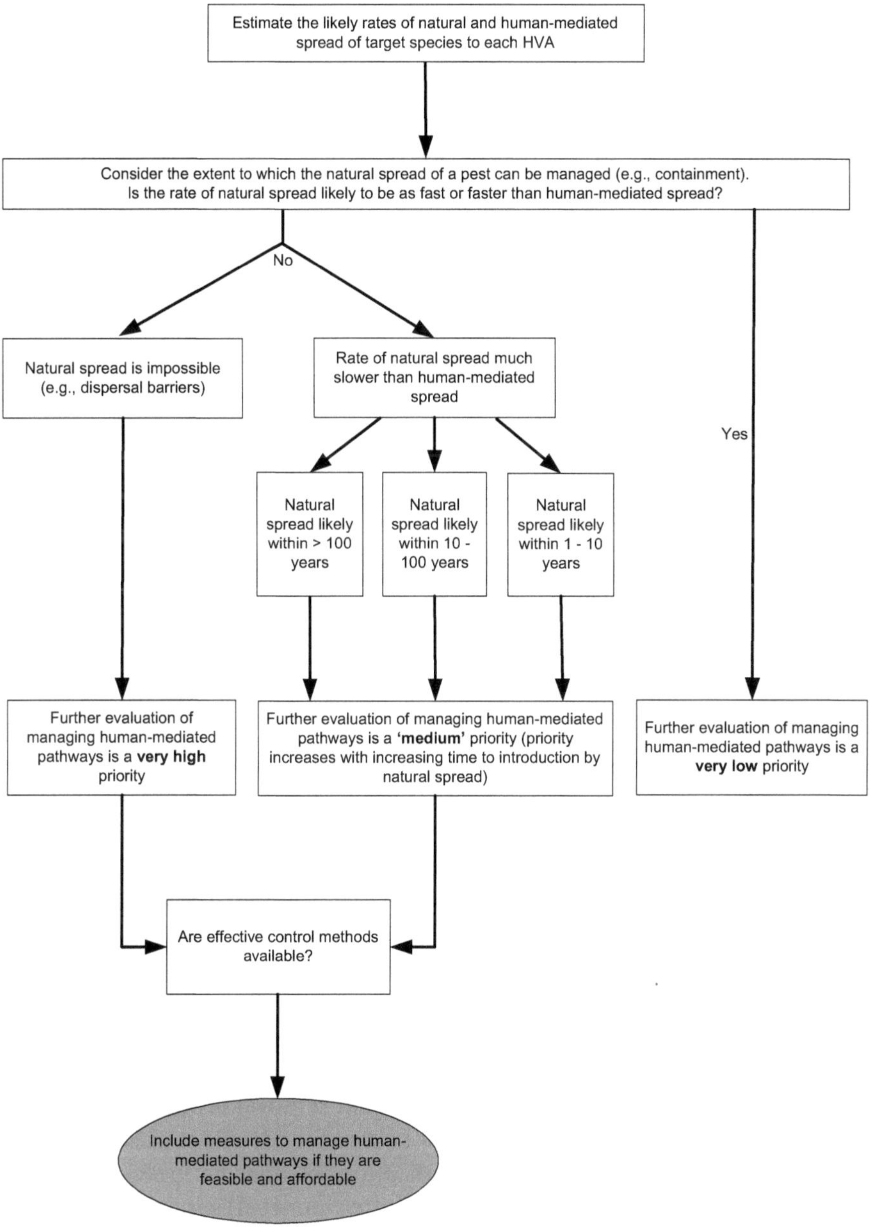

Fig. 25.2 Decision tree for evaluating whether, and to what extent management of human-mediated transfer mechanisms might be desirable for a given high-value area (HVA)

In terms of Eq. (25.4), P'_{Iij} is the residual risk that management measures will fail to prevent spread. Hence, the analyst will need to estimate the likely effectiveness of management for each pathway where feasible measures are available. This can be done by estimating each $P'_{I/Xij}$ directly (where X represents a pathway), or by estimating the proportion by which the measure would reduce risk of introduction via the pathway and multiplying this by $P_{I/Xij}$. For example, for a ballast water measure that reduced delivery of species *i* to site *j* by 50 %, $P'_{I/BWij}=P_{I/BWij}x0.5$. Using the examples of human-mediated (ballast water, hull fouling, aquaculture) and natural spread pathways given in Section 25.3.1, this will allow determination of P'_{Iij} as follows (this can be calculated as in Eq. 25.3):

$$P'_{Iij}=f\,(P'_{I/HFij}, P'_{I/BWij}, P'_{I/AQij}, P'_{I/NSij}) \tag{25.6}$$

The analyst also needs to provide an estimate for the term C_{SMij}, which is the sum of costs for feasible measures to manage spread. These estimates do not need to be precise, but need to be reasonably accurate relative to the cost estimates for other measures.

25.4.2 Surveillance and Incursion Response: P_{SCij}

Table 25.1 highlights features of marine environments and marine pests that affect the feasibility of traditional approaches to surveillance and incursion response (i.e., eradication, containment, or control of pest populations). Key challenges in marine systems lie in the early detection of target pests, and in the development of practical and cost-effective incursion response tools that have minimal adverse side effects (Wotton and Hewitt 2004; Thresher and Kuris 2004). In these respects, classical biological control is considered high risk (e.g., Secord 2003), and the mechanical and chemical treatment approaches commonly used in terrestrial and freshwater environments are not always applicable. Localized control of subtidal *Undaria* populations, for example, relies on diver detection and manual removal of the visible sporophyte stage of the seaweed, and is rarely successful (Hewitt et al. 2005). In the few situations where successful eradication of marine pests has been reported, there were usually particular (often unusual) circumstances that favored a positive outcome, as revealed by examples with *Undaria* on a sunken vessel near New Zealand's Chatham Islands (Wotton et al. 2004), and the black-striped mussel *Mytilopsis sallei* in a Darwin marina (McEnnulty et al. 2001).

Given the poor record of post-invasion management success in marine systems, highest priority should ideally be given to preventing new incursions (McEnnulty et al. 2001; Eno and Hamer 2002), but the lack of completely effective management measures for this purpose means that unwanted introduc-

Table 25.1 Key features of marine environments and marine pests that affect the feasibility of surveillance and incursion response

A. Receiving environment attributes

Relative ease of surveillance or response	Water clarity	Wave exposure	Bathymetric complexity	Biological complexity	Remoteness	Tidal state	Habitat availability
Easy	Clear	Sheltered	2D	Simple	Accessible	Intertidal	Limited
↓	↓	↓	↓	↓	↓	↓	↓
Difficult	Turbid	Exposed	3D	Complex	Remote	Deep subtidal	Unlimited

B. Invader attributes

Relative ease of surveillance or response	Invasiveness	Invader distribution	Invader conspicuousness	Habitat preferences	Propagule dispersal range
Easy	Low	Confined	Large or conspicuous	Specific	Short
↓	↓	↓	↓	↓	↓
Difficult	High	Widespread	Small or cryptic	Generalist	Distant

tions will continue. Hence, in the context of our framework, initial judgment is required as to whether surveillance and incursion response is likely to be worthwhile, such that the P_{SCij} term is retained in Eq. (25.4) for more detailed evaluation. Figure 25.3 provides a structured approach to assist with this decision, leading through a series of questions that relate to the feasibility of surveillance or incursion response based on the attributes of the pest and the receiving environment (e.g., Table 25.1), and on the availability of effective management options. A range of non-technical aspects also need to be considered as part of this process (McEnnulty et al. 2001; Wotton and Hewitt 2004).

The likelihood of successful incursion response, P_{SCij}, will clearly be species- and situation-specific, and driven to a large extent by the desired management endpoints (e.g., eradication vs. control). In considering options for managing *Undaria*, for example, Sinner et al. (2000) described the cumulative probability in a given year that (1) an infestation would occur, (2) the infestation would be detected while still at a 'level 1' stage (i.e., no reproductive plants), (3) the response to level 1 and 'level 2' (i.e., reproductive plants present) infestations failed, and (4) control efforts failed, leading to an uncontrolled infestation. These same elements will not always be appropriate for different species or situations. For example, it may be desirable to manage *Undaria* and other biofouling pests on aquaculture structures to a level that avoids adverse effects, even when repeated incursions are inevitable and eradication is not feasible.

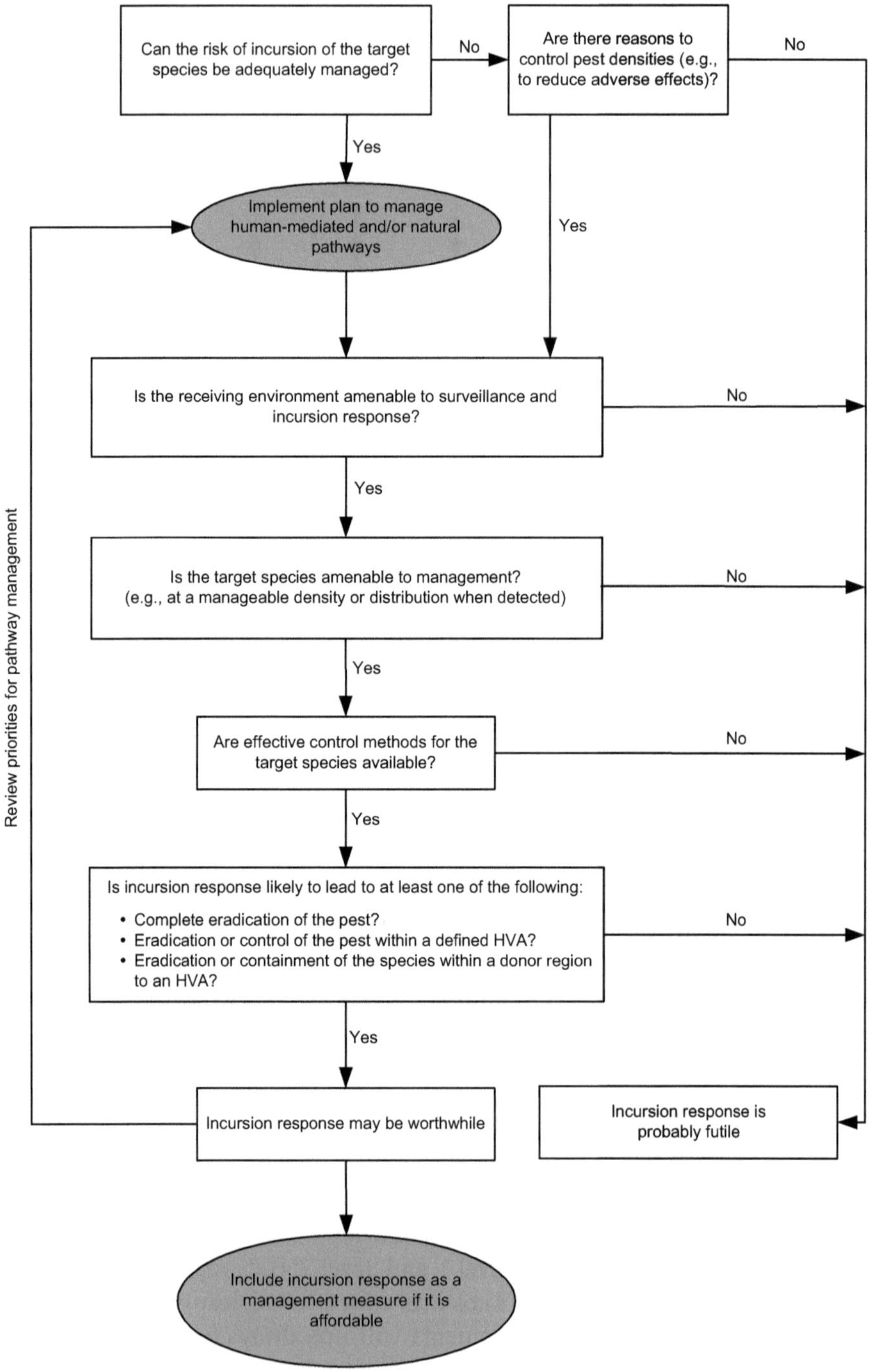

Fig. 25.3 Decision tree for considering whether incursion response for existing and potential pests is likely to be worthwhile

25.5 Risk Evaluation: Ranking Management Priorities

Risk evaluation involves comparing unmanaged risk with the risk after management, taking account of the costs of management. The most comprehensive analysis would be to determine priorities across all species and HVAs, and with respect to the full range of management measures. This involves estimating Eqs. (25.1) and (25.4) for each *ij* combination, and comparing the results – that is, for species *i* in area *j*:

$$R_{Uij}-R_{Mij}=\text{expected value of damage reduction} \quad (25.7)$$

To indicate relative priorities for management, the expected value of damage reduction represents the benefits (B) of management and, from the resulting matrix, one can identify the species–site combinations whose management would provide the greatest returns. This can be compared to the costs (C) of management in the form of a benefit–cost (B:C) ratio, as follows:

$$B_{ij}{:}C_{ij}=[R_{Uij}-R_{Mij}]/[C_{SMij}+(P'_{Iij}xP_{PDij}xC_{SCij})] \quad (25.8)$$

The B:C ratios can be ranked from highest to lowest to determine relative priorities, within a species–site matrix. Where V_j has been expressed in monetary terms, B:C>1 indicates a worthwhile expenditure, although when there are budget constraints only those actions with the highest returns would be implemented (see also Choquenot et al. 2004).

Another application of the framework would be to assess the relative return from management interventions that might be applied across all areas and species, e.g., a new hull fouling regulation. This requires estimating risk for the 'with' and 'without' management situations to calculate R_{Uij}, R_{Mij} and C_{ij}. The B:C ratio for measure X can be represented as follows:

$$B{:}C(X)=[SumR_{Uij}-SumR_{M(X)ij}]/SumC(X)_{ij} \quad (25.9)$$

A similar ratio can be estimated for alternative management interventions (or combinations thereof), and the ratios compared to see which delivers the greatest benefits (i.e., damage avoided) per dollar spent. For exercises such as these, one could use representative species or taxa (i.e., representing key attributes of interest for risk species), rather than a comprehensive list of target species, in order to keep the evaluation process manageable.

In the case of *Undaria*, Sinner et al. (2000) applied a simplified version of this framework to selected HVAs by estimating the parameters P'_IxP_{PD} (as a single parameter) and P_{SC} in order to obtain cost estimates for ranking a range of management options. V and I were assessed qualitatively, and used to describe the likely outcomes (i.e., benefits) of each option to inform decision makers in their selection of a preferred approach.

25.6 Conclusions

The aim of this framework is to provide an approach to setting priorities that caters for marine biosecurity threats (from existing or potential pests) to different types of coastal values or stakeholder sectors (e.g., aquaculture, conservation) at different scales of interest (e.g., national vs. internal border control). It is a framework that promotes forward planning to avoid poorly informed, ad hoc decision-making.

For full application, this approach would require a significant amount of data about particular pest species and the vulnerability of high-value areas to those species. In many circumstances, this information will not be available or the analyst might consider that it is not possible to identify the species that pose the greatest risk (e.g., given uncertainty regarding how an organism will behave in a new environment). However, the framework can be simplified to accommodate these situations, e.g., by using representative species or taxa, rather than a complete suite of target species. Furthermore, at least in certain situations, some of the parameters or even dimensions of the framework can be condensed if there is insufficient information, or deleted if the management question does not require their consideration.

The data for implementation of this framework can be accumulated and refined over time, and there is clearly scope to automate the assessment process. Rudimentary first applications covering a range of scales and values, if properly documented, will provide a useful platform for further applications and, given that many policy decisions will require consideration of similar parameters, the tool will become progressively more sophisticated. In further development of the framework, we emphasize the importance of information sharing among the various scientific disciplines and stakeholder groups involved in biosecurity, both in New Zealand and overseas, since many of the issues and needs raised in relation to the marine environment are common to all.

Acknowledgements Our sincere thanks to Chad Hewitt (Ministry of Fisheries), Geoff Hicks (Department of Conservation), and John Hellstrom (Biosecurity Limited) for their valuable comments on a draft manuscript. This project was funded by New Zealand Foundation for Research, Science and Technology, contract CAWX0202.

References

Aurand D, Walko L, Pond R (2000) Developing consensus ecological risk assessments: environmental protection in oil spill response planning – a guidebook. United States Coast Guard, Washington, DC

Choquenot D, Nicol SJ, Koehn JD (2004) Bioeconomic modelling in the development of invasive fish policy. NZ J Mar Freshwater 38:419–428

Coutts ADM (2002) A biosecurity investigation of a barge in the Marlborough Sounds. Cawthron Rep no 744. Cawthron Institute, Nelson, New Zealand

Coutts ADM, Taylor MD (2004) A preliminary investigation of biosecurity risks associated with biofouling of merchant vessels in New Zealand. NZ J Mar Freshwater 38:215–229

Coutts ADM, Moore KM, Hewitt CL (2003) Ships' sea-chests: an overlooked transfer mechanism for non-indigenous marine species? Mar Pollut Bull 46:1510–1513

Cranfield HJ, Gordon DP, Willan RC, Marshall BA, Battershill CN, Francis MP, Nelson WA, Glasby CJ, Read GB (1998) Adventive marine species in New Zealand. NIWA Tech Rep 34. Wellington, New Zealand

Eno C, Hamer JP (2002) Nature conservation implications of marine biological invasions. In: Leppäkoski E, Gollasch S, Olenin S (eds) Invasive aquatic species of Europe: distribution, impacts and management. Kluwer, Dordrecht, The Netherlands, pp 477–483

Floc'h J, Pajot R, Mouret V (1996) *Undaria pinnatifida* (Laminariales, Phaeophyta) 12 years after its introduction into the Atlantic Ocean. Hydrobiologia 326/327:217–222

Forrest BM, Blakemore KA (2002) Inter-regional marine farming pathways for the Asian kelp *Undaria pinnatifida*. Cawthron Rep no 726. Cawthron Institute, Nelson, New Zealand

Forrest BM, Taylor MD (2002) Assessing invasion impact: survey design considerations and implications for management of an invasive marine plant. Biol Invasions 4:375–386

Forrest BM, Taylor MD, Hay CH (1997) Foreign marine species in New Zealand: towards a risk assessment and management model. Cawthron Rep no 424. Cawthron Institute, Nelson, New Zealand

Groves RH, Panetta FD, Virtue JG (2001) Weed risk assessment. CSIRO, Victoria, Australia

Hay CH, Dodgshun TJ (1997) Ecosystem transplant? The case of the *Yefim Gorbenko*. In: Seafood New Zealand, May, pp 13–14

Hay CH, Handley S, Dodgshun TJ, Taylor M, Gibbs W (1997) Cawthron's ballast water research programme final report 1996–1997. Cawthron Rep no 417. Cawthron Institute, Nelson, New Zealand

Hayes KR (2002) Identifying hazards in complex ecological systems. Part 1: fault-tree analysis for biological invasions. Biol Invasions 4:235–249

Hayes KR, Hewitt CL (1998) Risk assessment framework for ballast water introductions. Tech Rep no 14. Centre for Research on Introduced Marine Pests, CSIRO Marine Research, Hobart, Tasmania

Hayes KR, Sliwa C (2003) Identifying potential marine pests – a deductive approach applied to Australia. Mar Pollut Bull 46:91–98

Hayes KR, McEnnulty FR, Sliwa C (2002) Identifying potential marine pests: an inductive approach. Final report for Environment Australia National Priority Pests Project, Centre for Research on Introduced Marine Pests, CSIRO Marine Research, Hobart, Tasmania

Hewitt CL, Hayes KR (2002) Risk assessment of marine biological invasions. In: Leppäkoski E, Gollasch S, Olenin S (eds) Invasive aquatic species of Europe: distribution, impacts and management. Kluwer, Dordrecht, The Netherlands, pp 456–466

Hewitt CL, Campbell ML, McEnnulty F, Moore KM, Murfet NB, Robertson B, Schaffelke B (2005) Efficacy of physical removal of a marine pest: the introduced kelp *Undaria pinnatifida* in a Tasmanian Marine Reserve. Biol Invasions 7:251–263

Inglis GJ (2001) Criteria for identifying and selecting high value locations and locations at risk of invasion by exotic marine organisms in New Zealand. Final research report for Ministry of Fisheries, NIWA, Wellington, New Zealand

Kareiva P, Parker IM, Pascual M (1996) Can we use experiments and models in predicting the invasiveness of genetically engineered organisms? Ecology 77:1670–1675

Lawton JH, Brown KC (1986) The population and community ecology of invading insects. Philos Trans R Soc Lond B Biol Sci 314:607–617

Lodge DM (1993) Biological invasions: lessons for ecology. Trends Ecol Evol 8:133–137

Mack RN (1996) Predicting the identity and fate of plant invaders: emergent and emerging approaches. Biol Conserv 78:107–121

McEnnulty FR, Bax NJ, Schaffelke B, Campbell ML (2001) A review of rapid response options for the control of ABWMAC listed introduced marine pest species and related taxa in Australian waters. Centre for Research on Introduced Marine Pests Tech Rep no 23. CSIRO Marine Research, Hobart, Tasmania

Nyberg CD, Wallentinus I (2005) Can species traits be used to prdict marine macroalgal introductions. Biol Invasions 7:265–279

Rhodes LL, Mackenzie AL, Kaspar HF, Todd KE (2001) Harmful algae and mariculture in New Zealand. ICES J Mar Sci 58:398–403

Ruiz GM, Hewitt CL (2002) Toward understanding patterns of coastal marine invasions: a prospectus. In: Leppäkoski E, Gollasch S, Olenin S (eds) Invasive aquatic species of Europe: distribution, impacts and management. Kluwer, Dordrecht, The Netherlands, pp 529–547

Secord D (2003) Biological control of marine invasive species: cautionary tales and land-based lessons. Biol Invasions 5:117–131

Simberloff D (1989) Which insect introductions succeed and which fail? In: Drake JA, Mooney HA, di Castri E, Groves RH, Kruger FJ, Rejmanek M, Williamson M (eds) Biological invasions: a global perspective. Wiley, New York, pp 61–75

Simberloff D (2003) How much information on population biology is needed to manage introduced species. Conserv Biol 17:83–92

Sinner J, Gibbs N (1998) A proposed framework for the management of biosecurity risks at the New Zealand border. Revised final report. Report to the Biosecurity Council, Wellington, New Zealand

Sinner J, Forrest BM, Taylor MD (2000) A strategy for managing the Asian kelp *Undaria*: final report. Cawthron Rep no 578. Cawthron Institute, Nelson, New Zealand

Smith LD, Wonham MJ, McCann L, Ruiz GM, Hines AH, Carlton JT (1999) Invasion pressure to a ballast-flooded estuary and an assessment of inoculant survival. Biol Invasions 1:67–87

Snedecor GW, Cochran WG (1980) Statistical methods, 7th edn. Iowa State Univ Press, Ames, IA

Taylor MD, Hay CH, Forrest BM (1999) Patterns of marine bioinvasion in New Zealand and mechanisms for internal quarantine. In: Pederson J (ed) Marine bioinvasions. Proc 1st National (US) Conf Marine Bioinvasions, 24–27 January 1999, MIT Sea Grant College Program, Massachusetts Institute of Technology, Cambridge, MA

Thresher RE, Kuris AM (2004) Options for managing invasive species. Biol Invasions 6:295–300

Vermeij GJ (1996) An agenda for invasion biology. Biol Conserv 78:3–9

Williamson M (2001) Can the impacts of invasive species be predicted? In: Groves RH, Panetta FD, Virtue JG (eds) Weed risk assessment. CSIRO, Collingwood, Australia, pp 20–33

Wotton DM, Hewitt CL (2004) Marine biosecurity post-border management: developing incursion response systems for New Zealand. NZ J Mar Freshwater 38:553–559

Wotton DM, O'Brien C, Stuart MD, Fergus DJ (2004) Eradication success down under: heat treatment of a sunken trawler to kill the invasive seaweed *Undaria pinnatifida*. Mar Pollut Bull 49:844–849

26 Does Commercial Harvesting of Introduced Wild Mammals Contribute to Their Management as Conservation Pests?

J.P. Parkes

26.1 Introduction

The dynamics of exploitation of wildlife as a commercial resource, and of control when they are pests share a common theoretical basis of renewable resource harvesting (e.g., Caughley and Sinclair 1994; Hone 1994). For most animal populations, per capita resources such as food decline once animal densities pass some threshold, resulting in reduced fecundity and increased mortality, until the population stops increasing when fecundity equals mortality at carrying capacity, K (see Chap. 12). At all densities up to K, the annual number of births exceeds the annual number of deaths, and the difference can be harvested or culled sustainably. The actual number that can be removed without affecting the population size depends on this density-dependent difference and the actual population size, resulting in a humped curve (Fig. 26.1) where the maximum sustainable harvest is taken from a population held at about K/2 – often a little higher for large mammals, as the density-dependent effects of resource limits on births and deaths are usually delayed (Fowler 1981; Barlow 1991). The different goals of commercial harvesters and pest managers require that they focus on different parts of Fig. 26.1 (Choquenot et al. 1995). The commercial harvester aims for a harvest size and achieves this by managing for a population density, whereas the pest controller aims for a population density and achieves this by managing for a sustained cull. As an aside, recreational hunters in New Zealand, largely hunting ungulates, have rather mixed aims. Some prefer to maximize their harvests, and have similar aims as commercial harvesters. Others prefer to maximize the number of trophy animals, which would be best achieved from populations at lower densities with more food per capita, and have aims more like those of pest managers (Nugent and Fraser 1993).

Ideally, a commercial harvester aims to take the economically optimal harvest of animals at the apex of Fig. 26.1, and so attempts to manage the popula-

Ecological Studies, Vol. 186
R.B. Allen and W.G. Lee (Eds.)
Biological Invasions in New Zealand

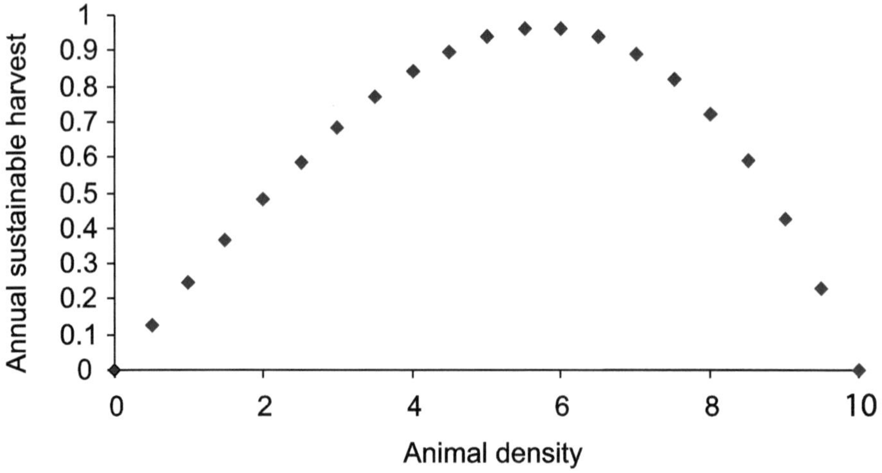

Fig. 26.1 Relationship between population density and annual recruitment. R=0.25, K=10, θ=2

tion density to the appropriate level. However, various equal, but economically suboptimal, harvests can be sustained from either lower or higher population sizes (Fig. 26.1). Sometimes, commercial harvesters may choose to not sustain their industry, depending on their perception of the present and future value of the animals (May 1976; Clark 1990). This is economically rational when it pays to kill as many as possible, and invest the profits elsewhere, i.e., before someone else does, or in cases where the population's rate of increase is less than the discount rate indexed by the interest rate a harvester can expect on any profits (Clark 1990).

Ideally, a pest controller aims to reduce animal populations to some target density at which their impact is absent or acceptable, and then (assuming eradication is not an option) hold them there by some sustained control strategy (e.g., Choquenot and Parkes 2001). Pest impacts are clearly likely to be higher and unacceptable above densities where the animals are competing for food – the food usually being the conservation or productive resource people value. Therefore, unlike a commercial harvester, a pest manager generally aims for a density below the apex in Fig. 26.1.

However, the biology and economics that underpin these general systems are rarely straightforward. Commercial harvesters have to deal with changes in the rates of population increase that are not simply due to the effects of food on the animal's rate of increase (e.g., Caughley and Sinclair 1994), uncertain market or discount rates, and the economic effects of any competing harvesters or of "by-catch" of other species. Pest controllers also have to deal with uncertainties about whether their target pest densities actually protect the conservation values impacted – notably, targets change as extrinsic events affect either the dynamics of the pest, the resource, or the interaction between the two.

In New Zealand, all 31 species of introduced wild or feral mammals are legally defined as pests, especially when they occur on land reserved in the public conservation estate (e.g., King 2001). New Zealand spends at least NZ$ 40 million per year managing the worst of these mammals to protect conservation values, and at least a further NZ$ 60 million per year on those that also affect agricultural production (Parkes and Murphy 2003). Several species are also sought by commercial hunters either alone (brushtail possums, *Trichosurus vulpecula*) or in competition with recreational hunters (red deer, *Cervus elaphus*; feral pigs, *Sus scrofa*; chamois, *Rupicapra rupicapra*; Himalayan tahr, *Hemitragus jemlahicus*) who also see the animals as a valuable resource. In fact, most New Zealanders have rather mixed views on the status of introduced wild mammals. Most see the rodents, mustelids, lagomorphs, and marsupials as pests, but only a minority agree with the law and view the larger ungulates as "only pests" (Fraser 2001). Dealing with the conflicting aims of these official and public views, and the competition between recreational and commercial hunters, forms a significant part of the management process for many species in New Zealand (e.g., for tahr; Hughey and Parkes 1996). Wild and feral animals are the property of the Crown on lands of all tenures, but become the property of any hunter who legally takes them. Hunters require only the permission of the landowner, and generally no limits are placed on when or how many animals may be taken. As the Crown (via the Department of Conservation) manages about 30% of New Zealand as public conservation estate, and as they see the animals as pests, there are usually few limits imposed on commercial or recreational hunters, i.e., no charges and few restrictions.

Where commercial harvesters do kill animals that others consider a pest, the question might be asked by the latter whether, and under what circumstances, the former can be of use as pest controllers. This chapter reviews that question by using some New Zealand examples of the commercial exploitation of red deer, Himalayan tahr, and brushtail possums.

26.2 What Does Pest Control Need to Deliver?

The best solution to deal with pests – apart from not introducing them – is to eradicate them. New Zealand management agencies have growing experience of extirpating pests on islands (Veitch and Bell 1990; Parkes and Murphy 2003), some isolated populations on the main islands (e.g., feral goats; Parkes 2001), and new populations of red deer (Fraser et al. 2003). However, generally the biological and social conditions that must be met before eradication is achievable, i.e., no immigration, all at risk, killed faster than they can replace their losses, and sufficient resources and political will to achieve success (Parkes 1990; Bomford and O'Brien 1995), cannot yet be achieved throughout the main islands of New Zealand, and sustained control is the only positive option to manage most vertebrate pest species (Parkes 1993a).

Unlike eradication, sustained control is not a simple strategy. At a single place, or in general, if funds are not limited, the optimal sustained control strategy to protect conservation values from pests is to reduce pest densities to as few as technology will allow, and then hold them there by a strategy of periodic maintenance control to kill the annual increment. This assumes that even low pest densities are sufficient to ensure the viability of the conservation resources affected. If the required pest density is lower than technically achievable, the appropriate strategy is to remove the threatened resource to a place of safety, e.g., as was done for the endangered parrot the kakapo (*Strigops habroptilus*) when threatened by feral cats on Stewart Island (Cresswell 1996).

A major issue for sustained control is therefore how to optimize the annual investment of $ 40 million by choosing the right pests to control in the right places at the right times, frequencies and intensities. The optimal strategy to control a pest species at a place where it affects a conservation resource depends mostly on the nature of their interaction. Generally, for pest species that have relatively stable populations affecting resource populations in some constant or chronic way, the best solution is to reduce the pest population to some target density at which the impact is absent or tolerable, and hold it there by some regular maintenance control regime (Parkes 1993b). However, for pest species whose density fluctuates widely, the optimal frequency and intensity of control depends more on identification of the timing of their acute, periodic impact on conservation values (Parkes 1993b). Especially for these acute-impact pests, decisions on the best intervention point are further complicated in New Zealand by the fact that their native prey also often fluctuate in response to events such as seed masting, which may also drive the pest's population dynamics such that their impacts may or may not be in phase with their density (e.g., Wilson et al. 1998). Fortunately for the simplicity of this review, the only mammals commercially harvested have relatively stable populations with chronic impacts on most affected conservation values.

26.3 What Does Commercial Harvesting Deliver as a Pest-Control Byproduct?

Commercial harvesting needs to deliver a profit to the harvester, and this can be potentially done by:

- a one-off harvest to actual extinction or to economic extinction if the population never recovers,
- an infrequent harvest if the population recovers from economic extinction to densities worth re-exploiting,
- a sustained annual harvest that may be constant or variable.

All of these options can potentially provide some benefit to pest managers under different circumstances.

Commercial harvesting to actual extinction (=eradication, in pest-control jargon) requires either that the last animal is so valuable that it is worth hunting (e.g., as with rhinoceroses in many parts of Africa), or that it is held at very low densities as secondary prey (a predator pit), i.e., where all individuals are vulnerable as by-catch – when some associated species remains profitably harvestable, the secondary prey is at risk of extinction due to the harvesting, or simply bad luck. This has not occurred with any modern introduced vertebrate in New Zealand, although it was arguably the result of some exploitation of native wildlife in the past (e.g., for the anthropogenic extinction of moa (Dinonithifoines; McGlone 1989). The only pest eradication program that at all paralleled the "rhinoceros" was the eradication of coypu (*Myocaster coypus*) from England where the value per animal, as a bounty rather than a commercial profit, was increased as the density of coypus decreased (Gosling and Baker 1989).

Exploitation to economic extinction, as seen in some fisheries (e.g., Walters and Maguire 1996), has not happened with any New Zealand introduced mammal populations. Still, the periodic harvests taken from Himalayan tahr might have fallen into this category if mortality from official and recreational hunting had been higher (see below). Sustained (but fluctuating) annual commercial harvests have been taken from red deer, feral pigs, chamois and possums in New Zealand.

In general, a commercial harvest might provide one of three sorts of conservation outcomes:

- Sustain a harvest sufficient to reduce the population density below some level at which the animals' impact on conservation resources is acceptable.
- Sustain a harvest that, although not reducing the pest animal enough to protect the conservation resource, saved the pest-control agency having to pay for that part of the control effort.
- Sustain a harvest that provides no benefit to conservation.

26.4 Case Study 1: Red Deer

Red deer are the most abundant of seven cervid species with wild populations in New Zealand, and form the bulk of the commercial harvest. The current population of about 200,000, 10 % of its size before the advent of commercial harvesting, covers a range of about 120,000 km^2 (Nugent et al. 2001a). Large-scale culling of deer as pests by government agencies began in the 1930s, with annual tallies averaging 28,000 animals until 1961 (Howard 1965), although since the 1960s this official control has been increasingly restricted to a few key conservation sites (e.g., to protect rare bird habitats; Maxwell 2001), and to remove new populations from previously deer-free areas (Fraser et al. 2003) with annual tallies of a few hundred animals at most. Deer are also a popular recreational hunting resource, about 40,000 being shot each year (Nugent 1992).

Commercial harvesting of venison for game meat, mostly exported to Europe, began in 1958 using ground hunting, often serviced by fixed-wing aircraft and helicopters to position the hunters and transport the carcasses (Challies 1990). Large-scale shooting from helicopters began in the mid-1960s, and has accounted for at least 2 million deer (Challies 1990; Parkes et al. 1996; Nugent et al. 2001a; Fig. 26.2). In the decade from ca. 1977, nearly 100,000 deer were also captured to stock deer farms (Challies 1990; Nugent et al. 2001a). The annual harvest of deer between 1988 and 1999 by commercial operators has varied between about 13,000 and 32,000, almost all of this variation being explained by the price of venison (Parkes et al. 1996; Nugent et al. 2001a). A potential constraint on the commercial harvest is the availability of processing factories. These usually also kill and process farmed deer, a NZ$ 210 million average per year industry, and require separate processing facilities if they are to handle wild-shot deer, a NZ$ 2.5 million average per year industry. In 2002, for example, a shipment of venison was withdrawn from export and tested for pesticide residues (none were found) after a few wild-deer harvesters took animals from areas where sodium monofluoroacetate (compound 1080) had been used to control possums. Processing factories refused to take any wild deer for about 6 months until the government, and they, could be assured that the rules for harvesting game animals from areas where pesticides had been used were being followed. As a consequence

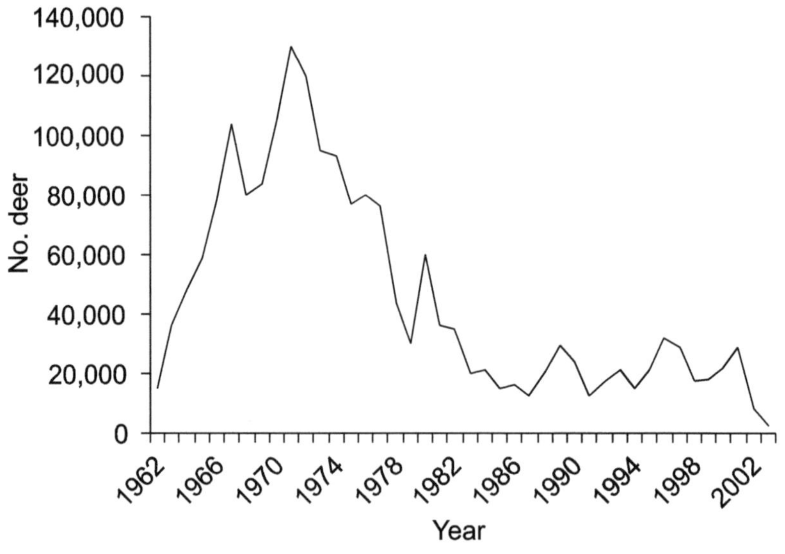

Fig. 26.2 Annual harvest of deer in New Zealand by commercial hunters, 1962–2003

of this, in 2002 and 2003 the harvests were only 8,305 and 2,203 deer, respectively (Fig. 26.2). The risks of pesticide contamination between farmed and wild venison in the separated processing facilities are very low, but market perceptions of such risks determined that the need to protect the farmed deer industry was paramount.

Overall, the annual commercial harvest of red deer has averaged about 20,000 animals. Assuming an exponential rate of increase of 0.33 and logistic growth (Nugent et al. 2001a), a population of 200,000 would increase by 78,000 per year if no deer were harvested, i.e., the commercial harvest takes about 25 % of the annual increment.

Before the advent of helicopter shooting, deer densities were greatest in grassland habitats, and least in forest habitats in catchments where the animals had a choice (Challies 1990). The use of helicopters has virtually extirpated deer from alpine and subalpine grasslands where basically all animals are at risk to the hunters. In forested habitats, deer are less vulnerable to helicopter hunting, and although densities have been substantially reduced – more where the proportion of grassland to forest in a catchment is highest (Nugent et al. 1987) – current populations appear to average about 3–4 animals km^{-2}, with maximum densities of about 15 animals km^{-2} in some North Island forests (Nugent et al. 2001a).

The benefit of the commercial exploitation of deer to conservation has been considerable in the alpine grasslands, especially in areas such as Fiordland National Park where deer are the main introduced herbivore present (e.g., Rose and Platt 1987). The conservation benefit in forest habitats has been minimal, despite much-reduced deer densities. This is because deer previously living in forests removed most of the palatable plants within browse range when they were at higher densities prior to the commercial harvesting. The current forest populations rely on leaf-fall from palatable canopy trees for a significant proportion of their diet, unless they have access to other habitats. This means that any seedlings that are more palatable than leaf-fall are in a "predator pit", and all accessible palatable seedlings are eaten across a wide range of residual deer densities (Nugent et al. 2001b). However, even the removal of all deer from New Zealand ecosystems (e.g., by fencing or eradication) does not necessarily reverse the changes they caused. Wardle et al. (2001) have shown that browse-induced changes in plant communities have altered litter quality and soil properties such that, even if deer herbivory is removed, the ecosystem (forest exclosures in their study) can follow a variety of trajectories. Similarly, Coomes et al. (2003) have noted that effects of competition from unpalatable plants, lack of seed sources, and responses of other herbivores to removal of deer may all preclude reversibility.

Irrespective of any benefits, the commercial harvest is sustainable because the majority of surviving deer now live in the forest and are less accessible to helicopters, unless they venture out into grasslands or other open sites – their preferred habitat.

26.5 Case Study 2: Himalayan Tahr

Himalayan tahr are restricted to about 4,300 km^2 of the Southern Alps (Fraser et al. 2000), with a current population of perhaps 10,000 animals (Forsyth and Tustin 2001). Government agencies culled about 25,000 tahr between 1937 and 1970, and currently cull between 1,000 and 2,000 animals per year, mostly from national parks (T. Farrell, DOC, personal communication). Recreational hunters kill about 750 tahr each year (Nugent 1992).

Commercial harvesting of tahr began in 1971, and until 1982 accounted for over 39,000 animals. Along with recreational hunting and government culling, this reduced the population from about 50,000 to fewer than 5,000 animals in 1982 (Parkes et al. 1996). The commercial harvest of tahr stopped in 1982 under pressure from recreational hunters, although it was by this stage only a by-catch of the larger venison industry, with a few hundred tahr being taken in the winter months when deer hunting was not usually conducted (Parkes et al. 1996). A tahr control plan was developed in 1993 (DOC 1993), in an attempt to hold tahr numbers below 10,000 animals by giving recreational hunters first harvesting rights in areas where they have legal access. In general, they have not been able to kill enough (most target adult males as trophy animals), and occasional commercial harvests (between 1995 and 2002; Fig. 26.3) and official culling have been necessary (Forsyth and Tustin 2001).

Tahr are the most obligate alpine ungulate in New Zealand, and as such most animals are always vulnerable to an aerial "predator" such as helicopter

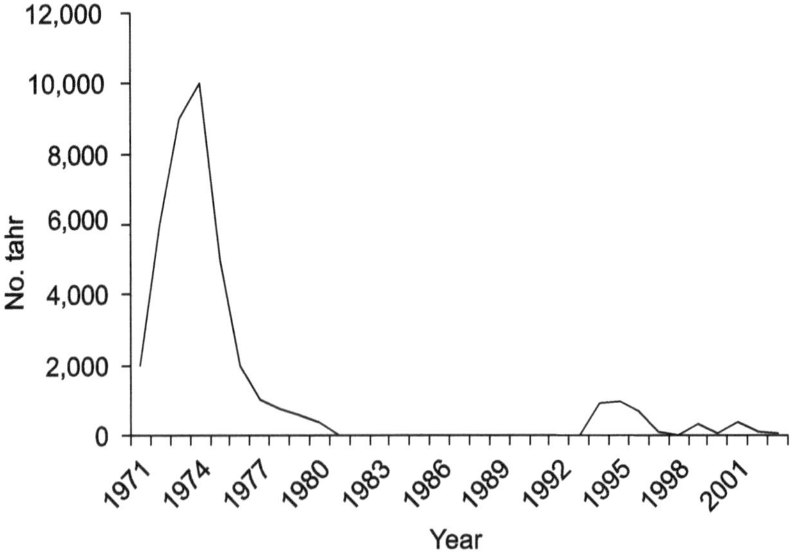

Fig. 26.3 Annual harvest of Himalayan tahr in New Zealand by commercial hunters, 1971–2003

hunters, especially in winter when the animals prefer to camp on steep, snow-free bluffs (Tustin and Parkes 1988). However, some tahr do live for at least part of their lives in the subalpine scrub and montane forests especially in Westland, and are not vulnerable to helicopter hunting (Forsyth and Tustin 2001).

Overall, the annual commercial harvest of tahr from 1994 to 2003 has averaged ca. 400 animals (Fig. 26.3). Assuming an exponential rate of increase of 0.128 (Caughley 1970), a population of 10,000 would increase by 1,136 in a year if all hunting stopped. The average commercial harvest would account for about 30 % of this potential increase.

The conservation benefits of the initial 90 % reduction in the 1970s were considerable, especially to their primary food of alpine snow tussocks (*Chionochloa* spp.; DOC 1993; Parkes and Thomson 1999). Surveys show a significant negative relationship between tahr density and tussock condition taken across their range, and catchments with very low densities (<1 tahr km^{-2}) continue to improve (Parkes et al. 2004).

26.6 Case Study 3: Possums

Possums are almost ubiquitous throughout New Zealand, and their population is probably over 50 million animals – the 70 million estimated by Batcheler and Cowan (1988, reported in Warburton et al. 2000), less those killed in the sustained control operations conducted by government agencies, although these estimates are rather speculative. Possums are subject to large-scale official control as vectors of bovine tuberculosis over about 40,000 km^2, and as conservation pests over about 10,650 km^2 of the public conservation estate (Parkes and Murphy 2003).

Possums were introduced to New Zealand from Australia in the 19th century to form an export fur industry, and have been harvested for that purpose since about 1912 (Warburton et al. 2000). Between 1921 and 1997, over 60 million possum skins were exported (Fig. 26.4), and, as with the red deer harvest, the annual harvest of furs (since 1967 at least) has been largely determined by the price the hunters receive per skin (Parkes et al. 1996; Warburton et al. 2000). Since the mid-1990s, a large number (e.g., about 2 million possums in 2002) have been harvested so that their fur can be used to mix with sheep wool to produce clothing (Warburton et al. 2000).

There have been few measures of the effect of possum harvesting for fur on possum densities. Brockie (1982) estimated commercial fur hunters reduced possum densities by 49–63 % in accessible areas over a period (1966–1982) with relatively high prices for possum skins (Warburton et al. 2000). Assuming an intrinsic rate of increase for possums of 0.3 (Barlow 2000), a population reduced to 50 million would increase by about 17 million per year in the absence of any hunting. The current average annual commercial harvest of

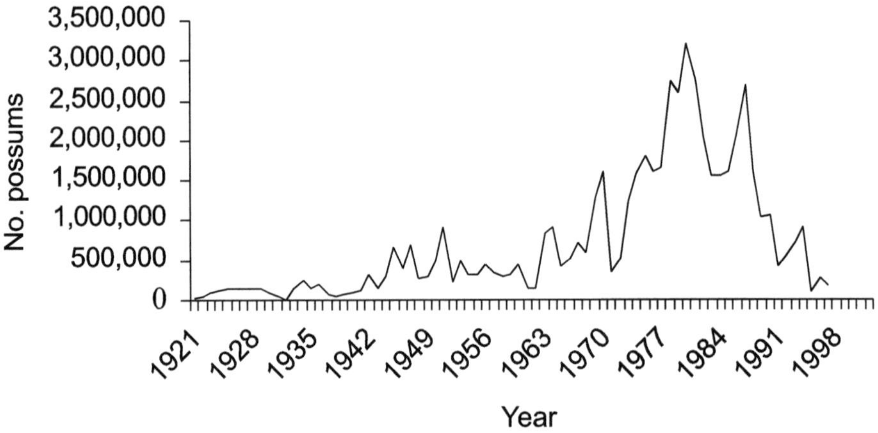

Fig. 26.4 Annual harvest of possums by commercial skin and fur harvesters in New Zealand, 1921–1997

less than 3 million (Warburton et al. 2000) would contribute little to maintaining these reduced densities – other than at accessible sites.

This level of reduction, if sustained, is probably not sufficient to eliminate bovine Tb (Barlow 2000), and is inadequate to protect vulnerable biota (e.g., Pekelharing et al. 1998; Sweetapple et al. 2002). Thus, the need continues for substantial government and private control efforts to protect conservation values and reduce bovine Tb risks at priority places.

26.7 Conclusions

The commercial harvesting industry in New Zealand is analogous to a predator–prey system involving primary and secondary prey with and without refugia (Holling 1959; Caughley and Sinclair 1994). Similarly, the herbivore–resource interaction that describes the benefits of the commercial harvest depends on whether the vegetation eaten by the animals is always available or not.

The current deer harvest is sustainable (extrinsic events excluded) because a significant proportion of the population finds refuge from the predator by living in forest habitats. Benefits to conservation are large, and probably directly related to deer density in the grasslands where deer have access to most of their food resource. However, benefits are less quantifiable in forests, at least in the short term. In many forests, that part of the food resource that is always accessible to deer (palatable seedlings and saplings in the forest understory) is a secondary prey. They are in a predator pit, as deer are sustained by the primary food of canopy litter fall and epicormic growth from the mature trees. In the long term, of course, the effects of deer, even at quite low densities, on regeneration will change the composition of forests toward unpalatable species. Furthermore, removal of deer after any long-term presence will not necessarily reverse this process (Wardle et al. 2001).

The tahr harvest is also potentially sustainable because it can be managed as a by-catch (secondary prey) of the larger deer harvesting industry. It is more at risk than the deer commercial industry because of the new requirement for exporters to test for pesticide residues in game meat. Commercial harvesting has provided direct conservation benefits, and has resulted in a general improvement in the condition of tahr's primary food – alpine snow tussocks. However, unless the commercial harvest is sustained, these benefits will only be maintained by increased official control (with an opportunity cost elsewhere), as recreational hunters have not by themselves been able to kill enough tahr to stop populations increasing.

The possum fur industry is sustainable, as it takes only a small proportion of the total possum population. However, it currently provides little conservation benefit because the residual densities are too high.

A significant difference between culls as pest control and commercial harvesting is that the former is, in practice, focused at sites where the pests' impacts are worst, whereas the latter is focused at places with the highest accessible animal densities. Thus, harvesting may give general benefits, as with the removal of alpine deer and reduction of tahr, but pest managers are still left with too many residual pests at their high-priority sites.

Some have argued that because commercial and/or recreational harvesting places a value on the animal, it can compromise optimal pest-control strategies and tactics. There is a degree of posturing among the protagonists in the debate over wild animal management in New Zealand. Pro-hunting groups in New Zealand lobby against some pest control, e.g., against culling male trophy tahr. Pro-control groups generally support commercial exploitation of pest species, presumably because they do not think the harvesters are organized enough to manage for higher animal densities from which to take their harvests. However, they generally oppose recreational harvesters other than as pest-control agents, presumably because they think the hunters might be effective in lobbying for higher animal densities. These conflicts will not go away. National legislation generally views introduced animals as pests, but national policy statements tend to be more pragmatic and countenance "spatial compromises", which set different densities at different places, and then partition the harvests required to keep the animals below these target densities among the competing harvesters (Department of Conservation 1991, 2001). In this case, commercial harvesting is seen by pest managers as, at best, a useful component of the tactical mix for pest control or, at worst, irrelevant. Hunters and conservation groups will usually have different views on what the target densities should be, but once decided it is, from the pest manager's point of view, the technical ability of the different harvesters to deliver that should influence who gets first pick – assuming such choices can be managed.

Acknowledgements I thank Rob Allen and Jacqueline Beggs for comments on drafts of this paper, and Peter van der Logt of the New Zealand Food Safety Authority for data on recent deer and tahr harvests.

References

Barlow ND (1991) Control of endemic bovine Tb in New Zealand possum populations: results from a simple model. J Appl Ecol 28:794–809

Barlow ND (2000) Models for possum management. In: Montague TL (ed) The brushtail possum: biology, impact and management of an introduced marsupial. Manaaki Whenua Press, Lincoln, New Zealand, pp 208–219

Bomford M, O'Brien P (1995) Eradication or control for vertebrate pests? Wildl Soc Bull 23:249–255

Brockie RE (1982) Effect of commercial hunters on the number of possums, *Trichosurus vulpecula*, in Orongorongo Valley, Wellington. NZ J Ecol 5:21–28

Caughley G (1970) Eruption of ungulate populations, with emphasis on Himalayan tahr in New Zealand. Ecology 51:53–72

Caughley G, Sinclair ARE (1994) Wildlife ecology and management. Blackwell, Cambridge, MA

Challies CN (1990) Red deer. In: King CM (ed) The handbook of New Zealand mammals. Oxford Univ Press, Auckland, New Zealand, pp 436–458

Choquenot D, Parkes J (2001) Setting thresholds for pest control: how does pest density affect resource viability? Biol Conserv 99:29–46

Choquenot D, O'Brien P, Hone J (1995) Commercial use of pests: can it contribute to conservation objectives? In: Grigg GC, Hale PT, Lunney D (eds) Conservation through sustainable use of wildlife. Centre for Conservation Biology, University of Queensland, Australia, pp 251–258

Clark CW (1990) Mathematical bioeconomics: the optimal management of renewable resources, 2nd edn. Wiley, New York

Coomes DA, Allen RB, Forsyth DM, Lee WG (2003) Factors preventing the recovery of New Zealand forests following control of invasive deer. Conserv Biol 17:450–459

Cresswell M (1996) Kakapo recovery plan 1996–2005. Threatened Species Recovery Plan 21. Department of Conservation, Wellington, New Zealand

Department of Conservation (1991) Himalayan thar management policy. Department of Conservation, Wellington, New Zealand

Department of Conservation (1993) Himalayan thar control plan. Canterbury Conservancy Conservation Management Planning Ser 3:1–47

Department of Conservation (2001) Department of Conservation policy statement on deer control. Department of Conservation, Wellington, New Zealand. www.doc.govt.nz/conservation(Animal pests)

Forsyth DM, Tustin KG (2001) Advances in New Zealand mammalogy 1990–2000: Himalayan tahr. J R Soc NZ 31:251–261

Fowler CW (1981) Density dependence as related to life history strategy. Ecology 62:602–610

Fraser W (2001) Introduced wildlife in New Zealand: a survey of general public views. Landcare Research Science Ser 23. Manaaki Whenua Press, Lincoln, New Zealand

Fraser KW, Cone JM, Whitford EJ (2000) A revision of the established ranges and new populations of 11 introduced ungulate species in New Zealand. J R Soc NZ 30:419–437

Fraser KW, Parkes JP, Thomson C (2003) Management of new deer populations in Northland and Taranaki. Science for Conservation 212. Department of Conservation, Wellington, New Zealand
Gosling LM, Baker SJ (1989) The eradication of muskrats and coypus from Britain. Biol J Linn Soc 38:39–51
Holling CS (1959) The components of predation as revealed by a study of small mammal predation of the European pine sawfly. Can Entomol 91:293–320
Hone J (1994) Analysis of vertebrate pest control. Cambridge Univ Press, Cambridge, UK
Howard WE (1965) Control of introduced mammals in New Zealand. DSIR Info Ser 45:1–96
Hughey KFD, Parkes JP (1996) Tahr management – planning and consultation under the Wild Animal Control Act. R Soc NZ Misc Ser 31:85–90
King CM (2001) Advances in New Zealand mammalogy, 1990–2000. Introduction. J R Soc NZ 31:1–5
Maxwell JM (2001) Fiordland takahe: population trends, dynamics and problems. In: Lee WG, Jamieson IG (eds) The takahe. Fifty years of conservation management and research. Univ Otago Press, Dunedin, New Zealand, pp 61–79
May RM (1976) Harvesting whale and fish populations. Nature 263:91–92
McGlone MS (1989) The Polynesian settlement of New Zealand in relation to environmental and biotic changes. NZ J Ecol suppl 12:115–129
Nugent G (1992) Big-game, small-game, and gamebird hunting in New Zealand: hunting effort, harvest, and expenditure in 1988. NZ J Zool 19:75–90
Nugent G, Fraser KW (1993) Pests or valued resources? Conflicts in management of deer. NZ J Zool 20:361–366
Nugent G, Parkes JP, Tustin KG (1987) Changes in the density and distribution of red deer and wapiti in northern Fiordland. NZ J Ecol 10:11–21
Nugent G, Fraser KW, Asher GW, Tustin KG (2001a) Advances in New Zealand mammalogy 1990–2000: Deer. J R Soc NZ 31:263–298
Nugent G, Fraser W, Sweetapple P (2001b) Top down or bottom up? Comparing the impacts of introduced arboreal possums and 'terrestrial' ruminants on native forests in New Zealand. Biol Conserv 99:65–79
Parkes JP (1990) Feral goat control in New Zealand. Biol Conserv 54:335–348
Parkes JP (1993a) Feral goats: designing solutions for a designer pest. NZ J Ecol 17:71–83
Parkes JP (1993b) The ecological dynamics of pest-resource-people systems. NZ J Zool 20:223–230
Parkes JP (2001) Advances in New Zealand mammalogy 1990–2000: feral livestock. J R Soc NZ 31:233–241
Parkes J, Murphy E (2003) Management of introduced mammals in New Zealand. NZ J Zool 30:335–359
Parkes JP, Thomson C (1999) Impact of Himalayan thar (*Hemitragus jemlahicus*) on snow tussock in the Southern Alps. Science for Conservation 132. Department of Conservation, Wellington, New Zealand
Parkes JP, Nugent G, Warburton B (1996) Commercial exploitation as a pest control tool for introduced mammals in New Zealand. Wildl Biol 2:171–177
Parkes JP, Thomson C, Newell CL, Forrester GJ (2004) Impacts of Himlayan thar (*Hemitragus jemlahicus*) on snow tussock grasslands in the Southern Alps. Landcare Research Contract Rep LC0304/077 for West Coast Conservancy, Department of Conservation, New Zealand
Pekelharing CJ, Parkes JP, Barker RJ (1998) Possum (*Trichosurus vulpecula*) densities and impacts on fuchsia (*Fuchsia excorticata*) in South Westland, New Zealand. NZ J Ecol 22:197–203

Rose AB, Platt KH (1987) Recovery of northern Fiordland alpine grasslands after reduction in the deer population. NZ J Ecol 10:23–33

Sweetapple PJ, Nugent G, Whitford J, Knightbridge PI (2002) Mistletoe (*Tupeia antarctica*) recovery and decline following possum control in a New Zealand forest. NZ J Ecol 26:61–71

Tustin KG, Parkes JP (1988) Daily movement and activity of female and juvenile Himalayan tahr (*Hemitragus jemlahicus*) in the eastern Southern Alps, New Zealand. NZ J Ecol 11:51–59

Veitch CR, Bell BD (1990) Eradication of introduced animals from the islands of New Zealand. In: Towns DR, Daugherty CH, Atkinson IAE (eds) Ecological restoration of New Zealand islands. Conservation Sciences Publ 2. Department of Conservation, Wellington, New Zealand, pp 137–146

Walters CJ, Maguire JJ (1996) Lessons for stock assessment from the northern cod collapse. Rev Fish Biol Fish 6:125–137

Warburton B, Tocher GH, Allan NS (2000) Possums as a resource. In: Montague TL (ed) The brushtail possum: biology, impact and management of an introduced marsupial. Manaaki Whenua Press, Lincoln, New Zealand, pp 251–261

Wardle DA, Barker GM, Yeates GW, Bonner KI, Ghani A (2001) Introduced browsing mammals in New Zealand natural forests: aboveground and belowground consequences. Ecol Monogr 71:587–614

Wilson PR, Karl BJ, Toft RJ, Beggs JR, Taylor RH (1998) The role of introduced predators and competitors in the decline of kaka (*Nestor meridionalis*) populations in New Zealand. Biol Conserv 83:175–185

27 Pest Control: Does the Answer Lie in New Biotechnologies?

J.A. Duckworth, A.E. Byrom, P. Fisher and C. Horn

27.1 Why Is a Biotechnological Approach Necessary?

The use of a recombinant oral vaccine to eliminate fox rabies in large areas of Western Europe, and more recently to control rabies in foxes, skunks and racoons in North America is a highly successful example of how new biotechnologies are being applied to solve old problems in wildlife management. This chapter focuses on three biotechnological areas currently being developed in New Zealand for improved vertebrate pest management: species-specific toxins, fertility control technologies for population control, and DNA genotyping for species identification and estimation of population parameters. We also address public acceptability of such technologies, which are sometimes perceived as potentially adding new problems.

Despite attempts at control, introduced mammals such as brushtail possums (*Trichosurus vulpecula*), stoats (*Mustela erminea*) as well as Norway (*Rattus norvegicus*) and ship (*R. rattus*) rats continue to pose major threats to conservation values and, in some instances, to the economic well-being of New Zealand (see Chaps. 16, 17). Annually, more than NZ$ 75 million is expended nationally on conventional methods for vertebrate pest control (mainly of possums), which include trapping and use of baits containing sodium fluoroacetate (1080), cyanide, brodifacoum or cholecalciferol. Control costs are likely to increase to over NZ$ 85 million per annum with renewed efforts to have New Zealand's livestock industries declared free of bovine Tb (*Mycobacterium bovis*) by 2013. The extensive field use of vertebrate pesticides in New Zealand reflects both the scale of the introduced pest problem and the unique array of native wildlife present. Ground or aerial application of 1080 baits is currently a mainstay of many regional strategies to control possums. However, there are increasing public and scientific concerns about the use of non-selective poisons for wildlife management (PCE 1994; Eason et al. 2000). Worldwide, regulatory scrutiny of the target specificity, environmental persistence, and humaneness of vertebrate pesticides has meant that some

Ecological Studies, Vol. 186
R.B. Allen and W.G. Lee (Eds.)
Biological Invasions in New Zealand

products are no longer available for use, and more stringent requirements are applied to the registration of toxic compounds for new applications. Reliance on poisons may become untenable in the future, as overseas markets become concerned about contamination of produce, or as target populations develop bait- or poison-shyness. In effect, the "tool box" of conventional pest control methods available to wildlife managers is shrinking, and so alternative methods are urgently required.

Rapid advances in biotechnology such as molecular biology, proteomics and genetic modification can improve our understanding of pest biology and allow development of more cost-effective and humane methods of pest control while reducing or avoiding some of the risks of current control methods. Genetic variation between species in biochemical and physiological processes permits the vulnerabilities of different pest species to be exploited. The use of genetic modification and biologically derived products to interfere with key physiological and reproductive processes should lead to more effective and "smarter" pest control. New technologies utilising genetic variations between individuals can also be applied to estimate pest numbers more accurately, and determine relationships between population density and pest impacts on the environment more precisely. Although conventional methods can provide effective control in the short term, integrated management using a combination of traditional and biotechnological approaches offers a more cost-effective and targeted approach to solving vertebrate pest problems in New Zealand.

27.2 Biotechnology and New-Generation Vertebrate Pesticides

Since the 1950s, ongoing developments have improved efficacy and minimized risk in the use of conventional pesticides, and only recently has biotechnology had any impact within this field. In New Zealand and Australia especially, where several introduced species are significant vertebrate pests, the focus between 1960 and 1990 was on optimising field pest control with broad-spectrum poisons such as 1080 and cyanide. Potent second-generation anticoagulant rodenticides (e.g. brodifacoum) replaced the first-generation anticoagulants, such as warfarin. Unfortunately, the unique persistence of second-generation anticoagulants combined improved effectiveness with increased risk to non-target species and food-chain contamination (Eason et al. 2002). Cholecalciferol was developed in the 1980s (Marsh and Tunberg 1986), and was the last "new" poison to reach the vertebrate pesticide market. In the past decade, partially target-selective control tools have been developed through reformulation of existing pesticides. Bait delivery systems designed to enhance efficacy and reduce non-target exposure have utilised new technological and research advances. For example, encapsulation of cyanide has offered improvements over older formulations in overcoming cyanide-shyness in possums (e.g. Eason et al. 2000), and the development of long-life bait formulations continues to build on the considerable body of research which underpins current pest control technologies. Today, reformulation of the "old" poison zinc phosphide for pest control (Morgan et al. 2001) is being pursued in New Zealand.

The approach of identifying and exploiting specific physiological or biochemical characteristics in target species to develop novel, species-specific toxins capitalizes on advances in biotechnology. Receptors and enzymes which play critical roles in physiological control systems are particularly suitable sites for exploitation in toxin intervention. Candidate species-specific toxicants have been identified, and other compounds with specific toxic mechanisms probably remain to be found. These need to be rigorously tested and their modes of action fully elucidated as part of the development of a "new generation" of vertebrate pesticides. Current New Zealand research in this area is focusing on rats (*Rattus* spp.), with the aim of extension to other pests such as possums and mustelids. The development of new-generation selective toxicants is an extremely important milestone in reducing environmental impact, and will complement the ongoing development of fertility control. These will offer significant improvements in species specificity, humaneness, and environmental safety, as replacements for the current suite of active ingredients and as additional control tools for integration within pest management programmes in the future.

27.3 Biotechnology and Fertility Control

Fertility control offers a humane and ethical method for managing pest species in New Zealand (Cowan 2000) and, when combined with conventional control, may provide a long-term and cost-effective solution to the possum problem (Bayliss and Choquenot 1999). Stoats, with their short life span, 9-month embryonic diapause, and high potential reproductive rate, are also excellent candidates for fertility control (N. Barlow, personal communication). Fertility control has the potential to improve operational efficiencies by increasing the time between applications of conventional control. Other advantages include decreasing the amount of toxin used, thereby reducing the risk of environmental contamination and threats to non-target wildlife. In New Zealand, new biotechnological approaches are being developed for possums and stoats: one involves immunologically mediated interference with fertility, known as immunocontraception (Cowan 2000), another uses gonadotrophin-releasing hormone-toxin complexes to attack key mechanisms in the hormonal control of reproduction (Eckery et al. 2001).

For possums, considerable research has sought to identify genes for target proteins essential for key reproductive processes (Duckworth et al. 1999; Cui et al. 2001; Eckery et al. 2001; Harris et al. 2001). One particularly promising area for immunocontraceptive control is disruption of the fertilization process. Immunizing possums against proteins from eggs and sperm reduces the number of possum young born (Duckworth et al. 1998, 1999). The zona pellucida (ZP), a glycoprotein coat around the egg, is a particularly attractive target because ZP proteins play an essential role during fertilization and early development of the embryo (Paterson et al. 2000). They are also strongly immunogenic, specific to reproductive function (Mate and McCartney 1998), and have the potential to provide possum- or marsupial-specific targets (Mate et al. 1998; Duckworth et al. 1999; Harris et al. 2001). Genes which code for possum ZP proteins have been cloned (Mate and McCartney 1998; McCartney and Mate 1999), and immunization with recombinant possum ZP3 and ZP2 proteins (Mate et al. 2003) reduces the fertility of female possums by 70–80 % (Duckworth et al. 1999; Molinia et al. 2001).

Molecules active during germ cell and embryonic development are also being investigated as fertility control targets. These include: a stem cell factor and its receptor, plus other local growth factors important for ovarian development (Eckery et al. 2001); vesicle-associated molecules (VAM) which act during oogenesis and cleavage (Frankenberg and Selwood 2001); coat proteins which are important in late cleavage and blastocyst development (Selwood 2000); and a leukaemia inhibitory factor (LIF) which acts during implantation (Cui et al. 2001).

27.3.1 Target Specificity of Fertility Control

Potential species specificity for fertility control is likely to arise from identifying regions of the target protein which show significant species variation. For instance, in mice one of the most divergent regions of the ZP3 protein is a putative sperm-binding site involved in fertilization (Millar et al. 1989). A potentially possum-specific peptide was identified from the same region of possum ZP3 (McCartney and Mate 1999). Immunization against this peptide reduced the number of embryos produced by female possums by more than 60 %, but had no effect on mouse fertility (J. Duckworth, unpublished data). Other potential contraceptive peptides are being identified in possum ZP proteins by linear epitope mapping and bacteriophage display libraries (Harris et al. 2001). Hormone analogues which demonstrate differing biological activity between species have also been investigated (Cowan 2000). Risks to non-target species will be further reduced by utilising formulation and deployment methods (e.g. bait stations) which maximize exposure of pest species to the control agent but which, through low acceptability or accessibility, minimize the exposure of native fauna.

27.3.2 Delivery Systems for Fertility Control

Oral vaccines for human and livestock disease prevention are a highly active area of biotechnology research, with direct application to vaccine delivery to wild animal populations. Options being assessed range from the use of non-disseminating, non-living vaccines in baits (Cowan 2000) to self-disseminating biological control vectors such as those being developed in Australia for the control of wild rabbits and mice (Kerr et al. 2001; Cowan et al. 2002).

Initially, fertility control for possums in New Zealand is likely to be delivered in baits. Contraceptive antigens for possums are being produced in transgenic plants and bacteria for further evaluation and testing (Cowan 2000). Transgenic plants and plant viruses are capable of producing large quantities of biologically active proteins which can be readily formulated into vaccine baits (Cowan et al. 1999). Possums have been shown to develop an immune response to foreign proteins expressed in genetically modified potatoes (Tacket et al. 1998; J. Duckworth and A. Walmsley, unpublished data). Plants expressing possum ZP antigens have been developed and are being tested. Another oral vaccine delivery system being investigated is non-living bacterial ghosts – empty cell envelopes resulting from membrane rupture following activation of a lysis gene (Szostak et al. 1996). Recombinant ghosts can express high levels of foreign proteins; they have inherent adjuvant properties, and can induce specific humoral and cellular immune responses against target components (Eko et al. 1999). Humoral immune responses of possums given ghosts expressing possum ZP3 antigens, by subcutaneous, oral and nasal immunization routes, indicate that bacterial ghosts show promise for bait or aerosol delivery (J. Duckworth, X. Cui and W. Lubitz, unpublished data). Microencapsulation and formulation methods to protect orally administered contraceptive agents from degradation in the gastrointestinal tract are being investigated (McLeod et al. 2001; Aldwell et al. 2003).

Alternatively, a non-infectious and replication-limited form of a microbial vector could be used which infects the host, produces the contraceptive protein, and stimulates an immune response (Catmull et al. 1999). Baits containing recombinant vaccinia virus expressing part of the rabies virus genome have been successfully used to combat rabies in populations of wild foxes (Pastoret and Brochier 1999). Such baits would need to be delivered to each possum, and offer potential advantages of increased effectiveness, species specificity, and public acceptability.

In the long term, a biocontrol agent which spreads by natural transmission from animal to animal causing infertility could provide a humane and more cost-effective solution to New Zealand's possum problem. Such an approach is being developed for rabbits, mice and foxes in Australia (Tyndale-Biscoe 1994; Robinson et al. 1999; Kerr et al. 2001; Cowan et al. 2002). A dissemination system requires a suitable possum-specific vector which can infect a high proportion of the target population, be genetically modified to include possum contraceptive antigens, and can present the antigens effectively to the host's immune system. Additional cell signal genes (such as cytokines and immunomodulators) may also be incorporated to enhance the immune response. A range of parasites and pathogens of possums have been considered (Cowan 2000); several viruses, including herpes viruses (Zheng et al. 2001) and adenoviruses (Thomson and Meers 2001), and the nematode *Parastrongyloides trichosuri* (Skinner et al. 2001) are being investigated as potential vectors. Once identified, potential vectors will require thorough evaluation of their biological impacts (e.g. their pattern and rate of spread, effects of competition between modified and wild field strains), and possible impacts of selection for resistance to infection will also need to be addressed (Cowan 2001; Williams 2002). As for all control methods, the risk profile and local and international acceptability concerns must be considered. For a disseminating delivery system, the selection of targets, the species specificity of a transmissible vector, and the mode of dissemination of that vector are critical. Any release of a transgenic organism will require regulatory approval, and depends on demonstrated biosafety, a clear understanding of biological consequences of releasing the agent, and public and international perceptions of the consequences and ongoing risk. Assessment for release is currently being undertaken for the immunocontraceptive murine cytomegalovirus vaccine developed to control mouse plagues in Australia (Williams 2002). Registration is a substantial process and may take several years. In New Zealand, most effort towards fertility control has targeted possums but the technology is applicable to many other species, and a combination fertility and disease control vaccine for wildlife (e.g. which would both render possums infertile and protect against infection from bovine Tb) is also an exciting possibility.

27.4 Biotechnologies for Improved Detection and Estimation of Mammalian Pest Populations

27.4.1 Use of DNA for Mark and Recapture: a Non-Invasive Approach to Population Estimation

For effective pest control, there is a need to accurately and precisely estimate animal abundance, either for monitoring success of control operations (Parkes 1996) or to relate a species' population density to its putative impacts (Choquenot and Parkes 2001). Estimating abundance using conventional methods (e.g. live-trapping for mark-recapture, kill-trapping, or indexing) can be problematic (Ruscoe et al. 2001), particularly for populations at low densities following control operations, and for those species which are naturally rare and/or difficult to monitor by traditional means. This includes species which are trap-shy, or where trappability is highly variable among individuals. In addition, traditional mark-recapture methods (e.g. live-trapping) can be expensive, and are therefore often avoided by pest managers.

One new approach for estimating animal abundance is to use DNA profiling to identify individuals in a population without having to physically capture and/or mark animals (Valderrama et al. 1999). Repeated surveys can be used to estimate survival and rates of change for a population (Mills et al. 2000). Recent developments with the use of microsatellite DNA markers, in combination with traditional statistical methods and novel tissue collection techniques, offer possibilities for measuring both population parameters and behaviour of individuals. The DNA markers are highly variable, and therefore well suited to comparing genetic variation among individuals. Microsatellite analysis, using polymerase chain reaction (PCR), can be performed using very small samples of DNA. Modern equipment allows repeatable analysis of large quantities of samples, making microsatellite markers particularly useful for consistently assigning identity to samples taken from unsighted individuals. Match statistics can be used to ensure that the available genetic data are able to resolve individuals, even when the study population contains many close relatives (Manel et al. 2002). It is also possible to use genetic markers to assign gender.

Those methods therefore offer new options for population estimation of species which are difficult to detect due to their habitat and behaviour. The technique has been used successfully internationally, particularly for endangered species (e.g. bears, Sloane et al. 2000; wombats, Woods et al. 1999) and also to detect wildlife poaching (Manel et al. 2002). The method has not previously been applied to pest species in New Zealand, but potentially has a number of useful applications for the management of animal pests.

27.4.2 Extraction of Genetic Material from Pest Species

For many species, DNA samples can be obtained from hair follicles or from epithelial cells sloughed off onto faeces (Valderrama et al. 1999; Wilson and Delahay 2001). Program MARK (White and Burnham 1999) has been used to analyse the resultant mark-recapture data for a range of species internationally.

To date, work in New Zealand has focused on possums and stoats. Extraction of DNA from possum hair is not possible because no genetic material is attached in this case. However, useful (amplifiable) quantities of mitochondrial DNA are recovered from faecal pellets collected by systematically walking predetermined transects in possum habitat. Rates of DNA recovery are very high from fresh pellets, and at least 50 % from pellets up to 27 days old which have not been rained upon (rainfall strongly reduces recovery rates of DNA from possum faecal pellets). DNA markers in possums are highly variable (individual possums have been separated by genotype on the basis of only eight loci). Preliminary results suggest that DNA sampling detects significant numbers of possums not captured by conventional trapping (G. Nugent, D. Gleeson, R. Howitt and C. Thomson, unpublished data).

For stoats, hair samples are obtained using adhesive gel on a rubber band stretched across both apertures of a 20-cm length of drainpipe (45-mm diameter) baited with rabbit meat. Pipes are placed 250–500 m apart in a grid pattern in forest habitats where stoats are most commonly found. Amplifiable quantities of DNA have been successfully extracted from a sample of three or more stoat hairs. Microsatellite DNA of stoats is highly variable, and eight loci have been successfully amplified and analysed. Using this method, individual stoats are often repeatedly identified ("recaptured") over a sampling period, and the method "captures" more stoats than would be captured by live-trapping. However, about 45 % of all samples are mixed, due to different individuals entering the tube, which affects the ability to accurately assign genotypes.

27.4.3 Is DNA Analysis a Useful Approach for Pest Management in New Zealand?

Science evolves rapidly in areas where "new technologies" are used. Protocols for field collection of viable DNA, extraction, and analysis of microsatellite data are being continuously refined. For both possums and stoats, recovery of viable DNA may be difficult if samples degrade before collection and, for stoats, mixed samples may make successful DNA extraction and analysis difficult. Disadvantages also include the lag time between collection of field samples and laboratory results (for genetic analysis), possible shyness around detection devices (Wilson and Delahay 2001), and a lack of population closure (Boulanger and McLellan 2001). Compared with conventional mark-recapture methods, however, data collection using DNA analysis has proven to be highly cost-effective for both stoats and possums, and is often more cost-effective than indexing methods. The technology has a number of other advantages. It not only provides high-quality data for mark-recapture analyses, but can also provide information on gender of individuals, family relationships, dispersal patterns, and genetic diversity among populations. The animals are never physically captured or restrained, and so negative responses to traps are reduced. The data can also be used for a simple presence/absence index or for home range delineation. Consequently, use of this non-invasive technique for accurate and precise estimation of population density has the potential to provide us with insights into species management which would not have been gained using conventional trapping or indexing methods. The methodology will be invaluable for a range of pest species in New Zealand, particularly those whose populations are maintained at low densities (possums) or which historically have been regarded as rare or difficult to trap (stoats).

27.5 The Public and New Biotechnologies

The public generally accept the need for controlling mammalian pests in New Zealand. They prefer manual control techniques such as trapping and shooting, which are generally seen to be humane and environmentally friendly (Fitzgerald et al. 1996, 2000, 2002). Fitzgerald and colleagues also found that the New Zealand public are least accepting of poisoning as a control technique. Forms of biological control are rated somewhere between manual control methods and poisons. However, this finding is complicated by the following factors.

- First, people rated themselves less knowledgeable about biological control methods than poisoning methods. Higher acceptance may be partly a function of lower levels of knowledge and debate, and a relatively low exposure in the media.
- Second, the public have not yet been faced as directly with the use of biotechnologies as they have with aerial application of 1080. Higher levels of acceptance may therefore come from lower levels of engagement, rather than purely from higher acceptability.
- Third, for people concerned with animal welfare, fertility control is rated as very humane, since it allows existing animals to live out their lives. Thus, a focus on fertility control increases the acceptability of some gene technologies.

In terms of public acceptance of genetic modification work for pest control, the Parliamentary Commissioner for the Environment (PCE 2000) notes that the public overall find gene technologies more or less acceptable based on both level of perceived risk *and* "health profile" (see Fig. 27.1).

Gene technologies perceived as high risk are less acceptable than those with a low perceived risk. Likewise, gene technologies placed at the "fit and healthy" end of the health profile are more acceptable than those at the "poor health" end. These findings were achieved through focus groups where much discussion occurs and people involved have weighed up their answers (not merely a public opinion poll), which indicates they are likely to hold over time. New Zealanders value their natural heritage, so the costs of using pest control techniques will have to be weighed carefully (*by the general public*, not only by experts) against factors such as the benefits of protecting native species (e.g. kiwi), and the efficacy and acceptability of available pest control methods.

An example of the "weighing up" process in pest control is illustrated by Horn and Kilvington (2003), who found that although Maori groups dislike using 1080 poison, some have allowed its aerial application. In those situations, Maori (indigenous New Zealanders) felt they had a good level of influence, good access to relevant information, and time to learn about and discuss the issues. This example suggests that public acceptance of biotechnologies

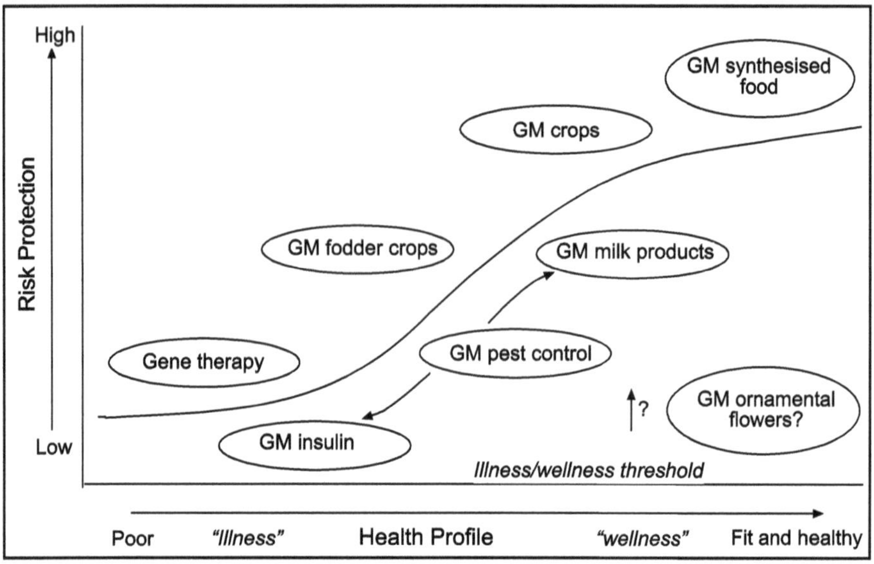

Fig. 27.1 Risk perception and health profile of different gene technologies. Gene technologies in the *top right corner* of the figure are least acceptable, whereas those in the *bottom left corner* are most acceptable (PCE 2000)

for pest control may depend as much on *how* these technologies are introduced to the public, and how much control people feel they have over possible long-term effects (i.e. how much risk is associated with the technology). A new form of pest control which can be contained within a limited area is more likely to be acceptable than a mechanism which spreads quickly and widely throughout the country and which cannot be stopped once it is released.

A corollary of considerable importance is the fact that all alien pests in New Zealand are native and valued elsewhere in the world. Possums, for example, are native to Australia and are protected there. Any mechanism which might spread possum infertility to Australia is likely to be viewed negatively by that country. Thus, debate over control technologies also has an international component. These issues require effective dialogue processes between scientists, managers, and other stakeholder groups on pest management using biotechnology.

27.6 Conclusions

New approaches to mammalian pest management in New Zealand are urgently required to complement ongoing incremental improvements in conventional control. A range of options are being explored, from non-invasive genotyping for estimating populations, to new pest control methods including fertility control and novel species-specific toxins. Developing new biotechnologies is a complex process both scientifically, in developing and demonstrating effective control and monitoring methods, and in ensuring the acceptance of the technology within New Zealand and overseas. The different risk profiles relating to each option, including the development of various delivery systems, mean that any discussion of ethical, social, or political acceptability needs to occur separately for each approach. A process of effective dialogue is required. If the risks, particularly those related to effectiveness, species specificity, and delivery systems, prove to be acceptable to the public, then the potential benefits from biotechnology to New Zealand's economy and for enhancing and protecting the natural environment are enormous.

Acknowledgements Dianne Gleeson and Robyn Howitt extracted DNA from stoat hair and possum faeces. Graham Nugent provided information on analysis of mark-recapture data obtained from possum DNA. Brian Hopkins provided information regarding the development of new-generation vertebrate pesticides. The Australian Government's Cooperative Research Centres Program, New Zealand Foundation for Research, Science and Technology, Landcare Research, Animal Health Board, Fletcher Challenge Forests, the New Zealand Department of Conservation, and New Zealand Lottery Grants Board supported research on possum fertility control, and development of methods for improved detection and estimation of stoats and possums.

References

Aldwell FE, Tucker IG, de Lisle GW, Buddle BM (2003) Oral delivery of *Mycobacterium bovis* BCG in a lipid formulation induces resistance to pulmonary tuberculosis in mice. Infect Immun 71:101–108

Bayliss P, Choquenot D (1999) Ecological modelling and possum biocontrol in New Zealand. In: Sutherland G (ed) Advances in the biological control of possums. R Soc NZ Misc Ser 56:19–23

Boulanger J, McLellan B (2001) Closure violation in DNA-based mark-recapture estimation of grizzly bear populations. Can J Zool 79:642–651

Catmull J, Wilson ME, Kirchhoff LV, Metwali A, Donelson JE (1999) Induction of specific cell-mediated immunity in mice by oral immunization with *salmonella* expressing *Onchocerca volvulus* glutathione S-transferase. Vaccine 17:31–39

Choquenot D, Parkes J (2001) Setting thresholds for pest control: how does pest density affect resource viability? Biol Conserv 99:29–46

Cowan PE (2000) Biological control of possums: prospects for the future. In: Montague TL (ed) The brushtail possum: biology, impact and management of an introduced marsupial. Manaaki Whenua Press, Lincoln, New Zealand, pp 262–270

Cowan P (2001) Immunologically based fertility control for possums – is non-response an issue? In: Proc Biological Management of Possums, 2–4 April 2001, Ministry of Agriculture and Forestry, Wellington, New Zealand, pp 14–16

Cowan P, Walmsley A, Kirk D, Lee SM, Young P (1999) Plant-derived vaccines for possum fertility control. In: Sutherland G (ed) Advances in the biological control of possums. R Soc NZ Misc Ser 56:24–27

Cowan P, Pech R, Curtis P (2002) Field applications of fertility control for wildlife management. In: Holt WV, Pickard AR, Rodger JC, Wildt DE (eds) Reproductive science and integrated conservation. Cambridge Univ Press, Cambridge, pp 305–318

Cui S, Griffith N, Nanayakkara K, Cruz Y, Selwood L (2001) Developmental targets for fertility control in the brushtail possum. In: Proc Biological Management of Possums, 2–4 April 2001, Ministry of Agriculture and Forestry, Wellington, New Zealand, pp 6–11

Duckworth JA, Buddle BM, Scobie S (1998) Fertility of brushtail possums (*Trichosurus vulpecula*) immunised against sperm. J Reprod Immunol 37:125–138

Duckworth J, Harris M, Mate K, McCartney C, Buist J, Scobie S, Jones D, Jones S (1999) Development of brushtail possum immunocontraception: targeting sperm and egg antigens. In: Sutherland G (ed) Advances in the biological control of possums. R Soc NZ Misc Ser 56:72–76

Eason CT, Warburton B, Henderson RJ (2000) Toxicants used for possum control. In: Montague TL (ed) The brushtail possum: biology, impact and management of an introduced marsupial. Manaaki Whenua Press, Lincoln, New Zealand, pp 154–163

Eason CT, Murphy EC, Wright GRG, Spurr EB (2002) Assessment of risks of brodifacoum to non-target birds and mammals in New Zealand. Ecotoxicology 11:35–48

Eckery DC, Juengel JL, Bauer E, Heath DA, Lawrence SB, Lun S, McLeod BJ, Moore LG, Thompson EG, Thomson BP, Western A, Whale L, McNatty KP (2001) Disabling reproductive function in possums: current strategies, research findings and future directions. In: Proc Biological Management of Possums, 2–4 April 2001, Ministry of Agriculture and Forestry, Wellington, New Zealand, pp 72–81

Eko FO, Witte A, Huter V, Kuen B, Furst-Ladani S, Haslberger A, Katinger A, Hensel A, Szostak MP, Resch S, Mader H, Raza P, Brand E, Marchart J, Jechlinger W, Haidinger W, Lubitz W (1999) New strategies for combination vaccines based on the extended recombinant bacterial ghost system. Vaccine 17:1643–1649

Fitzgerald G, Saunders L, Wilkinson R (1996) Public attitudes to the biological control of rabbits in New Zealand. MAF Policy Tech Pap 96/3. Ministry of Agriculture, Wellington, New Zealand, pp 1–26

Fitzgerald GP, Wilkinson RL, Saunders LS (2000) Public perceptions and issues in possum control. In: Montague TL (ed) The brushtail possum: biology, impact and management of an introduced marsupial. Manaaki Whenua Press, Lincoln, New Zealand, pp 187–197

Fitzgerald G, Fitzgerald N, Wilkinson R (2002) Social acceptability of stoats and stoat control methods: focus group findings. Sci Conserv 207:1–45

Frankenberg S, Selwood L (2001) Ultrastructure of oogenesis in the brushtail possum. Mol Reprod Dev 58:297–306

Harris MS, Cui X, Jones S, McCartney C, O'Rand MG (2001) Identification of new antigen targets and the molecular dissection of antigenicity for possum immunocontraception. In: Proc Biological Management of Possums, 2–4 April 2001, Ministry of Agriculture and Forestry, Wellington, New Zealand, pp 20–23

Horn C, Kilvington M (2003) Maori and 1080. 2002/2003. Protect (Summer), pp 17–21. www.landcareresearch.co.nz/research/social

Kerr P, van Leeuwen BH, Perkins H, Holland MK, Gu W, Jackson RJ, Williams CK, Robinson AJ (2001) Development of fertility control for wild rabbits in Australia using a virally-vectored immunocontraceptive. In: Vurro M, Gressel J, Butt T, Harman G, St Lagar R, Nuss D, Pilgeram A (eds) Enhancing biocontrol agents and handling risks. NATO Science Series. Life and Behavioural Sciences, vol 339. IOS Press, Amsterdam

Manel S, Berthier P, Luikart G (2002) Detecting wildlife poaching: identifying the origin of individuals with Bayesian assignment tests and multilocus genotypes. Conserv Biol 16:650–659

Marsh RE, Tunberg A (1986) Characteristics of cholecalciferol. Rodent control: other options. Pest Control Technol 14:43–45

Mate KE, McCartney CA (1998) Sequence and analysis of the zona pellucida 2 cDNA(ZP2) from a marsupial, the brushtail possum, *Trichosurus vulpecula*. Mol Reprod Dev 51:322–329

Mate KE, Molinia FC, Rodger JC (1998) Manipulation of the fertility of marsupials for conservation of endangered species and control of over-abundant populations. Anim Reprod Sci 53:65–76

Mate KE, Buist JM, Duckworth JA (2003) Expression in *Escherichia coli* and immunological characterization of three zona pellucida proteins (ZP1, ZP2, and ZP3) from a marsupial, the brushtail possum (*Trichosurus vulpecula*). Mol Reprod Dev 64:136–143

McCartney CA, Mate KE (1999) Cloning and characterisation of a zona pellucida 3 cDNA from a marsupial, the brushtail possum, *Trichosurus vulpecula*. Zygote 7:1–9

McLeod BJ, Wen J, Tucker IG, Davies N, Ledger R, Butt AG (2001) Oral delivery of bioactives to possums: protecting bioactives and enhancing their uptake. In: Proc Biological Management of Possums, 2–4 April 2001, Ministry of Agriculture and Forestry, Wellington, New Zealand, pp 28–31

Millar SE, Chamow SM, Baur AW, Oliver C, Robey F, Dean J (1989) Vaccination with a synthetic zona pellucida peptide produces long-term contraception in female mice. Science 246:935–938

Mills LS, Citta JJ, Lair KP, Schwartz MK, Tallmon DA (2000) Estimating animal abundance using noninvasive DNA sampling: promise and pitfalls. Ecol Appl 10:283–294

Molinia FC, Glazier AG, Myers JV, Sidhu KS, Mate KE, Berg DK et al. (2001) In vitro and in vivo models of key reproductive events as tools for possum biological control. In: Proc Biological Management of Possums, 2–4 April 2001, Ministry of Agriculture and Forestry, Wellington, New Zealand, pp 1–5

Morgan DR, Eason CT, Wickstrom ML (2001) Zinc phosphide as a poison for control of possums (*Trichosurus vulpecula*): the need for encapsulation. In: Pelz H-J, Cowan DP, Feare CJ (eds) Advances in Vertebrate Pest Management, II. Filander, Fürth, pp 337–344

Parkes JP (1996) Integrating the management of introduced mammal pests of conservation values in New Zealand. Wildl Biol 2:179–184

Pastoret PP, Brochier B (1999) Epidemiology and control of fox rabies in Europe. Vaccine 17:1750–1754

Paterson M, Jennings ZA, van Duin M, Aitken RJ (2000) Immunocontraception with zona pellucida proteins. Cells Tissues Org 166:228–232

PCE (1994) Possum management in New Zealand. Office of the Parliamentary Commissioner for the Environment, Wellington, New Zealand, pp 1–196

PCE (2000) Caught in the headlights: New Zealanders' reflections on possums, control options and genetic engineering. Office of the Parliamentary Commissioner for the Environment, Wellington, New Zealand, pp 1–149

Robinson AJ, Jackson RJ, Merchant JC, Kerr PJ, Holland MK, Lawson MA, Shellam GR (1999) Biocontrol of pest mammals in Australia – progress towards virally vectored immunocontraception for the mouse and rabbit. Rec Adv Microbiol 7:63–98

Ruscoe WA, Goldsmith R, Choquenot D (2001) A comparison of population estimates and abundance indices for house mice inhabiting beech forests in New Zealand. Wildl Res 28:173–178

Selwood L (2000) Marsupial egg and embryo coats. Cells Tissues Organs 166:208–219

Skinner SJM, Newton-Howes J, Ralston M, Heath DD, Grant WN, Shoemaker CB (2001) A nematode parasite with potential as a disseminating vector for possum biocontrol. In: Proc Biological Management of Possums, 2–4 April 2001, Ministry of Agriculture and Forestry, Wellington, New Zealand, pp 51–55

Sloane MA, Sunnucks D, Alpers D, Beheregaray LB, Taylor AC (2000) Highly reliable genetic identification of individual northern hairy-nosed wombats from single remotely collected hairs: a feasible censusing method. Mol Ecol 9:1233–1240

Szostak MP, Hensel A, Eko FO, Klein R, Auer T, Mader H, Haslberger A, Bunka S, Wanner G, Lubitz W (1996) Bacterial ghosts: non-living candidate vaccines. J Biotechnol 44:161–170

Tacket CO, Mason HS, Losonsky G, Clements JD, Levine MM, Arntzen CJ (1998) Immunogenicity in humans of a recombinant bacterial antigen delivered in a transgenic potato. Nat Med 4:607–609

Thomson D, Meers J (2001) Characterisation of an adenovirus in brushtail possums. In: Proc Biological Management of Possums, 2–4 April 2001, Ministry of Agriculture and Forestry, Wellington, New Zealand, pp 36–39

Tyndale-Biscoe CH (1994) Virus-vectored immunocontraception of feral mammals. Reprod Fertil Dev 6:281–287

Valderrama X, Karesh WB, Wildman DE, Melnick DJ (1999) Noninvasive methods for collecting fresh hair tissue. Mol Ecol 8:1749–1752

White GC, Burnham KP (1999) Program MARK: survival estimation from populations of marked animals. Bird Study 46 suppl S:120–139

Williams CK (2002) Risk assessment for release of genetically modified organisms: a virus to reduce the fertility of introduced wild mice, *Mus domesticus*. Reproduction suppl 60:81–88

Wilson GJ, Delahay RH (2001) A review of methods to estimate the abundance of terrestrial carnivores using field signs and observation. Wildl Res 28:151–164

Woods JG, Paetkau D, Lewis D, McLellan BN, Proctor M, Strobeck C (1999) Genetic tagging of free-ranging black and brown bears. Wildl Soc Bull 27:616–627

Zheng T, Dickie A, Lu G, Buddle B (2001) Marsupial herpes viruses for use in biological control of possums. In: Proc Biological Management of Possums, 2–4 April 2001, Ministry of Agriculture and Forestry, Wellington, New Zealand, pp 46–50

28 Updated Perspective on Biological Invasions in New Zealand

R.B. ALLEN, R.P. DUNCAN and W.G. LEE

28.1 Introduction

Since Darwin's visit (Darwin 1859), New Zealand has been considered an exemplar of island vulnerability to invasion by alien biota (e.g., Elton 1958; Carlquist 1974; Walter 1979). New Zealand comprises two main islands (with >600 adjacent smaller islands) and several outlying island groups with a latitudinal range of 29°16′ to 52°33′S and a total land area of ca. 270,000 km^2 (Wardle 1991). This land area has fluctuated considerably through geological and glacial events during the Tertiary, creating a suite of islands that are now geologically diverse, have been exposed for varying lengths of time, and with climates extending from subtropical to cool temperate, and habitats from thermal pools to perpetual snow. Consequently, the role that New Zealand's insularity has played in facilitating biological invasions must be viewed within a context of its historically dynamic and diverse landscapes and habitats (Chap. 2). In addition, increasingly questions are being raised about New Zealand's biotic isolation because there is growing evidence of significant, post-Gondwanic immigration from neighboring continents and islands, and high turnover in the biota (see Chap. 2).

Isolation and small land area are thought to promote a suite of biological characteristics that make islands vulnerable to invasions (e.g., Denslow 2003), and result in novel invader impacts (e.g., Simberloff 1995). In this chapter, we assess whether these widely cited island-specific characteristics are important determinants of island, and in particular New Zealand's, vulnerability to invasive species. Other, non-island-specific determinants we also consider include disturbance, human activity, resource availability, climate, assembly history and time, all of which potentially covary with island-specific characteristics. We then consider the potential for ongoing invasions, management of alien species, and finally the long-term consequences of biological invasions for the New Zealand landscape.

Ecological Studies, Vol. 186
R.B. Allen and W.G. Lee (Eds.)
Biological Invasions in New Zealand

28.2 Are New Zealand Islands Vulnerable to Invasion?

New Zealand is one of the most invaded countries in the world, as judged by the number and percentage of naturalized alien species (e.g., Vitousek et al. 1997; Clout 1999). Presently, 94, 43, 40, 33 and 31 % of New Zealand's terrestrial mammals, freshwater fish, vascular plants, amphibians, and terrestrial bird species, respectively, are alien species (Clout 1999). Nevertheless, while widely used, the number or percentage of alien species in a biota may not provide a useful measure of invasibility because it does not allow for differences in the number of attempted introductions (e.g., Simberloff 1995). Islands may have disproportionately more alien species simply because more alien species have been transported and introduced to islands (Blackburn and Duncan 2001). Indeed, studies of bird introductions, using historical records documenting both failed and successful attempts to establish populations, have consistently shown no difference between island and mainland locations in the likelihood that an alien species will establish (e.g., Sol 2000). The success of bird introductions to New Zealand and continental Australia illustrate this: 120 bird species have been introduced to New Zealand, with 34 (28 %) successfully established (Chap. 9), whereas 52 species have been introduced to Australia, with 19 successes (37 %; Duncan et al. 2001). Hence, the greater number (and percentage) of alien bird species in the New Zealand fauna, relative to Australia, results from more introduction attempts. New Zealand actually has a lower success rate, implying that it is harder to invade.

Similar outcomes may hold for other taxonomic groups. The New Zealand flora, with approximately 40 % alien species, appears much more invaded than the Australian flora, with approximately 15 % alien species. However, approximately 25,000 plant species are recorded as having been introduced to each country, of which ca. 1,800 (7 %) have naturalized in New Zealand, compared with ca. 2,800 (11 %) in Australia (Duncan and Williams 2002; Rod Randall, personal communication). Of the plant species introduced, a lower proportion have successfully invaded the New Zealand archipelago than continental Australia.

In summary, the number or percentage of alien species in a biota tells us little about the vulnerability of a location to invasion. What we require is information on the proportion of invasion attempts that have resulted in successful establishment. Where these data are available, there is currently no evidence to suggest that New Zealand islands are inherently more invasible than comparable mainland locations. This calls into question the widely held view that islands are easy to invade, and that certain island-specific biological characteristics enhance that invasibility (Table 28.1). In the following section, we consider in more detail the role of island-specific biological characteristics versus other non-island-specific determinants contributing to invasion vulnerability.

Table 28.1 Biological characteristics contributing to the invasibility of islands by alien species, and the mechanism responsible for their importance on islands

Characteristic	Mechanism
Low diversity on islands (Elton 1958)	Distance of islands from mainland sources results in a low immigration rate of new species, and island area limits viable population sizes (MacArthur and Wilson 1967)
Island biota are depauperate in certain characteristics (Carlquist 1974)	Novel environmental conditions and long-distance oceanic dispersal filter insular biota for certain taxonomic lineages and functional characteristics (Carlquist 1974)
Island species appear poor competitors (Darwin 1859; Carlquist 1974)	Evolution processes on islands are restricted by small populations along strong environmental gradients (Carlquist 1974; Denslow 2003)
Relatively few species at higher trophic levels on islands (e.g., Williamson 1981)	Dispersal to new and isolated habitats has led to the loss of enemies, and a biota without specialized pests

28.2.1 Assessment of Island-Specific Biological Characteristics Contributing to Invasibility

The apparent vulnerability of islands to invasion by alien species has often been attributed to lower species richness on islands relative to comparable areas of mainland (Denslow 2003). Species-poor islands may be less able to resist invaders due to fewer native competitors, predators or pathogens impacting on invading populations. For some biotic groups, the New Zealand biota is clearly species poor (e.g., social insects) but this is not always the case. With ca. 2,500 native vascular plant species, New Zealand is not particularly species poor when compared to continental landscapes of a similar surface area (Vitousek et al. 1997). New Zealand has rich marine floras and bryozoan assemblages (Chap. 8), and high fungal diversity with, for example, 10 and 7 % of the world's recorded ascomycete and basidiomycete species, respectively (Buchanan et al. 2004). Although considerable data exist, there have been few comparisons of local small-scale species richness between islands and mainlands for many taxonomic groups. Here at least, New Zealand native fungal communities appear relatively species rich (e.g., Allen et al. 2000).

At an island scale, Lonsdale (1999) found that, although island floras have a greater proportion of alien plant species than do mainland areas, this could not be attributed to lower native species richness because the islands he included were no less species rich than mainland areas of comparable size. Nevertheless, species richness at local patch scales may be more important because this is the scale at which ecological interactions such as competition and predation occur. However, at the patch scale, there is considerable debate over whether greater species richness confers resistance to invasion. In a New Zealand *Nothofagus* forest, the invasive herb *Hieracium lepidulum* more often invaded species-rich, rather than species-poor sites (Wiser et al. 1998), a pattern that has been observed elsewhere (e.g., Lonsdale 1999; Stohlgren et al. 2003). Such an outcome could arise through covariation of species richness with other factors that affect invasibility, particularly because experimental studies controlling for other factors have demonstrated negative species richness effects (e.g., Naeem et al. 2000), although these effects are often weak. Nevertheless, if species richness were a critical determinant of invasion vulnerability, which is a core argument for why islands are easy to invade, then an effect of species richness should be readily observed in natural systems. Instead, other factors appear to exert a much stronger influence (Hooper et al. 2005).

Due to their isolation, island biotas are often considered to be missing taxonomic and functional groups, creating opportunities for alien species to exploit those gaps (Table 28.1; Mack 2003). Functional gaps in the New Zealand fauna include no snakes, predatory terrestrial mammals, and few social insects or long-tongue bees and, in the flora, few grasses resistant to mammalian grazers or fire, few annuals that rapidly colonize disturbed sites, and few deciduous and cold-tolerant tall trees (Chap. 2). As taxonomic and functional groups can also be absent in continental biotas (Mack 2003), a remaining challenge is to quantify the degree to which island biotas are disproportionately lacking in such groups when compared to mainlands. Although the lack of entire groups may result from chance events (Simberloff 1995), it may also be the outcome of how specific environmental characteristics have shaped the biota. Distinctive features of the New Zealand terrestrial environment include high oceanicity, infrequent natural fire, a short history of continental climates, and an abundance of nutrient-poor soils (Chap. 11). McGlone et al. (2004) argue that the relatively low nutrient status of most New Zealand forest soils makes a deciduous phenology, with its high turnover of foliar nutrients, less competitive than a nutrient-conserving evergreen phenology. Nevertheless, there are now ca. 110 alien naturalized deciduous trees and shrubs in New Zealand, mostly growing where natural forests have been cleared (McGlone et al. 2004). In contrast, alien cold-tolerant conifers, such as *Pinus contorta*, appear to outperform native species in relatively unmodified habitats, having greater net carbon gain at current treeline than the native *Nothofagus* treeline species (Benecke and Nordmeyer 1982), and establishing and growing above current New Zealand treelines (Wardle 1991).

Darwin (1859) considered island species inherently less competitive than continental species, based on the apparent success of introduced species on islands, and the greater range of selection pressures thought to be present on continental areas. Denslow (2003) suggested that as steep environmental gradients occur over short distances on mountainous islands, this might then constrain the development of specialized adaptations, resulting in low competitive pressures and wide habitat breadths. In temperate forests, however, habitat breadth appears to be narrower, and plant species turnover greater, along elevation gradients in New Zealand forests than in Rocky Mountain forests at similar latitudes (Allen et al. 1991). Certainly, experimental manipulations of above- and below-ground competition suggest there is intense competition for light and nutrients among native plants even in New Zealand's relatively species-poor *Nothofagus* forests (e.g., Platt et al. 2004; Wiser et al. 2005). The assumption that island species are necessarily poor competitors is also questioned by the success of island species as pests in continental regions of the world. The percentage of New Zealand's native angiosperm flora that has naturalized overseas is of the same order of magnitude as the proportion of the world's angiosperm flora that has naturalized in New Zealand, and there are a growing number of examples of New Zealand native species that are serious environmental pests in Australia, South Africa, North America and Europe (e.g., Chap. 6). Maybe the concept of low competitive pressure in island biotas, like that of New Zealand, should be refined to allow for intense competition when there has been strong selective pressures for specific characteristics (e.g., dispersal and growth at low nutrients; Chap. 2).

On islands, few native species in higher trophic levels (see Table 28.1) may lower resistance to invading species, although demonstrating a lack of top-down control may be confounded by the fact that invading species often leave their natural enemies behind. Torchin and Mitchell (2004) considered that introduced animals and plants may escape most of the parasites and pathogens from their native range. Brushtail possums (*Trichosurus vulpecula*) in New Zealand, for example, have very few of their endoparasite fauna compared with those in their native Australia, and those parasites present in New Zealand are patchily distributed (Chap. 5). Naturalized herbaceous Asteraceae in New Zealand contain very few seed-eating insect larvae (less than 0.1 % of florets) compared with conspecifics in their native range (greater than 6 %; Fenner and Lee 2001), but this may be due to the relatively recent introduction of the plants (less than 100—200 years), their phylogenetic distance from the native flora, and the specialized nature of the bud-infesting habit of the insects. In general, only a few native species are known as pests on alien species, for example, native *Armillaria* fungi can kill radiata pine (*Pinus radiata*) seedlings and kiwifruit (*Actinidia*) vines when growing on forest soils (Buchanan et al. 2004). For some alien species, their predators are cosmopolitan, and therefore are already present in New Zealand on arrival. Most of the carnivorous (on nematodes) and entomoparasitic fungi identified in New Zealand are widely distributed elsewhere (Buchanan et al. 2004). If escape from pest pressure is a factor in the success of invaders, we should see subsequent demographic, or geographic, declines as pests catch up with alien species. There are records of widespread plant species declining for unknown reasons (Chap. 3), and also naturalized fungi are now being evaluated as biocontrol agents on alien plants (Buchanan et al. 2004). Although the release of invasive species from natural pests has been experimentally demonstrated elsewhere (e.g., Mitchell and Power 2003), few studies show the subsequent effects of a lagged pest invasion. There are world-wide examples of lagged animal declines occurring in New Zealand where the pathogen is known: rabbits (*Oryctolagus cuniculus*), upon the introduction (illegally) of rabbit hemorrhagic disease (Parkes et al. 2002), and honey bees upon the arrival in New Zealand of the varroa bee mite (*Varroa destructor*).

If top-down controls have not been an important evolutionary factor on islands, we would expect a native biota poorly adapted to pests. For example, it has long been argued that the New Zealand flora is poorly adapted to mammalian herbivory, but given native avian and invertebrate herbivory, this remains an area of considerable debate. One view is that common traits seen in the New Zealand flora, for example, heterophylly and the divaricate growth form, reflect specific defenses against browsing by the extinct moa (Greenwood and Atkinson 1977; Bond et al. 2004; cf. Howell et al. 2002). Bond et al. (2004) considered approximately 20 % of the woody trees and shrubs endemic to New Zealand exhibit these ratite-resistant structural strategies that would confer little resistance to introduced browsing mammals. However, such plants were not preferred by invading deer, and remain common in forests with a century of deer browsing (Wardle 1984), as well as sometimes increasing with deer browsing (e.g., Wardle et al. 2001). Certainly, there is considerable variability in the palatability of native plant species to alien mammal herbivores (Chap. 21). Forsyth et al. (2005) show, from a range of species-specific leaf traits, that foliar fiber is the best predictor of diet selection by red deer (*Cervus elaphus scoticus*) in the understory of a native forest. Although such patterns have been considered elsewhere, what remains unknown is whether the leaf traits of New Zealand plants are within the range for plants eaten by red deer in their native habitats. Clearly, the New Zealand flora has been subjected to various forms of herbivory, and there is an opportunity to collect comparative data to evaluate the extent to which traits exhibited by New Zealand plants differ systematically from those on other landmasses.

28.2.2 Importance of Non-Island-Specific Determinants

Numerous determinants, not necessarily characteristic of islands, have been implicated in the movement, release, establishment and spread of invasive species to New Zealand (e.g., see Chaps. 3, 8, 9 and 10). Most ecological studies focus on attributes affecting establishment (e.g., introduction effort and hunting for vertebrate animals) or spread (e.g., mutualisms, soil fertility, and disturbance for vascular plants) individually. Structured analyses are required to simultaneously show the relative importance of what are clearly island-specific attributes versus the wide range of other determinants controlling invasions (e.g., Lonsdale 1999). In one of the few such studies from New Zealand, Russell et al. (2004) related the species richness of small

(<10 kg) and large alien mammals on 297 offshore islands to a range of island attributes and other determinants of invasion. Species richness of large mammals was mostly related to human activities on islands (e.g., wharf presence) that are not exclusively island-specific attributes, whereas species richness of small mammals was in part related to island attributes (e.g., area) but also non-island-specific attributes (e.g., latitude). However, a limitation of the Russell et al. (2004) study is that it did not account for the importance of failures in assessing invasibility (Simberloff 1995). Overall, the New Zealand experience suggests island-specific biological characteristics and other attributes are often not the critical determinants of invasion success.

28.3 Do Invasive Species Have Novel Impacts on New Zealand Islands?

Alien species can have ecological consequences that are normally rare or absent from island ecosystems, and in that respect, are novel. However, several chapters show the nature of such impacts can be unclear when viewed at the ecosystem level, because there are many potential biotic and abiotic interactions and feedbacks (see also Simberloff 1995). For example, those impacts brought about by introduced mammalian herbivores are difficult to ascribe relative to the historical impacts of native herbivores (Chap. 11). As a consequence, it is often difficult to determine whether the island-specific factors leading to novel impacts on islands (Table 28.2) are the key reasons why New Zealand ecosystems appear to have been strongly impacted by alien species.

28.3.1 Assessment of Factors Contributing to Novel Impacts on Islands

The causal relationship between alien invasions and the extinction of native species is widely accepted, although supporting data are often sparse (Gurevitch and Padilla 2004). Dramatic extinctions in the New Zealand avifauna were associated with human arrival, but it has taken a long time for a consensus to emerge that most of these extinctions were caused principally by the introduction of mammalian predators (e.g., Fleming 1962; Worthy and Holdaway 2002). Thirty-one percent of native bird species, of a total of 131 species, have become extinct on the two main islands since human arrival. The pre-European colonization (before 1769 A.D.) extinctions appear largely explained by the selective extinction of large-bodied species, and the post-European extinctions by the absence of predator-escape responses, including the absence of flight (Duncan and Blackburn 2004). Although small population size is often considered a basis for the vulnerability of island taxa to extinc-

Table 28.2 Factors contributing to the novel impact of alien species on islands, and the associated mechanism responsible for their importance on islands

Factor	Mechanism
Extinctions are more numerous on islands (e.g., Atkinson 1989; Simberloff 1995)	Small and geographically restricted species populations on islands have limited refuges (e.g., Simberloff 2000)
Declines in population size are more pronounced on islands (e.g., Simberloff 2000)	Island endemics have small range sizes, and there is a positive correlation between range size and local abundance (Brown and Maurer 1984)
Island species prone to diseases vectored by alien species (e.g., Crosby 1986)	Probability of resistant genotypes on islands is lower, and they will evolve more slowly than in mainland situations (e.g., Carson 1981)
Alien species provide novel functional roles in island ecosystems (e.g., Vitousek et al. 1987)	Aliens that can alter ecosystem processes (e.g., decomposition or disturbance regimes) change the abilities of species to coexist (e.g., Vitousek et al. 1997)

tion (Table 28.2), geographical range, as a surrogate for population size, was not a predictor of extinction in New Zealand birds (Duncan and Blackburn 2004), and many extinct species were initially widely abundant. Rather, the case of New Zealand bird extinctions supports the importance of novel functional roles displayed by alien predators. Although the notion of native animal extinctions resulting from novel alien predators is compelling, introduced plant and animal species have not led to extinction in native plant species. Only three native plant species are thought to have become extinct during colonization, even though two-thirds of New Zealand land surfaces has been deforested and a wide range of herbivores introduced (Clout 1999). To some degree, the impact of an alien species on a native species population will depend upon the presence of spatial and temporal refugia, related to duration of susceptible life stages, species longevity, and habitat tolerance of the native species. Moreover, the impact of alien herbivores on many New Zealand tree species awaits resolution because of the long generation turnover times of many forest canopy tree species (Lee 1998).

Within New Zealand, the population size of endemic species does not appear a determinant of invasive species impacts. For example, apparent population imbalances in widely distributed canopy tree species have often been thought of as being a consequence of alien herbivores (e.g., brushtail possums, goats and deer). More specifically, some authors consider brushtail possums have caused dieback in some canopy tree species, and local extinctions, although others argue these apparent imbalances are potentially a response to factors other than introduced herbivores (Chaps. 17 vs. 21). Widely distributed, numerically dominant tree species appear to have been subjected, so far, to more catastrophic impacts of alien species in continental areas than in New Zealand. Since chestnut blight invaded the USA, chestnut (*Castanea dentata*) has been reduced from a common canopy dominant in eastern deciduous forests to an occasional understory plant maintained by sprouting (e.g., Greller 1988). Certainly, some New Zealand endemic birds (e.g., kakapo, *Strigopus habrotilus*) and reptiles (e.g., tuatara, *Sphenodon* spp.) were widely distributed and numerous in prehuman times, but now only occur as very small populations on adjacent islands. Giant weta (*Deinacrida* spp.) and snails (*Powelliphanta* spp.) that are preyed on by rats were also numerous in many types of habitat but are now largely absent in the lowlands (Daugherty et al. 1993). Conversely, some native taxa with small populations appear to maintain viable populations only in alien-dominated vegetation (e.g., several rare orchids).

Although alien species often carry reduced parasite and pathogen loads, diseases caused by parasites/pathogens can still be widely found in alien and native species in New Zealand (Chaps. 10 and 17). It has long been considered that the general decline in New Zealand's avifauna populations, which occurred in some areas before some alien predators (e.g., mustelids) arrived, may be a consequence of avian diseases (Myers 1923). Diseases such as avian malaria (*Plasmodium relictum*) do occur in alien and native birds, and that disease is likely to expand its geographical and host range due to the further introduction of alien mosquitoes, with likely consequences for bird population declines, including threatened species (Chap. 16). Alien parasitoids have also utilized native invertebrates and fish. Native galaxiid fish share many parasites with alien salmonid fish, and these are thought to explain why galaxiids have been lost from certain freshwater systems (Chap. 5). Certainly, it is necessary for a more careful characterization of diseases found in the native and alien biota, but in particular their consequences for native species population dynamics, before New Zealand studies can make a clearer contribution to debate about the importance of alien-vectored diseases.

Alien species with novel functional roles have the capacity to alter ecosystem processes, and change the ability of native species to coexist (Table 28.2; Vitousek et al. 1997). For example, introduced social bees (e.g., honey bees, *Apis mellifera*, and bumblebees, *Bombus* spp.) differ functionally from any New Zealand native bees in their temperature flight thresholds (aliens operate at lower temperatures), quantity of floral resource needed, and population dynamics (native bees are solitary; Newstrom and Robertson 2005). Exotic bees may then have unexpected effects on native plant pollination (e.g., in cool environments), as they are important flower visitors to some native plants (Chap. 15). Alien N-fixing plant species generally have higher fixation rates than do native N-fixing species, and alien N-fixing species have been observed to outcompete a wide range of native species through vigorous growth and rapid dispersal (Chap. 19). N-fixing aliens apparently alter successional trajectories away from those dominated by native species, and this for at least a century (Chap. 19). Alien mammalian predators (e.g., stoats) in New Zealand have strong olfactory capabilities, in contrast to top native predators (birds and reptiles) that principally use sight (Daugherty et al. 1993); thus, native avifauna traits such as cryptic coloring, freezing in presence of predators, and nocturnal habits, effective against diurnal aerial predators, have performed poorly against terrestrial predators with an acute sense of smell. The difficult question remains as to whether continental aliens have novel functional roles on islands more often than island aliens have on mainland landmasses. The New Zealand terrestrial planarians are devastating British Isles lumbricid earthworm populations, and subsequently modifying soil properties, which suggests New Zealand taxa can certainly fill functional gaps in the biota elsewhere (Chap. 6).

28.3.2 Importance of Non-Island-Specific Factors on Alien Impacts

There is increasing evidence that the nature of alien species impacts is context dependent, varying in space and time. It has long been known that stoat abundance and impacts on bird populations can be temporally variable (Wardle 1984). This variability is in part because of trophic interactions where stoat numbers wax and wane with mice numbers, which in turn depend on annually variable seed production in New Zealand *Nothofagus* forest (Chap. 16). Furthermore, other trends in beech forests may affect outcomes. For example, background decadal-level trends in beech tree seeding related to climate trends (Richardson et al. 2005), and the relatively recent arrival of alien wasps that now compete for food with adult birds and prey on nestlings (Chap. 18), are likely to cause dramatic shifts in the nature of these trophic interactions. Not only are the processes driving alien species impacts temporally complex, involving island-specific and other factors, but they also vary spatially. Although it has been experimentally demonstrated in one region of New Zealand that brown trout (*Salmo trutta*) have more profound impacts on the functioning of streams, across multiple levels of ecological organization, than do native galaxiid fish, these impacts may not apply elsewhere in New Zealand (Chap. 14). Similarly, although long-term herbivory by alien mammals in forests throughout New Zealand was expected to reduce plant species that produce high-quality litter and favor decomposer activity, and also cause an increase in plant species that produce litter that is slow to decompose, the effects of herbivores on soil carbon and nitrogen storage were idiosyncratic, with different responses in different parts of New Zealand (Chap. 20). Distinctive responses in the soil and soil biota at a location may be the result of spatially varying soil fertility, macroclimate, and mammal species present. Such examples show that demonstrating the impacts of some alien species has been challenging, let alone whether the impacts are novel. There is a need for more careful experimental testing of alien species impacts under a range of conditions, complemented by simulation models to overcome some of the timescale issues.

28.4 What Are the Ongoing Consequences of Invasions for the New Zealand Landscape?

28.4.1 Is the Invasion Over?

Recent historical trends give a first approximation of our expectations about the types of ongoing invasions and their consequences. All current alien mammal species were established in the wild >80 years ago, but there are ongoing distributional adjustments giving different combinations and population levels (Chap. 4). Some of these distributional expansions are a consequence of a legislative relaxation 20 years ago that allowed traditional pest species to be farmed outside their feral range. This process of range expansion will continue for many mammals, increasing the local diversity and potential impacts of introduced species. However, both importation and release of new mammal species are unlikely under current border-control security systems. With 2,200 alien plant species naturalized in New Zealand, the ongoing invasion of alien plants will also reflect distributional adjustments, but also the >25,000 species in cultivation around New Zealand provide a large species pool and multiple sources for ongoing naturalization, promoted by human activities such as disposal of gardening waste (e.g., Sullivan et al. 2005). In contrast to naturalization of established aliens, the arrival of new alien invertebrate species continues unabated in New Zealand (e.g., Charles and Henderson 2002; Brockerhoff et al. 2003; G.M. Barker, unpublished data), despite border controls refined in legislation over the last 150 years (Chap. 1) and a better knowledge of species invasiveness. However, our inability to early detect and control a significant proportion of cryptic alien invaders is likely to remain a major challenge. At a smaller scale again, the transfer of genetic material from alien species to native species, through spontaneous hybridization, has gained an increased profile recently because of the potential release of transgenic crops in New Zealand (Chap. 7). In an assessment of the future risk of transgenic escapes from genetically modified introduced crops, Armstrong et al. (2005) found that 66 (54%) of 123 widely grown temperate crops in New Zealand have reproductive compatibility with plant species, nearly all alien, in the wild, with the potential to acquire transgenes promoting resistance to pests, diseases and abiotic stresses. Historical trends may also be altered in the future through changes in the sources of alien species, their pathways to New Zealand, and pervasive changes on the New Zealand landscape. Certainly, the view that biological invasions will increase with increasing global trade (e.g., Levine and D'Antonio 2003) likely applies to New Zealand, and propagule pressure of small organisms from new trading partners (e.g., in Asia) will access a largely untapped pool of potentially new invasive species.

28.4.2 Reducing Alien Impacts?

New Zealand has acquired considerable expertise in the control of a wide range of alien taxa for both conservation and economic goals. The eradication, over the last decade, of an increasing number of mammal predator and herbivore species from offshore islands of increasing size has been a major conservation success. This has created unparalleled opportunities for the translocation and population recovery of threatened bird species (Chap. 24), as well as increased regeneration by some palatable plant species (e.g., Campbell and Atkinson 1999). The successes on offshore islands have increasingly inspired mammal eradication over small areas on the main islands, with re-immigration sometimes being addressed through fencing or ongoing alien species control (Chap. 24). So far, these attempts have focused on alien mammals and some plant species, but have not attempted the simultaneous eradication of a full range of alien biota, which may compromise the ability of some ecosystems to recover native biodiversity components. Furthermore, the past loss of native species may mean some native biodiversity components will never recover without additional human interventions other than alien eradication. For example, apparently ratite-resistant plants may now be disadvantaged by the extirpation of moa (Bond et al. 2004), and there are considerable challenges to re-establishing such selective pressure.

Eradication is not currently feasible for most of the alien species found over extensive areas on the main islands — the only option, apart from local predator-proof fences, is to determine what level of control is required. Reductions in alien mammal populations have been achieved, at a range of scales, through commercial harvesting and ongoing pest control operations. However, conservation gains have not always been consistent or convincing (Chap. 21). For example, although commercial harvesting of introduced deer has reduced their numbers nationally by 90 %, the benefits to palatable native plant species in forests appear limited (Chap. 26). This may be because conservation benefits are indeed limited, or else reflect the impacts of other alien herbivores, depletion of local seed sources, and the suppression of regeneration by the dominance of unpalatable species. The decline in North Island kokako (*Callaeas cinerea*) had been widely known, but the mechanism debated, before it became clear that alien nest predation (e.g., by brushtail possums) was a critical determinant of the population-level decline. More widely, studies attempting to show the benefits of alien mammal control for native species populations have often failed to yield the appropriate demographic data for understanding the consequences for native species populations. In addition, particularly for long-lived tree species, there is a need to model the long-term consequences of aliens in order to understand the benefits of management and to improve pest control strategies (Chap. 22). There are relatively few resources aimed at reducing the conservation impacts of alien taxa other than mammals, such as alien grasses and herbs. This is a major challenge because ca. 30 % of the New Zealand landscape is now primarily managed by the Government for conservation of native biodiversity (Chap. 1).

With an increasing number of alien species throughout the country, pest management increasingly becomes orientated around prioritization of control. This will become a complex task, as the dynamics and impacts of each species interact. Even small patches of relatively unmodified native forest may now contain six alien mammal herbivores, five alien mammal predators, two alien fish, numerous alien plants, and an unknown number of alien invertebrate, fungi and bacteria species. These organisms can have important trophic interactions and feedbacks with native species (e.g., Chaps. 16 and 18). Choosing which species to control, when, and where becomes rather more complex when it must be done within the context of ecosystem-level dynamics (e.g., see Chaps. 14 and 20). Management of multiple alien species may also be more complex because some will provide environmental (e.g., carbon storage by invasive trees), economic (e.g., trout fishing and tourism), recreational (e.g., trophy hunting opportunities), and cultural (e.g., wild pig for Maori festivities) benefits. The recreational freshwater fishery in New Zealand is based upon alien salmonids, is worth $ 800 million per year, and far exceeds the Government's allocation of ca. $ 250 million per year to fund the country's Department of Conservation, the principal agency responsible for biodiversity conservation.

28.4.3 Invasive Species as Determinants of the Future New Zealand Landscape

The significance of existing biological invasions to the New Zealand biota will gradually unfold in the long term, mediated by changes in land use and urban settlement across the landscape. Long-term scenarios are needed to motivate society to make better choices. These scenarios could, in the first instance, be based on a simple premise that there will be few, further knowledge gains to improve management, and also that a similar level of resources is allocated to the control of aliens. Under this scenario, over the next 100 years, we would expect the avifauna to be protected from predators on ca. 1 % of the landscape, providing habitat and supporting viable populations of most bird species but not covering a complete range of biota. This activity will fluctuate in response to varying economic conditions, changes in land ownership, and unpredictable natural events. We expect there will have been several major outbreaks of pathogens, widely affecting exotics and native forests. Within the next 1,000 years, perhaps 20 % of the New Zealand landscape would be dominated by alien conifers, particularly in dryland areas and above former treelines. In addition, 25 % of bird protection sites would have been lost through a massive central North Island volcanic eruption, with far-reaching consequences for New Zealand alien invasions at a time when New Zealand society is disrupted. In about 10,000 years, human activity would be the dominant evolutionary driver, with New Zealand's novel environmental conditions forming a new biota. Even now, many alien species are diverging from their mother populations. Surely such scenarios must be accommodated in our thinking about biological invasions today?

28.5 Conclusions

Because New Zealand has been colonized so recently, it is often possible to compile authoritative data on naturalized alien species, and for some taxa, definitive lists of failures (e.g., birds; Duncan 1997). This has facilitated some comprehensive analyses of factors controlling invasibility and the population dynamics of invading species in New Zealand. Many of the chapters in this book point to introduction effort, propagule pressure, and human activity as dominant determinants of alien species success on New Zealand islands, rather than island-specific attributes. Although novel impacts of alien species are a feature of some New Zealand invasions, it has not been shown that this is disproportionately so when compared with continental areas. Clearly, there is a need for ecosystem-level studies on the impacts of multiple alien species if we are to develop more effective management systems, including soundly based risk assessment systems for prioritization of control or eradication. The anticipated ongoing arrival and spread of a wide range of further alien species gives some urgency to these studies. Despite whatever we do, however, there is no doubt that we are still in the midst, if not at the beginning, of a major readjustment in the New Zealand biota.

Acknowledgements We thank Peter Bellingham and Matt McGlone for comments on a draft of this article. Preparation of this article was funded by the New Zealand Foundation for Research, Science and Technology.

References

Allen RB, Reif A, Hall GMJ (1991) Elevational distribution of conifer-broadleaved hardwood forests on South Island, New Zealand. J Veg Sci 2:323–330

Allen RB, Buchanan PK, Clinton PW, Cone AJ (2000) Composition and diversity of fungi on decaying logs in a New Zealand temperate beech (*Nothofagus*) forest. Can J For Res 30:1025–1033

Armstrong TT, Fitzjohn RG, Newstrom LE, Wilton AD, Lee WG (2005) Transgene escape: what potential for crop-wild hybridisation? Mol Ecol 14:2111–2132

Atkinson I (1989) Introduced animals and extinctions. In: Western D, Pearl M (eds) Conservation for the twenty first century. Oxford Univ Press, New York, pp 54–69

Benecke U, Nordmeyer AH (1982) Carbon uptake and allocation by *Nothofagus solandri* var *cliffortioides* (Hook. f.) Poole and *Pinus contorta* Douglas ex Loudon ssp. *contorta* at montane and subalpine altitudes. In: Waring RH (ed) Carbon uptake and allocation in subalpine ecosystems as a key to management. Forest Research Laboratory, Oregon State University, Corvallis, pp 9–21

Blackburn TM, Duncan RP (2001) Establishment patterns of exotic birds are constrained by non-random patterns in introduction. J Biogeogr 28:927–939

Bond WJ, Lee WG, Craine JM (2004) Plant structural defences against browsing birds: a legacy of New Zealand's extinct moas. Oikos 104:500–508

Brockerhoff EG, Kní_ek M, Bain J (2003) Checklist of indigenous and adventive bark and ambrosia beetles (Curculionidae: Scolytinae and Platypodinae) of New Zealand and interceptions of exotic species (1952–2000). NZ Entomol 26:29–44

Brown JH, Maurer B (1984) On the relationship between abundance and distribution of species. Am Nat 124:255–279

Buchanan PK, Beever RE, Galloway DJ, Glare TR, Johnston PR, McKenzie EHC, di Menna ME, Pennycook SR, Ridley GS, Smith JMB, Stephenson SL (2004) The fungi of New Zealand – an introduction. In: McKenzie EHC (ed) Introduction to fungi of New Zealand. Fungi of New Zealand, vol 1. Fungal Diversity Press, Hong Kong, pp 1–48

Campbell DJ, Atkinson IAE (1999) Effects of kiore (*Rattus exulans* Peale) on recruitment of indigenous coastal trees on northern offshore islands of New Zealand. J R Soc NZ 29:265–290

Carlquist S (1974) Island biology. Columbia Univ Press, New York

Carson HL (1981) Microevolution in an insular ecosystem. In: Mueller-Dombios D, Bridges KW, Carson HL (eds) Island ecosystems: biological organization in selected Hawaiian communities. Hutchinson-Ross, Stroudsburg, Pennsylvania

Charles JG, Henderson RC (2002) Catalogue of the exotic armoured scale insects (Hemiptera: Coccoidea: Diaspididae) in New Zealand. J R Soc NZ 32:587–615

Clout MN (1999) Biodiversity conservation and the management of invasive animals in New Zealand. In: Sandlund OT, Schei PJ, Viken Å (eds) Invasive species and biodiversity management. Kluwer, Dordrecht, pp 349–361

Crosby AW (1986) Ecological imperialism: the biological expansion of Europe, 900–1900. Cambridge Univ Press, Cambridge

Darwin CR (1859) The origin of species, 9th edn, 1979. Avenel Books, New York

Daugherty CH, Gibbs GW, Hitchmough RA (1993) Mega-island or micro-continent? New Zealand and its fauna. Trends Ecol Evol 8:437–442

Denslow JS (2003) Weeds in paradise: thoughts on the invasibility of tropical islands. Ann Miss Bot Gard 90:119–127

Duncan RP (1997) The role of competition and introduction effort in the success of passeriform birds introduced to New Zealand. Am Nat 149:903–915

Duncan RP, Blackburn TM (2004) Extinction and endemism in the New Zealand avifauna. Global Ecol Biogeogr 13:509–517

Duncan RP, Williams PA (2002) Darwin's naturalisation hypothesis challenged. Nature 417:608–609

Duncan RP, Bomford M, Forsyth DM, Conibear L (2001) High predictability in introduction outcomes and geographic range size of introduced Australian birds: a role for climate. J Anim Ecol 70:621–632

Elton CS (1958) The ecology of invasions by animals and plants. Methuen, London

Fenner M, Lee WG (2001) Lack of pre-dispersal seed predators in introduced Asteraceae in New Zealand. NZ J Ecol 25:95–100
Fleming CA (1962) The extinction of moas and other animals during the Holocene period. Notornis 10:113–117
Forsyth DM, Richardson SJ, Menchenton K (2005) Foliar fibre predicts diet selection by invasive red deer *Cervus elaphus scoticus* in a temperate New Zealand forest. Funct Ecol 19:495–504
Greenwood RM, Atkinson IAE (1977) Evolution of divaricating plants in New Zealand in relation to moa browsing. Proc NZ Ecol Soc 24:21–33
Greller AM (1988) Deciduous forests. In: Barbour MG, Billings WD (eds) North American terrestrial vegetation. Cambridge Univ Press, Cambridge
Gurevitch J, Padilla DK (2004) Are invasive species a major cause of extinction? Trends Ecol Evol 19:470–474
Hooper DU, Chapin FS III, Ewel JJ, Hector A, Inchausti P, Lavorel S, Lawton JH, Lodge DM, Loreau M, Naeem S, Schmid B, Setälä H, Symstad AJ, Vandermeer J, Wardle DA (2005) Effects of biodiversity on ecosystem functioning: a consensus of current knowledge. Ecol Monogr 75:3–35
Howell CJ, Kelly D, Turnbull MH (2002) Moa ghosts exorcised? New Zealand's divaricate shrubs avoid photoinhibition. Funct Ecol 16:232–240
Lee WG (1998) The vegetation of New Zealand – functional, spatial, and temporal gaps. R Soc NZ Misc Ser 48:91–101
Levine JM, D'Antonio CM (2003) Forecasting biological invasions with increasing international trade. Conserv Biol 17:322–326
Lonsdale WM (1999) Global patterns of plant invasions and the concept of invasibility. Ecology 80:1522–1536
MacArthur RH, Wilson EO (1967) The theory of island biogeography. Princeton Univ Press, Princeton, New Jersey
Mack RN (2003) Phylogenetic constraint, absent life forms, and preadapted alien plants: a prescription for biological invasions. Int J Plant Sci 164:S185–S196
McGlone MS, Dungan RJ, Hall GMJ, Allen RB (2004) Winter leaf loss in the New Zealand woody flora. NZ J Bot 42:1–19
Mitchell CE, Power AG (2003) Release of invasive plants from fungal and viral pathogens. Nature 421:625–627
Myers JG (1923) The present position of the endemic birds of New Zealand. NZ J Sci Technol 6:65–99
Naeem S, Knops JMH, Tilman D, Howe KM, Kennedy T, Gale S (2000) Plant diversity increases resistance to invasion in the absence of covarying extrinsic factors. Oikos 91:97–108
Newstrom L, Robertson A (2005) Progress in understanding pollination systems in New Zealand. NZ J Bot 43:1–59
Parkes JP, Norbury GL, Heyward RP, Sullivan G (2002) Epidemiology of rabbit haemorrhagic disease (RHD) in the South Island, New Zealand, 1997–2001. Wildl Res 29:543–555
Platt KH, Allen RB, Coomes DA, Wiser SK (2004) Mountain beech seedling responses to removal of below-ground competition and fertiliser. NZ J Ecol 28:289–293
Richardson SJ, Allen RB, Whitehead D, Carswell FE, Ruscoe WA, Platt KH (2005) Temperature and net carbon balance both determine seed production in a temperate tree species. Ecology 86(4):972–981
Russell JC, Clout MN, McArdle BH (2004) Island biogeography and the species richness of introduced mammals on New Zealand offshore islands. J Biogeogr 31:653–664
Simberloff D (1995) Why do introduced species appear to devastate islands more than mainland areas? Pac Sci 49:87–97

Simberloff D (2000) Extinction-proneness of island species – causes and management implications. Raffles Bull Zool 48:1–9

Sol D (2000) Are islands more susceptible to be invaded than continents? Birds say no. Ecography 23:687-692

Stohlgren TJ, Barnett DT, Kartesz JT (2003) The rich get richer: patterns of plant invasions in the United States. Front Ecol Environ 1:11–14

Sullivan J, Timmins S, Williams P (2005) Movement of invasive plant pests from coastal settlements into coastal native forest of northern New Zealand. NZ J Ecol 29:1–10

Torchin ME, Mitchell CE (2004) Parasites, pathogens, and invasions by plants and animals. Front Ecol Environ 2:183–190

Vitousek PM, Walker LR, Whiteaker LD, Mueller-Dombois D, Matson PA (1987) Biological invasion by *Myrica faya* alters ecosystem development in Hawaii. Science 238:802–804

Vitousek PM, D'Antonio CM, Loope LL, Rejmánek M, Westbrooks R (1997) Introduced species: a significant component of human-caused global change. NZ J Ecol 21:1–16

Walter HW (1979) Vegetation of the earth and ecological systems of the geo-biosphere, 2nd edn. Springer, Berlin Heidelberg New York

Wardle JA (1984) The New Zealand beeches: ecology, utilisation and management. New Zealand Forest Service, Wellington, New Zealand

Wardle P (1991) Vegetation of New Zealand. Cambridge Univ Press, Cambridge

Wardle DA, Barker GM, Yeates GW, Bonner KI, Ghani A (2001) Introduced browsing mammals in New Zealand natural forests: aboveground and belowground consequences. Ecol Monogr 71:587–614

Williamson M (1981) Island populations. Oxford Univ Press, Oxford

Wiser SK, Allen RB, Clinton PW, Platt KH (1998) Community structure and forest invasion by an exotic herb over 23 years. Ecology 79:2071–2081

Wiser SK, Allen RB, Benecke U, Baker G, Peltzer DA (2005) Tree growth and mortality after small-group harvesting in New Zealand old-growth *Nothofagus* forests. Can J For Res 35:2323–2331

Worthy TH, Holdaway RN (2002) The lost world of the moa – prehistoric life of New Zealand. Canterbury Univ Press, Christchurch, New Zealand

Subject Index

Ecological Studies

Volumes published since 2001

Volume 148
A Systems Analysis of the Baltic Sea (2001)
F.V. Wulff, L.A. Rahm, and P. Larsson (Eds.)

Volume 149
Banded Vegetation Patterning in Arid and Semiarid Environments (2001)
D. Tongway and J. Seghieri (Eds.)

Volume 150
Biological Soil Crusts: Structure, Function, and Management (2001, 2003)
J. Belnap and O.L. Lange (Eds.)

Volume 151
Ecological Comparisons of Sedimentary Shores (2001)
K. Reise (Ed.)

Volume 152
Global Biodiversity in a Changing Environment: Scenarios for the 21st Century (2001)
F.S. Chapin, O. Sala, and E. Huber-Sannwald (Eds.)

Volume 153
UV Radiation and Arctic Ecosystems (2002)
D.O. Hessen (Ed.)

Volume 154
Geoecology of Antarctic Ice-Free Coastal Landscapes (2002)
L. Beyer and M. Bölter (Eds.)

Volume 155
Conserving Biological Diversity in East African Forests: A Study of the Eastern Arc Mountains (2002)
W.D. Newmark

Volume 156
Urban Air Pollution and Forests: Resources at Risk in the Mexico City Air Basin (2002)
M.E. Fenn, L. I. de Bauer, and T. Hernández-Tejeda (Eds.)

Volume 157
Mycorrhizal Ecology (2002, 2003)
M.G.A. van der Heijden and I.R. Sanders (Eds.)

Volume 158
Diversity and Interaction in a Temperate Forest Community: Ogawa Forest Reserve of Japan (2002)
T. Nakashizuka and Y. Matsumoto (Eds.)

Volume 159
Big-Leaf Mahogany: Genetic Resources, Ecology and Management (2003)
A. E. Lugo, J. C. Figueroa Colón, and M. Alayón (Eds.)

Volume 160
Fire and Climatic Change in Temperate Ecosystems of the Western Americas (2003)
T. T. Veblen et al. (Eds.)

Volume 161
Competition and Coexistence (2002)
U. Sommer and B. Worm (Eds.)

Volume 162
How Landscapes Change: Human Disturbance and Ecosystem Fragmentation in the Americas (2003)
G.A. Bradshaw and P.A. Marquet (Eds.)

Volume 163
Fluxes of Carbon, Water and Energy of European Forests (2003)
R. Valentini (Ed.)

Volume 164
Herbivory of Leaf-Cutting Ants: A Case Study on *Atta colombica* in the Tropical Rainforest of Panama (2003)
R. Wirth, H. Herz, R.J. Ryel, W. Beyschlag, B. Hölldobler

Volume 165
Population Viability in Plants: Conservation, Management, and Modeling of Rare Plants (2003)
C.A Brigham, M.W. Schwartz (Eds.)

Volume 166
North American Temperate Deciduous Forest Responses to Changing Precipitation Regimes (2003)
P. Hanson and S.D. Wullschleger (Eds.)

Volume 167
Alpine Biodiversity in Europe (2003)
L. Nagy, G. Grabherr, Ch. Körner, D. Thompson (Eds.)

Volume 168
Root Ecology (2003)
H. de Kroon and E.J.W. Visser (Eds.)

Volume 169
Fire in Tropical Savannas: The Kapalga Experiment (2003)
A.N. Andersen, G.D. Cook, and R.J. Williams (Eds.)

Volume 170
Molecular Ecotoxicology of Plants (2004)
H. Sandermann (Ed.)

Volume 171
Coastal Dunes: Ecology and Conservation (2004)
M.L. Martínez and N. Psuty (Eds.)

Volume 172
Biogeochemistry of Forested Catchments in a Changing Environment: A German Case Study (2004)
E. Matzner (Ed.)

Volume 173
Insects and Ecosystem Function (2004)
W. W. Weisser and E. Siemann (Eds.)

Volume 174
Pollination Ecology and the Rain Forest: Sarawak Studies (2005)
D. Roubik, S. Sakai, and A.A. Hamid (Eds.)

Volume 175
Antarctic Ecosystems: Environmental Contamination, Climate Change, and Human Impact (2005)
R. Bargagli

Volume 176
Forest Diversity and Function: Temperate and Boreal Systems (2005)
M. Scherer-Lorenzen, Ch. Körner, and E.-D. Schulze (Eds.)

Volume 177
A History of Atmospheric CO_2 and its Effects on Plants, Animals, and Ecosystems (2005)
J.R. Ehleringer, T.E. Cerling, and M.D. Dearing (Eds.)

Volume 178
Photosynthetic Adaptation: Chloroplast to Landscape (2005)
W.K. Smith, T.C. Vogelmann, and C. Chritchley (Eds.)

Volume 179
Lamto: Structure, Functioning, and Dynamics of a Savanna Ecosystem (2005)
L. Abbadie et al. (Eds.)

Volume 180
Plant Ecology, Herbivory, and Human Impact in Nordic Mountain Birch Forests (2005)
F.E. Wielgolaski (Ed.) and P.S. Karlsson, S. Neuvonen, D. Thannheiser (Ed. Board)

Volume 181
Nutrient Acquisition by Plants: An Ecological Perspective (2005)
H. BassiriRad (Ed.)

Volume 182
Human Ecology: Biocultural Adaptations in Human Cummunities (2006)
H. Schutkowski

Volume 183
Growth Dynamics of Conifer Tree Rings: Images of Past and Future Environments (2006)
E.A. Vaganov, M.K. Hughes, and A.V. Shashkin

Volume 184
Reindeer Management in Northernmost Europe: Linking Practical and Scientific Knowledge in Social-Ecological Systems (2006)
B.C. Forbes, M. Bölter, L. Müller-Wille, J. Hukkinen, F. Müller, N. Gunslay, Y. Konstantinov (Eds.)

Volume 185
Ecology and Conservation of Neotropical Montane Oak Forests (2006)
M. Kappelle (Ed.)

Volume 186
Biological Invasions in New Zealand (2006)
R.B. Allen and W.G. Lee (Eds.)

MIX
Papier aus verantwortungsvollen Quellen
Paper from responsible sources
FSC® C105338

If you have any concerns about our products,
you can contact us on
ProductSafety@springernature.com

In case Publisher is established outside the EU,
the EU authorized representative is:
Springer Nature Customer Service Center GmbH
Europaplatz 3, 69115 Heidelberg, Germany

Printed by Libri Plureos GmbH
in Hamburg, Germany